AF581779

Sensor Signal and Information Processing III

Sensor Signal and Information Processing III

Editors

Wai Lok Woo
Bin Gao

MDPI • Basel • Beijing • Wuhan • Barcelona • Belgrade • Manchester • Tokyo • Cluj • Tianjin

Editors
Wai Lok Woo
Northumbria University
UK

Bin Gao
University of Electronic Science and Technology of China
China

Editorial Office
MDPI
St. Alban-Anlage 66
4052 Basel, Switzerland

This is a reprint of articles from the Special Issue published online in the open access journal *Sensors* (ISSN 1424-8220) (available at: https://www.mdpi.com/journal/sensors/special_issues/SSIP_III).

For citation purposes, cite each article independently as indicated on the article page online and as indicated below:

LastName, A.A.; LastName, B.B.; LastName, C.C. Article Title. *Journal Name* **Year**, *Volume Number*, Page Range.

ISBN 978-3-0365-0012-6 (Hbk)
ISBN 978-3-0365-0013-3 (PDF)

Contents

About the Editors

Wai Lok Woo is currently a Professor of Machine Learning at Northumbria University, UK. Previously, he was the Director of Research for the Newcastle Research and Innovation Institute and the Director of Operations for Newcastle University, UK, where he received the B.Eng. degree in electrical and electronics engineering and the M.Sc. and Ph.D. degrees in 1993, 1995, and 1998, respectively. His research interests include mathematical development of sensor signal processing and machine learning for anomaly detection, digital health, and digital sustainability. He is Associate Editor of several IEEE journals. He is interested in answering the global question of how the integration of smart sensors and machine learning advances humanity and sustains the ecosystem in the current digital transformation era. His research is funded by the UK Research and Innovation.

Bin Gao is currently a Professor at the School of Automation Engineering, University of Electronic Science and Technology of China (UESTC), Chengdu, China. He received his B.Sc. degree in communications and signal processing from Southwest Jiao Tong University (2001–2005), China, and received his M.Sc. degree in communications and signal processing with distinction and Ph.D. degree from Newcastle University, UK (2006–2011). He worked as a Research Associate (2011-2013) with Newcastle University on wearable acoustic sensor technology. His research interests include electromagnetic and thermographic sensing, supervised and unsupervised machine learning, wearable sensing, and nondestructive testing and evaluation, and he actively publishes in these areas. He has coordinated several National Natural Science Foundation of China research projects.

Editorial

Sensor Signal and Information Processing III

Wai Lok Woo [1,*] and Bin Gao [2]

1 Department of Computer and Information Sciences, Northumbria University, Newcastle upon Tyne NE1 8ST, UK

2 School of Automation Engineering, University of Electronic Science and Technology of China, Chengdu 611731, China; bin_gao@uestc.edu.cn

* Correspondence: wailok.woo@northumbria.ac.uk

Received: 24 November 2020; Accepted: 24 November 2020; Published: 26 November 2020

Sensors are one of the key factors in the success of the Internet of Things (IoT); however, these sensors are not conventional types that simply convert physical variables into electrical signals. They have substantially evolved into something more complex and sophisticated and perform a technically and economically viable role in most IoT applications. Sensors have traditionally been functionally simple devices that convert physical variables into electrical signals or create changes in electrical properties. While this functionality is a crucial starting point, sensors have extra functionalities such as very low power consumption, self-identification and self-verification, and they are wireless and physically small enough to "disappear" unobtrusively into any environment. In addition, information from multiple sensors can be combined and correlated to make conclusions about latent problems, for example, temperature sensor and vibration sensor data can be used to detect the onset of mechanical failure. In some cases, multiple sensor functions are available in one device, in others, the functions are combined in software to create a "soft" sensor. It has become evident that sensor intelligence, apart from facilitating connectivity, also has many other benefits including predictive maintenance, more flexible manufacturing, and improved productivity.

This is the third in a series of Special Issues dedicated to Sensor Signal and Information Processing (SSIP) [1]. The first two series of SSIP were published in 2018 and 2019, respectively [2–4]. SSIP has become an overarching field of research that is focused on the mathematical foundations and practical applications of signal processing algorithms that learn, reason, and act. It bridges the boundary between theory and application to develop novel, theoretically-inspired methodologies that target both longstanding and emergent signal processing applications. The core of SSIP lies in its use of nonlinear and non-Gaussian signal processing methodologies combined with convex and nonconvex optimization. SSIP encompasses new theoretical frameworks for statistical signal processing (e.g., deep learning, latent component analysis, tensor factorization, and Bayesian methods) coupled with information theoretical learning, and novel developments in these areas are specialized in the processing of a variety of signal modalities including audio, bio-signals, multi-physics signals, images, multispectral and video, among others. In recent years, many signal processing algorithms have incorporated some forms of computational intelligence as part of their core framework in problem solving. These algorithms have the capacity to generalize and discover knowledge for themselves and to learn new information whenever unseen data are captured. The focus of this Special Issue is on a broad range of sensors, signal and information processing involving the introduction and development of new advanced theoretical and practical algorithms. It includes twenty works focused on sensor signal and information processing based on diverse technologies for different applications.

Permutation entropy (PE) is a powerful complexity measure for analyzing time series and it has the advantage of easy implementation and high efficiency. In order to improve the performance of PE, Li et al. [5] have recently proposed improved methods by introducing amplitude information and distance information. Weighted-permutation entropy (W-PE) uses variance information to weight each arrangement pattern, which has good robustness and stability in the case of high noise level and can extract complexity information from data with spike features or abrupt amplitude changes. Dispersion entropy (DE) introduces amplitude information by using the normal cumulative distribution function. It is not only able to detect the change of simultaneous frequency and amplitude, but also is superior to the PE method in distinguishing different data sets. Reverse permutation entropy (RPE) is defined as the distance to white noise in the opposite trend to PE and W-PE, and it has high stability for time series with varying lengths. To further improve the performance of PE, the authors have proposed a new complexity measure known as reverse dispersion entropy (RDE), which takes PE as its theoretical basis and combines the advantages of DE and RPE by introducing amplitude information and distance information.

Along the same line of time series research, Parathai et al. [6] have proposed a solution based on complex non-negative matrix factorization (CMF) for events classification from a noisy mixture. It encodes the spectra pattern and estimates the phase of the original signals in the time-frequency representation through an adaptive L_1 sparsity CMF algorithm. The features enhance the efficiency of the temporal decomposition process. The support vector machine-based one-versus-one strategy was applied with a mean supervector to categorize the demixed sound into the matching sound-event class. The rapid development of sensor technology has given rise to the emergence of huge amounts of tensor data. For various reasons, such as sensor failures and communication loss, the tensor data may be corrupted by both small noises and gross corruptions. Fang et al. [7] have studied the Stable Tensor Principal Component Pursuit (STPCP), which aims to recover a tensor from its corrupted observations. The proposed model is based on the tubal nuclear norm, which has shown superior performance in comparison with other tensor nuclear norms. It is shown theoretically that the underlying tensor and the sparse corruption tensor can be stably recovered under tensor incoherence conditions.

Indoor positioning using Wi-Fi signals is an economic technique. Its main drawback is that multipath propagation distorts these signals, thus leading to inaccurate localization. One approach to improve the positioning accuracy consists of using fingerprints based on channel state information (CSI). Wang et al. [8] have proposed a new positioning method that consists of three stages. In the first stage, a model is built for the fingerprints of the environment. This model obtains a precise interpolation of fingerprints at positions where a fingerprint measurement is not available. In the second stage, the model is then used to obtain a preliminary position estimate based only on the fingerprint measured at the receiver's location. Finally, the preliminary estimation with the dynamical model of the receiver's motion are used to obtain the final estimation. The obtained experimental results from the proposed method show that it has a promising future.

The automatic sleep stage classification technique can facilitate the diagnosis of sleep disorders and free medical experts from labor-consuming work. Shen et al. [9] have proposed an improved model-based essence features that combines locality energy and dual state space models for automatic sleep stage detection on single-channel electroencephalograph signals. The experimental results have shown high classification accuracy compared with state-of-the-art methods. Similarly, motor imagery (MI)-based brain-computer interface (BCI) systems detect electrical brain activity patterns through electroencephalogram signals to forecast user intention while performing movement imagination tasks. As the microscopic details of individuals' brains are directly shaped by their rich experiences, musicians can develop certain neurological characteristics, such as improved brain plasticity following extensive musical training. Riquelme-Ros et al. [10] have developed a new approach to assess the performance of pianists as they interacted with an MI-based BCI system and compared it with that of a control group. The outcome indicates that musical training could enhance the performance of individuals using BCI systems.

In image processing, the recognition of scene changes plays an essential role in a variety of real-world applications, such as scene anomaly detection. Most scene understanding research has focused on static scenes and most existing scene change captioning methods detect scene changes from single-view color images, neglecting the underlying three-dimensional structures.

Previous three-dimensional scene change captioning methods have used simulated scenes consisting of geometry primitives, making them unsuitable for real-world applications. To solve these problems, Qiu et al. [11] have proposed an end-to-end framework for describing scene changes from various input modalities, namely, RGB images, depth images, and point cloud data, which are available in most robot applications. In a similar development, Kim and Kudo [12] have proposed a new class of nonlocal Total Variation (TV), in which the first derivative and the second derivative are mixed. Since most existing TV only considers the first-order derivative, it suffers from problems such as staircase artifacts and loss in smooth intensity changes for textures and low-contrast objects, which is a major limitation in improving image quality. The proposed nonlocal TV combines the first and second order derivatives to effectively preserve smooth intensity changes. The non-rigid multi-modal 3D medical image registration is highly challenging due to the difficulty in constructing a similarity measure and the solution of non-rigid transformation parameters. Yang et al. [13] have proposed a novel structural representation-based registration method to address these problems. Firstly, an improved modality independent neighborhood descriptor (MIND) based on the foveated nonlocal self-similarity is designed for effective structural representations. Subsequently, the foveated MIND-based spatial constraint is introduced into the Markov random field (MRF) optimization to reduce the number of transformation parameters and restrict the calculation of the energy function in the image region involving non-rigid deformation. Finally, the accurate and efficient 3D medical image registration is realized by minimizing the similarity measure-based MRF energy function. Experiments on real magnetic resonance and ultrasound images with unknown deformation were also done to demonstrate the practicality and superiority of the method.

Deformable image registration is still a challenge when the considered images have strong variations in appearance and large initial misalignment. A huge performance gap currently remains for fast-moving regions in videos or strong deformations of natural objects. Ha et al. [14] have combined a U-Net architecture that is weakly supervised with segmentation information to extract semantically meaningful features with multiple stages of non-rigid spatial transformer networks parameterized with low-dimensional B-spline deformations. The models are compact, very fast in inference, and demonstrate clear potential for a variety of challenging tracking and/or alignment tasks in computer vision and medical image analysis. Compressive sensing (CS) spectroscopy is well known for developing a compact spectrometer that consists of two parts: compressively measuring an input spectrum and recovering the spectrum using reconstruction techniques. Kim et al. [15] have proposed a residual convolutional neural network for reconstructing the spectrum from the compressed measurements. The proposed network comprises learnable layers and a residual connection between the input and the output of these learnable layers. The proposed network has produced stable reconstructions under noisy conditions.

Although deep learning has achieved great success in many applications, its usage in communication systems has not been well explored. Zha et al. [16] investigated algorithms for multi-signal detection and modulation classification, which are significant in many communication systems. In their work, a deep learning framework for multi-signal detection and modulation recognition is proposed. Compared to some existing methods, the signal modulation format, center frequency, and start-stop time can be obtained from the proposed deep learning scheme. The control effect of various intelligent terminals is affected by the data sensing precision. Usually, the filtering method is based on the typical soft computing method used to promote the sensing level. Bai et al. [17] have proposed a neuron-based Kalman filter to overcome limitations due to the difficult recognition of the practical system and the empirical parameter estimation in the traditional Kalman filter. The neuro units optimize the filtering process to reduce the effect of the unpractical system model and hypothetical parameters. It is shown that the neuro-filter is effective in noise elimination within the soft computing solution. IoT systems generate a large volume of data all the time. How to choose and transfer which data are essential for decision-making is a challenge. This is especially important for low-cost and low-power designs where data volume and frequency are constrained by the protocols.

Tsapparellas et al. [18] have presented an unsupervised learning approach using Laplacian scores to discover which types of sensors can be reduced without compromising the decision-making. Here, a type of sensor is a feature. A comparative study has shown that when fewer types of sensors are used; the accuracy of the decision-making remains at a satisfactory level.

In defect detection, the surface quality of aluminum ingot is crucial for subsequent products, so it is necessary to adaptively detect different types of defects on the surface of milled aluminum ingots. Liang et al. [19] have proposed a two-stage detection to quickly apply the calculations to a real production line. A mask gradient response-based threshold segmentation is developed to extract the target defects. An inception-v3 network with a data augmentation technology and the focal loss is further proposed to overcome the class imbalance problem and improve the classification accuracy. The gearbox is one of the most fragile parts of a wind turbine. Yin et al. [20] have developed a fault diagnosis method for wind turbine gearboxes based on optimized long short-term memory neural networks with cosine loss (Cos-LSTM). The loss is converted from Euclid space to angular space by cosine loss, thus eliminating the effect of signal strength and improving the diagnosis accuracy. The energy sequence features and the wavelet energy entropy of the vibration signals are used to evaluate the Cos-LSTM networks. The effectiveness of the method is verified with the fault vibration data collected on a gearbox fault diagnosis experimental platform. To suppress noise in signals, a denoising method on the basis of the singular value decomposition (SVD) and the Akaike information criterion (AIC) is proposed in [21], which is called AIC-SVD. To verify the effectiveness of AIC-SVD, it is compared with wavelet threshold denoising and empirical mode decomposition with a Savitzky–Golay filter. The proposed method is self-adaptable and robust while avoiding the occurrence of over-denoising.

In wireless communication, a frequency-hopping (FH)-based dual-function multiple-input multiple-output (MIMO) radar communications system enables simultaneous implementation of a primary radar operation and a secondary communication function. The set of transmit waveforms employed to perform the MIMO radar task is generated using FH codes. However, as the radar channel is time-variant, it is necessary for a successive waveform optimization scheme to continually obtain target feature information. Yao et al. [22] have developed a method to enhance target detection and feature estimation performance by maximizing the mutual information (MI) between the target response and the target returns, and then minimizing the MI between successive target-scattering signals. Chen et al. [23] propose a geometric calibration method using sparse recovery to remove the linear array push-broom sensor bias. By using the sparse recovery method, the number and distribution of ground control points needed are greatly reduced. Meanwhile, the proposed method effectively removes short-period errors by recognizing periodic wavy patterns in the first step of the process. Finally, Sun et al. [24] have proposed a blind modulation classification method based on compressed sensing using a high-order cumulant and cyclic spectrum combined with the decision tree-support vector machine classifier. The proposed method solves the problem of low identification accuracy under single-feature parameters and reduces the performance requirements of the sampling system. Through calculating the fourth-order, eighth-order cumulant and cyclic spectrum feature parameters by breaking through the traditional Nyquist sampling law in the compressed sensing framework, six different cognitive radio signals are effectively classified. The results indicate that accurate and effective modulation classification can be achieved, and this provides a theoretical basis and technical advance for the field of optical-fiber signal detection.

Funding: This research received no external funding.

Acknowledgments: The authors of the submissions have expressed their appreciation to the work of the anonymous reviewers and the Sensors editorial team for their cooperation, suggestions and advice. Likewise, the guest editors of this Special Issue thank the staff of Sensors for the trust shown and the good work done.

Conflicts of Interest: The authors declare no conflict of interest.

References

1. Woo, W.L.; Gao, B. Sensor Signal and Information Processing III, MDPI Sensors 2020. Available online: https://www.mdpi.com/journal/sensors/special_issues/SSIP_III (accessed on 23 November 2020).
2. Woo, W.L.; Gao, B. Sensor Signal and Information Processing, MDPI Sensors 2018. Available online: https://www.mdpi.com/journal/sensors/special_issues/SSIP (accessed on 23 November 2020).
3. Woo, W.L.; Gao, B. Sensor Signal and Information Processing II, MDPI Sensors 2019. Available online: https://www.mdpi.com/journal/sensors/special_issues/SSIP_II (accessed on 23 November 2020).
4. Woo, W.L.; Gao, B. Sensor Signal and Information Processing II. *Sensors* **2020**, *20*, 3751. [CrossRef]
5. Li, Y.; Gao, X.; Wang, L. Reverse Dispersion Entropy: A New Complexity Measure for Sensor Signal. *Sensors* **2019**, *19*, 5203. [CrossRef]
6. Parathai, P.; Tengtrairat, N.; Woo, W.L.; Abdullah, M.A.M.; Rafiee, G.; Alshabrawy, O. Efficient Noisy Sound-Event Mixture Classification Using Adaptive-Sparse Complex-Valued Matrix Factorization and OvsO SVM. *Sensors* **2020**, *20*, 4368. [CrossRef]
7. Fang, W.; Wei, D.; Zhang, R. Stable Tensor Principal Component Pursuit: Error Bounds and Efficient Algorithms. *Sensors* **2020**, *19*, 5335. [CrossRef]
8. Wang, W.; Marelli, D.; Fu, M. Fingerprinting-Based Indoor Localization Using Interpolated Preprocessed CSI Phases and Bayesian Tracking. *Sensors* **2020**, *20*, 2854. [CrossRef]
9. Shen, H.; Ran, F.; Xu, M.; Guez, A.; Li, A.; Guo, A. An Automatic Sleep Stage Classification Algorithm Using Improved Model Based Essence Features. *Sensors* **2020**, *20*, 4677. [CrossRef]
10. Riquelme-Ros, J.-V.; Rodríguez-Bermúdez, G.; Rodríguez-Rodríguez, I.; Rodríguez, J.-V.; Molina-García-Pardo, J.-M. On the Better Performance of Pianists with Motor Imagery-Based Brain-Computer Interface Systems. *Sensors* **2020**, *20*, 4452. [CrossRef]
11. Qiu, Y.; Satoh, Y.; Suzuki, R.; Iwata, K.; Kataoka, H. Indoor Scene Change Captioning Based on Multimodality Data. *Sensors* **2020**, *20*, 4761. [CrossRef]
12. Kim, Y.; Kudo, H. Nonlocal Total Variation Using the First and Second Order Derivatives and Its Application to CT image Reconstruction. *Sensors* **2020**, *20*, 3494. [CrossRef]
13. Yang, F.; Ding, M.; Zhang, X. Non-Rigid Multi-Modal 3D Medical Image Registration Based on Foveated Modality Independent Neighborhood Descriptor. *Sensors* **2019**, *19*, 4675. [CrossRef]
14. Ha, I.Y.; Wilms, M.; Heinrich, M. Semantically Guided Large Deformation Estimation with Deep Networks. *Sensors* **2020**, *20*, 1392. [CrossRef]
15. Kim, C.; Park, D.; Lee, H.-N. Compressive Sensing Spectroscopy Using a Residual Convolutional Neural Network. *Sensors* **2020**, *20*, 594. [CrossRef]
16. Zha, X.; Peng, H.; Qin, X.; Li, G.; Yang, S. A Deep Learning Framework for Signal Detection and Modulation Classification. *Sensors* **2019**, *19*, 4042. [CrossRef]
17. Bai, Y.-T.; Wang, X.-Y.; Jin, X.B.; Zhao, Z.Y.; Zhang, B.H. A Neuron-Based Kalman Filter with Nonlinear Autoregressive Model. *Sensors* **2020**, *20*, 299. [CrossRef]
18. Tsapparellas, G.; Jin, N.; Dai, X.; Fehringer, G. Laplacian Scores-Based Feature Reduction in IoT Systems for Agricultural Monitoring and Decision-Making Support. *Sensors* **2020**, *20*, 5107. [CrossRef]
19. Liang, Y.; Xu, K.; Zhou, P. Mask Gradient Response-Based Threshold Segmentation for Surface Defect Detection of Milled Aluminum Ingot. *Sensors* **2020**, *20*, 4519. [CrossRef]
20. Yin, A.; Yan, Y.; Zhang, Z.; Li, C.; Sánchez, R.-V. Fault Diagnosis of Wind Turbine Gearbox Based on the Optimized LSTM Neural Network with Cosine Loss. *Sensors* **2020**, *20*, 2339. [CrossRef]
21. Yin, X.; Xu, Y.; Sheng, X.; Shen, Y. Signal Denoising Method Using AIC–SVD and Its Application to Micro-Vibration in Reaction Wheels. *Sensors* **2019**, *19*, 5032. [CrossRef]
22. Yao, Y.; Li, X.; Wu, L. Cognitive Frequency-Hopping Waveform Design for Dual-Function MIMO Radar-Communications System. *Sensors* **2020**, *20*, 415. [CrossRef]

23. Chen, J.; Sha, Z.; Yang, J.; An, W. Proposal of a Geometric Calibration Method Using Sparse Recovery to Remove Linear Array Push-Broom Sensor Bias. *Sensors* **2019**, *19*, 4003. [CrossRef]
24. Sun, X.; Su, S.; Zuo, Z.; Guo, X.; Tan, X. Modulation Classification Using Compressed Sensing and Decision Tree–Support Vector Machine in Cognitive Radio System. *Sensors* **2020**, *20*, 1438. [CrossRef] [PubMed]

Publisher's Note: MDPI stays neutral with regard to jurisdictional claims in published maps and institutional affiliations.

Article

Reverse Dispersion Entropy: A New Complexity Measure for Sensor Signal

Yuxing Li [1,*], Xiang Gao [1] and Long Wang [2]

[1] School of Automation and Information Engineering, Xi'an University of Technology, Xi'an 710048, China; gaoxiang@xaut.edu.cn
[2] School of Marine Science and Technology, Northwestern Polytechnical University, Xi'an 710072, China; wanglongwl@mail.nwpu.edu.cn
* Correspondence: liyuxing@xaut.edu.cn

Received: 14 October 2019; Accepted: 20 November 2019; Published: 27 November 2019

Abstract: Permutation entropy (PE), as one of the powerful complexity measures for analyzing time series, has advantages of easy implementation and high efficiency. In order to improve the performance of PE, some improved PE methods have been proposed through introducing amplitude information and distance information in recent years. Weighted-permutation entropy (W-PE) weight each arrangement pattern by using variance information, which has good robustness and stability in the case of high noise level and can extract complexity information from data with spike feature or abrupt amplitude change. Dispersion entropy (DE) introduces amplitude information by using the normal cumulative distribution function (NCDF); it not only can detect the change of simultaneous frequency and amplitude, but also is superior to the PE method in distinguishing different data sets. Reverse permutation entropy (RPE) is defined as the distance to white noise in the opposite trend with PE and W-PE, which has high stability for time series with varying lengths. To further improve the performance of PE, we propose a new complexity measure for analyzing time series, and term it as reverse dispersion entropy (RDE). RDE takes PE as its theoretical basis and combines the advantages of DE and RPE by introducing amplitude information and distance information. Simulation experiments were carried out on simulated and sensor signals, including mutation signal detection under different parameters, noise robustness testing, stability testing under different signal-to-noise ratios (SNRs), and distinguishing real data for different kinds of ships and faults. The experimental results show, compared with PE, W-PE, RPE, and DE, that RDE has better performance in detecting abrupt signal and noise robustness testing, and has better stability for simulated and sensor signal. Moreover, it also shows higher distinguishing ability than the other four kinds of PE for sensor signals.

Keywords: permutation entropy (PE); weighted-permutation entropy (W-PE); reverse permutation entropy (RPE); reverse dispersion entropy (RDE); time series analysis; complexity; sensor signal

---

1. Introduction

Due to the continuous development of measurement technology and the constant updating of high sensitivity sensor equipment, the accuracy of measured time series is greatly improved, which is conducive to the further analysis and processing of time series [1,2]. The complexity of time series is one of the most important means to represent the characteristics of time series. Entropy, as an effective complexity measure of time series, has been widely developed and used in different fields. Classic examples include permutation entropy (PE) [3], sample entropy (SE) [4], approximate entropy (AE) [5], fuzzy entropy (FE) [6], and multi-scale ones. However, among these different kinds of entropy, PE has successfully attracted attention from academics and practitioners by virtue of its own advantages.

In 2002, PE was first suggested in a scientific article by Bandt and Pompe [7]. As a complexity measure, PE introduced arrangement into time series, and determined each arrangement pattern according to the neighboring values. PE has the characteristics of easy implementation and high computation efficiency. With its own advantages, PE has been widely used in different fields, including the medical field [8], mechanical engineering field [9,10], economic field [11,12], and underwater acoustic field [13,14]. Aiming at weaknesses of PE, many revised PE methods have been proposed to improve the performance of traditional PE.

In 2013, Fadlallah et al. brought forward weighted-permutation entropy (W-PE) and first applied it to electroencephalogram signal processing [15]. In order to solve the limitation of PE, W-PE introduced amplitude information to weight each arrangement pattern by using variance information. Compared with PE, W-PE responds better to the sudden change of amplitude, in addition, it has better robustness and stability than PE at low signal-to-noise ratio (SNR). As an improvement of PE, W-PE has important influence and status in different fields [16–18]. For example, W-PE can show a better performance than PE in distinguishing Alzheimer's disease patients from normal controls [19].

In 2016, Rostaghi and Azami proposed dispersion entropy (DE) to quantify the complexity of time series and first applied it to electroencephalograms and bearing fault diagnosis database [20]. Unlike W-PE, DE introduced amplitude information to map the original signal to the dispersion signal by using the normal cumulative distribution function (NCDF). Compared with PE, DE has a better ability to detect the change of simultaneous frequency and amplitude, and also has a better ability to distinguish different datasets and requires less computation time. In [21], PE, AE, and DE were compared, the results suggest that DE leads to more stable results in describing the state of rotating machinery, and it is more suitable for real-time applications.

In 2017, reverse PE (RPE) was put forward by Bandt and employed to identify different sleep stages by using electroencephalogram data [22]. Since RPE is defined as the distance from white noise, it has the opposite trend with PE, W-PE, and DE. In [23,24], RPE was used for feature extraction of underwater acoustic signals, compared with PE, RPE has more stable performance and a higher classification recognition rate.

To improve the performance of PE and integrate the advantages of DE and RPE, we propose a new complexity measure for analyzing time series in this paper, and term reverse dispersion entropy (RDE) through introducing amplitude information of DE and distance information of RPE. In the next section, RDE is described in detail through comparison with PE, W-PE, RPE, and DE. In Sections 3 and 4, simulation experiments are carried out to further compare and analyze five kinds of PE. Finally, we summarize the total research work in Section 5.

2. Reverse Dispersion Entropy

RDE, as a new complexity measure for analyzing time series, takes PE as its theoretical basis and combines the advantages of DE and RPE. The flow chart of PE and RDE are shown in Figure 1. As shown in Figure 1, all steps of PE and RDE are different except for phase space reconstruction.

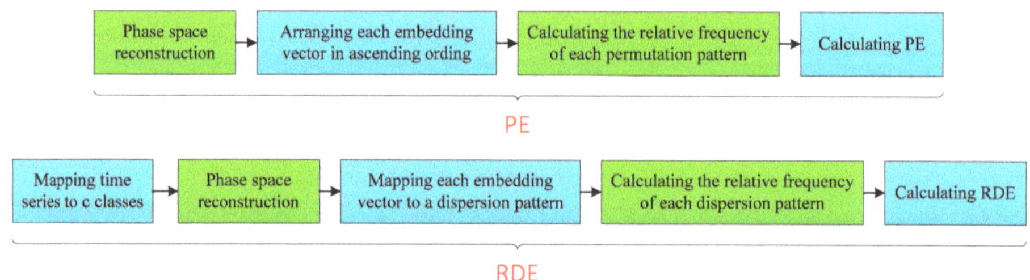

Figure 1. The flow chart of permutation entropy (PE) and reverse dispersion entropy (RDE).

The specific steps of RDE and detailed comparisons with the other four entropies are as follows [7,15,20,22]:

Step 1: mapping time series to c classes.

(1) Mapping by normal cumulative distribution function (NCDF)

For a time series $X = \{x(i),\ i = 1,\ 2,\ \cdots,\ T\}$ with T values, we map X to $Y = \{y(i),\ i = 1,\ 2,\ \cdots,\ T\}$ by NCDF, where $y(i)$ ranges from 0 to 1.

(2) Mapping by $round(c.y(i) + 0.5)$.

We map Y to $Z = \{z(i),\ i = 1,\ 2,\ \cdots,\ T\}$ by using $round(c.y(i) + 0.5)$, where c is the number of classes and $z(i)$ is a positive integer from 1 to c. There is no difference in this step between DE and RDE.

Step 2: phase space reconstruction.

We reconstruct Z into L. embedding vectors with the time delay τ and embedding dimension m, respectively. The matrix consisting of all embedding vectors can be represented as follows:

$$\begin{bmatrix} \{z(1),\ z(1+\tau),\ \cdots,\ z(1+(m-1)\tau)\} \\ \vdots \qquad \vdots \\ \{z(j),\ z(j+\tau),\ \cdots,\ z(j+(m-1)\tau)\} \\ \vdots \qquad \vdots \\ \{z(L),\ z(L+\tau),\ \cdots,\ z(L+(m-1)\tau)\} \end{bmatrix} \tag{1}$$

where the number of embedding vectors L is equal to $T - (m-1)\tau$. There is no difference in this step between PE, W-PE, RPE, and RDE.

Step 3: mapping each embedding vector to a dispersion pattern.

Since the embedding dimension and the number of classes are m and c, respectively, there exists c^m dispersion patterns, and each embedding vector can be mapped to a dispersion pattern π. For PE and W-PE, there exist $m!$ arrangement patterns, which is different from DE and RDE. However, there is also no difference in this step between DE and RDE.

Step 4: calculating the relative frequency of each dispersion pattern.

The relative frequency of i–th dispersion pattern can be expressed as follows:

$$P(\pi_i) = \frac{Number\{\pi_i\}}{N-(m-1)\tau}\ (1 \le i \le c^m) \tag{2}$$

In truth, $P(\pi_i)$ represents the proportion of the number of i–th dispersion patterns to the number of embedding vectors. The four kinds of entropy are the same in this step.

Step 5: calculating RDE.

Like RPE, RDE is defined as the distance to white noise by combining distance information. It can be expressed as:

$$H_{RDE}(X,m,c,\tau) = \sum_{i=1}^{c^m}\left(P(\pi_i) - \frac{1}{c^m}\right)^2 = \sum_{i=1}^{c^m} P(\pi_i)^2 - \frac{1}{c^m} \tag{3}$$

when $P(\pi_i) = 1/c^m$, the value of $H_{RDE}(X,m,c,\tau)$ is 0 (minimum value). In step 5, the calculation formulas of PE, W-PE, and DE are the same based on the definition of Shannon entropy, however, the calculation formula of RDE is the same as that of RPE by combining distance information.

When there is only one dispersion pattern, that is $P(\pi_i) = 1$, the value of $H_{RDE}(X,m,c,\tau)$ is $1 - \frac{1}{c^m}$ (maximum value). Therefore, the normalized RDE can be expressed as:

$$H_{RDE} = \frac{H_{RDE}(X,m,c,\tau)}{1-\frac{1}{c^m}} \tag{4}$$

Based on the test of simulation signals and real sensor signals, the recommended parameters of RDE are shown in Table 1. More details about PE, W-PE, DE, and RPE can be found in [7,15,20,22].

Table 1. The recommended parameters of RDE.

Parameters	τ	m	c	T
Values	1	2, 3	4, 5, 6, 7, 8	$T > c^m$

3. Simulations with Synthetic Signals

3.1. Simulation 1

To demonstrate the ability of RDE to detect mutation signals, we carried out a simulation experiment similar to [15]. The synthetic signals are as follows:

$$\begin{cases} y = x + s \\ x = \begin{cases} 50, & (t = 0.498) \\ 0, & (t >= 0\&t <= 1) \end{cases} \\ s = randn(t) \end{cases} \tag{5}$$

where the synthetic signal y with the sampling frequency of 1 kHz is composed of white Gaussian noise s and impulse signal x. The time domain waveform of y is shown in Figure 2. Five entropies are calculated by using a sliding window of 80 samples with 70 overlapped ones. For DE and RDE, the parameter c is 6. For all five entropies, the embedding dimension and time delay are 2 and 1. The five entropies of y are shown in Figure 3. Table 2 shows 5 entropies in the windows from 42 to 51. As shown in Figure 3 and Table 2, when the windows contain the impulse signal, the values of DE and RDE have a significant decrease and increase. For further comparison, the means of the five entropies and their variation ratios are shown in Table 3. A is the means of 82 windows without an impulse signal, B is the means of 8 windows with an impulse signal, and the variation ratio is the ratio of maximum to minimum of A and B. As shown in Table 3, for PE, W-PE, and RPE, A and B are very close, and the variation ratios are from 1.0002 to 1.04; for DE and RDE, there are obvious differences between A and B, the variation ratios are obviously greater than 1; DE has a variation ratio of 2.1503, and RDE has a variation ratio of up to 21.8034. The simulation results show that DE and RDE can detect mutation signals, and RDE with the highest variation ratio has better performance than the other four entropies in detecting mutation signals.

Table 2. The 5 entropies in the windows from 42 to 51.

Window	42	43	44	45	46	47	48	49	50	51
PE	0.995	0.993	0.995	0.997	0.995	0.996	0.997	0.998	0.997	0.996
W-PE	0.998	0.999	0.998	1.000	0.999	0.998	1.000	0.998	0.999	0.997
RPE	0.005	0.006	0.005	0.005	0.004	0.005	0.006	0.005	0.005	0.006
DE	0.934	0.432	0.434	0.435	0.436	0.435	0.433	0.434	0.436	0.935
RDE	0.012	0.265	0.257	0.252	0.248	0.252	0.256	0.256	0.255	0.011

Table 3. The means of the five entropies and their variation ratios.

Parameters	PE	W-PE	RPE	DE	RDE
A (Means of 82 windows)	0.9962	0.9980	0.0052	0.9345	0.0117
B (Means of 8 windows)	0.9964	0.9995	0.0050	0.4346	0.2551
Max(A, B)/Min(A, B)	1.0002	1.0015	1.0400	2.1503	21.8034

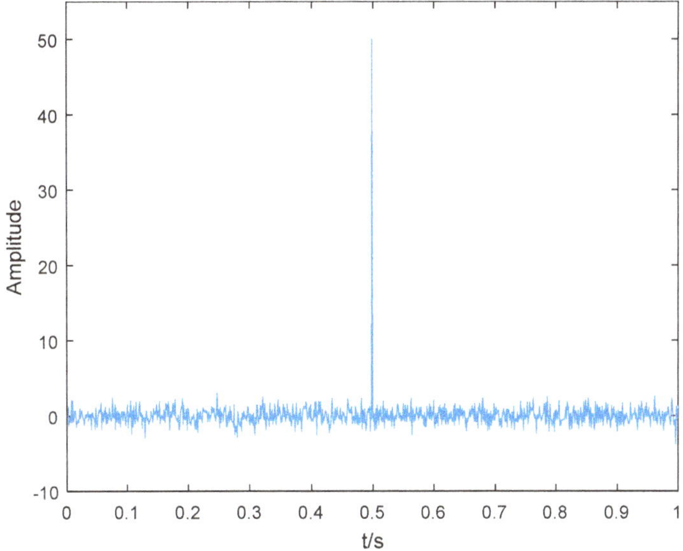

Figure 2. The time domain waveform of y.

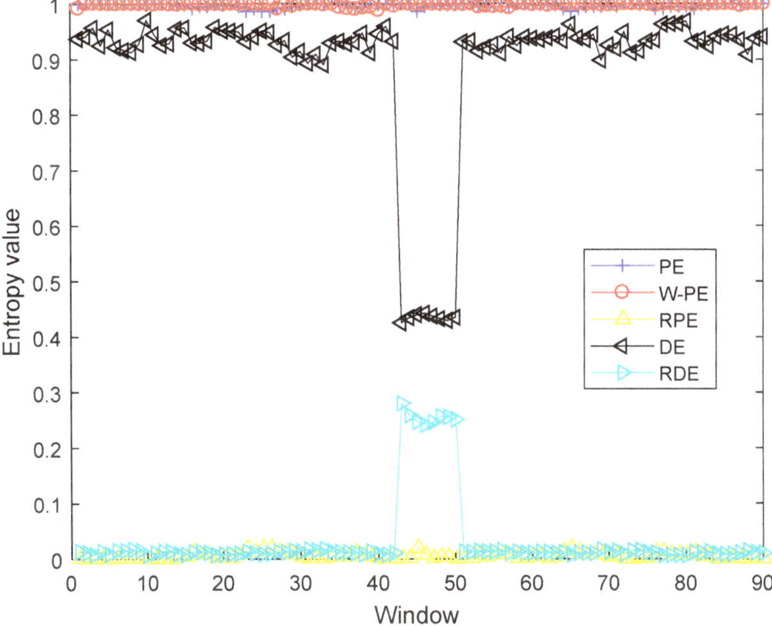

Figure 3. The five entropies of y. W-PE: weighted-permutation entropy; RPE: reverse permutation entropy; DE: dispersion entropy.

3.2. Simulation 2

Based on the recommended parameters range of RDE, we changed the embedding dimension of simulation 1 to 3. In view of $T > c^m$ and the sliding window of 80 samples, c can be set to four. Like simulation 1, simulation 2 was carried out with different parameters from simulation 1. The five entropies of y are shown in Figure 4. As shown in Figure 4, when the windows contain the impulse signal, the values of W-PE and DE have a significant decrease, and the values of RDE have a dramatical increase. Unlike simulation 1, W-PE with the embedding dimension of three can detect the mutation signal, the simulation results show that the value of embedding dimension can affect the capability of W-PE in detecting mutation signals.

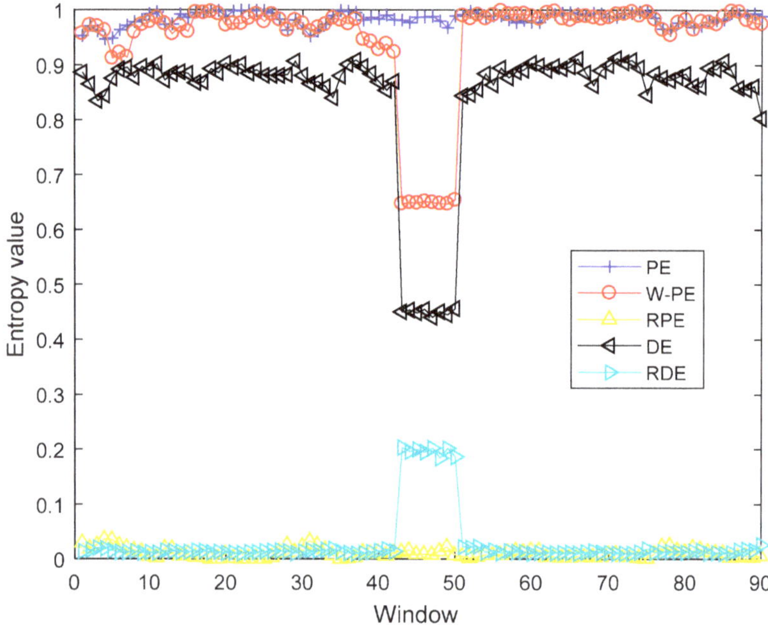

Figure 4. The five entropies of y.

Like Table 3, the means of the five entropies and their variation ratios are shown in Table 4. A is the means of 82 windows without an impulse signal, B is the means of eight windows with an impulse signal, and the variation ratio is the ratio of maximum to minimum of A and B. As shown in Table 4, for PE and RPE, A and B are close, and the variation ratios are almost 1 (1.0021 and 1.1468, respectively); for W-PE and DE, there are obvious differences between A and B, the variation ratios are obviously greater than 1 and less than 2 (1.5037 and 1.958, respectively); RDE has a variation ratio of up to 14.0143. The simulation results show that W-PE can detect mutation signals with the embedding dimension of three, DE and RDE can also detect mutation signals under different parameters, and RDE with the highest variation ratio have better performance than other four entropies in detecting mutation signal.

Table 4. The means of the five entropies and their variation ratios.

Parameters	PE	W-PE	RPE	DE	RDE
A (Means of 82 windows)	0.9844	0.9765	0.0109	0.8805	0.0140
B (Means of 8 windows)	0.9823	0.6494	0.0125	0.4497	0.1962
Max(A, B) / Min(A, B)	1.0021	1.5037	1.1468	1.9580	14.0143

3.3. Simulation 3

In order to verify the robustness of RDE to noise, we carried out noise robustness testing by using the same impulse signal x in simulation 1, and the synthetic signals y with different SNRs can be obtained by adding white Gaussian noise to x. All parameters are consistent with simulation 1. We set the embedding dimension and time delay to three and one for the five entropies and set *c* to six for DE and RDE.

The five entropies of synthetic signal under different SNRs from −10 dB to 10 dB are shown in Figure 5, and each entropy value is the mean under 1000 calculations. As shown in Figure 5, the values of PE and RPE have barely changed for different SNRs and are close to 1 and 0, respectively; the values of W-PE and DE monotonically decrease with the increase of SNR; the values of RDE monotonically increase with the increase of SNR. For further comparison of W-PE, DE, and RDE, the three entropies under −10 dB and 10 dB and their variation ratios are shown in Table 5. A and B are the entropies under 10 dB and −10 dB, and $\mathrm{Max(A, B)/Min(A, B)}$ is the ratio of maximum to minimum of A and B. As shown in Table 5, for W-PE and RDE, there are differences between A and B, the variation ratios are 1.3909 and 1.6993, respectively; RDE has a variation ratio of up to 94.3. Therefore, RDE can better reflect the difference under different SNRs than the other four entropies. The simulation results show that RDE with the highest variation ratio has better robustness to noise than the other four entropies.

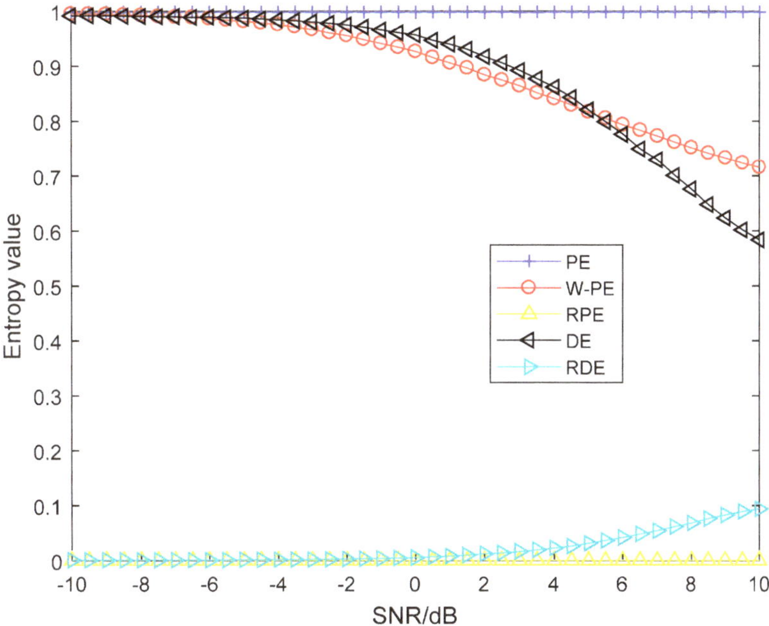

Figure 5. The five entropies of synthetic signal under different signal-to-noise ratios (SNRs).

Table 5. The three entropies under −10 dB and 10 dB and their variation ratios.

Parameters	W-PE	DE	RDE
A (10 dB)	0.7160	0.5839	0.0943
B (−10 dB)	0.9959	0.9922	0.0010
$\mathrm{Max(A, B)/Min(A, B)}$	1.3909	1.6993	94.3000

3.4. Simulation 4

In order to prove the stability of RDE for synthetic signal, we carried out stability testing by using the cosine signal of different lengths with the frequency of 100 Hz. For the five entropies, we set the embedding dimension and time delay to three and one and set *c* to six for DE and RDE. The five entropies of cosine signal with the frequency of 100 Hz are shown in Figure 6, the initial data length is 2000 sampling points, and 100 sampling points are added each time until the data length reached 12,000 sampling points.

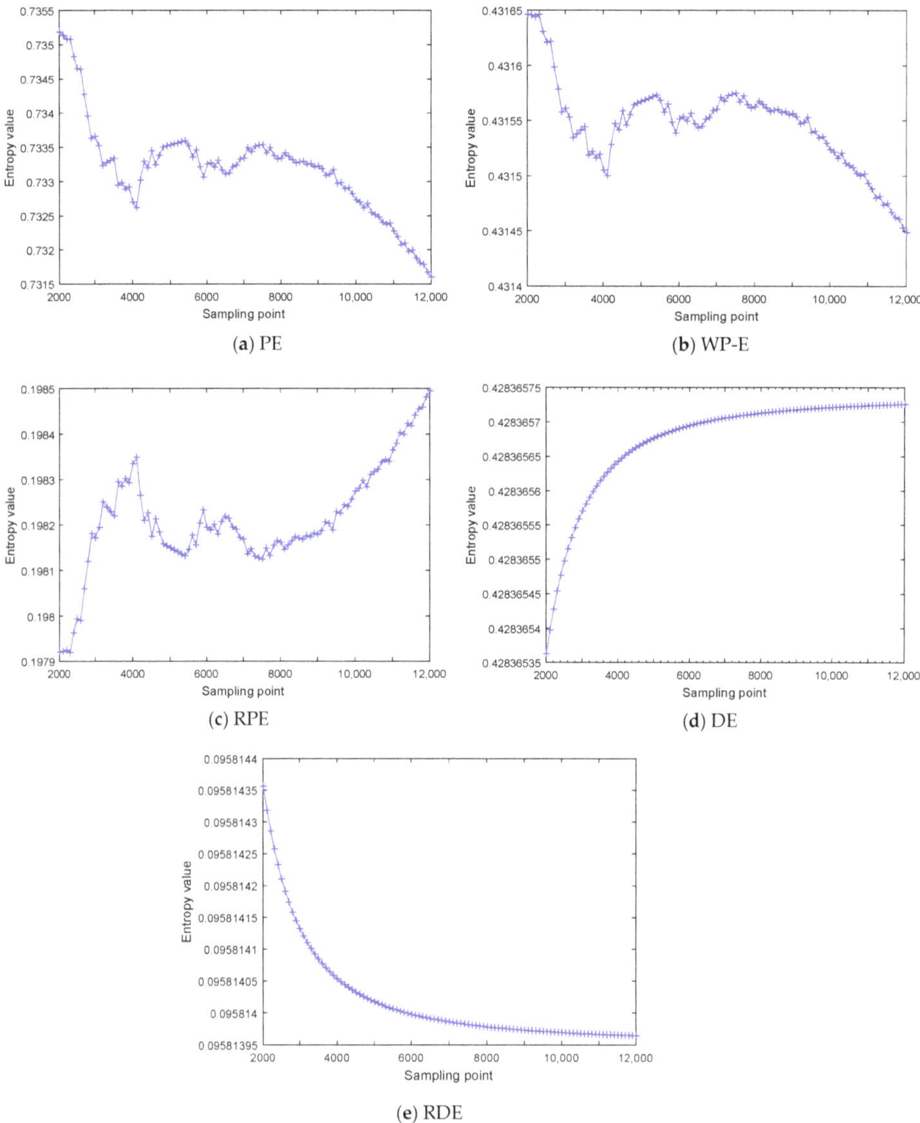

(**a**) PE (**b**) WP-E (**c**) RPE (**d**) DE (**e**) RDE

Figure 6. The five entropies of cosine signal with the frequency of 100 Hz.

As shown in Figure 6, the five entropies change in varying degrees with the increase of data length; the values of W-PE are from 0.43145 to 0.43165, the variation ranges of W-PE $(10-4)$ are one order of magnitude lower than ones of PE $(10-3)$ and RPE $(10-3)$; the values of DE are from 0.4283653 to 0.4283657, the values of RDE are from 0.095814 to 0.0958144, and the variation ranges of DE $((10-7)$ and RDE $(10-7)$ are smaller than ones of W-PE. The mean and standard deviation of five entropies for the cosine signal of different lengths are shown in Table 6. As shown in Table 6, RDE has the smallest standard deviation compared with the other four entropies. The stability testing results indicate that DE and RDE have better stability than the other three entropies under different length data.

Table 6. The mean and standard deviation of five entropies for the cosine signal of different lengths.

Parameters	PE	W-PE	RPE	DE	RDE
mean value	0.7334	0.4316	0.1982	0.4284	0.0958
standard deviation	5×10^{-4}	3×10^{-5}	9×10^{-5}	9×10^{-8}	8×10^{-8}

3.5. Simulation 5

In order to prove the stability of RDE for synthetic signal, we carried out stability testing by using the cosine signal $\cos(200\pi t)$ under 10 dB. The data length of each sample is 2000, we calculated the 5 entropies of 100 samples. The five entropies of cosine signal under 10 dB are shown in Figure 7. As shown in Figure 7, the entropy values of the same category are at the same level and have very little difference.

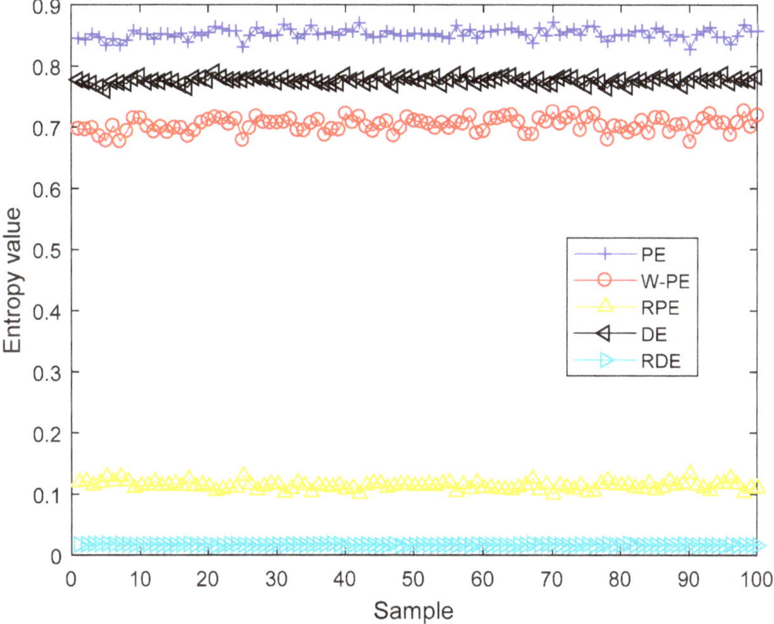

Figure 7. The five entropies of cosine signal under 10 dB.

To more intuitively compare the stability of the five entropies, the complexity feature boxplots of five entropies for cosine signal under 10 dB are shown in Figure 8. As shown in Figure 8, PE, W-PE, RPE, and DE have obvious fluctuations, however, RDE has the smallest fluctuation range compared to the other four entropies. The mean and standard deviation of the five entropies for the cosine signal under 10 dB are shown in Table 7. As shown in Table 7, RDE has the smallest standard deviation compared with the other four entropies. The experimental results show that RDE has better stability than the other four entropies under noisy conditions.

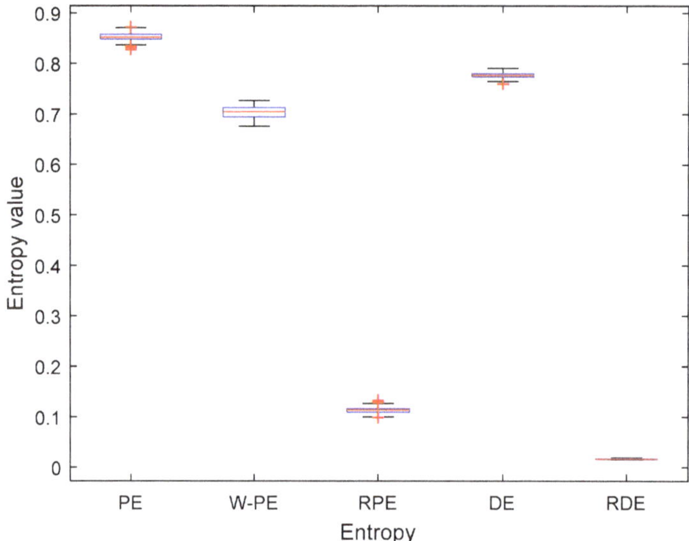

Figure 8. The complexity feature boxplots of five entropies for cosine signal under 10 dB.

Table 7. The mean and standard deviation of five entropies for the cosine signal under 10 dB.

Parameters	PE	W-PE	RPE	DE	RDE
mean value	0.8510	0.7034	0.1154	0.7770	0.0173
standard deviation	0.0073	0.0113	0.0056	0.0048	0.0006

4. Application for Real Sensor Signals

4.1. Simulation 1

In order to compare the ability of five entropies to distinguish real sensor signals, we carried out complexity testing by using three kinds of ship signals, termed as ship 1, ship 2, and ship 3. Each sample was 5000 points with a sampling frequency of 44.1 kHz. The five entropy distributions of three kinds of ship are shown in Figure 9, and each kind of ship includes 100 samples. As shown in Figure 9, compared with the distributions of PE, W-PE, and RPE, the distributions of DE and RDE were easier to distinguish the three kinds of ship signals.

The complexity feature boxplots of five entropies for three kinds of ship are shown in Figure 10, and the mean and standard deviation of five entropies for three kinds of ship are shown in Table 8. As shown in Figure 10 and Table 8, compared with the other four entropies, RDE had the smallest fluctuation range and standard deviation for each ship signal. The experimental results show that RDE has better stability for ship signals.

Table 8. The mean and standard deviation of five entropies for three kinds of ship.

	PE	W-PE	RPE	DE	RDE
mean value of ship 1	0.7325	0.4617	0.2039	0.5731	0.0637
standard deviation of ship 1	0.0096	0.0111	0.0069	0.0088	0.0033
mean value of ship 2	0.8259	0.5550	0.1347	0.7179	0.0284
standard deviation of ship 2	0.0065	0.0096	0.0049	0.0150	0.0030
mean value of ship 3	0.7862	0.5051	0.1646	0.6308	0.0472
standard deviation of ship 3	0.0112	0.0150	0.0083	0.0094	0.0028

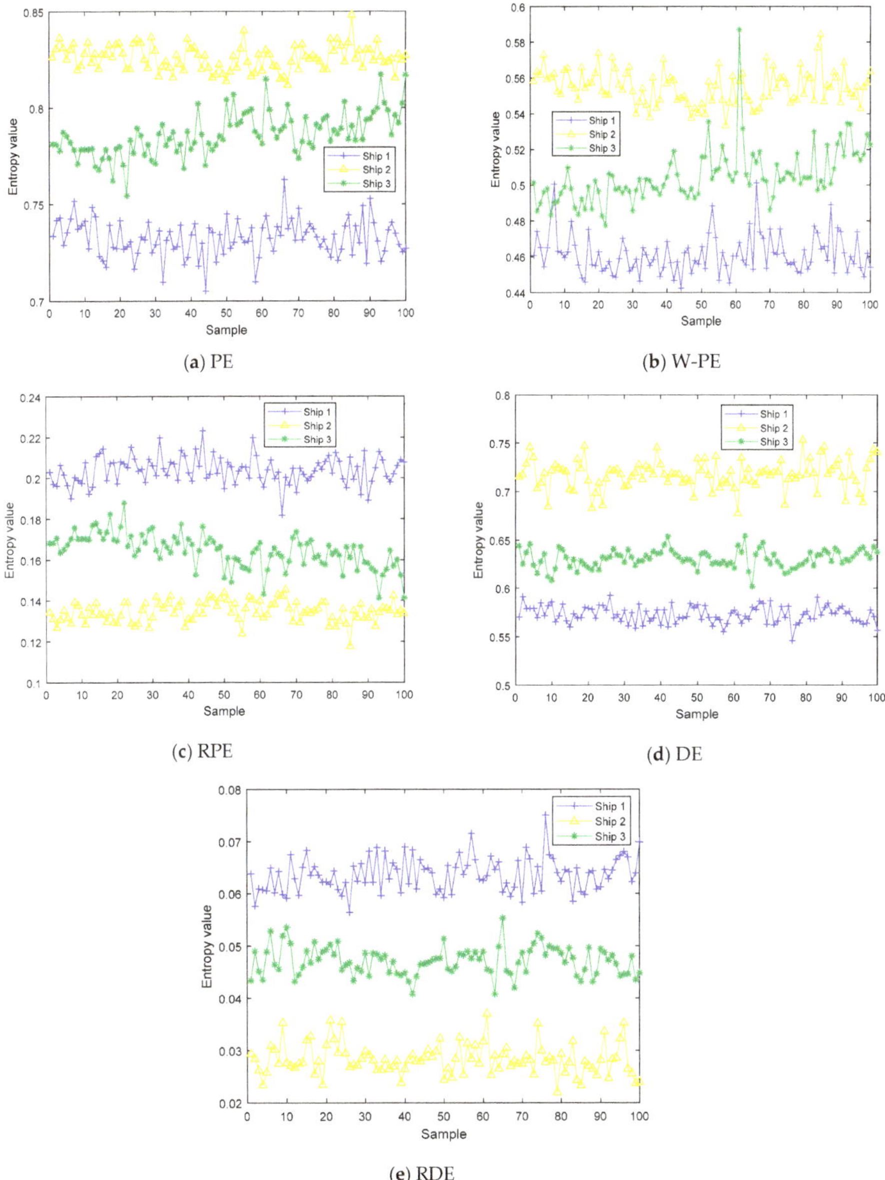

Figure 9. The five entropy distributions for three kinds of ship.

To further prove the distinguishing ability of RDE, we used a support vector machine (SVM) to distinguish the three kinds of ship signals; the classification results by five entropies for three kinds of ship are shown in Table 9. As seen in Table 9, PE and W-PE have a recognition rate of less than 95%; DE and RPE have a recognition rate of more than 95%; RDE has the highest recognition rate of up to 99%. The experimental results show that RDE has better distinguishing ability for ship signals.

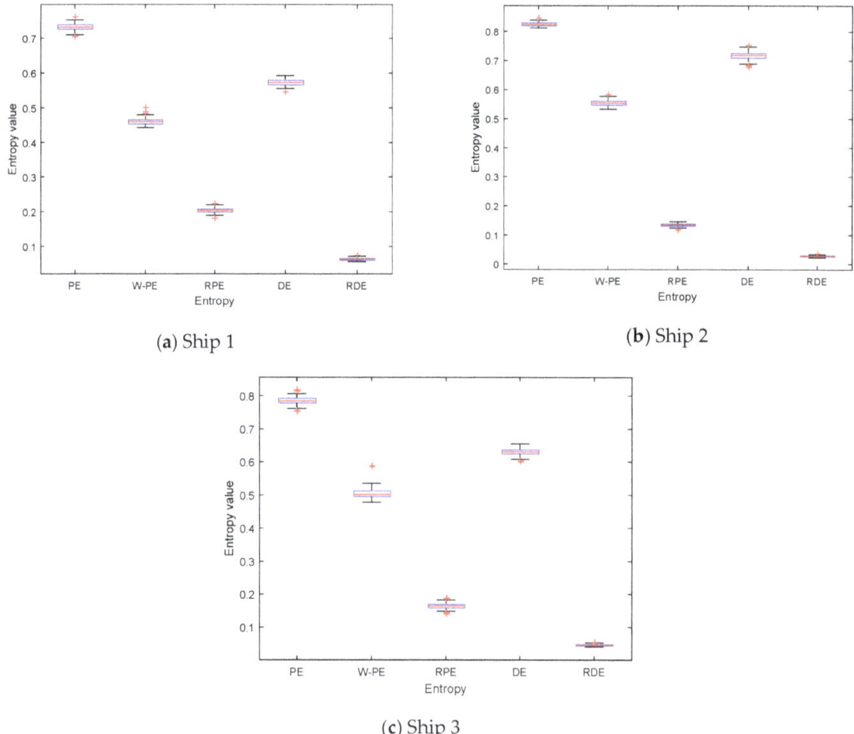

(**a**) Ship 1

(**b**) Ship 2

(**c**) Ship 3

Figure 10. The complexity feature boxplots of five entropies for three kinds of ship.

Table 9. The classification results by five entropies for three kinds of ship.

PE	W-PE	RPE	DE	RDE
92.67%	93.33%	96%	98.33%	99%

4.2. Simulation 2

Like simulation 1 in Section 4.1, we carried out complexity testing by using three kinds of rolling bearings signals, termed as fault 1, fault 2, and fault 3, which come from the Case Western Reverse Laboratory [25]. Each sample is 2000 points with a sampling frequency of 12 kHz. The mean and standard deviation of the five entropies for three kinds of fault are shown in Table 10, and each kind of fault includes 50 samples. As shown in Table 10, for PE, W-PPE, RPE, and DE, the mean values of fault 1 and fault 2 are very close, which makes it difficult to distinguish the two faults; for RDE, there are obvious differences in the mean values of the three faults, and it has the smallest standard deviation compared to the other four entropies. The experimental results show that RDE has better stability for rolling bearing signals.

Table 10. The mean and standard deviation of five entropies for three kinds of fault.

	PE	W-PE	RPE	DE	RDE
mean value of fault 1	0.7752	0.4932	0.1785	0.7378	0.0221
standard deviation of fault 1	0.0063	0.0040	0.0054	0.0045	0.0009
mean value of fault 2	0.9702	0.8820	0.0226	0.9480	0.0023
standard deviation of fault 2	0.0023	0.0069	0.0018	0.0028	0.0001
mean value of fault 3	0.9717	0.9045	0.0214	0.9227	0.0039
standard deviation of fault 3	0.0024	0.0073	0.0018	0.0135	0.0009

To further prove the distinguishing ability of RDE for rolling bearing signals, we used an SVM to distinguish three kinds of rolling bearing signals; the classification results by five entropies for three kinds of rolling bearing signals are in Table 11. As seen in Table 11, PE and W-PE have a recognition rate of less than 80%; RPE has a recognition rate of more than 80%; DE and RDE have a recognition rate of less than 95%; RDE has the highest recognition rate of up to 100%. The experimental results show that RDE has better distinguishing ability for rolling bearing signals.

Table 11. The classification results by five entropies for three kinds of rolling bearing signals.

PE	W-PE	RPE	DE	RDE
74.67%	77.33%	83.33%	96.67%	100%

5. Conclusions

This paper proposed a new complexity measure for analyzing time series and termed RDE. A large number of simulation experiments was carried out to verify the effectiveness of this complexity measure. Its main contributions are as follows:

1. Compared with PE, W-PE, RPE, and DE, RDE had better performance in detecting mutation signals under different embedding dimensions.
2. Compared with PE, W-PE, RPE, and DE, RDE had better robustness to noise and also better stability in the case of different length data and the presence of noise.
3. For real signals, RDE had better distinguishing ability and stability than PE, W-PE, RPE, and DE.

Overall, as an effective complexity metric, RDE could be used to analyze more real sensor signals in different fields.

Author Contributions: Development of theoretical approach, Y.L. and X.G.; numerical analysis, Y.L., X.G. and L.W.; writing—original draft preparation, Y.L.; writing—review and editing, Y.L. and X.G.

Funding: The authors gratefully acknowledge the support of the National Natural Science Foundation of China (No. 11574250).

Conflicts of Interest: The authors declare no conflicts of interest.

References

1. Olsson, A.; Persson, A.; Bartfai, A.; Boman, I. Sensor technology more than a support. *Scand. J. Occup. Ther.* **2018**, *25*, 79–87. [CrossRef]
2. Guerrero-Ibáñez, J.; Zeadally, S.; Contreras-Castillo, J. Sensor Technologies for Intelligent Transportation Systems. *Sensors* **2018**, *18*, 1212. [CrossRef]
3. Zanin, M.; Zunino, L.; Rosso, O.A.; Papo, D. Permutation Entropy and Its Main Biomedical and Econophysics Applications: A Review. *Entropy* **2012**, *14*, 1553–1577. [CrossRef]
4. Li, Y.; Chen, X.; Yu, J.; Yang, X.; Yang, H. The Data-Driven Optimization Method and Its Application in Feature Extraction of Ship-Radiated Noise with Sample Entropy. *Energies* **2019**, *12*, 359. [CrossRef]
5. Pincus, S. Approximate entropy (ApEn) as a complexity measure. *Chaos* **1995**, *5*, 110–117. [CrossRef] [PubMed]

6. Ishikawa, A.; Mieno, H. The fuzzy entropy concept and its application. *Fuzzy Sets Syst.* **1979**, *2*, 113–123. [CrossRef]
7. Bandt, C.; Pompe, B. Permutation entropy: A natural complexity measure for time series. *Phys. Rev. Lett.* **2002**, *88*, 174102. [CrossRef] [PubMed]
8. Nicolaou, N.; Georgiou, J. Detection of epileptic electroencephalogram based on Permutation Entropy and Support Vector Machines. *Expert Syst. Appl.* **2012**, *39*, 202–209. [CrossRef]
9. Zhang, X.; Liang, Y.; Zhou, J.; Zang, Y. A novel bearing fault diagnosis model integrated permutation entropy, ensemble empirical mode decomposition and optimized SVM. *Measurement* **2015**, *69*, 164–179. [CrossRef]
10. Yan, R.; Liu, Y.; Gao, R. Permutation entropy: A nonlinear statistical measure for status characterization of rotary machines. *Mech. Syst. Signal Process.* **2012**, *29*, 474–484. [CrossRef]
11. Zunino, L.; Zanin, M.; Tabak, B.; Perez, D.; Rosso, O. Forbidden patterns, permutation entropy and stock market inefficiency. *Phys. A* **2009**, *388*, 2854–2864. [CrossRef]
12. Hou, Y.; Liu, F.; Gao, J.; Cheng, C.; Song, C. Characterizing Complexity Changes in Chinese Stock Markets by Permutation Entropy. *Entropy* **2017**, *19*, 514. [CrossRef]
13. Li, Y.-X.; Li, Y.-A.; Chen, Z.; Chen, X. Feature Extraction of Ship-Radiated Noise Based on Permutation Entropy of the Intrinsic Mode Function with the Highest Energy. *Entropy* **2016**, *18*, 393. [CrossRef]
14. Li, Y.; Li, Y.; Chen, X.; Yu, J. A Novel Feature Extraction Method for Ship-Radiated Noise Based on Variational Mode Decomposition and Multi-Scale Permutation Entropy. *Entropy* **2017**, *19*, 342.
15. Fadlallah, B.; Chen, B.; Keil, A.; Principe, J. Weighted-permutation entropy: A complexity measure for time series incorporating amplitude information. *Phys. Rev. E* **2013**, *87*, 022911. [CrossRef]
16. Zhou, S.; Qian, S.; Chang, W.; Xiao, Y.; Cheng, Y. A Novel Bearing Multi-Fault Diagnosis Approach Based on Weighted Permutation Entropy and an Improved SVM Ensemble Classifier. *Sensors* **2018**, *18*, 1934. [CrossRef]
17. Bian, Z.; Ouyang, G.; Li, Z.; Li, Q.; Wang, L.; Li, X. Weighted-Permutation Entropy Analysis of Resting State EEG from Diabetics with Amnestic Mild Cognitive Impairment. *Entropy* **2016**, *18*, 307. [CrossRef]
18. Li, Y.; Wang, L.; Li, X.; Yang, X. A Novel Linear Spectrum Frequency Feature Extraction Technique for Warship Radio Noise Based on Complete Ensemble Empirical Mode Decomposition with Adaptive Noise, Duffing Chaotic Oscillator, and Weighted-Permutation Entropy. *Entropy* **2019**, *21*, 507. [CrossRef]
19. Deng, B.; Liang, L.; Li, S.; Wang, R.; Yu, H.; Wang, J.; Wei, X. Complexity extraction of electroencephalograms in Alzheimer's disease with weighted-permutation entropy. *Chaos Interdiscip. J. Nonlinear Sci.* **2015**, *25*, 043105. [CrossRef]
20. Rostaghi, M.; Azami, H. Dispersion Entropy: A Measure for Time Series Analysis. *IEEE Signal Process. Lett.* **2016**, *23*, 610–614. [CrossRef]
21. Mostafa, R.; Reza, A.; Hamed, A. Application of Dispersion Entropy to Status Characterization of Rotary Machines. *J. Sound Vib.* **2018**, *438*, 291–308.
22. Bandt, C. A New Kind of Permutation Entropy Used to Classify Sleep Stages from Invisible EEG Microstructure. *Entropy* **2017**, *19*, 197. [CrossRef]
23. Li, Y.; Li, Y.; Chen, X.; Yu, J. Denoising and Feature Extraction Algorithms Using NPE Combined with VMD and Their Applications in Ship-Radiated Noise. *Symmetry* **2017**, *9*, 256. [CrossRef]
24. Li, Y.; Chen, X.; Yu, J.; Yang, X. A Fusion Frequency Feature Extraction Method for Underwater Acoustic Signal Based on Variational Mode Decomposition, Duffing Chaotic Oscillator and a Kind of Permutation Entropy. *Electronics* **2019**, *8*, 61. [CrossRef]
25. Available online: http://csegroups.case.edu/bearingdatacenter/pages/download-data-file (accessed on 25 November 2019).

Article

Efficient Noisy Sound-Event Mixture Classification Using Adaptive-Sparse Complex-Valued Matrix Factorization and OvsO SVM

Phetcharat Parathai [1], Naruephorn Tengtrairat [1], Wai Lok Woo [2,*], Mohammed A. M. Abdullah [3], Gholamreza Rafiee [4] and Ossama Alshabrawy [2]

1 School of Software Engineering, Payap University, Chiang Mai 50000, Thailand; phetcharat@payap.ac.th (P.P.); naruephorn_t@payap.ac.th (N.T.)
2 Department of Computer and Information Sciences, Northumbria University, Newcastle upon Tyne NE1 8ST, UK; ossama.alshabrawy@northumbria.ac.uk
3 Computer and Information Engineering Department, Ninevah University, Mosul 41002, Iraq; mohammed.abdulmuttaleb@uoninevah.edu.iq
4 School of Electronics, Electrical Engineering and Computer Science, Queen's University Belfast, Belfast BT9 5BN, UK; g.rafiee@qub.ac.uk
* Correspondence: wailok.woo@northumbria.ac.uk

Received: 20 June 2020; Accepted: 4 August 2020; Published: 5 August 2020

Abstract: This paper proposes a solution for events classification from a sole noisy mixture that consist of two major steps: a sound-event separation and a sound-event classification. The traditional complex nonnegative matrix factorization (CMF) is extended by cooperation with the optimal adaptive L_1 sparsity to decompose a noisy single-channel mixture. The proposed adaptive L_1 sparsity CMF algorithm encodes the spectra pattern and estimates the phase of the original signals in time-frequency representation. Their features enhance the temporal decomposition process efficiently. The support vector machine (SVM) based one versus one (OvsO) strategy was applied with a mean supervector to categorize the demixed sound into the matching sound-event class. The first step of the multi-class MSVM method is to segment the separated signal into blocks by sliding demixed signals, then encoding the three features of each block. Mel frequency cepstral coefficients, short-time energy, and short-time zero-crossing rate are learned with multi sound-event classes by the SVM based OvsO method. The mean supervector is encoded from the obtained features. The proposed method has been evaluated with both separation and classification scenarios using real-world single recorded signals and compared with the state-of-the-art separation method. Experimental results confirmed that the proposed method outperformed the state-of-the-art methods.

Keywords: audio signal processing; sound event classification; nonnegative matric factorization; blind signal separation; support vector machines

---

1. Introduction

Surveillance systems have become increasingly ubiquitous in our living environment. These systems have been used in various applications including CCTV in traffic and site monitoring, and navigation. Automated surveillance is currently based on video sensory modality and machine intelligence. Recently, intelligent audio analysis has been taken into account in surveillance to improve the monitoring system via detection, classification, and recognition sound in a scenario. However, in a real-world situation, background noise has interfered in both the image and sound of a surveillance system. This will hinder the performance of a surveillance system. Hence, an automatic signal separation and event classification algorithm was proposed to improve the surveillance system by classifying the observed sound-event in noisy scenarios. The proposed noisy sound separation and event classification method is based on two approaches (i.e., blind signal separation and sound classification, which are introduced in the sections to follow, respectively).

The classical problem of blind source separation (BSS), the so-called "cocktail party problem", is a psycho-acoustic spectacle that alludes to the significant human-auditory capability to selectively focus on and identify the sound-source speaker from the scenarios. The interference is produced by competing speech sounds or a variety of noises that are often assumed to be independent of each other. In the case of only a single microphone being available, this reduces to the single channel blind source separation (SCBSS) [1–4]. The majority of SCBSS algorithms work in time-frequency domain, for example, binary masking [5–7] or nonnegative matrix factorization (NMF) [8–11]. NMF has been continuously developed with great success for decomposing underlying original signals when a sole sensor is available. NMF was developed using the multiplicative update (MU) algorithm to solve its parametrical optimization based on a cost function such as the Kullback–Liebler divergence and the least square distance. Later, other families of cost functions have been continuously proposed, for example, the Beta divergence [12], Csiszar's divergences, and Itakura–Saito divergence [13]. Additionally, iterative gradient update was presented where a sparsity constraint can be included into the optimizing function through regularization by minimizing penalized least squares [14] and using different sparsity constraints for dictionary and code [15]. The complex nonnegative matrix factorization (CMF) spreads the NMF model by combining a sparsity representation with the complex-spectrum domain to improve the audio separability. The CMF can extract the recurrent patterns of the phase estimates and magnitude spectra of constituent signals [16–18]. Nevertheless, the CMF lacks the generalized mechanics used for controlling the sparseness of the code. However, the sparsity parameter is manually determined for the above proposed methods. Approximate sparsity is an important consideration as they represent important information. Many sparse solutions have been proposed in the last decade [19–25]. Nonetheless, the optimal sparse solution remains an open issue.

Sound event classification (SEC) has vastly been exploited by many researchers. Sound can be categorized into speech, music, noise, environmental sound, or daily living sound [26]. Sound events are available in all classes, for example, car horn, traffic, walking, or knocking, etc. [27,28]. Sound-events contain significant information that can be used to describe what has happened or to predict what will happen next in the future. Most algorithms of the SEC methods are conveyed from sound classification approaches such as sparse coding, deep learning, and support vector machine (SVM). These approaches have been exploited to categorize a sound event in both indoor and outdoor scenarios. In recent years, the deep learning approach has been used to classify the sound-event. A deep learning framework can be established with two convolutional neural networks (CNNs) and a deep multi-layer perceptron (MLP) with rectified linear units (ReLU) as the activation function [29,30]. A Softmax function that is the final activation function is used to classify the sound into its corresponding class. The Softmax function is considered as the generalization of the logistic function, which aims to avoid overfitting. One of the advantages of deep learning is that it does not require feature extraction for the input sound. However, a deep neural network requires large training samples and despite a plethora of research, there is a general consensus that deep neural networks are still difficult to fine tune and generalize to test data. Moreover, it does not lend itself to the explanation as to why a certain decision is being made. Separate from the deep learning framework, another SEC approach is support vector machines [31,32], which has been practically presented to solve the classifier problem in various fields. The SVM algorithm relies on supervised learning by using the fundamental concept of statistic and risk minimization. The main process of the SVM is to draw the optimal separating hyperplane as the decision boundary located in such a way that the margin of separation between classes is maximized. The SVM approach is considered as supervised learning algorithm that is comprised of two sections: (1) a training section to model feature space and an optimal hyperplane, and (2) a testing section to use the SVM model for separating the observed data. The margin denotes the distance of the closest instance and the hyperplane. SVM has the desirable properties in that it requires only two differentiating factors to categorize two classes and a hyperplane that can be constructed to suit for an individual problem, even in the nonlinear case by selecting a kernel. Second, SVM provides a unique solution, since it is a convex optimization problem.

The rest of this paper is organized as follows. Section 2 presents the proposed noisy sound separation and event classification method, respectively. Next, Section 3 demonstrates and analyzes the performance of the proposed method. Finally, conclusions are drawn up in Section 4.

2. Background

Noisy mixed signals observed via a recording device can be stated as: $y(t) = s_1(t) + s_2(t) + n(t)$ where $s_1(t)$ and $s_2(t)$ denote the original sounds, and $n(t)$ is noise. This research is focused on two sound events in a single recorded signal. The proposed method consists of two steps: noisy sound separation and sound event classification, which is illustrated in Figure 1, where $y(t)$ and $\mathbf{Y}(\omega, t)$ denote a sound-event mixture in the time domain and time-frequency domain, respectively. The terms $\mathbf{W}^k(\omega), \mathbf{H}^k(t)$, $\boldsymbol{\phi}^k(\omega, t)$ are spectral basis, temporal code or weight matrix, and phase information, respectively. The term $\lambda^k(t)$ represents sparsity and $\hat{s}_j(t)$ is an estimated sound event source. The abbreviations MFCC, STE, and STZCR stand for Mel frequency cepstral coefficients, short-time energy, and short-time zero-crossing rate, respectively. The proposed method is consecutively elaborated in the following parts.

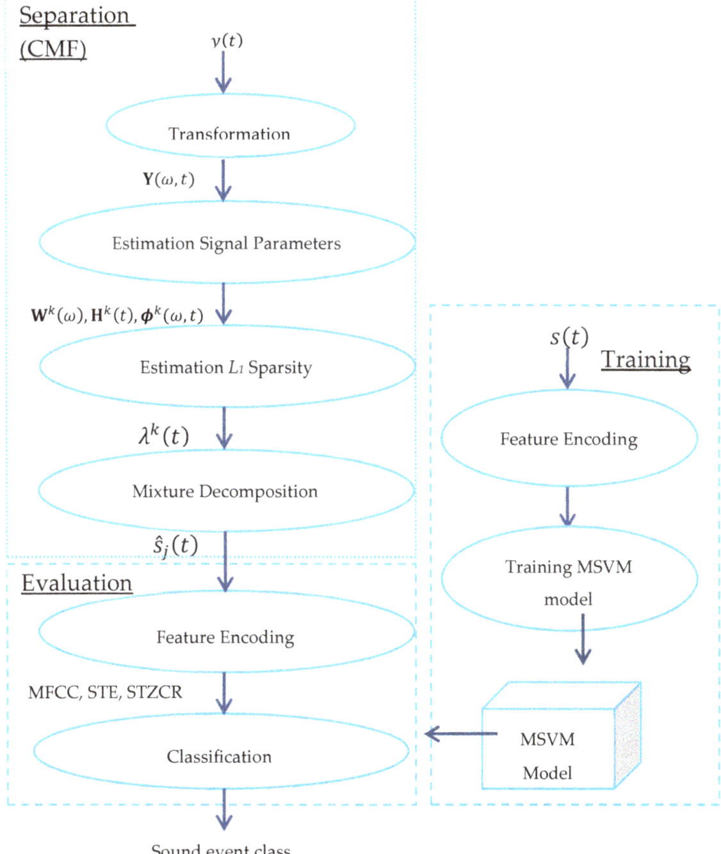

Figure 1. Signal flow of the proposed method.

2.1. Single-Channel Sound Event Separation

The problem formulation in time-frequency (TF) representation is given by an observed complex spectrum, $\mathbf{Y}_{f,t} \in \mathbb{C}$, to estimate the optimal parameters $\theta = \{\mathbf{W}, \mathbf{H}, \boldsymbol{\phi}\}$ of the model. A new factorization algorithm named as the adaptive L_1-sparse complex non-negative matrix factorization (adaptive L_1-SCMF) is derived in the following section. The generative model is given by

$$\mathbf{Y}(\omega,t) = \sum_{k=1}^{K} \mathbf{W}^k(\omega)\mathbf{H}^k(t)\mathbf{Z}^k(\omega,t) = \mathbf{X}(\omega,t) + \epsilon(\omega,t) \tag{1}$$

where $\mathbf{Z}^k(\omega,t) = e^{j\boldsymbol{\phi}^k(\omega,t)}$ and the reconstruction error $\boldsymbol{\epsilon}(\omega,t) \sim \mathcal{N}_{\mathbb{C}}\left(0,\sigma^2\right)$ is assumed to be independently and identically distributed (i.i.d.) with white noise having zero mean and variance σ^2. The term $\boldsymbol{\epsilon}(\omega,t)$ is used to denote a modeling error for each source. The likelihood of $\theta = \{\mathbf{W}, \mathbf{H}, \boldsymbol{\phi}\}$ is thus written as

$$P(\mathbf{Y}|\theta) = \prod_{f,t} \frac{1}{\pi\sigma^2} \exp\left(-\frac{\left|\mathbf{Y}(\omega,t) - \mathbf{X}(\omega,t)\right|^2}{\sigma^2}\right) \tag{2}$$

It is assumed that the prior distributions for $\mathbf{W}, \mathbf{H}$, and $\boldsymbol{\phi}$ are independent, which yields

$$P(\theta|\boldsymbol{\lambda}) = P(\boldsymbol{W})P(\boldsymbol{H}|\boldsymbol{\lambda})P(\boldsymbol{\phi}) \tag{3}$$

The prior $P(\mathbf{H}|\boldsymbol{\lambda})$ corresponds to the sparsity cost, for which a natural choice is a generalized Gaussian prior. When $p = 1$, $P(\mathbf{H}|\boldsymbol{\lambda})$ promotes the L_1-norm sparsity. L_1-norm sparsity has been shown to be probabilistically equivalent to the pseudo-norm, L_0, which is the theoretically optimum sparsity [33,34]. However, L_0-norm is non-deterministic polynomial-time (NP) hard and is not useful in large datasets such as audio. Given Equation (3), the posterior density [35,36] is defined as the maximum a posteriori probability (MAP) estimation problem, which leads to minimizing the following optimization problem with respect to θ. Equations of Gaussian prior and maximum a posteriori probability (MAP) estimation are expressed in Appendix A.

The CMF parameters have been upgraded by using an efficient auxiliary function for an iterative process. The auxiliary function for $f(\theta)$ can be expressed as the following: for any auxiliary variables with $\sum_k \overline{\mathbf{Y}}^k(\omega,t) = \mathbf{Y}(\omega,t)$, for any $\beta^k(\omega,t) > 0$, $\sum_k \beta^k(\omega,t) = 1$, for any $\mathbf{H}^k(t) \in \mathcal{R}, \overline{\mathbf{H}}^k(t) \in \mathcal{R}$, and $p = 1$. The term $f(\theta) \le f^+\left(\theta,\overline{\theta}\right)$ with an auxiliary function was defined as:

$$f^+\left(\theta,\overline{\theta}\right) \equiv \sum_{f,k,t} \frac{\left|\overline{\mathbf{Y}}^k(\omega,t) - \mathbf{W}^k(\omega)\mathbf{H}^k(t)\cdot e^{j\phi^k(\omega,t)}\right|^2}{\beta^k(\omega,t)} + \sum_{k,t}\left[\left(\lambda^k(t)\right)^p\left(p\left|\overline{\mathbf{H}}^k(t)\right|^{p-2}\mathbf{H}^k(t)^2 + (2-p)\left|\overline{\mathbf{H}}^k(t)\right|^p\right) - \log\lambda^k(t)\right] \tag{4}$$

where $\overline{\theta} = \left\{\overline{\mathbf{Y}}^k(\omega,t), \overline{\mathbf{H}}^k(t) \middle| 1 \le f \le F,\ 1 \le t \le T,\ 1 \le k \le K\right\}$. The function $f^+\left(\theta,\overline{\theta}\right)$ is minimized w.r.t. $\overline{\theta}$ when

$$\overline{\mathbf{Y}}^k(\omega,t) = \mathbf{W}^k(\omega)\overline{\mathbf{H}}^k(t)\cdot e^{j\phi^k(\omega,t)} + \beta^k(\omega,t)(\mathbf{Y}(\omega,t) - \mathbf{X}(\omega,t)) \tag{5}$$

$$\overline{\mathbf{H}}^k(t) = \mathbf{H}^k(t) \tag{6}$$

2.2. Formulation of Proposed CMF Based Adaptive Variable Regularization Sparsity

2.2.1. Estimation of the Spectral Basis and Temporal Code

In Equation (4), the update rule for θ is derived by differentiating $f^{+}\left(\theta,\overline{\theta}\right)$ partially w.r.t. $\mathbf{W}^{k}(\omega)$ and $\mathbf{H}^{k}(t)$, and setting them to zero, which yields the following:

$$\mathbf{W}^{k}(\omega) = \frac{\sum_{t}\frac{\mathbf{H}^{k}(t)}{\beta^{k}(\omega,t)}Re\left[\overline{\mathbf{Y}}^{k}(\omega,t)^{*}\cdot e^{j\phi^{k}(\omega,t)}\right]}{\sum_{t}\frac{\mathbf{H}^{k}(t)^{2}}{\beta^{k}(\omega,t)}} \tag{7}$$

$$\mathbf{H}^{k}(t) = \frac{\sum_{f}\frac{\mathbf{W}^{k}(\omega)}{\beta^{k}(\omega,t)}Re\left[\overline{\mathbf{Y}}^{k}(\omega,t)^{*}\cdot e^{j\phi^{k}(\omega,t)}\right]}{\sum_{f}\frac{\mathbf{W}^{k}(\omega)^{2}}{\beta^{k}(\omega,t)} + \left(\lambda^{k}(t)\right)^{p} p\left|\overline{\mathbf{H}}^{k}(t)\right|^{p-2}} \tag{8}$$

The update rule for the phase, $\phi^{k}(\omega,t)$, can be derived by reformulating Equation (4) as follows:

$$\begin{aligned} f^{+}\left(\theta,\overline{\theta}\right) &= \sum_{k,f,t}\frac{\left|\overline{\mathbf{Y}}^{k}(\omega,t)\right|^{2} - 2\mathbf{W}^{k}(\omega)\mathbf{H}^{k}(t)\mathrm{Re}\left[\overline{\mathbf{Y}}^{k}(\omega,t)\cdot e^{-j\phi^{k}(\omega,t)}\right] + \mathbf{W}^{k}(\omega)^{2}\mathbf{H}^{k}(t)^{2}}{\beta^{k}(\omega,t)} + \sum_{k,t}\lambda^{k}(t)\left(\left|\overline{\mathbf{H}}^{k}(t)\right|^{-1}\mathbf{H}^{k}(t)^{2} - \overline{\mathbf{H}}^{k}(t)\right) - \sum_{k,t}\log\lambda^{k}(t) \\ &= A - 2\sum_{k,f,t}\left|\mathbf{B}^{k}(\omega,t)\right|\cos\left(\phi^{k}(\omega,t) - \mathbf{\Omega}^{k}(\omega,t)\right) \end{aligned} \tag{9}$$

where A denotes the terms that are irrelevant with $\phi^{k}(\omega,t)$, $\mathbf{B}^{k}(\omega,t) = \frac{\mathbf{W}^{k}(\omega)\mathbf{H}^{k}(t)\overline{\mathbf{Y}}^{k}(\omega,t)}{\beta^{k}(\omega,t)}$, $\cos\Omega^{k}(\omega,t) = \frac{Re\left[\overline{\mathbf{Y}}^{k}(\omega,t)\right]}{\left|\overline{\mathbf{Y}}^{k}(\omega,t)\right|}$, and $\sin\mathbf{\Omega}^{k}(\omega,t) = \frac{\mathrm{Im}\left[\overline{\mathbf{Y}}^{k}(\omega,t)\right]}{\left|\overline{\mathbf{Y}}^{k}(\omega,t)\right|}$. Derivation of (9) is elucidated in Appendix B. The auxiliary function, $f^{+}\left(\theta,\overline{\theta}\right)$ in Equation (4) is minimized when $\cos\left(\phi^{k}(\omega,t) - \Omega^{k}(\omega,t)\right) = \cos\phi^{k}(\omega,t)\cos\Omega^{k}(\omega,t) + \sin\phi^{k}(\omega,t)\sin\Omega^{k}(\omega,t) = 1$, namely, $\cos\phi^{k}(\omega,t) = \cos\Omega^{k}(\omega,t)$ and $\sin\phi^{k}(\omega,t) = \sin\Omega^{k}(\omega,t)$. The update formula for $e^{j\phi^{k}(\omega,t)}$ eventually leads to

$$\begin{aligned} e^{j\phi^{k}(\omega,t)} &= \cos\phi^{k}(\omega,t) + j\sin\phi^{k}(\omega,t) \\ &= \frac{\overline{\mathbf{Y}}^{k}(\omega,t)}{\left|\overline{\mathbf{Y}}^{k}(\omega,t)\right|} \end{aligned} \tag{10}$$

The update formula for $\beta^{k}(\omega,t)$ and $\mathbf{H}^{k}(t)$ for projection onto the constraint space is set to

$$\beta^{k}(\omega,t) = \frac{\mathbf{W}^{k}(\omega)\mathbf{H}^{k}(t)}{\sum_{k}\mathbf{W}^{k}(\omega)\mathbf{H}^{k}(t)} \tag{11}$$

$$\mathbf{H}^{k}(t) \leftarrow \frac{\mathbf{H}^{k}(t)}{\sum_{k}\mathbf{H}^{k}(t)} \tag{12}$$

2.2.2. Estimation of L_1-Optimal Sparsity Parameter $\lambda^{k}(t)$

This section aims to facilitate spectral dictionaries with adaptive sparse coding. First, the CMF model is defined as the following terms:

$$\overline{\mathbf{W}} = \left[\mathbf{I} \otimes \mathbf{W}^1(\omega) \vdots \mathbf{I} \otimes \mathbf{W}^2(\omega) \vdots \cdots \vdots \mathbf{I} \otimes \mathbf{W}^K(\omega) \right],$$

$$e^{j\underline{\Phi}(t)} = \left[e^{j\underline{\Phi}^1(t)} \vdots \cdots \vdots e^{j\underline{\Phi}^K(t)} \right]$$

$$\underline{\mathbf{y}} = \mathrm{vec}(\mathbf{Y}) = \begin{bmatrix} \underline{\mathbf{Y}}^1(:) \\ \cdots \\ \underline{\mathbf{Y}}^2(:) \\ \cdots \\ \vdots \\ \cdots \\ \underline{\mathbf{Y}}^K(:) \end{bmatrix}, \underline{\mathbf{h}} = \begin{bmatrix} \mathbf{H}^1(t) \\ \cdots \\ \mathbf{H}^2(t) \\ \cdots \\ \vdots \\ \cdots \\ \mathbf{H}^K(t) \end{bmatrix}, \underline{\boldsymbol{\lambda}} = \begin{bmatrix} \lambda^1(t) \\ \cdots \\ \lambda^2(t) \\ \cdots \\ \vdots \\ \cdots \\ \lambda^K(t) \end{bmatrix}, \underline{\boldsymbol{\Phi}} = \begin{bmatrix} \boldsymbol{\Phi}^1(:,t) \\ \cdots \\ \boldsymbol{\Phi}^2(:,t) \\ \cdots \\ \vdots \\ \cdots \\ \boldsymbol{\Phi}^K(:,t) \end{bmatrix} \tag{13}$$

$$\overline{\mathbf{A}} = \begin{bmatrix} \overline{\mathbf{W}} \circ e^{j\underline{\Phi}(t)} & 0 & \cdots & 0 \\ 0 & \overline{\mathbf{W}} \circ e^{j\underline{\Phi}(t)} & 0 & \vdots \\ \vdots & 0 & \overline{\mathbf{W}} \circ e^{j\underline{\Phi}(t)} & 0 \\ 0 & \cdots & 0 & \overline{\mathbf{W}} \circ e^{j\underline{\Phi}(t)_t} \end{bmatrix}$$

where "$\otimes$" and "$\circ$" are the Kronecker product and the Hadamard product, respectively. The term $\mathrm{vec}(\cdot)$ denotes the column vectorization and the term $\mathbf{I}$ is the identity matrix. The goal is then set to compute the regularization parameter $\lambda^k(t)$ related to each $\mathbf{H}^k(t)$. To achieve the goal, the parameter p in Equation (4) is set to 1 to acquire a linear expression (in $\lambda^k(t)$). In consideration of the noise variance σ^2, Equation (4) can concisely be rewritten as:

$$F(\underline{\mathbf{h}}, \underline{\lambda}) = \frac{1}{2\sigma^2} \parallel \underline{\mathbf{y}} - \overline{\mathbf{A}}\underline{\mathbf{h}} \parallel_F^2 + \underline{\boldsymbol{\lambda}}^{\mathbf{T}}\underline{\mathbf{h}} - (\log \underline{\boldsymbol{\lambda}})^{\mathbf{T}}\underline{\mathbf{1}} \tag{14}$$

where the $\underline{\mathbf{h}}$ and $\underline{\boldsymbol{\lambda}}$ terms indicate vectors of dimension $R \times 1$ (i.e., $R = F \times T \times K$), and the superscript '$\mathbf{T}$' is used to denote complex Hermitian transpose (i.e., vector (or matrix) transpose followed by complex conjugate). The Expectation–Maximization (EM) algorithm will be used to determine $\underline{\boldsymbol{\lambda}}$ and $\underline{\mathbf{h}}$ is the hidden variable where the log-likelihood function can be optimized with respect to $\underline{\lambda}$. The log-likelihood function satisfies the following [12]:

$$\ln p\left(\underline{\mathbf{y}} \,\middle|\, \underline{\lambda}, \overline{\mathbf{A}}, \sigma^2\right) \geq \int Q(\underline{\mathbf{h}}) \ln\left(\frac{p\left(\underline{\mathbf{y}}, \underline{\mathbf{h}} \middle| \underline{\lambda}, \overline{\mathbf{A}}, \sigma^2\right)}{Q(\underline{\mathbf{h}})}\right) d\underline{\mathbf{h}} \tag{15}$$

by applying the Jensen's inequality for any distribution $Q(\underline{\mathbf{h}})$. The distribution can simply verify the posterior distribution of $\underline{\mathbf{h}}$, which maximizes the right-hand side of Equation (15), is given by $Q(\underline{\mathbf{h}}) = p\left(\underline{\mathbf{h}} \middle| \underline{\mathbf{y}}, \underline{\lambda}, \overline{\mathbf{A}}, \sigma^2\right)$. The posterior distribution in the form of the Gibbs distribution is proposed as follows:

$$Q(\underline{\mathbf{h}}) = \frac{1}{Z_h} \exp[-F(\underline{\mathbf{h}})] \text{ where } Z_h = \int \exp[-F(\underline{\mathbf{h}})] d\underline{\mathbf{h}} \tag{16}$$

The term $F(\underline{\mathbf{h}})$ in Equation (16) as the function of the Gibbs distribution is essential for simplifying the adaptive optimization of $\underline{\lambda}$. The maximum-likelihood (ML) estimation of $\underline{\lambda}$ can be decomposed as follows:

$$\underline{\lambda}^{ML} = \arg\max_{\underline{\lambda}} \int Q(\underline{\mathbf{h}}) \ln p(\underline{\mathbf{h}} | \underline{\lambda}) d\underline{\mathbf{h}} \tag{17}$$

In the same way,

$$\sigma^2{}_{ML} = \arg\max_{\sigma^2} \int Q(\underline{\mathbf{h}}) \ln p\left(\underline{\mathbf{y}} \,\middle|\, \underline{\mathbf{h}}, \overline{\mathbf{A}}, \sigma^2\right) d\underline{\mathbf{h}} \tag{18}$$

Individual element of **H** is required to be exponentially distributed with independent decay parameters that delivers $p(\underline{\mathbf{h}}|\underline{\lambda}) = \prod_g \lambda_g \exp(-\lambda_g h_g)$, thus Equation (17) obtains

$$\underline{\lambda}^{ML} = \arg\max_{\underline{\lambda}} \int Q(\underline{\mathbf{h}})\left(\ln \lambda_g - \lambda_g h_g\right) d\underline{\mathbf{h}} \tag{19}$$

The term $\underline{\mathbf{h}}$ denotes the dependent variable of the distribution $Q(\underline{\mathbf{h}})$, whereas other parameters are assumed to be constant. As such, the $\underline{\lambda}$ optimization in Equation (19) is derived by differentiating the parameters within the integral with respect to $\underline{\mathbf{h}}$. As a result, the functional optimization [37] of $\underline{\lambda}$ then obtains

$$\lambda_g = \frac{1}{\int h_g Q(\underline{\mathbf{h}}) d\underline{\mathbf{h}}} \tag{20}$$

where $g = 1, 2, \ldots, R$, λ_g denotes the gth element of $\underline{\lambda}$. Notice that the solution $\underline{\mathbf{h}}$ naturally splits its elements into distinct subsets $\underline{\mathbf{h}}_M$ and $\underline{\mathbf{h}}_P$, consisting of components $\forall_m \in M$ so that $h_m > 0$ and components $\forall_p \in P$ so that $h_P = 0$. The sparsity parameter is then obtained as presented in Equation (21):

$$\lambda_g = \begin{cases} \frac{1}{\int h_g Q_M(\underline{\mathbf{h}}_M) d\underline{\mathbf{h}}_M} = \frac{1}{h_g^{\text{MAP}}} \text{ if } g \in M \\ \frac{1}{\int h_g \hat{Q}_P(\underline{\mathbf{h}}_P) d\underline{\mathbf{h}}_P} = \frac{1}{u_g} \text{ if } g \in P \end{cases} \tag{21}$$

and its covariance X is given by

$$X_{ab} = \begin{cases} \left(\overline{\mathbf{C}}_P^{-1}\right)_{ab}, \text{ if } a, b \in M \\ u_p^2 \delta_{ab}, \text{ Otherwise.} \end{cases} \tag{22}$$

where $\hat{Q}_P(\underline{\mathbf{h}}_P \geq 0) = \prod_{p \in P} \frac{1}{u_p} \exp\left(\frac{-h_p}{u_p}\right)$, $\overline{\mathbf{C}}_P = \frac{1}{\sigma^2} \overline{\mathbf{A}}_P^{\mathrm{T}} \overline{\mathbf{A}}_P$ and $u_p \leftarrow u_p \frac{-\hat{b}_p + \sqrt{\hat{b}_p^2 + 4\frac{(\hat{\mathbf{C}}\underline{\mathbf{u}})_p}{\overline{u}_p}}}{2(\hat{\mathbf{C}}\underline{\mathbf{u}})_p}$. The function $Q_M(\underline{\mathbf{h}}_M)$ will be expressed as the unconstrained Gaussion with mean $\underline{\mathbf{h}}_M^{\text{MAP}}$ and covariance $\overline{\mathbf{C}}_M^{-1}$ based on a multivariate Gaussian distribution. Similarly, the inference for σ^2 can be computed as

$$\sigma^2 = \frac{1}{N_0} \int Q(\underline{\mathbf{h}})\left(\|\underline{\mathbf{y}} - \overline{\mathbf{A}}\underline{\mathbf{h}}\|^2\right) d\underline{\mathbf{h}} \tag{23}$$

where

$$\hat{h}_g = \begin{cases} h_g^{\text{MAP}} \text{ if } g \in M \\ u_g \text{ if } g \in P \end{cases}$$

The core procedure of the proposed CMF method is based on L_1-optimal sparsity parameters. The estimated sources are discovered by multiplying the respective rows of the $W^k(\omega)$ components with the corresponding columns of the $H^k(t)$ weights and time-varying phrase spectrum $e^{j\phi^k(\omega,t)}$. The separated source $\hat{s}_j(t)$ is obtained by converting the time-frequency represented sources into the time domain. Derivation of L_1-optimal sparsity parameter, is elucidated in the Appendix C.

2.3. Sound Event Classification

Once the separated sound signal is obtained, the next step is to identify the sound event. A multiclass support vector machine (MSVM) is employed to achieve the goal. The MSVM is comprised of two phases: the learning phase and the evaluation phase. The MSVM is based on one versus one strategy (OvsO) that splits observed c classes into $\frac{c(c-1)}{2}$ binary classification sub-problems. To train the $\sqsupset^{th}$ MSVM model, the MSVM will construct hyperplanes for discriminating each observed data into its corresponding class by executing the series of binary classification. Starting from the learning phase, sound signatures are extracted from the training dataset in the time-frequency domain. The sound signatures that were studied in this research were the Mel frequency cepstral coefficients (MFCC: MF), short-time energy (STE: $E(t)$), and short term zero-crossing rate (STZCR: $STZ(t)$), which can be orderly expressed as: $MF = 2525 \times \log[1 + (f/7)]$, $E(t) = \sum_{\tau=-\infty}^{\infty} [y(t) \cdot f_w(t-\tau)]^2$, $Z(t) = \sum_{\tau=-\infty}^{\infty} \left|sgn[s(\tau)] - sgn[s(\tau-1)]\right| \cdot f_w(t-\tau)$ where $f_w(t)$ denotes the windowing function. The training signals are segmented into small blocks, then the individual block is extracted to the three signature parameters. The mean supervector is then computed as an average of individual feature of all blocks for each sound event input. Thus, the mean feature supervector (O) with a corresponding sound-event-label vector ((w)) is paired together (i.e., ($\psi(O,w)$)) and supplied to the MSVM model. The discriminant formula can be expressed as:

$$
\begin{aligned}
\{\hat{\sqsupset}, \beta\} &= \arg\max_{\sqsupset} \left\{ \alpha_{\sqsupset}^{T} \psi(O, w; \beta) \right\} \\
&= \arg\max_{\sqsupset} \left\{ \max_{\beta} \sum_{i=1}^{|w|} \alpha_{\sqsupset}^{T} \psi\left(O_{i|\beta}, w_i\right) \right\}
\end{aligned}
\tag{24}
$$

where $\left(O_{i|\beta}, w_i\right), i = 1, \ldots, M$ represents the ith separated sound signals; the weight vector $\alpha_{\sqsupset}$ is employed for individual class $\sqsupset$ to compute a discriminant score for the O; the i term is the index of the block order (β); and the function $\alpha_{\sqsupset}^{T}\psi(O, w; \beta)$ measures a linear discriminant distance of the hyperplane with the extracted feature vector from the observed data. The MSVM based OvsO strategy for class $\sqsupset^{th}$ and other, the hyperplane, can be maximized as $\alpha_{\sqsupset}^{T}\psi(O, w; \beta) + b_{\sqsupset}$ and can then be learned via the following equation as

$$
\min_{\alpha_{\sqsupset}, \xi_{\sqsupset}} \frac{1}{2} ||\alpha_{\sqsupset}||^2 + C \sum_{i=1}^{M} \xi_i^{\sqsupset}
\tag{25}
$$

where $\xi_i^{\sqsupset} \geq 0$, $b_{\sqsupset}$ is a constant. The term $\sum_{i=1}^{M} \xi_i^{\sqsupset}$ denotes a penalty function for tradeoff between a large margin and a small error penalty. The optimal hyperplane can be determined by minimizing $\frac{1}{2}||\alpha_{\sqsupset}||^2$ to maximize the condition (i.e., $\alpha_{\sqsupset}^{T}\psi(O, w; \beta) + b_{\sqsupset} \geq 1 - \xi_i^{\sqsupset}$). If the conditional term is greater than $1 - \xi_i^{\sqsupset}$, then the estimated sound event belongs to the $\sqsupset^{th}$ class. Otherwise, the estimated sound event classifies into other classes.

The overview of the proposed algorithm is presented in the following table as Algorithm 1.

Algorithm 1 Overview of the proposed algorithm.

(1) Compute $\mathbf{Y}(\omega,t) = \text{STFT}(y(t))$ from the noisy single-channel mixture $y(t)$.

(2) Initial values $\mathbf{W}^k(\omega), \mathbf{H}^k(t),\ \beta^k(\omega,t)$, fixing the value of $\boldsymbol{\phi}^k(\omega,t)$ at $e^{j\boldsymbol{\phi}^k(\omega,t)} = \frac{\mathbf{Y}(\omega,t)}{|\mathbf{Y}(\omega,t)|}$, and calculate $\lambda^k(t)$ and σ^2.

(3) Update $\overline{\theta} = \left\{\overline{\mathbf{X}}, \overline{\mathbf{H}}\right\}, \theta = \left\{\mathbf{W}^k(\omega), \mathbf{H}^k(t),\ \boldsymbol{\phi}^k(\omega,t)\right\}, \beta^k(\omega,t)$.

(4) Update parameters (21) and (23) until convergence is reached as determined by the rate of change of the parameters update falling within a pre-determined threshold.

(5) Estimation of each source by multiplying the respective rows of the spectral components $\boldsymbol{W}^k(\omega)$ with the corresponding columns of the mixture weights $\boldsymbol{H}^k(t)$ and time-varying phrase spectrum $e^{j\boldsymbol{\phi}^k(\omega,t)}$. (i.e., $\left|\hat{\mathbf{S}}_i\right|^{.2} = \sum_{k=1}^{K_i} \mathbf{W}_{i_f}^k \mathbf{H}_{i_t}^k . e^{j\phi_{i_{f,t}}^{(k)}}$ and construct the binary TF mask for the ith source

$$M_i(f,t_s) := \begin{cases} 1, & \text{if} \left|\hat{\mathbf{S}}_i(f,t_s)\right|^{.2} > \left|\hat{\mathbf{S}}_j(f,t_s)\right|^{.2},\ i \neq j \\ 0, & otherwise \end{cases}).$$

(6) Convert the time-frequency represented sources into time domain to obtain the separated sources $\hat{s}_j(t)$ i.e., $\hat{s}_j(t) = STFT^{-1}\left(\left|\hat{\mathbf{S}}_i\right|^2\right)$.

(7) Classify the $\beth^{th}$ sound event by computing the optimal hyperplane $\alpha_\beth^T \psi(O,w;\beta) + b_\beth$ by minimizing the following equation: $\min\limits_{\alpha_\beth,\xi_\beth} \frac{1}{2}\|\alpha_\beth\|^2 + C\sum_{i=1}^{M} \xi_i^\beth$.

3. Experimental Results and Analysis

The performance was evaluated on recorded sound-event signals in a low noisy environment at 20 signal-to-noise ratios (SNRs). The sound-event database had a total of 500 recorded signals containing four event classes: speech (SP), door open (DO), door knocking (DK), and footsteps (FS). An overview of the experimental setup is given as the following: all signals had a 16-bit resolution and a sampling frequency of 44.1 KHz. A 2048 length of Hanning window with 50% overlap was used for signal processing. Nonlinear SVM with a Gaussian RBF kernel was used for constructing the MSVM learning model. Other kernels such as polynomials, sigmoid, and even linear function were tested, but the best performance was delivered by the Gaussian kernel. A 4-fold cross-validation strategy was used in the training phase for tuning the classifier parameters when using 80% of the recorded signals (n = 400) from the sound-event database.

The performance of the proposed noisy sound separation and event classification (NSSEC) method was demonstrated and presented into the following two sections: (1) the separating performance, and (2) the MSVM classifier.

3.1. Sound-Event Separation and Classification Performance

Event mixtures consist of two sound-event signals in low noisy environment at 20 dB SNRs. A hundred sound-event signals of four classes were randomly selected and then mixed to generate 120 mixtures of six types (i.e., DO + DK, DO + FS, DO + SP, DK + FS, DK + SP, and FS + SP). The separation performance measured the signal-to-distortion ratio (SDR) (i.e., $SDR = 10\ log_{10}\left(\|s_{target}\|^2 / \|e_{interf} + e_{noise} + e_{artif}\|^2\right)$ where e_{interf}, e_{noise}, and e_{artif}). The SDR represents the ratio of the magnitude distortion of the original signal by the interference from other sources. The proposed separation method was compared with the state-of-the-art NMF approach (i.e., CMF [38], NMF-ISD [14,39], and SNMF [40–42] methods). The cost function was the least squares with 500 maximum number of iterations.

3.1.1. Variational Sparsity Versus Fixed Sparsity

In this implementation, several experiments were conducted to investigate the effect of sparsity regularization on source separation performance. The proposed separation method was evaluated by variational sparsity in the case of (1) uniform constant sparsity with low sparseness e.g., $\lambda_t^k = 0.01$ and (2) uniform constant sparsity with high sparseness (e.g., $\lambda_t^k = 100$). The hypothesis is that the proposed variational sparsity will significantly yield improvement of the audio source separation when compared with fixed sparsity.

To investigate the impact of uniform sparsity parameter, the set of sparsity regularization values from 0 to 10 with a 0.5 interval were determined for each experiment of 60 mixtures of six types. Results of the uniform regularization given by various sparsity (i.e., $\lambda_t^k = 0, 0.5, \ldots, 10$) is illustrated in Figure 2.

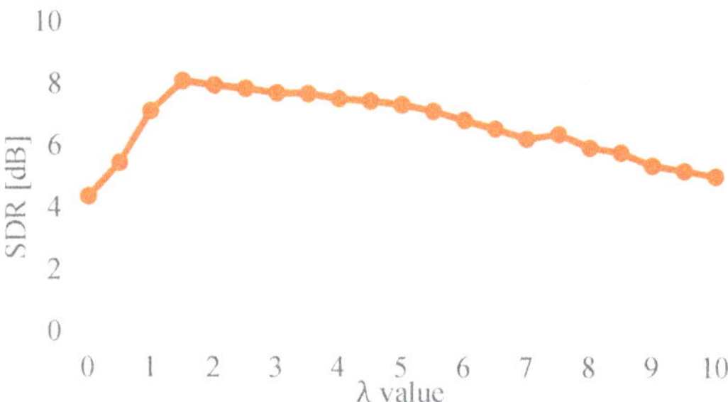

Figure 2. Separation results of the proposed method by using different uniform regularization.

Figure 2 illustrates that the best performance of the unsupervised CMF was in a range of 1.5–3, which yielded the highest SDR of over 8dB. When the term λ_t^k was set too high, the low spectral values of sound-event signals were overly sparse. This overfitting sparsity $\boldsymbol{H}^k(t)$ caused the separation performance toward a tendency to degrade. Conversely, the underfitting sparsity $\boldsymbol{H}^k(t)$ occurred when the term λ_t^k was set too low. The coding parameter $\boldsymbol{H}^k(t)$ could not distinguish between the two sound-event signals. It was also noticed that if the factorization is non-regularized, this will cause the separation results to contain a mixed sound. According to the uniform sparsity results in Figure 2, the separation performance of the proposed method varies depending on the assigned sparsity values. Thus, it is challenging to find a solution for the indistinctness among the sound-event sources in the TF representation to determine the optimal value of sparseness. Thus, this introduces the importance of determining the optimal λ for separation. Table 1 presents the essential sparsity value on the separation performance by comparing the proposed method given by variational sparsity against the uniform sparsity scheme. The average performance improvement of the proposed adaptive CMF method against the uniform constant sparsity was 1.32 dB SDR. The SDR results clearly indicate that the adaptive sparsity yielded the surpass separation performance over the constant sparsity scheme. Hence, the proposed variational sparsity improves the performance of the discovered original sound-event signals by adaptively selecting the appropriate sparsity parameters to be individually adapted for each element code (i.e., $\lambda_g = \begin{cases} \frac{1}{\int h_g Q_M(\underline{\mathbf{h}}_M) d\underline{\mathbf{h}}_M} = \frac{1}{h_g^{\mathrm{MAP}}} \text{ if } g \in M \\ \frac{1}{\int h_g \hat{Q}_P(\underline{\mathbf{h}}_P) d\underline{\mathbf{h}}_P} = \frac{1}{u_g} \text{ if } g \in P \end{cases}$ and $\sigma^2 = \frac{1}{N_0} \int Q(\underline{\mathbf{h}}) \left(\| \underline{\mathbf{y}} - \overline{\mathbf{A}} \underline{\mathbf{h}} \|^2 \right) d\underline{\mathbf{h}}$ where $\hat{h}_g = \begin{cases} h_g^{\mathrm{MAP}} \text{ if } g \in M \\ u_g \text{ if } g \in P \end{cases}$). Consequently, the optimal sparsity facilitates the estimated spectral dictionary via the estimated temporal code. The quantitative measures of separation performance were performed to assess the proposed single-channel sound event separation method. The overall average signal-to-distortion ratio (SDR) was 8.62 dB as illustrated in Figure 3.

Table 1. Comparison of average SDR performance on three types of mixtures between uniform regularization methods and the proposed method.

Mixtures	Methods	SDR
DO + DK	Proposed method	7.63
	(Best) Uniform regularization sparsity	6.59
DO + FS	Proposed method	9.06
	(Best) Uniform regularization sparsity	8.74
DO + SP	Proposed method	8.45
	(Best) Uniform regularization sparsity	6.91
DK + FS	Proposed method	7.04
	(Best) Uniform regularization sparsity	6.35
DK + SP	Proposed method	9.72
	(Best) Uniform regularization sparsity	7.78
FS + SP	Proposed method	9.81
	(Best) Uniform regularization sparsity	7.42

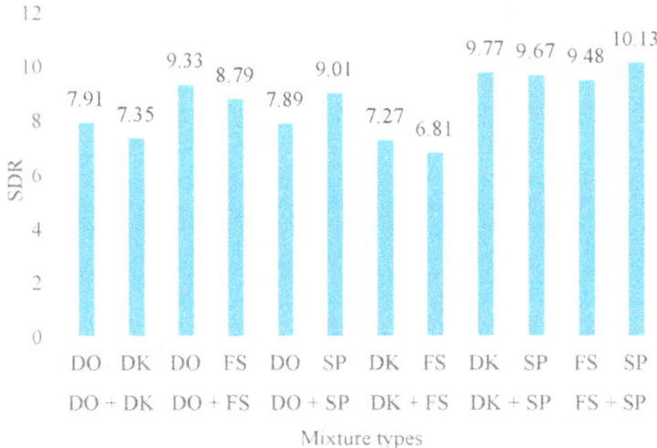

Figure 3. Average SDR results of six-mixture types.

Each sound-event signal has its own temporal pattern that can be clearly noticed in TF representation. Examples of sound-event signals in the TF domain are illustrated in Figure 4. Through the adaptive L_1-SCMF method, the proposed single-channel separation method can generate complex temporal patterns such as speech. Thus, the separation results clearly indicate that the performances of noisy source separation perform with high SDR values.

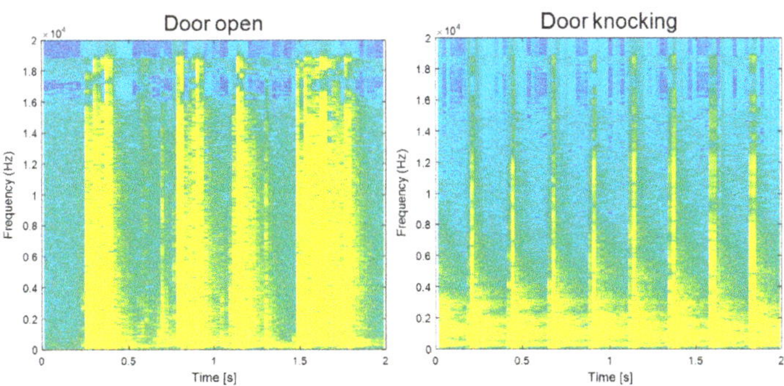

Figure 4. *Cont.*

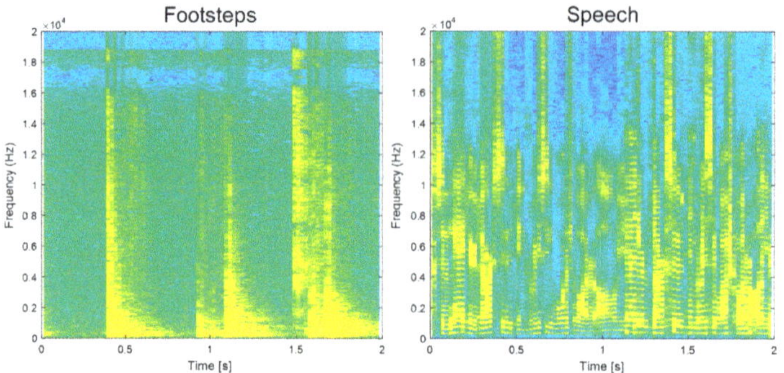

Figure 4. Example of time-frequency representation of four sound event classes.

3.1.2. Comparison of the Proposed Adaptive CMF with Other SCBSS Methods Based on NMF

This section presents the adaptive CMF separating performance against the state-of-the-art NMF methods (i.e., CMF, SNMF, and NMF-ISD). In the compared methods, the experimental variables such as the normalizing time-frequency domain were computed by using the short-time Fourier transform (i.e., 1024-point Hanning window with 50% overlap). The number of factors was two, with a sparsity weight of 1.5. One hundred random realizations of twenty second-event mixtures were executed. As a result, the average SDRs are presented in Table 2. The proposed adaptive CMF method yielded the best separating performance over the CMF, SNMF, and NMF_ISD methods with the average improvement SDR at 2.13 dB. The estimated door open signals obtained the highest SDR among the four event categories.

Table 2. Comparison of average SDR and SIR performance on three types of mixtures between SCICA, NMF-ISD, SNMF, CMF, and the proposed method.

Mixtures	Methods	SDR
Door Open	Proposed method	8.38
	CMF	7.11
	SNMF	6.23
	NMF-ISD	6.17
Door Knocking	Proposed method	8.13
	CMF	7.06
	SNMF	6.52
	NMF-ISD	6.55
Footsteps	Proposed method	8.36
	CMF	7.89
	SNMF	6.62
	NMF-ISD	6.06
Speech	Proposed method	9.60
	CMF	6.73
	SNMF	5.61
	NMF-ISD	5.32

The sparsity parameter was carefully adapted using the proposed adaptive L_1-SCMF method exploiting the phase information and temporal code of the sources, which is inherently ignored by SNMF and NMF-ISD and has led to an improved performance of about 2 dB in SDR. On the other hand, the parts decomposed by the CMF, SNMF, and NMF-ISD methods were unable to capture the phase spectra and the temporal dependency of the frequency patterns within the audio signal.

Additionally, the CMF and NMF-ISD are unique when the signal adequately spans the positive octant. Thus, the rotation of **W** and opposite **H** can obtain the same results. The CMF method can easily be over or under sparse resolution of the factorization due to manually determining the sparsity value.

3.2. *Performance of Event Classification Based on MSVM Algorithm*

This section elucidates the features and performance of the MSVM-learning model. The MSVM-learning model was investigated to obtain the optimal size of the sliding window and then determine the significant features that led to the classification performance. Finally, the efficiency of the MSVM model was evaluated. These topics are presented in order in the following parts.

3.2.1. Determination Optimal Window Length for Feature Encoding

For the MSVM method, sound-event signals are segmented into small blocks for encoding feature parameters by using a fixed-length of the sliding window. The sets of feature vectors are computed using the mean supervector and then loaded to the MSVM model for learning and constructing the hyperplane. The size of blocks can affect the information of the feature vectors, which leads to the classifier performance. The block's size will affect the $\alpha_{\beth}$, hence modifying the block size will mark the learning efficiency of the MSVM model. Therefore, in order to obtain the optimal value of $\alpha_{\beth}$, the optimal block size was exploited by training the MSVM model given various lengths of window sizes (i.e., 0.5, 1, 1.5, and 2.0 s) to learn the 400 noisy sound-event signals of four event classes with cross-validation.

The experimental results are plotted in Figure 5, where the block size varies from 0.5 to 2.0 with 0.5 increments. The MSVM model of the 1.5 s block size yielded the best sound-event classification at 100% accuracy. The sliding window function benefits from SVM to learn an unknown sound event by generating the set of blocks from the observed event, regarded as a number of observed events. As a result, a set of sound event characteristics were computed for each block (i.e., $O_{i|\beta}, w_i$ in Equation (24)).

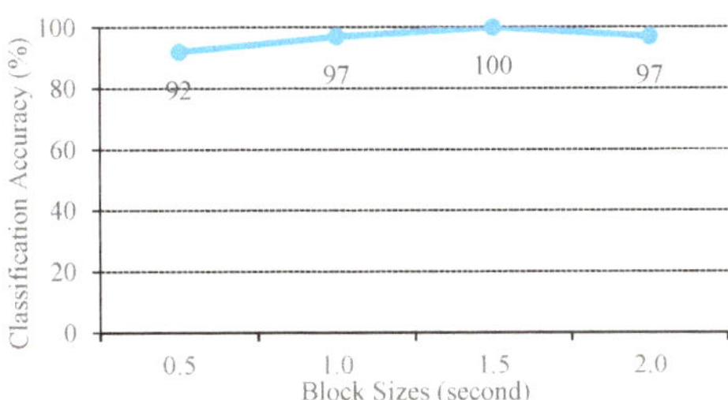

Figure 5. Classification performance of the original and combination source MSVM with various block sizes.

The optimal length of the window size can capture the signature of the sound event. If the window length is too short, the encoded features will then deviate from the character of the sound event. In addition, the mean supervector is computed from the set of features of all blocks, which can be regarded as the mean of the probability distribution of the features. This mean supervector advantages the MSVM to reduce misclassifications when compared to the conventional SVM. Hence, the STFT of all experiments set the window function at 1.5 s.

3.2.2. Determination of Sound-Event Features

Each sound-event signal was encoded with three features: Mel frequency cepstral coefficients (MFCCs), short-time energy (STE), and short-time zero-crossing rate (STZCR). MFCCs are represented as a frequency domain feature that is evaluated in a similar assembly to the human ear (i.e., logarithmic frequency perception). STE is the total spectrum power of an observed event.

The STZCR denotes the number of times that the signal amplitude interval satisfies the condition (i.e., $STZCR = (1/T-1)\sum_{t=1}^{T-1}[[\{s_t s_{t-1} < 0\}$ where $[[\{s_t s_{t-1} < 0\}$ is 1 if the condition is true and 0 otherwise). The STZCR features of four sound-event classes are illustrated in Figure 6.

(**a**) Door open (**b**) Door knocking

(**c**) Footstep (**d**) Speech

Figure 6. STZCR patterns of four sound-events (**a**–**d**).

The STZCR feature represents unique patterns of four sound-event classes. The four sound-event patterns are different in shape and data range. Similarly, the MCFFs and STE features extract distinctive patterns of all event classes, except for the patterns between door knocking and footstep, as illustrated in Figure 7.

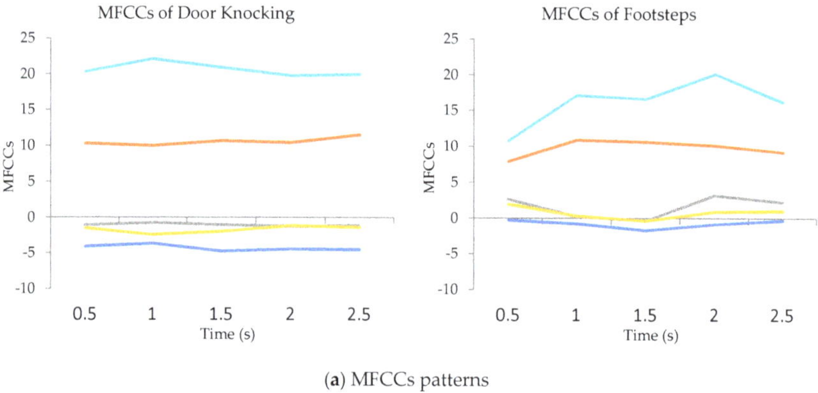

(**a**) MFCCs patterns

Figure 7. *Cont.*

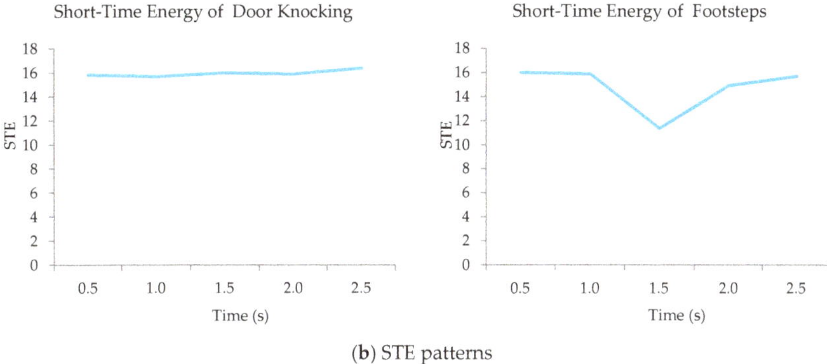

(**b**) STE patterns

Figure 7. MFCCs (**a**) and STE (**b**) patterns of door knocking and footstep.

Figure 7 aims to compare the characteristics of similar sound events such as door knocking and footsteps. Thus, MFCCs and STE features were used to illustrate the patterns of sound events. Figure 7a represents the five orders of MFCC features to compare patterns between door knocking and walking while the STE features are shown in Figure 7b.

The proposed method separated the six categories of mixtures, then classified each estimated sound event signal into its corresponding class. Classified results of the six categories are presented as confusion matrixes below:

Actual

Predict	DO	DK
DO	19	3
DK	3	15

	DO	FS
DO	12	8
FS	4	16

	DO	SP
DO	19	5
SP	3	13

	DK	FS
DK	12	4
FS	9	15

	DK	SP
DK	16	2
SP	5	17

	FS	SP
FS	14	6
SP	3	17

The classification of the proposed method was measured by Precision = TP/(TP + FP), Recall = TP/(TP + FN), and F1-score = 2 × (Precision × Recall)/(Precision + Recall). The TP and TN terms refer to the true positive and true negative, while the FP and FN terms mean false positive and false negative. The scores of Precision, Recall, and F1-score were 0.7667, 0.7731, and 0.7699, respectively.

Each feature represents unique characteristics of an individual sound event. Thus, features were matched into seven cases for exploiting their influence on the MSVM classifiers (i.e., {(MFCC), (STE), (STZCR), (MFCC, STE), (MFCC, STZCR), (STE, STZCR), (MFCC, STE, STZCR)}).

As shown in Figure 8, the MSVM model given by MFCCs and STZCR yielded the best classified accuracy at 100%, with less deviation among the other cases. Therefore, the separated signals were then classified by the proposed MSVM method given by the MFCC and STZCR vectors and the 1.5 s window function. The computational complexity of the proposed method was analyzed by two steps. First, the adaptive L1-SCMF method was NP-hard. Big-O of the adaptive L_1-SCMF method consists of spectral basis (m), temporal code (n), and phase information that rely on components (k). Thus, Big-O of the separation step is $(mn)^{O(k^2)}$. For MSVM steps, it is a polynomial algorithm where Big-O is $O(n^3)$. Therefore, the computational complexity of the proposed method based on Big-O is $(mn)^{O(k^2)}$. All experiments were performed by a PC with Intel® Core™ i7-4510U CPU2.00 GHz and 8 GB RAM. MATLAB was used as the programming platform.

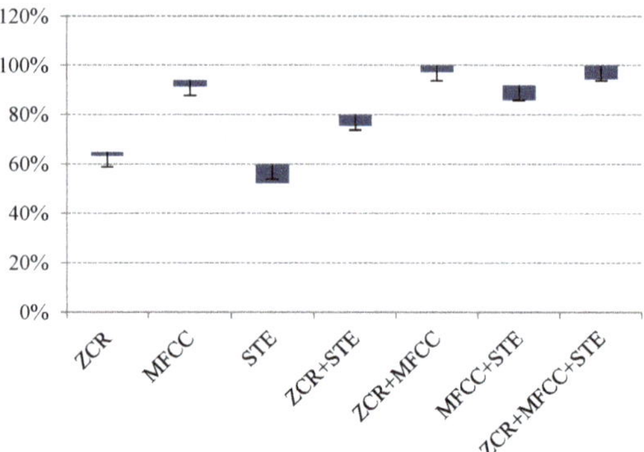

Figure 8. Classification performances of multi-class MSVM of various sets of features and length of event signal.

3.2.3. Performance of MSVM Classifier

The MSVM-classifier performance is presented in terms of percentage of the corrected sound-event classification. The 240 separated signals of four classes from the proposed separation method were individually identified by the MSVM classifier.

Figure 9 compares the classification performance on the four classes of individual sound events. The best classification accuracy was door open, followed by footstep, door knocking, and speech. On the other hand, the classification results based on the mixed sound events are illustrated in Figure 10. The MSVM model delivered the highest performance of the door-open event with 84% accuracy.

From the above experiments, the proposed method yields an average classification accuracy of 76.67%. The MSVM method can well discriminate and classify the mixed event signals with high classification accuracy (i.e., the mixture of door open with door knocking and door knocking with speech were correctly classified with above 80% accuracy). Due to the MFCC and STZCR features in the individual event, these signals had obvious distinguishable patterns, as shown in the example of STZCR plots in Figure 6. Despite the SDR scores of the separated signals between door open and door knocking being relatively low (as given in Figure 3), the MSVM yielded the highest classification accuracy for the door open with door knocking mixture (DO + DK). This is attributed to the fact that interference remaining in the separated event signals causes the extracted MFCC and STZCR vectors to deviate from their original sound event vectors.

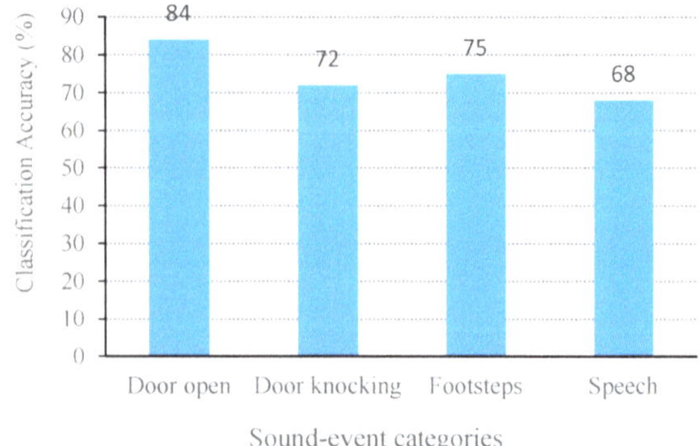

Figure 9. Average percentage of classification accuracy from the perspective of event group of the proposed NSSEC method.

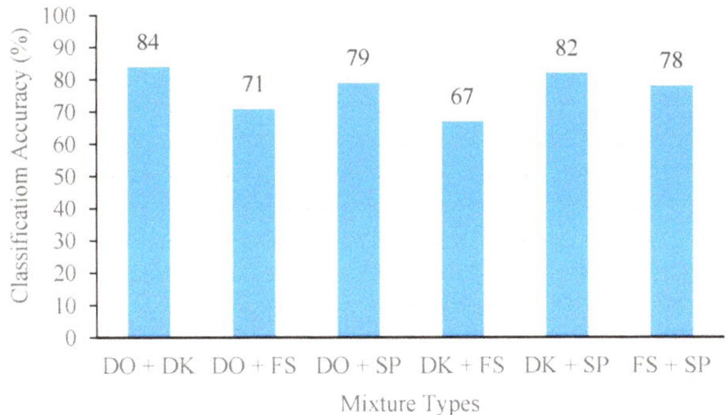

Figure 10. Classification performance of NSSEC model with 1.5 s block size.

4. Conclusions

A novel solution for classification of the noisy mixtures using a single microphone was presented. The complex matrix factorization was proposed and extended by adaptively tuning the sparse regularization. Thus, the desired L_1-optimal sparse decomposition was obtained. In addition, the phase estimates of the CMF could extract the recurrent pattern of the magnitude spectra. The updated equation was derived through an auxiliary function. For classification, the multiclass support vector was used as the mean supervector for encoding the sound-event signatures. The proposed noisy sound separation and event classification method was demonstrated by using four sets of noisy sound-event mixtures, which were door open, door knocking, footsteps, and speech. Based on the experimental results, first, the optimal window length of STFT was found where 1.5 s of the sliding window yielded the best separation performance. The second was two significant features that were ZCR and MFCCs. These parameters were set for examining the proposed method. The proposed method achieved outstanding results in both separation and classification. In future work, the proposed method will be evaluated on a public dataset such as the DCASE 2016, alongside the comparison with other machine learning algorithms.

Author Contributions: Conceptualization, P.P. and W.L.W.; Methodology, P.P. and N.T.; Software, P.P.; Validation, N.T. and W.L.W.; Investigation, P.P. and N.T.; Writing—original draft preparation, P.P. and W.L.W.; Writing—review and editing, N.T., M.A.M.A., O.A., and G.R.; Visualization, M.A.M.A.; Supervision, W.L.W.; Project administration, N.T., M.A.M.A., and O.A.; Funding acquisition, W.L.W. All authors have read and agreed to the published version of the manuscript.

Funding: This research was funded by the UK Global Challenge Research Fund, the National Natural Science Foundation of China (No. 61971093, No. 61401071, No. 61527803), and supported by the NSAF (Grant No. U1430115) and EPSRC IAA Phase 2 funded project: "3D super-fast and portable eddy current pulsed thermography for railway inspection (EP/K503885/1).

Conflicts of Interest: The authors declare no conflict of interest.

Appendix A. Single-Channel Sound Event Separation

The prior $P(\mathbf{H}|\boldsymbol{\lambda})$ corresponds to the sparsity cost, for which a natural choice is a generalized Gaussian prior:

$$P(\mathbf{H}|\boldsymbol{\lambda}) = \prod_{k,t} \frac{p\lambda^k(t)}{2\Gamma(1/p)} \exp\left(-\left(\lambda^k(t)\right)^p \left|\mathbf{H}^k(t)\right|^p\right) \quad \text{(A1)}$$

where $\lambda^k(t)$ and p are the shape parameters of the distribution. When $p = 1$, $P(\mathbf{H}|\boldsymbol{\lambda})$ promotes the L_1-norm sparsity. L_1-norm sparsity has been shown to be probabilistically equivalent to the pseudo-norm, L_0, which is the theoretically optimum sparsity [29,30]. However, L_0-norm is non-deterministic polynomial-time (NP) hard and is not useful in large datasets such as audio. Given Equation (3), the posterior density is defined as

$$P(\theta|\mathbf{Y},\boldsymbol{\lambda}) \propto P(\mathbf{Y}|\theta)P(\mathbf{H}|\boldsymbol{\lambda}) \tag{A2}$$

The maximum a posteriori probability (MAP) estimation problem leads to minimizing the following optimization problem with respect to θ:

$$f(\theta) = \sum_{f,t} \left|\mathbf{Y}(\omega,t) - \mathbf{X}(\omega,t)\right|^2 + \sum_{k,t}\left[\left(\lambda^k(t)\right)^p \left|\mathbf{H}^k(t)\right|^p - \log\lambda^k(t)\right] \tag{A3}$$

subject to $\sum_f \mathbf{W}^k(\omega) = 1\ (k = 1,\ldots,K)$.

The CMF parameters has been upgraded by using an efficient auxiliary function for an iterative process. The auxiliary function for $f(\theta)$ can be expressed as the following: for any auxiliary variables with $\sum_k \overline{\mathbf{Y}}^k(\omega,t) = \mathbf{Y}(\omega,t)$, for any $\beta^k(\omega,t) > 0$, $\sum_k \beta^k(\omega,t) = 1$, for any $\mathbf{H}^k(t) \in \mathcal{R}, \overline{\mathbf{H}}^k(t) \in \mathcal{R}$, and $p = 1$. The term $f(\theta) \leq f^+\left(\theta,\overline{\theta}\right)$ with an auxiliary function was defined as

$$f^+\left(\theta,\overline{\theta}\right) \equiv \sum_{f,k,t} \frac{\left|\overline{\mathbf{Y}}^k(\omega,t) - \mathbf{W}^k(\omega)\mathbf{H}^k(t)\cdot e^{j\phi^k(\omega,t)}\right|^2}{\beta^k(\omega,t)} + \sum_{k,t}\left[\left(\lambda^k(t)\right)^p\left(p\left|\overline{\mathbf{H}}^k(t)\right|^{p-2}\mathbf{H}^k(t)^2 + (2-p)\left|\overline{\mathbf{H}}^k(t)\right|^p\right) - \log\lambda^k(t)\right] \tag{A4}$$

where $\overline{\theta} = \left\{\overline{\mathbf{Y}}^k(\omega,t), \overline{\mathbf{H}}^k(t) \,\middle|\, 1 \leq f \leq F,\ 1 \leq t \leq T,\ 1 \leq k \leq K\right\}$. The function $f^+\left(\theta,\overline{\theta}\right)$ is minimized w.r.t. $\overline{\theta}$ when

$$\overline{\mathbf{Y}}^k(\omega,t) = \mathbf{W}^k(\omega)\overline{\mathbf{H}}^k(t)\cdot e^{j\phi^k(\omega,t)} + \beta^k(\omega,t)(\mathbf{Y}(\omega,t) - \mathbf{X}(\omega,t)) \tag{A5}$$

$$\overline{\mathbf{H}}^k(t) = \mathbf{H}^k(t) \tag{A6}$$

Appendix B. Estimation of the Spectral Basis and Temporal Code

In Equation (4), the update rule for θ is derived by differentiating $f^+\left(\theta,\overline{\theta}\right)$. partially w.r.t. $\mathbf{W}^k(\omega)$ and $\mathbf{H}^k(t)$, and setting them to zero, which yields the following:

$$\mathbf{W}^k(\omega) = \frac{\sum_t \frac{\mathbf{H}^k(t)}{\beta^k(\omega,t)} Re\left[\overline{\mathbf{Y}}^k(\omega,t)^* \cdot e^{j\phi^k(\omega,t)}\right]}{\sum_t \frac{\mathbf{H}^k(t)^2}{\beta^k(\omega,t)}} \tag{A7}$$

$$\mathbf{H}^k(t) = \frac{\sum_f \frac{\mathbf{W}^k(\omega)}{\beta^k(\omega,t)} Re\left[\overline{\mathbf{Y}}^k(\omega,t)^* \cdot e^{j\phi^k(\omega,t)}\right]}{\sum_f \frac{\mathbf{W}^k(\omega)^2}{\beta^k(\omega,t)} + \left(\lambda^k(t)\right)^p p\left|\overline{\mathbf{H}}^k(t)\right|^{p-2}} \tag{A8}$$

The update rule for the phase, $\boldsymbol{\phi}^k(\omega,t)$, can be derived by reformulating Equation (A1) as follows:

$$\begin{aligned}
&f^+\left(\theta,\overline{\theta}\right)\\
&= \sum_{k,f,t} \frac{\left|\overline{\mathbf{Y}}^k(\omega,t)\right|^2 - 2\mathbf{W}^k(\omega)\mathbf{H}^k(t)\mathrm{Re}\left[\overline{\mathbf{Y}}^k(\omega,t)\cdot e^{-j\phi^k(\omega,t)}\right] + \mathbf{W}^k(\omega)^2\mathbf{H}^k(t)^2}{\beta^k(\omega,t)} + \sum_{k,t}\lambda^k(t)\left(\left|\overline{\mathbf{H}}^k(t)\right|^{-1}\mathbf{H}^k(t)^2 - \overline{\mathbf{H}}^k(t)\right) - \sum_{k,t}\log\lambda^k(t)\\
&= A - 2\sum_{k,f,t} \frac{\mathbf{W}^k(\omega)\mathbf{H}^k(t)\left|\overline{\mathbf{Y}}^k(\omega,t)\right|}{\beta^k(\omega,t)}\left(\frac{Re\left[\overline{\mathbf{Y}}^k(\omega,t)\cdot e^{-j\phi^k(\omega,t)}\right]}{\left|\overline{\mathbf{Y}}^k(\omega,t)\right|}\right)\\
&= A - 2\sum_{k,f,t}\left|\mathbf{B}^k(\omega,t)\right| \frac{Re\left[\left(\overline{\mathbf{Y}}^k(\omega,t)^{(r)} + j\overline{\mathbf{Y}}^k(\omega,t)^{(i)}\right)\left(\cos\phi^k(\omega,t) - j\sin\phi^k(\omega,t)\right)\right]}{\left|\overline{\mathbf{Y}}^k(\omega,t)\right|}\\
&= A - 2\sum_{k,f,t}\left|\mathbf{B}^k(\omega,t)\right|\cos\boldsymbol{\phi}^k(\omega,t)\cos\Omega^k(\omega,t) + \sin\boldsymbol{\phi}^k(\omega,t)\sin\Omega^k(\omega,t)\\
&= A - 2\sum_{k,f,t}\left|\mathbf{B}^k(\omega,t)\right|\cos\left(\boldsymbol{\phi}^k(\omega,t) - \Omega^k(\omega,t)\right)
\end{aligned} \tag{A9}$$

where A denotes the terms that are irrelevant with $\phi^k(\omega,t)$, $\mathbf{B}^k(\omega,t) = \frac{\mathbf{W}^k(\omega)\mathbf{H}^k(t)\overline{\mathbf{Y}}^k(\omega,t)}{\beta^k(\omega,t)}$, $\cos\Omega^k(\omega,t) = \frac{Re\left[\overline{\mathbf{Y}}^k(\omega,t)\right]}{\left|\overline{\mathbf{Y}}^k(\omega,t)\right|}$, and $\sin\Omega^k(\omega,t) = \frac{\mathrm{Im}\left[\overline{\mathbf{Y}}^k(\omega,t)\right]}{\left|\overline{\mathbf{Y}}^k(\omega,t)\right|}$. The auxiliary function, $f^+\left(\theta,\overline{\theta}\right)$ in (A4) is minimized when $\cos\left(\phi^k(\omega,t) - \Omega^k(\omega,t)\right) = \cos\phi^k(\omega,t)\cos^k(\omega,t) + \sin\phi^k(\omega,t)\sin\Omega^k(\omega,t) = 1$, namely, $\cos\phi^k(\omega,t) = \cos\Omega^k(\omega,t)$ and $\sin\phi^k(\omega,t) = \sin\Omega^k(\omega,t)$. The update formula for $e^{j\phi^k(\omega,t)}$ eventually leads to

$$\begin{aligned} e^{j\phi^k(\omega,t)} &= \cos\phi^k(\omega,t) + j\sin\phi^k(\omega,t) \\ &= \frac{Re\left[\overline{\mathbf{Y}}^k(\omega,t)\right] + \mathrm{Im}\left[\overline{\mathbf{Y}}^k(\omega,t)\right]}{\left|\overline{\mathbf{Y}}^k(\omega,t)\right|} \\ &= \frac{\overline{\mathbf{Y}}^k(\omega,t)}{\left|\overline{\mathbf{Y}}^k(\omega,t)\right|} \end{aligned} \tag{A10}$$

The update formula for $\beta^k(\omega,t)$ and $\mathbf{H}^k(t)$ for projection onto the constraint space is set to

$$\beta^k(\omega,t) = \frac{\mathbf{W}^k(\omega)\mathbf{H}^k(t)}{\sum_k \mathbf{W}^k(\omega)\mathbf{H}^k(t)} \tag{A11}$$

$$\mathbf{H}^k(t) \leftarrow \frac{\mathbf{H}^k(t)}{\sum_k \mathbf{H}^k(t)} \tag{A12}$$

Appendix C. Estimation of L_1-Optimal Sparsity Parameter $\lambda^k(t)$

This section aims to facilitate spectral dictionaries with adaptive sparse coding. First, the CMF model is defined as the following terms:

$$\begin{gathered} \overline{\mathbf{W}} = \left[\mathbf{I}\otimes\mathbf{W}^1(\omega)\vdots\mathbf{I}\otimes\mathbf{W}^2(\omega)\vdots\cdots\vdots\mathbf{I}\otimes\mathbf{W}^K(\omega)\right], \\ e^{j\underline{\Phi}(t)} = \left[e^{j\underline{\Phi}^1(t)}\vdots\cdots\vdots e^{j\underline{\Phi}^K(t)}\right] \\ \underline{\mathbf{y}} = \mathrm{vec}(\mathbf{Y}) = \begin{bmatrix} \underline{\mathbf{Y}}^1(:) \\ \cdots \\ \underline{\mathbf{Y}}^2(:) \\ \cdots \\ \vdots \\ \cdots \\ \underline{\mathbf{Y}}^K(:) \end{bmatrix}, \underline{\mathbf{h}} = \begin{bmatrix} \mathbf{H}^1(t) \\ \cdots \\ \mathbf{H}^2(t) \\ \cdots \\ \vdots \\ \cdots \\ \mathbf{H}^K(t) \end{bmatrix}, \underline{\lambda} = \begin{bmatrix} \lambda^1(t) \\ \cdots \\ \lambda^2(t) \\ \cdots \\ \vdots \\ \cdots \\ \lambda^K(t) \end{bmatrix}, \underline{\Phi} = \begin{bmatrix} \Phi^1(:,t) \\ \cdots \\ \Phi^2(:,t) \\ \cdots \\ \vdots \\ \cdots \\ \Phi^K(:,t) \end{bmatrix} \\ \overline{\mathbf{A}} = \begin{bmatrix} \overline{\mathbf{W}}\circ e^{j\underline{\Phi}(t)} & 0 & \cdots & 0 \\ 0 & \overline{\mathbf{W}}\circ e^{j\underline{\Phi}(t)} & 0 & \vdots \\ \vdots & 0 & \overline{\mathbf{W}}\circ e^{j\underline{\Phi}(t)} & 0 \\ 0 & \cdots & 0 & \overline{\mathbf{W}}\circ e^{j\underline{\Phi}(t)_t} \end{bmatrix} \end{gathered} \tag{A13}$$

where "⊗" and "∘" are the Kronecker product and the Hadamard product, respectively. The term vec(·) denotes the column vectorization and the term $\mathbf{I}$ is the identity matrix. The goal is then set to compute the regularization parameter $\lambda^k(t)$ related to each $\mathbf{H}^k(t)$. To achieve the goal, the parameter p in Equation (A3) was set at 1 to acquire a linear expression (in $\lambda^k(t)$). In consideration of the noise variance σ^2, Equation (A3) can concisely be rewritten as:

$$F(\underline{\mathbf{h}},\underline{\lambda}) = \frac{1}{2\sigma^2}\underline{\mathbf{y}} - \overline{\mathbf{A}}\underline{\mathbf{h}}_F^2 + \underline{\lambda}^{\mathbf{T}}\underline{\mathbf{h}} - (\log\underline{\lambda})^{\mathbf{T}}\underline{\mathbf{1}} \tag{A14}$$

where the $\underline{\mathbf{h}}$ and $\underline{\boldsymbol{\lambda}}$ terms indicate vectors of dimension $R \times 1$ (i.e., $R = F \times T \times K$), and the superscript '**T**' is used to denote complex Hermitian transpose (i.e., vector (or matrix) transpose), followed by complex conjugate. The Expectation–Maximization (EM) algorithm is used to determine $\underline{\boldsymbol{\lambda}}$ and $\underline{\mathbf{h}}$ is the hidden variable, where the log-likelihood function can be optimized with respect to $\underline{\lambda}$. The log-likelihood function satisfies the following [12]:

$$\ln p\left(\underline{\mathbf{y}} \,\middle|\, \underline{\lambda}, \overline{\mathbf{A}}, \sigma^2\right) \geq \int Q(\underline{\mathbf{h}}) \ln\left(\frac{p\left(\underline{\mathbf{y}}, \underline{\mathbf{h}} \middle| \underline{\lambda}, \overline{\mathbf{A}}, \sigma^2\right)}{Q(\underline{\mathbf{h}})}\right) d\underline{\mathbf{h}} \tag{A15}$$

by applying the Jensen's inequality for any distribution $Q(\underline{\mathbf{h}})$. The distribution can simply verify the posterior distribution of $\underline{\mathbf{h}}$ that maximizes the right-hand side of Equation (A19) is given by $Q(\underline{\mathbf{h}}) = p\left(\underline{\mathbf{h}} \middle| \underline{\mathbf{y}}, \underline{\lambda}, \overline{\mathbf{A}}, \sigma^2\right)$. The posterior distribution in the form of the Gibbs distribution is proposed as follows:

$$Q(\underline{\mathbf{h}}) = \frac{1}{Z_h} \exp[-F(\underline{\mathbf{h}})] \text{ where } Z_h = \int \exp[-F(\underline{\mathbf{h}})] d\underline{\mathbf{h}} \tag{A16}$$

The term $F(\underline{\mathbf{h}})$ in Equation (A16) as the function of the Gibbs distribution is essential for simplifying the adaptive optimization of $\underline{\lambda}$. The maximum-likelihood (ML) estimation of $\underline{\lambda}$ can be decomposed as follows:

$$\begin{aligned} \underline{\lambda}^{ML} &= \arg\max_{\underline{\lambda}} \ln p\left(\underline{\mathbf{y}} \,\middle|\, \underline{\lambda}, \overline{\mathbf{A}}, \sigma^2\right) \\ &= \arg\max_{\underline{\lambda}} \int Q(\underline{\mathbf{h}}) \left(\ln p\left(\underline{\mathbf{y}} \,\middle|\, \underline{\mathbf{h}}, \overline{\mathbf{A}}, \sigma^2\right) + \ln p(\underline{\mathbf{h}} | \underline{\lambda})\right) d\underline{\mathbf{h}} \\ &= \arg\max_{\underline{\lambda}} \int Q(\underline{\mathbf{h}}) \ln p(\underline{\mathbf{h}} | \underline{\lambda}) d\underline{\mathbf{h}} \end{aligned} \tag{A17}$$

In the same way,

$$\begin{aligned} {\sigma^2}_{ML} &= \arg\max_{\sigma^2} \ln p\left(\underline{\mathbf{y}} \,\middle|\, \underline{\lambda}, \overline{\mathbf{A}}, \sigma^2\right) \\ &= \arg\max_{\sigma^2} \int Q(\underline{\mathbf{h}}) \left(\ln p\left(\underline{\mathbf{y}} \,\middle|\, \underline{\mathbf{h}}, \overline{\mathbf{A}}, \sigma^2\right) + \ln p(\underline{\mathbf{h}} | \underline{\lambda})\right) d\underline{\mathbf{h}} \\ &= \arg\max_{\sigma^2} \int Q(\underline{\mathbf{h}}) \ln p\left(\underline{\mathbf{y}} \,\middle|\, \underline{\mathbf{h}}, \overline{\mathbf{A}}, \sigma^2\right) d\underline{\mathbf{h}} \end{aligned} \tag{A18}$$

Individual element of **H** is required to be exponentially distributed with independent decay parameters that delivers $p(\underline{\mathbf{h}} | \underline{\lambda}) = \prod_g \lambda_g \exp\left(-\lambda_g h_g\right)$, thus Equation (20) obtains

$$\underline{\lambda}^{ML} = \arg\max_{\underline{\lambda}} \int Q(\underline{\mathbf{h}}) \left(\ln \lambda_g - \lambda_g h_g\right) d\underline{\mathbf{h}} \tag{A19}$$

The term $\underline{\mathbf{h}}$ denotes the dependent variable of the distribution $Q(\underline{\mathbf{h}})$ whereas other parameters are assumed to be constant. As such, the $\underline{\lambda}$ optimization in (A19) is derived by differentiating the parameters within the integral with respect to $\underline{\mathbf{h}}$. As a result, the functional optimization of $\underline{\lambda}$ then obtains

$$\lambda_g = \frac{1}{\int h_g Q(\underline{\mathbf{h}}) d\underline{\mathbf{h}}} \tag{A20}$$

where $g = 1, 2, \ldots, R$, λ_g denotes the gth element of $\underline{\lambda}$. The iterative update for σ^2_{ML} is given by

$$\begin{aligned} \sigma^2_{ML} &= \arg\max_{\sigma^2} \int Q(\underline{\mathbf{h}}) \left(\tfrac{-N_0}{2} \ln\left(\pi\sigma^2\right) - \tfrac{1}{2\sigma^2} \|\underline{\mathbf{y}} - \overline{\mathbf{A}}\underline{\mathbf{h}}\|^2\right) d\underline{\mathbf{h}} \\ &= \tfrac{1}{N_0} \int Q(\underline{\mathbf{h}}) \left(\|\underline{\mathbf{y}} - \overline{\mathbf{A}}\underline{\mathbf{h}}\|^2\right) d\underline{\mathbf{h}} \end{aligned} \tag{A21}$$

where $p\left(\underline{\mathbf{y}}\,\middle|\underline{\mathbf{h}},\overline{\mathbf{A}},\sigma^2\right)=\left(\pi\sigma^2\right)^{-N_0/2}\exp\left(-\left(1/2\sigma^2\right)\|\underline{\mathbf{y}}-\overline{\mathbf{A}}\underline{\mathbf{h}}\|^2\right)$ and $N_0=K\times T$. However, the integral forms in Equations (A20) and (A21) are complex to compute and analyzed analytically. Thus, an approximation to $Q(\underline{\mathbf{h}})$ is exploited. Notice that the solution $\underline{\mathbf{h}}$ naturally splits its elements into distinct subsets $\underline{\mathbf{h}}_M$ and $\underline{\mathbf{h}}_p$ consisting of components $\forall_m\in M$ such that $h_m>0$ and components $\forall_p\in P$ such that $h_P=0$. Hence, this can be derived as follows:

$$F(\underline{\mathbf{h}},\underline{\lambda})=F\left(\underline{\mathbf{h}}_M,\underline{\lambda}_M\right)+F\left(\underline{\mathbf{h}}_P,\underline{\lambda}_P\right)+G \tag{A22}$$

Defined $F\left(\underline{\mathbf{h}}_M,\underline{\lambda}_M\right)=\frac{1}{2\sigma^2}\|\underline{\mathbf{y}}-\overline{\mathbf{A}}_M\underline{\mathbf{h}}_M\|^2+\underline{\lambda}_M^{\mathbf{T}}\underline{\mathbf{h}}_M-(\log\underline{\lambda})_M^{\mathbf{T}}\mathbf{1}M$, $F\left(\underline{\mathbf{h}}_P,\underline{\lambda}_P\right)=\frac{1}{2\sigma^2}\|\underline{\mathbf{y}}-\overline{\mathbf{A}}_P\underline{\mathbf{h}}_P\|^2+\underline{\lambda}_P^{\mathbf{T}}\underline{\mathbf{h}}_P-(\log\underline{\lambda})_P^{\mathbf{T}}\mathbf{1}P$, and $G=\frac{1}{2\sigma^2}\left[2\left(\overline{\mathbf{A}}_M\underline{\mathbf{h}}_M\right)^{\mathbf{T}}\left(\overline{\mathbf{A}}_P\underline{\mathbf{h}}_P\right)-\|\underline{\mathbf{y}}\|^2\right]$. Here, the term $\|\underline{\mathbf{y}}\|^2$ is a constant and the cross-term $\left(\overline{\mathbf{A}}_M\underline{\mathbf{h}}_M\right)^{\mathbf{T}}\left(\overline{\mathbf{A}}_P\underline{\mathbf{h}}_P\right)$ measures the orthogonality between $\overline{\mathbf{A}}_M\underline{\mathbf{h}}_M$ and $\overline{\mathbf{A}}_P\underline{\mathbf{h}}_P$, where $\overline{\mathbf{A}}_M$ and $\overline{\mathbf{A}}_P$ denote the sub-matrix of $\overline{\mathbf{A}}$ that corresponds to $\underline{\mathbf{h}}_M$ and $\underline{\mathbf{h}}_P$. To obtain a simplified expression in Equation (A22), the $F(\underline{\mathbf{h}})$ function can be approximated as $F(\underline{\mathbf{h}},\underline{\lambda})\approx F\left(\underline{\mathbf{h}}_M,\underline{\lambda}_M\right)+F\left(\underline{\mathbf{h}}_P,\underline{\lambda}_P\right)$ and the G can be safely discounted since its value is typically much smaller than $F\left(\underline{\mathbf{h}}_M,\underline{\lambda}_M\right)$ and $F\left(\underline{\mathbf{h}}_P,\underline{\lambda}_P\right)$. Thus, the approximation of $Q(\underline{\mathbf{h}})$ can be expressed as

$$\begin{aligned}Q(\underline{\mathbf{h}},\underline{\lambda}) &= \tfrac{1}{Z_h}\exp[-F(\underline{\mathbf{h}},\underline{\lambda})]\\ &\approx \tfrac{1}{Z_h}\exp\left[-\left(F\left(\underline{\mathbf{h}}_M,\underline{\lambda}_M\right)+F\left(\underline{\mathbf{h}}_P,\underline{\lambda}_P\right)\right)\right]\\ &= \tfrac{1}{Z_M}\exp\left[-F\left(\underline{\mathbf{h}}_M,\underline{\lambda}_M\right)\right]\\ \tfrac{1}{Z_P}\exp[-&\ F\left(\underline{\mathbf{h}}_P,\underline{\lambda}_P\right)]\\ &= Q_M\left(\underline{\mathbf{h}}_M\right)Q_P\left(\underline{\mathbf{h}}_P\right)\end{aligned} \tag{A23}$$

Defining $Z_M=\int\exp\left[-F\left(\underline{\mathbf{h}}_M,\underline{\lambda}_M\right)\right]d\underline{\mathbf{h}}_M$ and $Z_P=\int\exp\left[-F\left(\underline{\mathbf{h}}_P,\underline{\lambda}_P\right)\right]d\underline{\mathbf{h}}_P$. With the purpose of characterizing $Q_P\left(\underline{\mathbf{h}}_P\right)$, some positive deviation to $\underline{\mathbf{h}}_P$ is needed to be allowed for, whereas the $\underline{\mathbf{h}}_P$ values will reject all negative values due to CMF only accepting zero and positive values. Thus, $\underline{\mathbf{h}}_P$ admits zero and positive values in $Q_P\left(\underline{\mathbf{h}}_P\right)$. The approximation of the distribution $Q_P\left(\underline{\mathbf{h}}_P\right)$ is then utilized in the Taylor expansion as the *maximum a posterior probability* (MAP) estimate. Therefore, with $\underline{\mathbf{h}}^{\mathrm{MAP}}$, one obtains

$$\begin{aligned}Q_P\left(\underline{\mathbf{h}}_P\geq 0\right) &\propto \exp\left\{-\left[\left(\frac{\partial F}{\partial \underline{\mathbf{h}}}\right)\Bigg|_{\underline{\mathbf{h}}^{\mathrm{MAP}}}\right]_P^{\mathbf{T}}\underline{\mathbf{h}}_P-\tfrac{1}{2}\underline{\mathbf{h}}_P^{\mathbf{T}}\overline{\mathbf{C}}_P\underline{\mathbf{h}}_P\right\}\\ &= \exp\left[-\left(\overline{\mathbf{C}}\underline{\mathbf{h}}^{\mathrm{MAP}}-\tfrac{1}{\sigma^2}\overline{\mathbf{A}}^{\mathbf{T}}\underline{\mathbf{y}}+\underline{\lambda}\right)_P^{\mathbf{T}}\underline{\mathbf{h}}_P-\tfrac{1}{2}\underline{\mathbf{h}}_P^{\mathbf{T}}\overline{\mathbf{C}}_P\underline{\mathbf{h}}_P\right]\end{aligned} \tag{A24}$$

where $\overline{\mathbf{C}}_P=\frac{1}{\sigma^2}\overline{\mathbf{A}}_P^{\mathbf{T}}\overline{\mathbf{A}}_P$ and $\overline{\mathbf{C}}=\frac{1}{\sigma^2}\overline{\mathbf{A}}^{\mathbf{T}}\overline{\mathbf{A}}$. The integration of the term $Q_P\left(\underline{\mathbf{h}}_P\right)$ in Equation (A24) is hard to derive in its closed form expression for analytical evaluation, which subsequently prohibits inference of the sparsity parameters. A fixed form distribution is employed for computing variational approximate $Q_P\left(\underline{\mathbf{h}}_P\right)$. As a result, the closed form expression is obtained. Subsequently, the term $\underline{\mathbf{h}}_P$ only takes on nonnegative values, so a suitable fixed form distribution is to use the factorized exponential distribution given by

$$\hat{Q}_P\left(\underline{\mathbf{h}}_P\geq 0\right)=\prod_{p\in P}\frac{1}{u_p}\exp\left(\frac{-h_p}{u_p}\right) \tag{A25}$$

By minimizing the Kullback–Leibler divergence between Q_P and $\hat{Q}_P$, the variational parameters $\underline{\mathbf{u}}=\left\{u_p\right\}$ where $\forall p\in P$ can be derived as:

$$\begin{aligned}\underline{\mathbf{u}} &= \arg=\min_{\underline{\mathbf{u}}}\hat{Q}_P\left(\underline{\mathbf{h}}_P\right)\ln\frac{\hat{Q}_P\left(\underline{\mathbf{h}}_P\right)}{Q_P\left(\underline{\mathbf{h}}_P\right)}d\underline{\mathbf{h}}_P\\ &= \arg=\min_{\underline{\underline{\mathbf{u}}}}\hat{Q}_P\left(\underline{\mathbf{h}}_P\right)\left[\ln\hat{Q}_P\left(\underline{\mathbf{h}}_P\right)-\ln Q_P\left(\underline{\mathbf{h}}_P\right)\right]d\underline{\mathbf{h}}_P\end{aligned} \tag{A26}$$

Solving Equation (A26) for u_p leads to the following update [37]:

$$u_p \leftarrow u_p \frac{-\hat{b}_p + \sqrt{\hat{b}_p^2 + 4\frac{(\hat{\mathbf{C}}\underline{\mathbf{u}})_p}{\overline{u}_p}}}{2(\hat{\mathbf{C}}\underline{\mathbf{u}})_p} \tag{A27}$$

The approximate distribution for components $\underline{\mathbf{h}}_M$ can be obtained by substituting $F(\underline{\mathbf{h}}_M, \underline{\boldsymbol{\lambda}}_M)$ into $Q_M(\underline{\mathbf{h}}_M)$ as follows:

$$\begin{aligned} Q_M(\underline{\mathbf{h}}_M) &= \frac{1}{Z_M} \exp\left[-F(\underline{\mathbf{h}}_M, \underline{\boldsymbol{\lambda}}_M)\right] \\ &\propto \exp\left[-\left(\frac{1}{2}\underline{\mathbf{h}}_M^{\mathrm{T}} \overline{\mathbf{C}}_M \underline{\mathbf{h}}_M - \frac{1}{\sigma^2}\underline{\mathbf{y}}^{\mathrm{T}} \overline{\mathbf{A}}_M \underline{\mathbf{h}}_M + \underline{\boldsymbol{\lambda}}_M \underline{\mathbf{h}}_M\right)\right] \end{aligned} \tag{A28}$$

In Equation (A28), the function $Q_M(\underline{\mathbf{h}}_M)$ will be expressed as the unconstrained Gaussion with mean $\underline{\mathbf{h}}_M^{\mathrm{MAP}}$ and covariance $\overline{\mathbf{C}}_M^{-1}$ based on a multivariate Gaussian distribution. The term $\overline{\mathbf{C}}_M$ denotes the sub-matrix of $\overline{\mathbf{C}}$. The sparsity parameter is then obtained by substituting Equations (A24), (A25), and (A28) into Equation (A20) as presented in Equation (A29):

$$\lambda_g = \begin{cases} \frac{1}{\int h_g Q_M(\underline{\mathbf{h}}_M) d\underline{\mathbf{h}}_M} = \frac{1}{h_g^{\mathrm{MAP}}} \text{ if } g \in M \\ \frac{1}{\int h_g \hat{Q}_P(\underline{\mathbf{h}}_P) d\underline{\mathbf{h}}_P} = \frac{1}{u_g} \text{ if } g \in P \end{cases} \tag{A29}$$

and its covariance X is given by

$$X_{ab} = \begin{cases} \left(\overline{\mathbf{C}}_P^{-1}\right)_{ab}, & \text{if } a, b \in M \\ u_p^2 \delta_{ab}, & \text{Otherwise.} \end{cases} \tag{A30}$$

Similarly, the inference for σ^2 can be computed from Equation (24) as

$$\sigma^2 = \frac{1}{N_0} \int Q(\underline{\mathbf{h}})\left(\|\underline{\mathbf{y}} - \overline{\mathbf{A}}\underline{\mathbf{h}}\|^2\right) d\underline{\mathbf{h}} \tag{A31}$$

where

$$\hat{h}_g = \begin{cases} h_g^{\mathrm{MAP}} \text{ if } g \in M \\ u_g \text{ if } g \in P \end{cases}$$

The core procedure of the proposed CMF method is based on L_1-optimal sparsity parameters. The estimated sources are discovered by multiplying the respective rows of the $\mathbf{W}^k(\omega)$ components with the corresponding columns of the $\boldsymbol{H}^k(t)$ weights and time-varying phrase spectrum $e^{j\phi^k(\omega,t)}$. The separated sources $\hat{s}_j(t)$ are obtained by converting the time-frequency represented sources into time domain.

References

1. Wang, Q.; Woo, W.L.; Dlay, S. Informed single-channel speech separation using hmm–gmm user-generated exemplar source. *IEEE/ACM Trans. Audio Speech Lang. Process.* **2014**, *22*, 2087–2100. [CrossRef]
2. Gao, B.; Bai, L.; Woo, W.L.; Tian, G.; Cheng, Y. Automatic defect identification of eddy current pulsed thermography using single channel blind source separation. *IEEE Trans. Instrum. Meas.* **2013**, *63*, 913–922. [CrossRef]
3. Yin, A.; Gao, B.; Tian, G.; Woo, W.L.; Li, K. Physical interpretation and separation of eddy current pulsed thermography. *J. Appl. Phys.* **2013**, *113*, 64101. [CrossRef]

4. Cheng, L.; Gao, B.; Tian, G.; Woo, W.L.; Berthiau, G. Impact damage detection and identification using eddy current pulsed thermography through integration of PCA and ICA. *IEEE Sens. J.* **2014**, *14*, 1655–1663. [CrossRef]
5. Cholnam, O.; Chongil, G.; Chol, R.K.; Gwak, C.; Rim, K.C. Blind signal separation method and relationship between source separation and source localisation in the TF plane. *IET Signal Process.* **2018**, *12*, 1115–1122. [CrossRef]
6. Tengtrairat, N.; Woo, W.L.; Dlay, S.S.; Gao, B. Online noisy single-channel blind separation by spectrum amplitude estimator and masking. *IEEE Trans. Signal Process* **2016**, *64*, 1881–1895. [CrossRef]
7. Tengtrairat, N.; Gao, B.; Woo, W.L.; Dlay, S.S. Single-Channel Blind Separation Using Pseudo-Stereo Mixture and Complex 2-D Histogram. *IEEE Trans. Neural Netw. Learn. Syst.* **2013**, *24*, 1722–1735. [CrossRef]
8. Koundinya, S.; Karmakar, A. Homotopy optimisation based NMF for audio source separation. *IET Signal Process.* **2018**, *12*, 1099–1106. [CrossRef]
9. Kim, M.; Smaragdis, P. Single channel source separation using smooth Nonnegative Matrix Factorization with Markov Random Fields. In Proceedings of the 2013 IEEE International Workshop on Machine Learning for Signal Processing (MLSP), Southapmton, UK, 22–25 September 2013; pp. 1–6.
10. Yoshii, K.; Itoyama, K.; Goto, M. Student's T nonnegative matrix factorization and positive semidefinite tensor factorization for single-channel audio source separation. In Proceedings of the 2016 IEEE International Conference on Acoustics, Speech and Signal Processing (ICASSP), Shanghai, China, 20–25 March 2016; pp. 51–55.
11. Al-Tmeme, A.; Woo, W.L.; Dlay, S.; Gao, B. Underdetermined convolutive source separation using gem-mu with variational approximated optimum model order NMF2D. *IEEE/ACM Trans. Audio, Speech, Lang. Process.* **2016**, *25*, 35–49. [CrossRef]
12. Woo, W.L.; Gao, B.; Bouridane, A.; Ling, B.W.-K.; Chin, C.S. Unsupervised learning for monaural source separation using maximization–minimization algorithm with time–frequency deconvolution. *Sensors* **2018**, *18*, 1371. [CrossRef]
13. Gao, B.; Woo, W.L.; Dlay, S.S. Unsupervised single channel separation of non-stationary signals using Gammatone filterbank and Itakura-Saito nonnegative matrix two-dimensional factorizations. *IEEE Trans. Circuits Syst. I* **2013**, *60*, 662–675. [CrossRef]
14. Févotte, C.; Bertin, N.; Durrieu, J.-L. Nonnegative matrix factorization with the itakura-saito divergence: With application to music analysis. *Neural Comput.* **2009**, *21*, 793–830. [CrossRef] [PubMed]
15. Pu, X.; Yi, Z.; Zheng, Z.; Zhou, W.; Ye, M. Face recognition using fisher non-negative matrix factorization with sparseness constraints. *Comput. Vis.* **2005**, *3497*, 112–117. [CrossRef]
16. Magron, P.; Virtanen, T. Towards complex nonnegative matrix factorization with the beta-divergence. In Proceedings of the 2018 16th International Workshop on Acoustic Signal Enhancement (IWAENC), Tokyo, Japan, 17–20 September 2018; pp. 156–160.
17. King, B. New Methods of Complex Matrix Factorization for Single-Channel Source Separation and Analysis. Ph.D. Thesis, University of Washington, Seattle, WA, USA, 2012.
18. Parathai, P.; Tengtrairat, N.; Woo, W.L.; Gao, B. Single-channel signal separation using spectral basis correlation with sparse nonnegative tensor factorization. *Circuits Syst. Signal Process.* **2019**, *38*, 5786–5816. [CrossRef]
19. Woo, W.L.; Dlay, S.; Al-Tmeme, A.; Gao, B. Reverberant signal separation using optimized complex sparse nonnegative tensor deconvolution on spectral covariance matrix. *Digit. Signal Process.* **2018**, *83*, 9–23. [CrossRef]
20. Tengtrairat, N.; Parathai, P.; Woo, W.L. Blind 2D signal direction for limited-sensor space using maximum likelihood estimation. *Asia-Pac. J. Sci. Technol.* **2017**, *22*, 42–49.
21. Gao, B.; Woo, W.L.; Tian, G.Y.; Zhang, H. Unsupervised diagnostic and monitoring of defects using waveguide imaging with adaptive sparse representation. *IEEE Trans. Ind. Inform.* **2016**, *12*, 405–416. [CrossRef]
22. Gao, B.; Woo, W.L.; He, Y.; Tian, G.Y. Unsupervised sparse pattern diagnostic of defects with inductive thermography imaging system. *IEEE Trans. Ind. Inform.* **2016**, *12*, 371–383. [CrossRef]
23. Tengtrairat, N.; Woo, W.L. Single-channel separation using underdetermined blind autoregressive model and lest absolute deviation. *Neurocomputing* **2015**, *147*, 412–425. [CrossRef]
24. Gao, B.; Woo, W.; Ling, B.W.-K. Machine learning source separation using maximum a posteriori nonnegative matrix factorization. *IEEE Trans. Cybern.* **2013**, *44*, 1169–1179. [CrossRef]

25. Tengtrairat, N.; Woo, W. Extension of DUET to single-channel mixing model and separability analysis. *Signal Process.* **2014**, *96*, 261–265. [CrossRef]
26. Zhou, Q.; Feng, Z.; Benetos, E. Adaptive noise reduction for sound event detection using subband-weighted NMF. *Sensors* **2019**, *19*, 3206. [CrossRef] [PubMed]
27. Yan, L.; Zhang, Y.; He, Y.; Gao, S.; Zhu, D.; Ran, B.; Wu, Q. Hazardous traffic event detection using markov blanket and sequential minimal optimization (MB-SMO). *Sensors* **2016**, *16*, 1084. [CrossRef] [PubMed]
28. Chen, Y.-L.; Chiang, H.-H.; Chiang, C.-Y.; Liu, J.; Yuan, S.-M.; Wang, J.-H. A vision-based driver nighttime assistance and surveillance system based on intelligent image sensing techniques and a heterogamous dual-core embedded system architecture. *Sensors* **2012**, *12*, 2373–2399. [CrossRef] [PubMed]
29. McLoughlin, I.V.; Zhang, H.; Xie, Z.; Song, Y.; Xiao, W. Robust sound event classification using deep neural networks. *IEEE/ACM Trans. Audio. Speech. Lang. Process.* **2015**, *23*, 540–552. [CrossRef]
30. Noh, K.; Chang, J.-H. Joint optimization of deep neural network-based dereverberation and beamforming for sound event detection in multi-channel environments. *Sensors* **2020**, *20*, 1883. [CrossRef]
31. Hsu, C.W.; Lin, C.J. A comparison of methods for multi-class support vector machines. *IEEE Trans. Neural Netw.* **2002**, *13*, 415–425.
32. Martin-Morato, I.; Cobos, M.; Ferri, F.J. A case study on feature sensitivity for audio event classification using support vector machines. In Proceedings of the 2016 IEEE 26th International Workshop on Machine Learning for Signal Processing (MLSP), Salerno, Italy, 13–16 September 2016; pp. 1–6.
33. Candès, E.J.; Romberg, J.K.; Tao, T. Stable signal recovery from incomplete and inaccurate measurements. *Commun. Pure Appl. Math.* **2006**, *59*, 1207–1223. [CrossRef]
34. Selesnick, I. Resonance-based signal decomposition: A new sparsity-enabled signal analysis method. *Signal Process.* **2011**, *91*, 2793–2809. [CrossRef]
35. Al-Tmeme, A.; Woo, W.L.; Dlay, S.; Gao, B. Single channel informed signal separation using artificial-stereophonic mixtures and exemplar-guided matrix factor deconvolution. *Int. J. Adapt. Control. Signal Process.* **2018**, *32*, 1259–1281. [CrossRef]
36. Gao, B.; Woo, W.L.; Dlay, S.S. Single channel blind source separation using EMD-subband variable regularized sparse features. *IEEE Trans. Audio. Speech Lang. Process.* **2011**, *19*, 961–976. [CrossRef]
37. Bertsekas, D.P. *Nonlinear Programming*, 2nd ed.; Athena Scientific: Belmont, MA, USA, 1999.
38. Kameoka, H.; Ono, N.; Kashino, K.; Sagayama, S. Complex NMF: A new sparse representation for acoustic signals. In Proceedings of the 2009 IEEE International Conference on Acoustics, Speech and Signal Processing, Taipei, Taiwan, 19–24 April 2009; pp. 3437–3440. [CrossRef]
39. Parathai, P.; Woo, W.L.; Dlay, S.; Gao, B. Single-channel blind separation using L1-sparse complex non-negative matrix factorization for acoustic signals. *J. Acoust. Soc. Am.* **2015**, *137*, 124–129. [CrossRef] [PubMed]
40. Zdunek, R.; Cichocki, A. Nonnegative matrix factorization with constrained second-order optimization. *Signal Process.* **2007**, *87*, 1904–1916. [CrossRef]
41. Yu, K.; Woo, W.L.; Dlay, S. Variational regularized two-dimensional nonnegative matrix factorization. *IEEE Trans. Neural Netw. Learn. Syst.* **2012**, *23*, 703–716.
42. Gao, B.; Woo, W.L.; Dlay, S. Adaptive sparsity non-negative matrix factorization for single-channel source separation. *IEEE J. Sel. Top. Signal Process.* **2011**, *5*, 989–1001. [CrossRef]

Article

Stable Tensor Principal Component Pursuit: Error Bounds and Efficient Algorithms

Wei Fang [1,*], Dongxu Wei [2] and Ran Zhang [3]

1 Department of Computer Science and Technology, Huaibei Vocational and Technical College, Huaibei 235000, China
2 School of Physics and Electronic Electrical Engineering, Huaiyin Normal University, Huaian 223300, China; weidx@hytc.edu.cn
3 Mathematics Teaching and Research Group, Nanjing No.9 High School, Nanjing 210018, China; exjzhang@163.com
* Correspondence: fangweixyz@sina.com

Received: 13 November 2019; Accepted: 29 November 2019; Published: 3 December 2019

Abstract: The rapid development of sensor technology gives rise to the emergence of huge amounts of tensor (i.e., multi-dimensional array) data. For various reasons such as sensor failures and communication loss, the tensor data may be corrupted by not only small noises but also gross corruptions. This paper studies the Stable Tensor Principal Component Pursuit (STPCP) which aims to recover a tensor from its corrupted observations. Specifically, we propose a STPCP model based on the recently proposed tubal nuclear norm (TNN) which has shown superior performance in comparison with other tensor nuclear norms. Theoretically, we rigorously prove that under tensor incoherence conditions, the underlying tensor and the sparse corruption tensor can be stably recovered. Algorithmically, we first develop an ADMM algorithm and then accelerate it by designing a new algorithm based on orthogonal tensor factorization. The superiority and efficiency of the proposed algorithms is demonstrated through experiments on both synthetic and real data sets.

Keywords: tensor principal component pursuit; stable recovery; tensor SVD; ADMM

1. Introduction

In recent years, different types of tensor data have emerged with the significant progress of modern sensor technology, such as color images [1], videos [2], functional MRI data [3], hyper-spectral images [4], point could data [5], traffic stream data [6], etc. Thanks to its multi-way nature, tensor-based methods have natural superiority over vector and matrix-based methods in analyzing and processing ubiquitous modern multi-way data, and have found extensive applications in computer vision [1,7], data mining [5], machine learning [2], signal processing [8], to name a few. In real applications, the acquired tensor data may often suffer from noises and gross corruptions owing to many different reasons such as sensor failure, lens pollution, communication interference, occlusion in videos, or abnormalities in a sensor network [9], etc. At the same time, many real-world tensor data, such as face images or videos, have been shown to have some low-dimensional structure and can be well approximated by a smaller number of "principal components" [8]. Then, a question naturally arises: how to pursue the principal components of an observed tensor data in the presence of both noises and gross corruptions? We will answer this question in this paper and refer to the proposed methodology as Stable Tensor Principal Component Pursuit (STPCP).

The tensor low-rankness is an ideal model of the property that a tensor data can be well approximated by a small number of principal components [8]. In the last decade, low-rank tensor models have attracted much attention in many fields [10]. There are multiple low-rank tensor models since there exist different definitions of tensor rank. Among these models, the low CP rank model [11] and the low Tucker rank model [1] should be the most famous two. The low CP rank model approximates the underlying tensor by the sum of a small number of rank-1 tensors, whereas the low Tucker rank model assumes the unfolding matrix along each mode are low rank. To estimate an unknown low-rank tensor from corrupted observations, it is a natural option to consider the rank minimization problem which chooses the tensor of lowest rank as the solution from a certain feasible set. However, tensor rank minimization, even in its 2-way (matrix) case, is generally NP-hard [12] and even harder in higher-way cases [13]. For tractable solutions, researchers turn to a variety of convex surrogates for tensor rank [1,14–18] to replace the tensor rank in rank minimization problem. Methods based on surrogates for the tensor CP rank and Tucker rank have been extensively explored in both the theoretical side and the application side [14,17,19–24].

Recently, the low-tubal-rank model [16,25] has shown better performance than traditional tensor low-rank models in many tensor recover tasks such as image/video inpainting/denoising/sensing [2,25,26], moving object detection [27], multi-view learning [28], seismic data completion [29], WiFi fingerprint [30], MRI imaging [16], point cloud data inpainting [31], and so on. The tubal rank is a new complexity measure of tensor defined through the framework of tensor singular value decomposition (t-SVD) [32,33]. At the core of existing low-tubal-rank models is the tubal nuclear norm (TNN) which is a convex surrogate for the tubal rank. In contrast to CP-based tensor nuclear norms or Tucker-based tensor nuclear norms which models low-rankness in the original domain, TNN models low-rankness in the Fourier domain. It is pointed out in [25,34,35] that TNN has superiority over traditional tensor nuclear norms in exploiting the ubiquitous "spatial-shifting" property in real-world tensor data.

Inspired by the superior performance of TNN, this paper adopts TNN as a low-rank regularizer in the proposed STPCP model. Specifically, the proposed STPCP aims to estimate the underlying tensor data $\underline{\mathbf{L}}_0 \in \mathbb{R}^{n_1 \times n_2 \times n_3}$ from an observation tensor $\underline{\mathbf{M}}$ polluted by both small dense noises and sparse gross corruptions as follows:

$$\underline{\mathbf{M}} = \underline{\mathbf{L}}_0 + \underline{\mathbf{S}}_0 + \underline{\mathbf{E}}_0, \tag{1}$$

where $\underline{\mathbf{S}}_0$ is a tensor denoting the sparse corruptions and $\underline{\mathbf{E}}_0$ is a tensor representing small dense noises. Model (1) is also known as robust tensor decomposition in [36,37].

Our STPCP model is first formulated as a TNN-based convex problem. Then, our theoretical analysis gives upper bound on the estimation error of $\underline{\mathbf{L}}_0$ and $\underline{\mathbf{S}}_0$. In contrast to the analysis in [37], the proposed STPCP can exactly recovery the underlying tensor $\underline{\mathbf{L}}_0$ and the sparse corruption tensor $\underline{\mathbf{S}}_0$ when the noise term $\underline{\mathbf{E}}_0$ vanishes. For efficient solution of the proposed STPCP model, we develop two algorithms with extensions to a more challenging scenario where missing observations are also considered. The first algorithm is an ADMM algorithm and the second algorithm accelerates it using tensor factorization. Experiments show the effectiveness and the efficiency of the designed algorithms.

We organize the rest of this paper as follows. In Section 2, we briefly introduce basic preliminaries for t-SVD and some related works. The proposed STPCP model is formulated and analyzed theoretically in Section 3. We design two algorithms in Section 4 and report experimental results in Section 5. This work is concluded in Section 6. The proofs of theorems, propositions, and lemmas are given in the appendix.

2. Preliminaries and Related Works

In this section, some preliminaries of t-SVD are first introduced. Then, the related works are presented.

Notations. We denote vectors by bold lower-case letters, e.g., $\mathbf{a} \in \mathbb{R}^n$, matrices by bold upper-case letters, e.g., $\mathbf{A} \in \mathbb{R}^{n_1 \times n_2}$, and tensors by underlined upper-case letters, e.g., $\underline{\mathbf{A}} \in \mathbb{R}^{n_1 \times n_2 \times n_3}$. For a given 3-way tensor, we define its fiber as a vector given through fixing all indices but one, and its *slice* as a matrix obtained by fixing all indices but two. For a given 3-way tensor $\underline{\mathbf{A}}$, we use $\underline{\mathbf{A}}_{ijk}$ to denote its (i,j,k)-th element; $\mathbf{A}^{(k)} := \underline{\mathrm{A}}(:,:,k)$ is used to denote its k-th frontal slice. $\underline{\widetilde{\mathrm{A}}}$ is used to denote the tensor after performing 1D Discrete Fourier Transformation (DFT) on all tube fibers $\underline{\mathrm{A}}(i,j,:)$ of $\underline{\mathrm{T}}$, $\forall i = 1,2,\cdots,n_1,\ j = 1,2,\cdots,n_2$, which can be efficiently computed by the Matlab command $\underline{\widetilde{\mathrm{A}}} = \mathrm{fft}(\underline{\mathrm{A}},[],3)$. We use $\mathrm{dft3}(\cdot)$ and $\mathrm{idft3}(\cdot)$ to represent the 1D DFT and inverse DFT along the tube fibers of 3-way tensors, i.e., $\mathrm{dft3}(\underline{\mathrm{A}}) := \mathrm{fft}(\underline{\mathrm{A}},[],3)$, $\quad \mathrm{idft3}(\underline{\mathrm{A}}) := \mathrm{ifft}(\underline{\mathrm{A}},[],3)$.

For a given matrix $\mathbf{M} \in \mathbb{R}^{n_1 \times n_2}$, define the nuclear norm and spectral norm of $\mathbf{M}$ respectively as:

$$\|\mathbf{M}\|_* := \sum_{i=1}^{p} \sigma_i(\mathbf{M}), \quad \text{and} \quad \|\mathbf{M}\|_{\mathrm{sp}} := \max\{\sigma_i(\mathbf{M})\},$$

where $p = \min\{n_1, n_2\}$, and $\sigma_1(\mathbf{M}) \geq \cdots \geq \sigma_p(\mathbf{M})$ are the singular values of $\mathbf{M}$ in a non-ascending order. The l_0-norm, l_1-norm, Frobenius norm, l_∞-norm of a tensor $\underline{\mathrm{A}} \in \mathbb{R}^{n_1 \times n_2 \times n_3}$ is defined as:

$$\|\underline{\mathrm{A}}\|_0 := \sum_{ijk} 1(\underline{\mathrm{A}}_{ijk} \neq 0), \quad \|\underline{\mathrm{A}}\|_1 := \sum_{ijk} |\underline{\mathrm{A}}_{ijk}|, \quad \|\underline{\mathrm{A}}\|_{\mathrm{F}} := \sqrt{\sum_{ijk} \underline{\mathrm{A}}_{ijk}^2}, \quad \|\underline{\mathrm{A}}\|_\infty := \max_{ijk} |\underline{\mathrm{A}}_{ijk}|,$$

where $1(C)$ is an indicator function whose value is 1 if the condition C is true, and 0 otherwise.

Given two matrices $\mathbf{A} = (a_{ij}) \in \mathbb{C}^{n_1 \times n_2}, \mathbf{B} = (b_{ij}) \in \mathbb{C}^{n_1 \times n_2}$, we define their inner product as follows:

$$\langle \mathbf{A}, \mathbf{B} \rangle = \mathrm{tr}(\mathbf{A}^{\mathsf{H}} \mathbf{B}) = \sum_{ij} \bar{a}_{ij} b_{ij},$$

where $\mathbf{A}^{\mathsf{H}}$ denotes conjugate transpose of matrix $\mathbf{A}$ and $\bar{a}_{ij}$ denotes the conjugation of complex number a_{ij}. Given two 3-way tensors $\underline{\mathrm{A}}, \underline{\mathrm{B}} \in \mathbb{R}^{n_1 \times n_2 \times n_3}$, we define their inner product as follows:

$$\langle \underline{\mathrm{A}}, \underline{\mathrm{B}} \rangle := \sum_{ijk} \underline{\mathrm{A}}_{ijk} \underline{\mathrm{B}}_{ijk}.$$

2.1. Tensor Singular Value Decomposition

We first define 3 operators based on block matrices which are introduced in [33]. For a given 3-way tensor $\underline{\mathrm{A}} \in \mathbb{R}^{n_1 \times n_2 \times n_3}$, we define its block vectorization $\mathrm{bvec}(\cdot)$ and the inverse operation $\mathrm{bvfold}(\cdot)$ in the following equation:

$$\mathrm{bvec}(\underline{\mathrm{A}}) := \begin{bmatrix} \mathbf{A}^{(1)} \\ \mathbf{A}^{(2)} \\ \vdots \\ \mathbf{A}^{(n_3)} \end{bmatrix} \in \mathbb{R}^{n_1 n_3 \times n_2}, \quad \mathrm{bvfold}(\mathrm{bvec}(\underline{\mathrm{A}})) = \underline{\mathrm{A}}.$$

We further define the block circulant matrix $\mathrm{bcirc}(\cdot)$ of any 3-way tensor $\underline{\mathrm{A}} \in \mathbb{R}^{n_1 \times n_2 \times n_3}$ as follows:

$$\mathrm{bcirc}(\underline{\mathrm{A}}) := \begin{bmatrix} \mathbf{A}^{(1)} & \mathbf{A}^{(n_3)} & \cdots & \mathbf{A}^{(2)} \\ \mathbf{A}^{(2)} & \mathbf{A}^{(1)} & \cdots & \mathbf{A}^{(3)} \\ \vdots & \ddots & \ddots & \vdots \\ \mathbf{A}^{(n_3)} & \mathbf{A}^{(n_3-1)} & \cdots & \mathbf{A}^{(1)} \end{bmatrix} \in \mathbb{C}^{n_1 n_3 \times n_2 n_3}$$

Equipped with above defined operators, we are now in a position to define the t-product of 3-way tensors.

Definition 1 (t-product [33]). *Given two tensors $\underline{\mathrm{A}} \in \mathbb{R}^{n_1 \times n_2 \times n_3}$ and $\underline{\mathrm{B}} \in \mathbb{R}^{n_2 \times n_4 \times n_3}$, the t-product of $\underline{\mathrm{A}}$ and $\underline{\mathrm{B}}$ is a new 3-way tensor $\underline{\mathrm{C}}$ with size $n_1 \times n_4 \times n_3$:*

$$\underline{\mathrm{C}} = \underline{\mathrm{A}} * \underline{\mathrm{B}} =: \mathrm{bvfold}\left(\mathrm{bcirc}(\underline{\mathrm{A}})\mathrm{bvec}(\underline{\mathrm{B}})\right). \tag{2}$$

A more intuitive interpretation of t-SVD is as follows [33]. If we treat a 3-way tensor $\underline{\mathrm{A}} \in \mathbb{R}^{n_1 \times n_2 \times n_3}$ as a matrix of size $n_1 \times n_2$ whose entries are the tube fibers, then the tensor t-product can be analogously understood as the "matrix multiplication" where the standard scalar product is replaced with the vector circular convolution between the tubes (i.e., vectors):

$$\underline{\mathrm{C}} = \underline{\mathrm{A}} * \underline{\mathrm{B}} \Leftrightarrow \underline{\mathrm{C}}(i,j,:) = \sum_{k=1}^{n_2} \underline{\mathrm{A}}(i,k,:) \star \underline{\mathrm{B}}(k,j,:), \quad \forall i = 1,2,\cdots,n_1,\ j = 1,2,\cdots,n_4, \tag{3}$$

where $\star$ represent the operation of circular convolution [33] of two vectors $\mathbf{a}, \mathbf{b} \in \mathbb{R}^{n_3}$ defined as $(\mathbf{a} \star \mathbf{b})_j = \sum_{k=1}^{n_3} \mathbf{a}_k \mathbf{b}_{1+(j-k)\mathrm{mod} n_3}$.

We also define the block diagonal matrix $\mathrm{bdiag}(\cdot)$ of any 3-way tensor $\underline{\mathrm{A}} \in \mathbb{R}^{n_1 \times n_2 \times n_3}$ and its inverse $\mathrm{bdfold}(\cdot)$ as follows

$$\mathrm{bdiag}(\underline{\mathrm{A}}) := \begin{bmatrix} \mathbf{A}^{(1)} & & \\ & \ddots & \\ & & \mathbf{A}^{(n_3)} \end{bmatrix} \in \mathbb{R}^{n_1 n_3 \times n_2 n_3}, \quad \mathrm{bdfold}(\mathrm{bdiag}(\underline{\mathrm{A}})) = \underline{\mathrm{A}}.$$

We also use $\overline{\mathbf{A}}$ (or $\overline{\underline{\mathrm{A}}}$) to denote the block diagonal matrix of tensor $\widetilde{\underline{\mathrm{A}}} = \mathrm{dft3}(\underline{\mathrm{A}})$ (i.e., the Fourier version of $\underline{\mathrm{A}}$) i.e.,

$$\overline{\mathbf{A}} = \mathrm{bdiag}(\widetilde{\underline{\mathrm{A}}}) := \begin{bmatrix} \widetilde{\mathbf{A}}^{(1)} & & \\ & \ddots & \\ & & \widetilde{\mathbf{A}}^{(n_3)} \end{bmatrix} \in \mathbb{C}^{n_1 n_3 \times n_2 n_3}.$$

Then the relationship between DFT and circular convolution further indicates that the conducting t-product in the original domain is equivalent to performing standard matrix product on the Fourier block diagonal matrices [33]. Since matrix product on the Fourier block diagonal matrices can be parallel written as matrix product of all the frontal slices in the Fourier domain, we have the following relationships:

$$\underline{\mathrm{C}} = \underline{\mathrm{A}} * \underline{\mathrm{B}} \Leftrightarrow \overline{\mathbf{C}} = \overline{\mathbf{A}\mathbf{B}} \Leftrightarrow \widetilde{\mathbf{C}}^{(k)} = \widetilde{\mathbf{A}}^{(k)} \widetilde{\mathbf{B}}^{(k)}, \quad k = 1,2,\cdots,n_3. \tag{4}$$

The relationship between the t-product and FFT also indicates that the inner product of two 3-way tensors $\underline{\mathrm{A}}, \underline{\mathrm{B}} \in \mathbb{R}^{n_1 \times n_2 \times n_3}$ and the inner product of their corresponding Fourier block diagonal matrices $\overline{\underline{\mathrm{A}}}, \overline{\underline{\mathrm{B}}} \in \mathbb{C}^{n_1 n_3 \times n_2 n_3}$ satisfy the following relationship:

$$\langle \underline{\mathrm{A}}, \underline{\mathrm{B}} \rangle = \frac{1}{n_3} \left\langle \widetilde{\underline{\mathrm{A}}}, \widetilde{\underline{\mathrm{B}}} \right\rangle = \frac{1}{n_3} \left\langle \overline{\mathbf{A}}, \overline{\mathbf{B}} \right\rangle.$$

When $\underline{\mathrm{A}} = \underline{\mathrm{B}} = \underline{\mathrm{X}}$, one has:

$$\|\underline{\mathrm{X}}\|_{\mathrm{F}} = \frac{1}{\sqrt{n_3}} \|\overline{\underline{\mathrm{X}}}\|_F.$$

We further define the concepts of tensor transpose, identity tensor, f-diagonal tensor and orthogonal tensor as follows.

Definition 2 (tensor transpose [33]). *Given a tensor $\underline{\mathrm{A}} \in \mathbb{R}^{n_1 \times n_2 \times n_3}$, then define its transpose tensor $\underline{\mathrm{A}}^\top$ of size $n_2 \times n_1 \times n_3$ which can be formed through first transposing all the frontal slices of $\underline{\mathrm{A}}$ and then exchanging each k-th transposed frontal slice with the $(n_3 + 2 - k)$-th transposed frontal slice for all $k = 2,3,\cdots,n_3$.*

For example, consider 3-way tensor $\underline{\mathbf{A}} = [\mathbf{A}^{(1)}|\mathbf{A}^{(2)}|\mathbf{A}^{(3)}|\mathbf{A}^{(4)}] \in \mathbb{R}^{n_1 \times n_2 \times 4}$ with 4 frontal slices, the tensor transpose $\underline{\mathbf{A}}^\top$ of $\underline{\mathbf{A}}$ is:

$$\underline{\mathbf{A}}^\top = [(\mathbf{A}^{(1)})^\top|(\mathbf{A}^{(4)})^\top|(\mathbf{A}^{(3)})^\top|(\mathbf{A}^{(2)})^\top] \in \mathbb{R}^{n_2 \times n_1 \times 4}.$$

Definition 3 (identity tensor [33]). *The identity tensor $\underline{\mathbf{I}} \in \mathbb{R}^{n \times n \times n_3}$ is a tensor whose first frontal slice is the n-by-n identity matrix with all other frontal slices are zero matrices.*

Definition 4 (f-diagonal tensor [33]). *We call a 3-way tensor f-diagonal if all the frontal slices of it are diagonal matrices.*

Definition 5 (orthogonal tensor [33]). *We call a tensor $\underline{\mathbf{Q}} \in \mathbb{R}^{n \times n \times n_3}$ an orthogonal tensor if the following equations hold:*

$$\underline{\mathbf{Q}}^\top * \underline{\mathbf{Q}} = \underline{\mathbf{Q}} * \underline{\mathbf{Q}}^\top = \underline{\mathbf{I}}.$$

Then, the tensor singular value decomposition (t-SVD) can be given as follows.

Definition 6 (Tensor singular value decomposition, and Tensor tubal rank [38]). *Given any 3-way tensor $\underline{\mathbf{X}} \in \mathbb{R}^{n_1 \times n_2 \times n_3}$, then it has the following factorization called tensor singular value decomposition (t-SVD):*

$$\underline{\mathbf{X}} = \underline{\mathbf{U}} * \underline{\Sigma} * \underline{\mathbf{V}}^\top, \tag{5}$$

where the left and right factor tensors $\underline{\mathbf{U}} \in \mathbb{R}^{n_1 \times n_1 \times n_3}$ and $\underline{\mathbf{V}} \in \mathbb{R}^{n_2 \times n_2 \times n_3}$ are orthogonal, and the middle tensor $\underline{\Sigma} \in \mathbb{R}^{n_1 \times n_2 \times n_3}$ is a rectangular f-diagonal tensor.

A visual illustration for the t-SVD is shown in Figure 1. It can be computed efficiently by FFT and IFFT in the Fourier domain according to Equation (4). For more details, see [2].

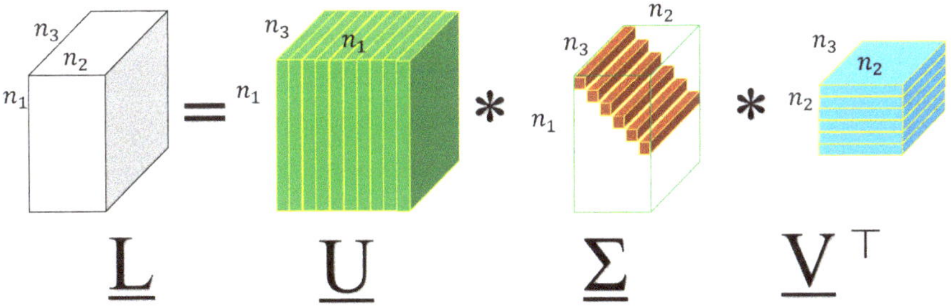

Figure 1. A visual illustration of t-SVD.

Definition 7 (Tensor tubal rank [38]). *The tensor tubal rank of any 3-way tensor $\underline{\mathbf{X}} \in \mathbb{R}^{n_1 \times n_2 \times n_3}$ is defined as the number of non-zero tubes of $\underline{\Sigma}$ in its t-SVD shown in Equation (5), i.e.,*

$$r_{tubal}(\underline{\mathbf{A}}) := \sum_i 1(\underline{\Sigma}(i,i,:) \neq \mathbf{0}). \tag{6}$$

Definition 8 (Tubal average rank [38]). *The tubal average rank $r_a(\underline{\mathbf{A}})$ of any 3-way tensor $\underline{\mathbf{A}} \in \mathbb{R}^{n_1 \times n_2 \times n_3}$ is defined as the averaged rank of all frontal slices of $\underline{\widetilde{\mathbf{A}}}$ as follows,*

$$r_a(\underline{\mathbf{A}}) := \frac{1}{n_3} \sum_{k=1}^{n_3} rank\left(\widetilde{\mathbf{A}}^{(k)}\right). \tag{7}$$

Definition 9 (Tensor operator norm [2,38]). *The tensor operator norm* $\|\underline{\mathrm{F}}\|_{\mathrm{op}}$ *of any 3-way tensor* $\underline{\mathrm{F}} \in \mathbb{R}^{n_1 \times n_2 \times n_3}$ *is defined as follows:*

$$\|\underline{\mathrm{F}}\|_{\mathrm{op}} := \sup_{\|\underline{\mathrm{A}}\|_{\mathrm{F}} \le 1} \|\underline{\mathrm{F}} * \underline{\mathrm{A}}\|_{\mathrm{F}}. \tag{8}$$

The relationship between t-product and FFT indicates that

$$\|\underline{\mathrm{F}}\|_{\mathrm{op}} := \sup_{\|\underline{\mathrm{A}}\|_{\mathrm{F}} \le 1} \|\underline{\mathrm{F}} * \underline{\mathrm{A}}\|_{\mathrm{F}} = \sup_{\|\overline{\mathbf{A}}\|_{\mathrm{F}} \le \sqrt{n_3}} \|\overline{\mathbf{F}} \cdot \overline{\underline{\mathrm{A}}}\|_{\mathrm{F}} = \|\overline{\mathbf{A}}\|_{\mathrm{sp}}. \tag{9}$$

Definition 10 (Tensor spectral norm [38]). *The tensor spectral norm* $\|\underline{\mathrm{A}}\|_{\mathrm{sp}}$ *of any 3-way tensor* $\underline{\mathrm{F}} \in \mathbb{R}^{n_1 \times n_2 \times n_3}$ *is defined as the matrix spectral norm of* $\overline{\mathbf{A}}$, *i.e.,*

$$\|\underline{\mathrm{A}}\|_{\mathrm{sp}} := \|\overline{\mathbf{A}}\|_{\mathrm{sp}}. \tag{10}$$

We further define the tubal nuclear norm.

Definition 11 (Tubal nuclear norm [2]). *For any tensor* $\underline{\mathrm{A}} \in \mathbb{R}^{n_1 \times n_2 \times n_3}$ *with t-SVD* $\underline{\mathrm{A}} = \underline{\mathrm{U}} * \underline{\Sigma} * \underline{\mathrm{V}}^{\top}$, *the tubal nuclear norm (TNN) of* $\underline{\mathrm{A}}$ *is defined as:*

$$\|\underline{\mathrm{A}}\|_{\mathrm{TNN}} := \langle \underline{\Sigma}, \underline{\mathrm{I}} \rangle = \sum_{i=1}^{r} \underline{\Sigma}(i,i,1), \tag{11}$$

where $r = r_{tubal}(\underline{\mathrm{A}})$.

To understand the tubal nuclear norm, first note that

$$r_{\mathrm{tubal}}(\underline{\mathrm{A}}) = \textstyle\sum_i 1(\underline{\Sigma}(i,i,:) \neq \mathbf{0}) \overset{(i)}{=} \sum_i 1(\widetilde{\underline{\Sigma}}(i,i,:) \neq \mathbf{0}) \overset{(ii)}{=} \sum_i 1(\|\widetilde{\underline{\Sigma}}(i,i,:)\|_1 \neq 0) \overset{(iii)}{=} \sum_i 1(\underline{\Sigma}(i,i,1) \neq 0), \tag{12}$$

where (i) holds because of the definition of DFT [2], (ii) holds by the property of l_1-norm, and (iii) is a result of DFT [2]. Thus, the tubal rank of $\underline{\mathrm{A}}$ is also the number of non-zero diagonal elements of $\underline{\Sigma}(i,i,1)$, i.e., the first frontal slice of tensor $\underline{\Sigma}$ in the t-SVD of $\underline{\mathrm{A}}$. Similar to the matrix singular values, the values $\underline{\Sigma}(i,i,1), i = 1,2,\cdots,n_3$ are also called the singular values of tensor $\underline{\mathrm{A}}$. As the matrix nuclear norm is the sum of matrix singular values, the tubal nuclear norm can be similarly understood as the sum of tensor singular values.

One can also verify by the property of DFT [2] that:

$$\|\underline{\mathrm{A}}\|_{\mathrm{TNN}} = \sum_{i=1}^{r} \underline{\Sigma}(i,i,1) = \sum_{k=1}^{n_3} \sum_{i=1}^{r} \widetilde{\underline{\Sigma}}(i,i,k) = \frac{1}{n_3} \sum_{k=1}^{n_3} \|\widetilde{\underline{\mathbf{A}}}^{(k)}\|_* = \frac{1}{n_3} \|\overline{\mathbf{A}}\|_*, \tag{13}$$

which indicates that the TNN of $\underline{\mathrm{A}} \in \mathbb{R}^{n_1 \times n_2 \times n_3}$ is also the averaged nuclear norm all frontal slices of $\widetilde{\underline{\mathrm{A}}}$. Thus, TNN indeed models the low-rankness of Fourier domain.

Now, we will show that the low-tubal-rank model is ideal to some real-world tensor data, such as color images and videos.

First, we consider a natural image of size $256 \times 256 \times 3$, shown in Figure 2a. In Figure 2b, we plot the distribution of its singular values, i.e., the values of $\underline{\Sigma}(i,i,1)$ along with the index i. As can be seen from Figure 2b, there are only a small number of singular values with large magnitude, and most of the singular values are close to 0. Then, we can say that some natural color images are approximately low tubal rank.

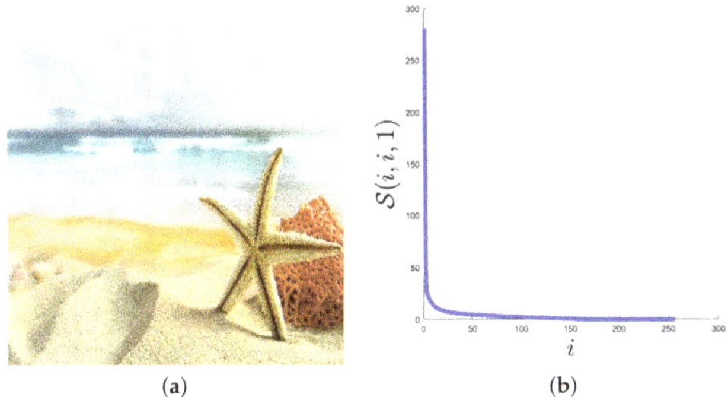

(a) (b)

Figure 2. The distribution of tensor singular values $\underline{\Sigma}(i,i,1)$ in a natural color image. (**a**) the sample image, (**b**) the distribution of $\underline{\Sigma}(i,i,1)$.

Then, consider a commonly used YUV sequence *Mother-daughter_qcif* (These data can be download from the following link https://sites.google.com/site/subudhibadri/fewhelpfuldownloads.) whose first frame is shown in Figure 3a. We use the Y components of the first 30 frames, and get a tensor of size $144 \times 176 \times 30$ and show the distribution of tensor singular values in Figure 3b. We can see from Figure 3b that similar to Figure 2b, there are only a small number of singular values with large magnitude, and most of the singular values are close to 0. Then, we can say that some videos can be well approximately low tubal rank.

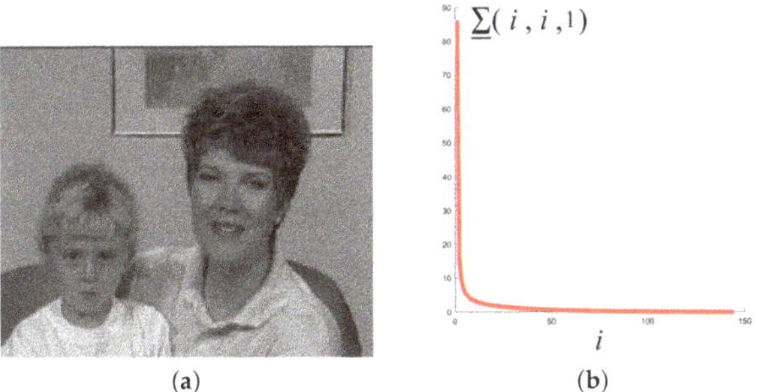

(a) (b)

Figure 3. The distribution of tensor singular values $\underline{\Sigma}(i,i,1)$ in a video sequence. (**a**) the first frame of the video, (**b**) the distribution of $\underline{\Sigma}(i,i,1)$.

For TNN and tensor spectral norm, we highlight the following two lemmas.

Lemma 1. *[2] TNN is the convex envelop of the tensor average rank in the unit ball of tensor spectral norm* $\{\underline{T} \in \mathbb{R}^{n_1 \times n_2 \times n_3} | \|\underline{T}\|_{sp} \leq 1\}$.

Lemma 2. *[2] The TNN and the tensor spectral norm are dual norms to each other.*

2.2. Related Works

In this subsection, we briefly introduce some related works. The proposed STPCP is tightly related to the Tensor Robust Principal Component Analysis (TRPCA) which aims to recover a low-rank tensor $\underline{L}_0$ and a sparse tensor $\underline{S}_0$ from their sum $\underline{M} = \underline{L}_0 + \underline{S}_0$. This is a special case of our measurement Model (1) where the noise tensor $\underline{E}_0$ is a zero tensor.

In [39], the SNN-based TRPCA model is proposed by modeling the underlying tensor as a low Tucker rank one

$$\min_{\underline{L},\underline{S}} \|\underline{L}\|_{\mathrm{SNN}} + \|\underline{S}\|_1 \quad \text{s.t.} \quad \underline{L} + \underline{S} = \underline{M}, \tag{14}$$

where SNN (Sum of Nuclear Norms) is defined as $\|\underline{L}\|_{\mathrm{SNN}} := \sum_{i=1}^{K} \alpha_k \|\mathbf{L}_{(k)}\|_*$, where $\alpha_k > 0$ and $\mathbf{L}_{(k)}$ is the mode-k matricization of $\underline{L}$ [40].

Model (14) indeed assumes the underlying tensor to be low Tucker rank, which can be too strong for some real tensor data. The TNN-based TRPCA model uses TNN to impose low-rankness in the final solution $\underline{L}$ as follows

$$\min_{\underline{L},\underline{S}} \|\underline{L}\|_{\mathrm{TNN}} + \lambda \|\underline{S}\|_1 \quad \text{s.t.} \quad \underline{L} + \underline{S} = \underline{M}. \tag{15}$$

As shown in [2], when the underlying tensor $\underline{L}_0$ satisfy the tensor incoherent conditions, by solving Problem (15), one can exactly recover the underlying tensor $\underline{L}_0$ and $\underline{S}_0$ with high probability with parameter $\lambda = 1/\sqrt{\max\{n_1, n_2\} n_3}$.

When the noise tensor $\underline{E}_0$ is not zero, the robust tensor decomposition based on SNN is proposed in [36] as follows:

$$\min_{\underline{L},\underline{S}} \frac{1}{2} \|\underline{M} - \underline{L} - \underline{S}\|_{\mathrm{F}} + \lambda_1 \|\underline{L}\|_{\mathrm{SNN}} + \lambda_2 \|\underline{S}\|_1, \tag{16}$$

where λ_1 and λ_2 are positive regularization parameters. The estimation error on $\underline{L}$ and $\underline{S}$ is analyzed with an upper bound in [36].

In [37], the TNN-based RTD model is proposed as follows:

$$\min_{\underline{L},\underline{S}} \frac{1}{2} \|\underline{M} - \underline{L} - \underline{S}\|_{\mathrm{F}} + \lambda_1 \|\underline{L}\|_{\mathrm{TNN}} + \lambda_2 \|\underline{S}\|_1, \quad \text{s.t.} \|\underline{L}\|_\infty \le \alpha, \tag{17}$$

where α is an upper estimate of l_∞-norm of the underlying tensor $\underline{L}_0$. An upper bound on the estimation error is also established. However, in the analysis of Model (17), the error does not vanish as the noise tensor $\underline{E}_0$ vanishes which means the analysis cannot guarantee exact recovery in the noiseless setting (which can be provided by the analysis of TNN-based TRPCA (15) by Lu et al. [2]).

The Bayesian approach is also used for robust tensor recovery. The CP decomposition under sparse corruption and small dense noise is considered [41], and tensor rank estimation is achieved using Bayesian approach. In [42], CP decomposition under missing value and small dense noise is considered with rank estimation similar to [41]. A sparse Bayesian CP model is proposed in [43] to recover a tensor with missing value, outliers and noises. In [44], a fully Bayesian treatment is proposed to recover a low-tubal-rank tensor corrupted by both noises and outliers.

3. Theoretical Guarantee for Stable Tensor Principal Component Pursuit

In this section, we formulate the proposed STPCP model and give the main theoretical result which upper bounds the estimation error and guarantees exact recovery in the noiseless setting.

3.1. The Proposed STPCP

As for the measurement Model (1), we further assume that the noise tensor $\underline{E}_0$ has bounded energy measured in F-norm, i.e., $\|\underline{E}_0\|_{\mathrm{F}} \le \delta$. Please note that the limited energy assumption is very mild, since most signals are of limited energy.

To recover the low-rank tensor $\underline{L}_0$ and the sparse tensor $\underline{S}_0$, we first produce the following optimization problem:

$$(\hat{\underline{L}}, \hat{\underline{S}}) = \operatorname*{argmin}_{\underline{L},\underline{S}} \|\underline{L}\|_{\mathrm{TNN}} + \lambda \|\underline{S}\|_1, \quad \text{s.t.} \ \|\underline{M} - \underline{L} - \underline{S}\|_{\mathrm{F}} \le \delta, \tag{18}$$

where λ is a positive parameter balancing the two regularizers. The motivation is to use TNN as a low-rank regularization term to exploit the low-dimensional structure in the signal tensor, whereas tensor l_1-norm is used to impose sparsity in the corruption tensor (since we assumes it to be sparse).

The relationship between Model (18) and existing models are discussed in Remark 1 and Remark 2.

Remark 1. *The following models can be seen as special cases as the proposed STPCP Model (18);*

(I). When $\delta = 0$, i.e., in the noiseless case, the proposed model degenerates to the TRPCA Model (15) [2].

(II). When $n_3 = 1$, then the stable tensor PCP Model (18) degenerates to the Stable Principal Component Pursuit (SPCP) [45] which aims to pursuit the principal components modeled by low-rank matrix $\underline{\mathbf{L}}_0$ from it observation $\mathbf{M}$ corrupted by both noises $\mathbf{E}_0$ and sparse corruptions $\mathbf{S}_0$. The SPCP is formulated as follows:

$$\min_{\mathbf{L},\mathbf{S}} \|\mathbf{L}\|_* + \lambda\|\mathbf{S}\|_1, \quad \text{s.t. } \|\mathbf{M}-\mathbf{L}-\mathbf{S}\|_{\mathrm{F}} \leq \delta. \tag{19}$$

(III). When $n_3 = 1$ and $\delta = 0$, the proposed STPCP further degenerates to Robust Principal Component Analysis (RPCA) [46] given as follows

$$\min_{\mathbf{L},\mathbf{S}} \|\mathbf{L}\|_* + \lambda\|\mathbf{S}\|_1, \quad \text{s.t. } \mathbf{L}+\mathbf{S} = \mathbf{M}. \tag{20}$$

Remark 2. *The differences from the proposed Model (18) and TNN-based RTD Model ((17) [37]) is as follows. First, our model does not need to upper estimate the l_∞-norm of the underlying tensor. Second, our model is a constrained optimization problem, whereas Model (17) is an unconstrained optimization problem.*

3.2. A Theorem for Stable Recovery

To analyze the statistical performance of Model (18), we should assume on the underlying low-rank tensor $\underline{\mathbf{L}}_0$ that it is not sparse. Only by this assumption, $\underline{\mathbf{L}}_0$ can be identified from its mixture with sparse $\underline{\mathbf{S}}_0$. Such an assumption can be described by the tensor incoherence condition [2,47], which is used to provide an identifiablility for low-rank $\underline{\mathbf{L}}_0$.

Definition 12 (Tensor incoherence condition [2,47]). *Given a 3-way tensor $\underline{\mathbf{T}} \in \mathbb{R}^{n_1\times n_2\times n_3}$ with tubal rank r, suppose it has the skinny t-SVD $\underline{\mathbf{T}} = \underline{\mathbf{U}} * \underline{\Lambda} * \underline{\mathbf{V}}^\top$, where $\underline{\mathbf{U}} \in \mathbb{R}^{n_1\times r\times n_3}, \underline{\mathbf{V}} \in \mathbb{R}^{r\times n_2\times n_3}$ are orthogonal tensors, and $\underline{\Lambda} \in \mathbb{R}^{r\times r\times n_3}$ is an f-diagonal tensor. Then, $\underline{\mathbf{T}}$ is said to satisfy the tensor incoherent condition (TIC) with parameter $\mu(\underline{\mathbf{T}})$ if the following inequalities hold:*

$$\max_{i\in[n_1]} \|\underline{\mathbf{U}}^\top * \mathring{\mathfrak{e}}_i\|_{\mathrm{F}} \leq \sqrt{\frac{r\mu(\underline{\mathbf{T}})}{n_1 n_3}}, \tag{21}$$

$$\max_{j\in[n_2]} \|\underline{\mathbf{V}}^\top * \mathring{\mathfrak{e}}_j\|_{\mathrm{F}} \leq \sqrt{\frac{r\mu(\underline{\mathbf{T}})}{n_2 n_3}}, \tag{22}$$

$$\|\underline{\mathbf{U}} * \underline{\mathbf{V}}^\top\|_\infty \leq \sqrt{\frac{r\mu(\underline{\mathbf{T}})}{n_1 n_2 n_3}}. \tag{23}$$

where $\mathring{\mathfrak{e}}_i \in \mathbb{R}^{n_1\times 1\times n_3}$ is a tensor column basis with only the $(i,1,1)$-th element being 1 and all the others being 0, and $\mathring{\mathfrak{e}}_j \in \mathbb{R}^{n_2\times 1\times n_3}$ is also a tensor column basis with only the $(j,1,1)$-th element being 1 and all the others being 0.

Assumption 1. *Suppose the true tensor $\underline{\mathbf{L}}_0$ in the measurement model (1) satisfies tensor incoherence condition with parameter μ.*

Assumption 1 intrinsically ensures that the row bases and column bases of $\underline{\mathbf{L}}_0$ do not align well with the canonical row and column bases. Thus, the low-rank $\underline{\mathbf{L}}_0$ is not sparse, which avoids the ambiguity that low-rank component can also be sparse in the measurement Model (1).

We should also force the sparse component in Model (1) is not low rank.

Assumption 2. *Assume the support Ω of $\underline{\mathbf{S}}_0$ is drawn uniformly at random.*

Now we can establish an upper bound on the estimation error of $\hat{\underline{\mathbf{L}}}$ and $\hat{\underline{\mathbf{S}}}$ in Problem (18).

Theorem 1 (An Upper Bound on the Estimation Error). *Suppose $\underline{\mathbf{L}}_0$ and $\underline{\mathbf{S}}_0$ satisfy Assumption 1 and Assumption 2, respectively. If the tubal rank r of $\underline{\mathbf{L}}_0$ and the sparsity (i.e., the l_0-norm) s of $\underline{\mathbf{S}}_0$ are respectively upper bounded as follows:*

$$r \leq \frac{c_r \min\{n_1, n_2\}}{\mu \log^2(n_3 \max\{n_1, n_2\})}, \quad \text{and} \quad s \leq c_s n_1 n_2 n_3 \tag{24}$$

where c_l and c_s are two sufficiently small numerical constants independent on the dimensions n_1, n_2 and n_3. Then the estimator defined in Model (18) satisfy the following inequalities:

$$\begin{aligned} \|\hat{\underline{\mathbf{L}}} - \underline{\mathbf{L}}_0\|_{\mathrm{F}} &\leq \left(\sqrt{1 + \frac{1}{\max\{n_1, n_2\}}} + 8(1 + 2\sqrt{2})\sqrt{\min\{n_1, n_2\} n_3}\right) \delta \\ \|\hat{\underline{\mathbf{S}}} - \underline{\mathbf{S}}_0\|_{\mathrm{F}} &\leq \left(\sqrt{1 + \max\{n_1, n_2\}} + 8(1 + 2\sqrt{2})\sqrt{n_1 n_2 n_3}\right) \delta, \end{aligned} \tag{25}$$

with probability at least $1 - c_1(n_3 \max\{n_1, n_2\})^{-c_2}$ (over the choice of support of $\underline{\mathbf{S}}_0$), where c_1 and c_2 are positive constants independent on the dimensions n_1, n_2 and n_3.

The proof of Theorem 1 are given in the appendix. In Theorem 1, estimation errors on $\underline{\mathbf{L}}_0$ and $\underline{\mathbf{S}}_0$ are separately established. It indicates that the estimation error scales linearly with the noise level δ, which is in consistence with the result in [37].

Remark 3. *A significant progress over [37] is that in the noiseless setting where $\underline{\mathbf{E}}_0$ vanishes, our analysis can provide exact recovery guarantee of $\underline{\mathbf{L}}_0$ and $\underline{\mathbf{S}}_0$. This is because the tensor incoherence condition adopted in our analysis intrinsically ensures that the low-rank tensor $\underline{\mathbf{L}}_0$ is not sparse and thus can be separated from the sparse corruption tensor, whereas the non-spiky condition adopted in [37] fails to provide identifiability in the measurement Model (1).*

For Theorem 1, we also give the following remark.

Remark 4. *The error bounds established in Theorem 1 are consistent with the theoretical analysis for the special cases shown in Remark 1.*

(I). When $\delta = 0$, i.e., in the noiseless case, the error bounds in Theorem 1 will vanish, which means exact recovery of $\underline{\mathbf{L}}_0$ and $\underline{\mathbf{S}}_0$ can be guaranteed. This result is consistent with the analysis in [2] for TNN-based TRPCA Model (15).

(II). When $n_3 = 1$, the error bound on the sparse component in Theorem 1 is consistent with the error bound shown in Equation (8) of [45]. The upper bound on error of the low-rank component in Theorem 1 is sharper than that given in Equation (8) of [45].

(III). When $n_3 = 1$ and $\delta = 0$, the proposed STPCP has consistent theoretical guarantee with the analysis of RPCA [46].

4. Algorithms

In this section, we design two algorithms. The first algorithm is based on the framework of ADMM [48] which has been extensively used in convex optimization with good convergence behavior. However, ADMM requires full SVDs on large matrices in each iteration which is high computational burden in high-dimensional settings. Thus, the second algorithm is proposed to solve this issue by using a factorization trick which can instead conducting SVDs on much smaller matrices.

4.1. An ADMM Algorithm

The proposed estimator (18) is equivalent to the following unconstrained problem:

$$\min_{\underline{L},\underline{S}} \frac{1}{2}\|\underline{L}+\underline{S}-\underline{M}\|_F^2 + \gamma(\|\underline{L}\|_{\mathrm{TNN}} + \lambda\|\underline{S}\|_1), \tag{26}$$

where γ is a positive parameter balancing the data fidelity term and the regularization term.

Besides being corrupted by noises and outliers, the observed tensor $\underline{M}$ may also suffer from missing entries which can be taken as outliers with known positions in many applications. Thus, it is more practical to consider the recovery of $\underline{L}_0$ against outliers $\underline{S}_0$, noises $\underline{E}_0$ and missing entries shown in the following measurement model:

$$\underline{M} = \underline{B} \odot (\underline{L}_0 + \underline{S}_0 + \underline{E}_0), \tag{27}$$

where tensor $\underline{B} \in \mathbb{R}^{n_1 \times n_2 \times n_3}$ denote the missing mask where $\underline{B}_{ijk} = 1$, if the (i,j,k)-th entry of $\underline{L}$ is observed and $\underline{B}_{ijk} = 0$ otherwise, and $\odot$ denotes element-wise multiplication. Taking into consideration of missing entries, Model (26) can be further modified as:

$$\min_{\underline{L},\underline{S}} \frac{1}{2}\|\underline{B}\odot(\underline{L}+\underline{S}-\underline{M})\|_F^2 + \gamma(\|\underline{L}\|_{\mathrm{TNN}} + \lambda\|\underline{S}\|_1). \tag{28}$$

By adding auxiliary variables to Problem (28), we obtain:

$$\begin{aligned} &\min_{\underline{K},\underline{L},\underline{R},\underline{S}} \frac{1}{2}\|\underline{B}\odot(\underline{L}+\underline{S}-\underline{M})\|_F^2 + \gamma\|\underline{K}\|_{\mathrm{TNN}} + \gamma\lambda\|\underline{R}\|_1 \\ &\text{s.t.} \ \ \underline{K}=\underline{L},\ \underline{R}=\underline{S}. \end{aligned} \tag{29}$$

The Augmented Lagrangian (AL) of Problem (29) is given as follows:

$$\begin{aligned} L_\rho(\underline{L},\underline{S},\underline{K},\underline{R},\underline{Y}_1,\underline{Y}_2) = &\frac{1}{2}\|\underline{B}\odot(\underline{L}+\underline{S}-\underline{M})\|_F^2 + \gamma\|\underline{K}\|_{\mathrm{TNN}} + \gamma\lambda\|\underline{R}\|_1 \\ &+ \langle \underline{Y}_1, \underline{K}-\underline{L}\rangle + \frac{\rho}{2}\|\underline{K}-\underline{L}\|_F^2 + \langle \underline{Y}_2, \underline{R}-\underline{S}\rangle + \frac{\rho}{2}\|\underline{R}-\underline{S}\|_F^2, \end{aligned} \tag{30}$$

where $\underline{Y}_1, \underline{Y}_2 \in \mathbb{R}^{n_1 \times n_2 \times n_3}$ are Lagrangian multipliers and ρ is a penalty parameter.

According the strategy of ADMM, we update prime variables $(\underline{L},\underline{S})$ and $(\underline{K},\underline{R})$ by alternative minimization of AL in Problem (29) as follows:

- Update $(\underline{L},\underline{S})$. We update $(\underline{L},\underline{S})$ by minimizing L_ρ with other variables fixed as follows:

$$\begin{aligned} &(\underline{L}^{t+1},\underline{S}^{t+1}) \\ &= \mathrm{argmin}_{(\underline{L},\underline{S})}\ L_\rho(\underline{L},\underline{S},\underline{K}^t,\underline{R}^t,\underline{Y}_1^t,\underline{Y}_2^t) \\ &= \mathrm{argmin}_{(\underline{L},\underline{S})}\ \tfrac{1}{2}\|\underline{B}\odot(\underline{L}+\underline{S}-\underline{M})\|_F^2 + \langle \underline{Y}_1^t, \underline{K}^t-\underline{L}\rangle + \tfrac{\rho}{2}\|\underline{K}^t-\underline{L}\|_F^2 + \langle \underline{Y}_2^t, \underline{R}^t-\underline{S}\rangle + \tfrac{\rho}{2}\|\underline{R}^t-\underline{S}\|_F^2. \end{aligned} \tag{31}$$

Taking derivatives of the right-hand side of Equation (31) with respect to $\underline{L}$ and $\underline{S}$ respectively, and setting the results zero, we obtain:

$$\begin{aligned} \underline{B}\odot(\underline{L}^{t+1}+\underline{S}^{t+1}) - \underline{B}\odot\underline{M} - \underline{Y}_1^t + \rho(\underline{L}^{t+1}-\underline{K}^t) = \underline{0} \\ \underline{B}\odot(\underline{L}^{t+1}+\underline{S}^{t+1}) - \underline{B}\odot\underline{M} - \underline{Y}_2^t + \rho(\underline{S}^{t+1}-\underline{R}^t) = \underline{0}. \end{aligned} \tag{32}$$

Resolving the above equation group yields:

$$\begin{aligned}\underline{\mathbf{L}}^{t+1} &= \left(\rho(\underline{\mathbf{B}}+\rho\underline{1})\odot\underline{\mathbf{K}}^{t}+\rho\underline{\mathbf{B}}\odot\underline{\mathbf{M}}+(\underline{\mathbf{B}}+\rho\underline{1})\odot\underline{\mathbf{Y}}_1^{t}-\underline{\mathbf{B}}\odot\underline{\mathbf{Y}}_2^{t}-\rho\underline{\mathbf{B}}\odot\underline{\mathbf{R}}^{t}\right)\oslash\left(\rho(2\underline{\mathbf{B}}+\rho\underline{1})\right),\\ \underline{\mathbf{S}}^{t+1} &= \left(\rho(\underline{\mathbf{B}}+\rho\underline{1})\odot\underline{\mathbf{R}}^{t}+\rho\underline{\mathbf{B}}\odot\underline{\mathbf{M}}+(\underline{\mathbf{B}}+\rho\underline{1})\odot\underline{\mathbf{Y}}_2^{t}-\underline{\mathbf{B}}\odot\underline{\mathbf{Y}}_1^{t}-\rho\underline{\mathbf{B}}\odot\underline{\mathbf{K}}^{t}\right)\oslash\left(\rho(2\underline{\mathbf{B}}+\rho\underline{1})\right),\end{aligned} \tag{33}$$

where $\oslash$ denotes entry-wise division and $\underline{1}$ denotes the tensor all whose entries are 1.

- Update $(\underline{\mathbf{K}}, \underline{\mathbf{R}})$. We update $(\underline{\mathbf{K}}, \underline{\mathbf{R}})$ by minimizing L_ρ with other variables fixed as follows:

$$\begin{aligned}&(\underline{\mathbf{K}}^{t+1}, \underline{\mathbf{R}}^{t+1})\\ &= \operatorname{argmin}_{(\underline{\mathbf{K}},\underline{\mathbf{R}})} L_\rho(\underline{\mathbf{L}}^{t+1}, \underline{\mathbf{S}}^{t+1}, \underline{\mathbf{K}}, \underline{\mathbf{R}}, \underline{\mathbf{Y}}_1^t, \underline{\mathbf{Y}}_2^t)\\ &= \operatorname{argmin}_{(\underline{\mathbf{K}},\underline{\mathbf{R}})} \gamma\|\underline{\mathbf{K}}\|_{\mathrm{TNN}} + \gamma\lambda\|\underline{\mathbf{R}}\|_1 + \left\langle \underline{\mathbf{Y}}_1^t, \underline{\mathbf{K}} - \underline{\mathbf{L}}^{t+1}\right\rangle + \tfrac{\rho}{2}\|\underline{\mathbf{K}} - \underline{\mathbf{L}}^{t+1}\|_{\mathrm{F}}^2 + \left\langle \underline{\mathbf{Y}}_2^t, \underline{\mathbf{R}} - \underline{\mathbf{S}}^{t+1}\right\rangle + \tfrac{\rho}{2}\|\underline{\mathbf{R}} - \underline{\mathbf{S}}^{t+1}\|_{\mathrm{F}}^2.\end{aligned} \tag{34}$$

Please note that Problem (34) can further be solved separately as follows:

$$\begin{aligned}\underline{\mathbf{K}}^{t+1} &= \operatorname*{argmin}_{\underline{\mathbf{K}}} \gamma\|\underline{\mathbf{K}}\|_{\mathrm{TNN}} + \left\langle \underline{\mathbf{Y}}_1^t, \underline{\mathbf{K}} - \underline{\mathbf{L}}^{t+1}\right\rangle + \frac{\rho}{2}\|\underline{\mathbf{K}} - \underline{\mathbf{L}}^{t+1}\|_{\mathrm{F}}^2\\ &= \mathfrak{S}_{\gamma\rho^{-1}}^{\|\cdot\|_{\mathrm{TNN}}}\left(\underline{\mathbf{L}}^{t+1} - \rho^{-1}\underline{\mathbf{Y}}_1^t\right).\end{aligned} \tag{35}$$

and

$$\begin{aligned}\underline{\mathbf{R}}^{t+1} &= \operatorname*{argmin}_{\underline{\mathbf{R}}} \gamma\lambda\|\underline{\mathbf{R}}\|_1 + \left\langle \underline{\mathbf{Y}}_1^t, \underline{\mathbf{R}} - \underline{\mathbf{S}}^{t+1}\right\rangle + \frac{\rho}{2}\|\underline{\mathbf{R}} - \underline{\mathbf{S}}^{t+1}\|_{\mathrm{F}}^2\\ &= \mathfrak{S}_{\gamma\lambda\rho^{-1}}^{\|\cdot\|_1}\left(\underline{\mathbf{S}}^{t+1} - \rho^{-1}\underline{\mathbf{Y}}_2^t\right),\end{aligned} \tag{36}$$

where $\mathfrak{S}_\tau^{\|\cdot\|_{\mathrm{TNN}}}(\cdot)$ is the proximity operator of TNN [5]. and $\mathfrak{S}_\tau^{\|\cdot\|_1}(\cdot)$ is the proximity operator of tensor l_1-norm given as follows [49]:

$$\mathfrak{S}_\tau^{\|\cdot\|_1}(\underline{\mathbf{A}}) := \operatorname*{argmin}_{\underline{\mathbf{X}}} \tau\|\underline{\mathbf{X}}\|_1 + \frac{1}{2}\|\underline{\mathbf{X}} - \underline{\mathbf{A}}\|_{\mathrm{F}}^2 = \operatorname{sign}(\underline{\mathbf{A}}) \odot \max\{(|\underline{\mathbf{A}}| - \tau, 0\},$$

In [5], a closed-form expression of $\mathfrak{S}_\tau(\cdot)$ is given as follows:

Lemma 3 (Proximity operator of TNN [5])**.** *For any 3D tensor $\underline{\mathbf{A}} \in \mathbb{R}^{n_1 \times n_2 \times n_3}$ with reduced t-SVD $\underline{\mathbf{A}} = \underline{\mathbf{U}} * \underline{\Lambda} * \underline{\mathbf{V}}^\top$, where $\underline{\mathbf{U}} \in \mathbb{R}^{n_1 \times r \times n_3}$ and $\underline{\mathbf{V}} \in \mathbb{R}^{n_2 \times r \times n_3}$ are orthogonal tensors and $\underline{\Lambda} \in \mathbb{R}^{r \times r \times n_3}$ is the f-diagonal tensor of singular tubes, the proximity operator $\mathfrak{S}_\tau^{\|\cdot\|_{\mathrm{TNN}}}(\underline{\mathbf{A}})$ at $\underline{\mathbf{A}}$ can be computed by:*

$$\mathfrak{S}_\tau^{\|\cdot\|_{\mathrm{TNN}}}(\underline{\mathbf{A}}) := \operatorname*{argmin}_{\underline{\mathbf{X}}} \tau\|\underline{\mathbf{X}}\|_{\mathrm{TNN}} + \frac{1}{2}\|\underline{\mathbf{X}} - \underline{\mathbf{A}}\|_{\mathrm{F}}^2 = \underline{\mathbf{U}} * \mathit{ifft3}(\max(\mathit{fft3}(\underline{\Lambda}) - \tau, 0)) * \underline{\mathbf{V}}^\top,$$

- Update $(\underline{\mathbf{Y}}_1, \underline{\mathbf{Y}}_2)$. The Lagrangian multipliers are updated by gradient ascent as follows:

$$\begin{aligned}\underline{\mathbf{Y}}_1^{t+1} &= \underline{\mathbf{Y}}_1^t + \rho(\underline{\mathbf{K}}^{t+1} - \underline{\mathbf{L}}^{t+1}),\\ \underline{\mathbf{Y}}_2^{t+1} &= \underline{\mathbf{Y}}_2^t + \rho(\underline{\mathbf{R}}^{t+1} - \underline{\mathbf{S}}^{t+1}).\end{aligned} \tag{37}$$

The algorithm is summarized in Algorithm 1. The convergence analysis of Algorithm 1 is established in Theorem 2.

Algorithm 1 Solving Problem (29) using ADMM.

Input: The observed tensor $\underline{\mathrm{M}}$, the parameters $\gamma, \lambda, \rho, \delta$.
1: Initialize $t = 0$, $\underline{\mathrm{L}}^0 = \underline{\mathrm{S}}^0 = \underline{\mathrm{K}}^0 = \underline{\mathrm{R}}^0 = \underline{\mathrm{Y}}_1^0 = \underline{\mathrm{Y}}_2^0 = \underline{0} \in \mathbb{R}^{n_1 \times n_2 \times n_3}$
2: **for** $t = 0, \cdots, T_{\max}$ **do**
3: Update $(\underline{\mathrm{L}}^{t+1}, \underline{\mathrm{S}}^{t+1})$ by Equation (33);
4: Update $(\underline{\mathrm{K}}^{t+1}, \underline{\mathrm{R}}^{t+1})$ by Equations (35)–(36);
5: Update $(\underline{\mathrm{Y}}_1^{t+1}, \underline{\mathrm{Y}}_2^{t+1})$ by Equation (37);
6: Check the convergence criteria:
(i) convergence of variables: $\|\underline{\mathrm{A}}^{t+1} - \underline{\mathrm{A}}^t\|_\infty \leq \delta, \ \forall \underline{\mathrm{A}} \in \{\underline{\mathrm{L}}, \underline{\mathrm{S}}, \underline{\mathrm{K}}, \underline{\mathrm{R}}\}$,
(ii) convergence of constraints: $\max\{\|\underline{\mathrm{K}}^{t+1} - \underline{\mathrm{L}}^t\|_\infty, \|\underline{\mathrm{R}}^{t+1} - \underline{\mathrm{S}}^{t+1}\|_\infty\} \leq \delta$.
7: **end for**
Output: $(\hat{\underline{\mathrm{L}}}, \hat{\underline{\mathrm{S}}}) = (\underline{\mathrm{L}}^{t+1}, \underline{\mathrm{S}}^{t+1})$.

Theorem 2 (Convergence of Algorithm 1)**.** *For any $\rho > 0$, if the unaugmented Lagrangian $L(\underline{\mathrm{L}}, \underline{\mathrm{S}}, \underline{\mathrm{K}}, \underline{\mathrm{R}}, \underline{\mathrm{Y}}_1, \underline{\mathrm{Y}}_2)$ has a saddle point, then the iterations $L(\underline{\mathrm{L}}^t, \underline{\mathrm{S}}^t, \underline{\mathrm{K}}^t, \underline{\mathrm{R}}^t, \underline{\mathrm{Y}}_1^t, \underline{\mathrm{Y}}_2^t)$ in Algorithm 1 satisfy the residual convergence, objective convergence and dual variable convergence of Problem (29) as $t \to \infty$.*

The proof of Theorem 2 is given in the Appendix A.

In a single iteration of Algorithm 1, the main cost comes from updating $\underline{\mathrm{L}}^t$ which involves computing FFT, IFFT and n_3 SVDs of $n_1 \times n_2$ matrices [47]. Hence Algorithm 1 has per-iteration complexity of order $O\big(n_1 n_2 n_3 (n_1 \wedge n_2 + \log n_3)\big)$. Thus, if the total iteration number is T, then the total computational complexity is:

$$O\Big(T n_1 n_2 n_3 (\min\{n_1, n_2\} + \log n_3)\Big). \tag{38}$$

4.2. A Faster Algorithm

To reduce the cost of computing TNN which is a main cost of Algorithm 1, we propose the following lemma which indicates that TNN is orthogonal invariant.

Lemma 4. *Given a tensor $\underline{\mathrm{X}} \in \mathbb{R}^{r \times r \times n_3}$, let $\underline{\mathrm{Q}} \in \mathbb{R}^{n_1 \times r \times n_3}$ a two semi-orthogonal tensors, i.e., $\underline{\mathrm{Q}}^\top * \underline{\mathrm{Q}} = \underline{\mathrm{I}} \in \mathbb{R}^{r \times r \times n_3}$ and $r \leq \min\{n_1, n_2\}$. Then, we have the following relationship:*

$$\|\underline{\mathrm{Q}} * \underline{\mathrm{X}}\|_{\mathrm{TNN}} = \|\underline{\mathrm{X}}\|_{\mathrm{TNN}}.$$

The proof of Lemma 4 can be found in the appendix. Equipped with Lemma 4, we decompose the low-rank component in Problem (28) as follows:

$$\underline{\mathrm{L}} = \underline{\mathrm{Q}} * \underline{\mathrm{X}}, \quad \text{s.t. } \underline{\mathrm{Q}}^\top * \underline{\mathrm{Q}} = \underline{\mathrm{I}}_r,$$

where $\underline{\mathrm{I}}_r \in \mathbb{R}^{r \times r \times n_3}$ is an identity tensor. The similar strategy has been used in low-rank matrix recovery from gross corruptions by [50]. Furthermore, we propose the following model for Problem (28):

$$\begin{aligned} &\min_{\underline{\mathrm{Q}}, \underline{\mathrm{X}}, \underline{\mathrm{S}}} \frac{1}{2}\|\underline{\mathrm{B}} \odot (\underline{\mathrm{Q}} * \underline{\mathrm{X}} + \underline{\mathrm{S}} - \underline{\mathrm{M}})\|_{\mathrm{F}}^2 + \gamma(\|\underline{\mathrm{X}}\|_{\mathrm{TNN}} + \lambda\|\underline{\mathrm{S}}\|_1) \\ &\text{s.t. } \underline{\mathrm{Q}}^\top * \underline{\mathrm{Q}} = \underline{\mathrm{I}}_r, \end{aligned} \tag{39}$$

where r is an upper estimation of tubal rank of the underlying tensor $r^* = r_{\mathrm{tubal}}(\underline{\mathrm{L}}_0)$.

In contrast to Model (28), the proposed Model (39) is a non-convex optimization problem. That means Model (39) may have many local minima. We establish a connection between the proposed Model (39) with Model (28) in the following theorem.

Theorem 3 (Connection between Model (39) and Model (28)). *Let $(\underline{Q}_*, \underline{X}_*, \underline{S}_*)$ be a global optimal solution to Problem (39). Furthermore, let $(\underline{L}^\star, \underline{S}^\star)$ be the solution to Problem (28), and $r_{tubal}(\underline{L}^\star) \leq r$, where r is the initialized tubal rank. Then $(\underline{Q}_* * \underline{X}_*, \underline{S}_*)$ is also the optimal solution to Problem (28).*

The proof of Theorem 3 can be found in the appendix. Theorem 3 states that the global optimal point of the (non-convex) Model (39) coincides with solution of the (convex) Model (28). It further indicates that the accuracy of Model (39) cannot exceed Model (28), which can be validated numerically in the experiment section.

To solve Model (39), we also use the ADMM framework.

First, by adding auxiliary variables, we have the following problem:

$$\begin{aligned} \min_{\underline{L},\underline{S},\underline{R},\underline{Q},\underline{X}} \ & \frac{1}{2}\|\underline{B} \odot (\underline{L} + \underline{S} - \underline{M})\|_F^2 + \gamma(\|\underline{X}\|_{\text{TNN}} + \lambda\|\underline{R}\|_1) \\ \text{s.t.} \ & \underline{Q} * \underline{X} = \underline{L}; \quad \underline{R} = \underline{S}; \quad \underline{Q}^\top * \underline{Q} = \underline{I}_r. \end{aligned} \tag{40}$$

The augmented Lagrangian of Problem (40) is:

$$\begin{aligned} L_2'(\underline{L}, \underline{S}, \underline{R}, \underline{Q}, \underline{X}) = & \frac{1}{2}\|\underline{B} \odot (\underline{L} + \underline{S} - \underline{M})\|_F^2 + \gamma(\|\underline{X}\|_{\text{TNN}} + \lambda\|\underline{R}\|_1) \\ & + \langle \underline{Y}_1, \underline{Q} * \underline{X} - \underline{L}\rangle + \frac{\rho}{2}\|\underline{Q} * \underline{X} - \underline{L}\|_F^2 + \langle \underline{Y}_2, \underline{R} - \underline{S}\rangle + \frac{\rho}{2}\|\underline{R} - \underline{S}\|_F^2 \\ \text{s.t.} \ \underline{Q}^\top * \underline{Q} = \underline{I}_r. & \end{aligned} \tag{41}$$

According the strategy of ADMM, we update prime variables $(\underline{L}, \underline{S})$ and $(\underline{Q}, \underline{X}, \underline{R})$ by alternative minimization of AL in Problem (41) as follows

- Update $(\underline{L}, \underline{S})$: We update $(\underline{L}, \underline{S})$ by minimizing L_ρ' with other variables fixed as follows

$$\begin{aligned} & (\underline{L}^{t+1}, \underline{S}^{t+1}) \\ & = \operatorname{argmin}_{(\underline{L},\underline{S})} L_\rho'(\underline{L}, \underline{S}, \underline{Q}^t, \underline{X}^t, \underline{R}^t, \underline{Y}_1^t, \underline{Y}_2^t) \\ & = \operatorname{argmin}_{(\underline{L},\underline{S})} \tfrac{1}{2}\|\underline{B} \odot (\underline{L} + \underline{S} - \underline{M})\|_F^2 + \langle \underline{Y}_1^t, \underline{Q}^t * \underline{X}^t - \underline{L}\rangle + \tfrac{\rho}{2}\|\underline{Q}^t * \underline{X}^t - \underline{L}\|_F^2 + \langle \underline{Y}_2^t, \underline{R}^t - \underline{S}\rangle + \tfrac{\rho}{2}\|\underline{R}^t - \underline{S}\|_F^2. \end{aligned} \tag{42}$$

Taking derivatives of the right-hand side with respect to $\underline{L}$ and $\underline{S}$ respectively, and setting the results zero, we obtain:

$$\begin{aligned} \underline{B} \odot (\underline{L}^{t+1} + \underline{S}^{t+1}) - \underline{B} \odot \underline{M} - \underline{Y}_1^t + \rho(\underline{L}^{t+1} - \underline{Q}^t * \underline{X}^t) = \underline{0} \\ \underline{B} \odot (\underline{L}^{t+1} + \underline{S}^{t+1}) - \underline{B} \odot \underline{M} - \underline{Y}_2^t + \rho(\underline{S}^{t+1} - \underline{R}^t) = \underline{0}, \end{aligned} \tag{43}$$

Resolving the above equation group yields:

$$\begin{aligned} \underline{L}^{t+1} &= \left((1+\rho)\underline{Q}^t * \underline{X}^t + \underline{B} \odot \underline{M} + \underline{Y}_1^t - \underline{R}^t\right) \oslash (2\underline{B} + \rho\underline{1}), \\ \underline{S}^{t+1} &= \left((1+\rho)\underline{R}^t + \underline{B} \odot \underline{M} + \underline{Y}_2^t - \underline{Q}^t * \underline{X}^t\right) \oslash (2\underline{B} + \rho\underline{1}). \end{aligned} \tag{44}$$

- Update $\underline{Q}$. We update $\underline{Q}$ by minimizing L'_ρ with other variables fixed as follows:

$$
\begin{aligned}
&\min_{\underline{Q}^\top * \underline{Q} = \underline{I}_r} L_\rho(\underline{L}^{t+1}, \underline{S}^{t+1}, \underline{Q}, \underline{X}^t, \underline{R}^t, \underline{Y}_1^t, \underline{Y}_2^t) \\
&= \min_{\underline{Q}^\top * \underline{Q} = \underline{I}_r} \left\langle \underline{Y}_1^t, \underline{Q} * \underline{X}^t - \underline{L}^{t+1} \right\rangle + \frac{\rho}{2} \|\underline{Q} * \underline{X}^t - \underline{L}^{t+1}\|_F^2. \\
&= \min_{\underline{Q}^\top * \underline{Q} = \underline{I}_r} \frac{\rho}{2} \|\underline{Q} * \underline{X}^t - (\underline{L}^{t+1} - \rho^{-1} \underline{Y}_1^t)\|_F^2 \\
&= \mathfrak{P}\left((\underline{L}^{t+1} - \rho^{-1} \underline{Y}_1^t) * (\underline{X}^t)^\top \right),
\end{aligned}
\tag{45}
$$

where operator $\mathfrak{P}(\cdot)$ is defined in Lemma 5 as follows.

Lemma 5 ([51])**.** *Given any tensors* $\underline{A} \in \mathbb{R}^{r \times n_2 \times n_3}, \underline{B} \in \mathbb{R}^{n_1 \times n_2 \times n_3}$, *suppose tensor* $\underline{B} * \underline{A}^\top$ *has t-SVD* $\underline{B} * \underline{A}^\top = \underline{U} * \underline{\Lambda} * \underline{V}^\top$, *where* $\underline{U} \in \mathbb{R}^{n_1 \times r \times n_3}$ *and* $\underline{V} \in \mathbb{R}^{r \times r \times n_3}$. *Then, the problem:*

$$
\min_{\underline{Q}^\top * \underline{Q} = \underline{I}_r} \|\underline{P} * \underline{A} - \underline{B}\|_F^2
\tag{46}
$$

has a closed-form solution as:

$$
\underline{Q} = \mathfrak{P}(\underline{B} * \underline{A}^\top) := \underline{U} * \underline{V}^\top.
\tag{47}
$$

- Update $(\underline{X}, \underline{R})$:We update $(\underline{X}, \underline{S})$ by minimizing L'_ρ with other variables fixed as follows:

$$
\begin{aligned}
&\min_{(\underline{X}, \underline{R})} L_\rho(\underline{L}^{t+1}, \underline{S}^{t+1}, \underline{Q}^{t+1}, \underline{X}, \underline{R}, \underline{Y}_1^t, \underline{Y}_2^t) \\
&= \min_{(\underline{X}, \underline{R})} \gamma \|\underline{X}\|_{\mathrm{TNN}} + \gamma\lambda \|\underline{R}\|_1 + \left\langle \underline{Y}_1^t, \underline{Q}^{t+1} * \underline{X} - \underline{L}^{t+1} \right\rangle + \frac{\rho}{2} \|\underline{Q}^{t+1} * \underline{X} - \underline{L}^{t+1}\|_F^2 \\
&\qquad + \left\langle \underline{Y}_2^t, \underline{R} - \underline{S}^{t+1} \right\rangle + \frac{\rho}{2} \|\underline{R} - \underline{S}^{t+1}\|_F^2.
\end{aligned}
\tag{48}
$$

Please note that Problem (48) can further be solved separately as follows:

$$
\begin{aligned}
\underline{K}^{t+1} &= \operatorname*{argmin}_{\underline{X}} \gamma \|\underline{X}\|_{\mathrm{TNN}} + \left\langle \underline{Y}_1^t, \underline{Q}^{t+1} * \underline{X} - \underline{L}^{t+1} \right\rangle + \frac{\rho}{2} \|\underline{Q}^\top * \underline{X} - \underline{L}^{t+1}\|_F^2 \\
&= \operatorname*{argmin}_{\underline{X}} \gamma \|\underline{X}\|_{\mathrm{TNN}} + \frac{\rho}{2} \|\underline{Q}^{t+1} * \underline{X} - (\underline{L}^{t+1} - \rho^{-1} \underline{Y}_1^t)\|_F^2 \\
&\overset{(i)}{=} \operatorname*{argmin}_{\underline{X}} \gamma \|\underline{X}\|_{\mathrm{TNN}} + \frac{\rho}{2} \|\underline{X} - (\underline{Q}^{t+1})^\top * (\underline{L}^{t+1} - \rho^{-1} \underline{Y}_1^t)\|_F^2 \\
&= \mathfrak{S}_{\gamma\rho^{-1}}^{\|\cdot\|_{\mathrm{TNN}}} \left((\underline{Q}^{t+1})^\top * (\underline{L}^{t+1} - \rho^{-1} \underline{Y}_1^t). \right)
\end{aligned}
\tag{49}
$$

and

$$
\begin{aligned}
\underline{R}^{t+1} &= \operatorname*{argmin}_{\underline{R}} \gamma\lambda \|\underline{R}\|_1 + \left\langle \underline{Y}_1^t, \underline{R} - \underline{S}^{t+1} \right\rangle + \frac{\rho}{2} \|\underline{R} - \underline{S}^{t+1}\|_F^2 \\
&= \mathfrak{S}_{\gamma\lambda\rho^{-1}}^{\|\cdot\|_1} \left(\underline{K}^{t+1} - \rho^{-1} \underline{Y}_2^t \right).
\end{aligned}
\tag{50}
$$

The equality (i) in Equation (49) holds because according to $\underline{Q}^\top * \underline{Q} = \underline{I}$, we have:

$$\begin{aligned}\min_{\underline{X}} \|\underline{Q} * \underline{X} - \underline{Y}\|_F^2 &= \min_{\overline{\mathbf{X}}} \frac{1}{n_3} \|\overline{\mathbf{Q}} \cdot \overline{\mathbf{X}} - \overline{\mathbf{Y}}\|_F^2 \\ &= \min_{\overline{\mathbf{X}}} \frac{1}{n_3} \|\overline{\mathbf{Y}}\|_F^2 - \frac{2}{n_3} \langle \overline{\mathbf{Q}} \cdot \overline{\mathbf{X}}, \overline{\mathbf{Y}} \rangle + \frac{1}{n_3} \|\overline{\mathbf{Q}} \cdot \overline{\mathbf{X}}\|_F^2 \\ &= \min_{\overline{\mathbf{X}}} \frac{1}{n_3} \|\overline{\mathbf{Y}}\|_F^2 - \frac{2}{n_3} \left\langle \overline{\mathbf{X}}, \overline{\mathbf{Q}}^{\mathsf{H}} \overline{\mathbf{Y}} \right\rangle + \frac{1}{n_3} \|\overline{\mathbf{X}}\|_F^2 \\ &= \min_{\overline{\mathbf{X}}} \frac{1}{n_3} \|\overline{\mathbf{X}} - \overline{\mathbf{Q}}^{\mathsf{H}} \overline{\mathbf{Y}}\|_F^2 \\ &= \min_{\underline{X}} \frac{\rho}{2} \|\underline{X} - \underline{Q}^\top * \underline{Y}\|_F^2. \end{aligned} \tag{51}$$

- Update $(\underline{Y}_1, \underline{Y}_2)$. The Lagrangian multipliers are updated by gradient ascent as follows:

$$\begin{aligned}\underline{Y}_1^{t+1} &= \underline{Y}_1^t + \rho(\underline{Q}^{t+1} * \underline{X}^{t+1} - \underline{L}^{t+1}), \\ \underline{Y}_2^{t+1} &= \underline{Y}_2^t + \rho(\underline{R}^{t+1} - \underline{S}^{t+1}).\end{aligned} \tag{52}$$

The algorithmic steps are summarized in Algorithm 2. The complexity analysis is given as follows. In each iteration of Algorithm 2, the update of $\underline{L}$ requires FFT/IFFT, and n_3 multiplications of n_1-by-r and r-by-n_2 matrices, which costs $O((n_1n_2 + rn_1 + r_n2)n_3 \log n_3 + rn_1n_2n_3)$; updating $\underline{S}$ costs $O(n_1n_2n_3)$; updating of $\underline{Q}$ involves FFT/IFFT and n_3 SVDs of n_1-by-r matrices, which costs $O(rn_1n_3 \log n_3 + r^2n_1n_3)$; updating $\underline{X}$ involves FFT/IFFT and n_3 SVDs of r-by-n_2, which costs $O(rn_2n_3 \log n_3 + r^2n_2n_3))$. Then, the per-iteration computational complexity of Algorithm 2 is dominated by:

$$O\left(\max\left\{n_1n_2n_3 \log n_3, r^2(n_1 + n_2)n_3\right\}\right).$$

Since the low-tubal-rank assumption $r \ll \min\{n_1, n_2\}$ is adopted in this paper, the per-iteration of Algorithm 2 is much lower than Algorithm 1.

Algorithm 2 Solving Problem (40) using ADMM.

Input: The observed tensor $\underline{M}$, an upper estimation r of $r_{\text{tubal}}(\underline{L}_0)$, the parameters $\gamma, \lambda, \rho, \delta$.
1: Initialize $t = 0$, $\underline{L}^0 = \underline{S}^0 = \underline{R}^0 = \underline{Y}_1^0 = \underline{Y}_2^0 = \underline{0} \in \mathbb{R}^{n_1 \times n_2 \times n_3}$, $\underline{Q}^0 = \underline{0} \in \mathbb{R}^{n_1 \times r \times n_3}$, $\underline{X}^0 = \underline{0} \in \mathbb{R}^{r \times n_2 \times n_3}$.
2: **for** $t = 0, \cdots, T_{\max}$ **do**
3: Update $(\underline{L}^{t+1}, \underline{S}^{t+1})$ by Equation (42);
4: Update $\underline{Q}^{t+1}$ by Equation (45);
5: Update $(\underline{X}^{t+1}, \underline{R}^{t+1})$ by Equations (49)–(50);
6: Update $(\underline{Y}_1^{t+1}, \underline{Y}_2^{t+1})$ by Equation (52);
7: Check the convergence criteria:
(i) convergence of variables: $\|\underline{A}^{t+1} - \underline{A}^t\|_\infty \leq \delta, \ \forall \underline{A} \in \{\underline{L}, \underline{S}, \underline{R}, \underline{Q}, \underline{X}\}$
(ii) convergence of constraints: $\max\{\|\underline{Q}^{t+1} * \underline{X}^{t+1} - \underline{L}^t\|_\infty, \|\underline{R}^{t+1} - \underline{S}^{t+1}\|_\infty\} \leq \delta$.
8: **end for**
Output: $(\hat{\underline{L}}, \hat{\underline{S}}) = (\underline{L}^{t+1}, \underline{S}^{t+1})$.

5. Experiments

5.1. Synthetic Data

We first verify the correctness of Theorem 1. Specifically, we check whether the following two statements indicated in Theorem 1 hold in experiments on synthetic data sets:

(I). (Exact recovery in the noiseless setting.) Our analysis guarantees that the underlying low-rank tensor $\underline{L}_0$ and sparse tensor $\underline{S}_0$ can be exactly recovered in the noiseless setting. This statement will be checked in Section 5.1.1.

(II). (Linear scaling of errors with the noise level.) In Theorem 1, the estimation errors on $\underline{L}_0$ and $\underline{S}_0$ scales linearly with the noise level δ. This statement will be checked in Section 5.1.2.

Signal Generation. With a given tubal rank r_0, we first generate the underlying tensor $\underline{L}_0 \in \mathbb{R}^{n_1 \times n_2 \times n_3}$ by $\underline{L}_0 = \underline{A} * \underline{B}/n_3$, where tensors $\underline{A} \in \mathbb{R}^{n_1 \times r_0 \times n_3}$ and $\underline{B} \in \mathbb{R}^{r_0 \times n_2 \times n_3}$ are generated with *i.i.d.* standard Gaussian elements. Then, the sparse corruption tensor $\underline{S}_0$ is generated by choosing its support uniformly at random. The non-zero elements of $\underline{S}_0$ will be *i.i.d.* sampled from a certain distribution that will be specified afterwards. Furthermore, the noise tensor $\underline{E}_0$ is generated with entries sampled *i.i.d.* from $\mathcal{N}(0, \sigma^2)$ with $\sigma = c\|\underline{L}_0\|_F / \sqrt{n_1 n_2 n_3}$, where we set constant c is to control the signal noise ratio. Finally, the observed tensor $\underline{M}$ is formed by $\underline{M} = \underline{L}_0 + \underline{S}_0 + \underline{E}_0$.

5.1.1. Exact Recovery in the Noiseless Setting

We first check *Statement (I)*, i.e., exact recovery in the noiseless setting. Specifically, we will show that Algorithm 1 and Algorithm 2 can exactly recover the underlying tensor $\underline{L}_0$ and the sparse corruption $\underline{S}_0$. We first test the recovery performance of different tensor sizes by setting $n = n_1 = n_2 \in \{100, 160, 200\}$ and $n_3 = 20$, with $(r_{\text{tubal}}(\underline{L}_0), \|\underline{S}_0\|_0) = (0.05n, 0.05n^2 n_3)$. The non-zero elements of tensor $\underline{S}_0$ is sampled from *i.i.d.* symmetric Bernoulli distribution, i.e., the possibility of being 1 or -1 are $1/2$. The results are shown in Table 1. It can be seen that both Algorithm 1 and Algorithm 2 can obtain relative standard error (RSE) smaller than $1\text{e}-5$ by which we can say that $\underline{L}_0$ and $\underline{S}_0$ are exact recovered. We can also see that Algorithm 2 runs much faster than Algorithm 1.

Table 1. Performance of Algorithm 1 and Algorithm 2 in both accuracy and speed for different tensor sizes when the gross corruption. Outliers from symmetric Bernoulli, observation tensor $\underline{M} \in \mathbb{R}^{n \times n \times n_3}$, $n_3 = 30$, $r_{\text{tubal}}(\underline{L}_0) = 0.05n$, $\|\underline{S}_0\|_1 = 0.05n^2 n_3$, noise level $c = 0$, $r = \max\{\lfloor 2r_{\text{tubal}}(\underline{L}_0) \rfloor, 15\}$.

n	$r_{\text{tubal}}\underline{L}_0$	$\|\underline{S}_0\|_0$	Method	$r_{\text{tubal}}\hat{\underline{L}}$	$\frac{\|\hat{\underline{L}}-\underline{L}_0\|_F}{\|\underline{L}_0\|_F}$	$\frac{\|\hat{\underline{S}}-\underline{S}_0\|_F}{\|\underline{S}_0\|_F}$	time/s
100	5	1×10^4	Algorithm 1	5	5.13×10^{-6}	5.27×10^{-6}	3.63
			Algorithm 2	5	4.92×10^{-6}	5.12×10^{-6}	**1.76**
160	8	2.56×10^4	Algorithm 1	8	3.86×10^{-6}	3.52×10^{-6}	9.52
			Algorithm 2	8	4.48×10^{-6}	4.08×10^{-6}	**4.42**
200	10	4×10^4	Algorithm 1	10	3.46×10^{-6}	3.59×10^{-6}	14.16
			Algorithm 2	10	4.12×10^{-6}	4.63×10^{-6}	**7.44**

We then test whether the recovery performance can be affected by the distribution of the corruptions. This is done by choosing the non-zeros elements of $\underline{S}_0$ from *i.i.d.* standard Gaussian distribution. The experimental results are reported in Table 2. We can find that both Algorithm 1 and Algorithm 2 can exactly recover the true $\underline{L}_0$ and $\underline{S}_0$ and Algorithm 2 runs much faster than Algorithm 1.

We also conduct STPCP by Algorithm 1 and Algorithm 2 with missing entries. After generating $\underline{L}_0$, $\underline{S}_0$ and $\underline{E}_0$, we get the observation by Model (27). We choose the support of $\underline{B}$ uniformly at random with possibility 0.8 and then set elements in the chosen support to be 1. Thus, %20 of the entries are missing. The corrupted observation M is then formed by $\underline{M} = \underline{B} \odot (\underline{L}_0 + \underline{S}_0 + \underline{E}_0)$. We show the recover results in Table 3. We can see that the underlying low-rank tensor $\underline{L}_0$ can be exactly recovered and the observed part of the corruption tensor $\underline{B} \odot \underline{S}_0$ can also be exactly recovered (Please note that it is impossible to recover the unobserved entries of a sparse tensor $\underline{S}_0$ [52]).

Table 2. Performance of Algorithm 1 and Algorithm 2 in both accuracy and speed for different tensor sizes when the gross corruption. Outliers from standard Gaussian distribution, observation tensor $\underline{\mathbf{M}} \in \mathbb{R}^{n \times n \times n_3}$, $n_3 = 30$, $r_{\text{tubal}}(\underline{\mathbf{L}}_0) = 0.05n$, $\|\underline{\mathbf{S}}_0\|_1 = 0.05n^2 n_3$, noise level $c = 0$, $r = \max\{\lfloor 2r_{\text{tubal}}(\underline{\mathbf{L}}_0)\rfloor, 15\}$.

n	$r_{\text{tubal}}\underline{\mathbf{L}}_0$	$\|\underline{\mathbf{S}}_0\|_0$	**Method**	$r_{\text{tubal}}\hat{\underline{\mathbf{L}}}$	$\frac{\|\hat{\underline{\mathbf{L}}}-\underline{\mathbf{L}}_0\|_F}{\|\underline{\mathbf{L}}_0\|_F}$	$\frac{\|\hat{\underline{\mathbf{S}}}-\underline{\mathbf{S}}_0\|_F}{\|\underline{\mathbf{S}}_0\|_F}$	**time/s**
100	5	1×10^4	Algorithm 1	5	2.7×10^{-6}	2.6×10^{-6}	4.43
			Algorithm 2	5	2.9×10^{-6}	3.2×10^{-6}	**1.82**
160	8	2.56×10^4	Algorithm 1	8	4.76×10^{-6}	4.08×10^{-6}	10.45
			Algorithm 2	8	4.24×10^{-6}	4.05×10^{-6}	**5.15**
200	10	4×10^4	Algorithm 1	10	3.78×10^{-6}	3.64×10^{-6}	18.97
			Algorithm 2	10	3.78×10^{-6}	3.63×10^{-6}	**8.04**

Table 3. Performance of Algorithm 1 and Algorithm 2 in both accuracy and speed for different tensor sizes when the gross corruption. Outliers from symmetric Bernoulli, observation tensor $\underline{\mathbf{M}} \in \mathbb{R}^{n \times n \times n_3}$, $n_3 = 30$, $r_{\text{tubal}}(\underline{\mathbf{L}}_0) = 0.05n$, $\|\underline{\mathbf{S}}_0\|_1 = 0.05n^2 n_3$, noise level $c = 0$, $r = \max\{\lfloor 2r_{\text{tubal}}(\underline{\mathbf{L}}_0)\rfloor, 15\}$, with %20 random missing entries.

n	$r_{\text{tubal}}\underline{\mathbf{L}}_0$	$\|\underline{\mathbf{B}} \odot \underline{\mathbf{S}}_0\|_0$	**Method**	$r_{\text{tubal}}\hat{\underline{\mathbf{L}}}$	$\frac{\|\hat{\underline{\mathbf{L}}}-\underline{\mathbf{L}}_0\|_F}{\|\underline{\mathbf{L}}_0\|_F}$	$\frac{\|\hat{\underline{\mathbf{S}}}-\underline{\mathbf{B}}\odot\underline{\mathbf{S}}_0\|_F}{\|\underline{\mathbf{B}}\odot\underline{\mathbf{S}}_0\|_F}$	**time/s**
100	5	8×10^3	Algorithm 1	5	7.52×10^{-6}	5.97×10^{-6}	3.87
			Algorithm 2	5	7.50×10^{-6}	5.96×10^{-6}	**1.69**
160	8	2.048×10^4	Algorithm 1	8	4.46×10^{-6}	5.17×10^{-6}	9.64
			Algorithm 2	8	5.60×10^{-6}	4.71×10^{-6}	**4.46**
200	10	3.2×10^4	Algorithm 1	10	4.78×10^{-6}	4.04×10^{-6}	14.78
			Algorithm 2	10	5.13×10^{-6}	4.20×10^{-6}	**7.77**

5.1.2. Linear Scaling of Errors with the Noise Level

We then verify *Statement (II)* that the estimation errors have linear scale behavior with respect to the noise level. The estimation errors are measured using the mean-squared-error (MSE):

$$\text{MSE}(\hat{\underline{\mathbf{L}}}) = \frac{\|\hat{\underline{\mathbf{L}}} - \underline{\mathbf{L}}_0\|_F^2}{n_1 n_2 n_3}, \quad \text{MSE}(\hat{\underline{\mathbf{S}}}) = \frac{\|\hat{\underline{\mathbf{S}}} - \underline{\mathbf{S}}_0\|_F^2}{n_1 n_2 n_3},$$

for the low rank component $\underline{\mathbf{L}}_0$ and the sparse component $\underline{\mathbf{S}}_0$, respectively. We test tensors of 3 different size by choosing $n \in \{60, 80, 100\}$ and $n_3 = 20$. The tubal rank $r_{\text{tubal}}(\underline{\mathbf{L}}_0)$ of $\underline{\mathbf{L}}_0$ and sparsity s of $\underline{\mathbf{S}}_0$ are set as $(r_{\text{tubal}}(\underline{\mathbf{L}}_0), s) = (5, 0.1n^2 n_3)$. We vary the signal noise ratio $c = 0.03 : 0.03 : 0.6$ which is in proportional of the noise level δ. We run the proposed Algorithm 1, test 50 trials, and report the averaged MSEs. The MSEs of $\hat{\underline{\mathbf{L}}}$ and $\underline{\mathbf{S}}_0$ versus c^2 are shown in sub-figures (a) and (b) in Figure 4. We can see that the estimation error has linear scaling behavior along with the noise level as Theorem 1 indicates. Since the results for $n = 80$ and $n = 100$ are quite similar to the case of $n = 60$, they are simply omitted.

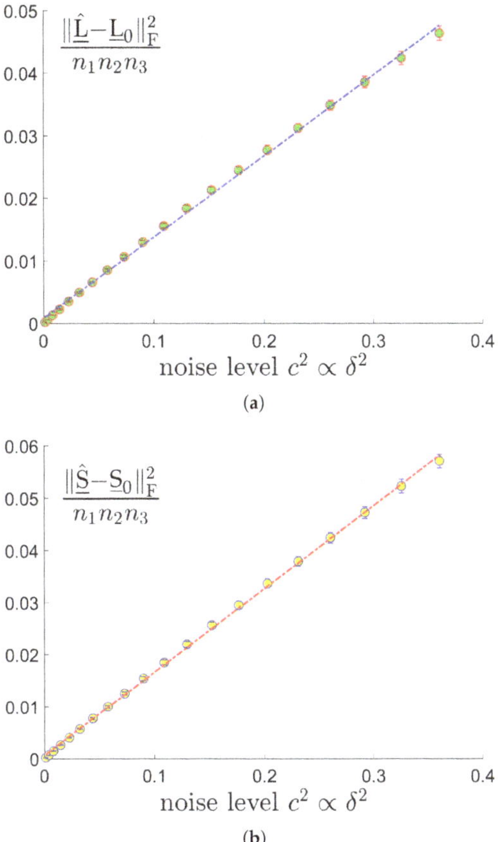

Figure 4. The MSEs of $\hat{\underline{\mathbf{L}}}$ and $\underline{\mathbf{S}}_0$ versus c^2 for tensors of size $60 \times 60 \times 20$ where the tubal rank $r_{\text{tubal}}(\underline{\mathbf{L}}_0)$ of $\underline{\mathbf{L}}_0$ and sparsity s of $\underline{\mathbf{S}}_0$ are set as $(r_{\text{tubal}}(\underline{\mathbf{L}}_0), s) = (5, 0.1n^2n_3)$. (**a**): MSE of $\hat{\underline{\mathbf{L}}}$ vs c^2. (**b**): MSE of $\hat{\underline{\mathbf{S}}}$ vs c^2.

5.2. Real Data Sets

In this section, experiments on real data sets (color images and videos) are carried out to evaluate the effectiveness and efficiency of the proposed Algorithms 1 and 2. Besides noises and sparse corruptions, we also consider missing values which is more challenging. The proposed algorithms are compared with the following typical models:

(I). NN-I: tensor recovery based on matrix nuclear norms of frontal slices formulated as follows:

$$\min_{\underline{\mathbf{L}},\underline{\mathbf{S}}} \frac{1}{2}\|\underline{\mathbf{B}} \odot (\underline{\mathbf{M}} - \underline{\mathbf{L}} - \underline{\mathbf{E}})\|_F + \gamma \sum_{k=1}^{n_3} (\|\mathbf{L}_{(k)}\|_* + \lambda \|\mathbf{S}^{(k)}\|_1). \tag{53}$$

This model will be used for image restoration in Section 5.2.1. Please note that Model (53) is equivalent to parallel matrix recovery on each frontal slice.

(II). NN-II: tensor recovery based on matrix nuclear norm formulated as follows:

$$\min_{\underline{\mathbf{L}},\underline{\mathbf{S}}} \frac{1}{2}\|\underline{\mathbf{B}} \odot (\underline{\mathbf{M}} - \underline{\mathbf{L}} - \underline{\mathbf{E}})\|_F + \gamma\|\mathbf{L}\|_* + \gamma\lambda\|\underline{\mathbf{S}}\|_1, \tag{54}$$

where $\mathbf{L} = [\mathbf{l}_1, \mathbf{l}_2, \cdots, \mathbf{l}_{n_3}] \in \mathbb{R}^{n_1 n_2 \times n_3}$ with $\mathbf{l}_k := \mathrm{vec}(\mathbf{L}^{(k)}) \in \mathbb{R}^{n_1 n_2}$ defined as the vectorization [40] of frontal slices $\mathbf{L}^{(k)}$, for all $k = 1, 2, \cdots, n_3$. This model will be used for video restoration in Section 5.2.2.

(III). SNN: tensor recovery based on SNN formulated as follows:

$$\min_{\underline{\mathbf{L}}, \underline{\mathbf{S}}} \frac{1}{2} \|\underline{\mathbf{B}} \odot (\underline{\mathbf{M}} - \underline{\mathbf{L}} - \underline{\mathbf{E}})\|_{\mathrm{F}} + \gamma \sum_{i=1}^{3} \alpha_m \|\mathbf{L}_{(i)}\|_* + \gamma \|\underline{\mathbf{S}}\|_1, \tag{55}$$

where $\mathbf{L}_{(i)} \in \mathbb{R}^{n_i \times \prod_{j \neq i} n_j}$ is the mode-i matriculation of tensor $\underline{\mathbf{L}} \in \mathbb{R}^{n_1 \times n_2 \times n_3}$, for all $i = 1, 2, 3$.

We solve the above Model (53)–(55) using ADMM implemented by ourselves in Matlab. The effectiveness of the algorithms is measured by Peaks Signal Noise Ratio (PSNR):

$$\mathrm{PSNR} := 10 \log_{10} \left(\frac{n_1 n_2 n_3 \|\underline{\mathbf{L}}_0\|_\infty^2}{\|\hat{\underline{\mathbf{L}}} - \underline{\mathbf{L}}_0\|_{\mathrm{F}}^2} \right).,$$

Please note that a larger PSNR value indicates higher quality of $\hat{\underline{\mathbf{L}}}$.

5.2.1. Color Images

Color images are the most commonly used 3-way tensors. We test the twenty 256-by-256-by-3 color images which have been used in [37], and carry out robust tensor recovery with missing entries (see Figure 5). Following [37], for a color image $\underline{\mathbf{L}}_0 \in \mathbb{R}^{n \times n \times 3}$, we choose its support uniformly at random with ratio ρ_s and fill in the values with *i.i.d.* symmetric Bernoulli variables to generate $\underline{\mathbf{S}}_0$. The noise tensor $\underline{\mathbf{E}}_0$ is generated with *i.i.d.* zero-mean Gaussian entries whose standard deviation is given by $\sigma = 0.05\|\underline{\mathbf{L}}_0\|_{\mathrm{F}} / \sqrt{3n^2}$. Then, we form the binary observation mask $\underline{\mathbf{B}}$ by choosing its support uniformly at random with ratio ρ_{obs}. Finally, the partially observed corruption $\underline{\mathbf{M}} = \underline{\mathbf{B}} \odot (\underline{\mathbf{L}}_0 + \underline{\mathbf{S}}_0 + \underline{\mathbf{E}}_0)$ are formed.

Figure 5. The 20 color images used.

We consider two scenarios by setting $(\rho_{\mathrm{obs}}, \rho_s) \in \{(0.9, 0.1), (0.8, 0.2)\}$. For NN (Model (53)), we set the regularization parameters $\lambda = 1/\sqrt{n\rho_{\mathrm{obs}}}$ (suggested by [46]), and set parameter $\gamma = \|\underline{\mathbf{E}}_0\|_{\mathrm{sp}}$ where $\|\underline{\mathbf{E}}_0\|_{\mathrm{sp}}$ is estimated as $6.5\sigma\sqrt{3\rho_{\mathrm{obs}} n \log(6n)}$ (suggested by [5]). For SNN, the parameters are chosen to satisfy $\gamma = 0.05$, $\alpha_1 = \alpha_2 = \sqrt{3n\rho_{\mathrm{obs}}}$, $\alpha_3 = 0.01\sqrt{3n\rho_{\mathrm{obs}}}$. For Algorithm 1 and Algorithm 2, we set $\gamma = 0.3\|\underline{\mathbf{E}}_0\|_{\mathrm{sp}}$, and $\lambda = 1/\sqrt{3n\rho_{\mathrm{obs}}}$. The initialized rank r in Algorithm 2 is set as 60. In each setting, we test each color image for 10 times and report the averaged PSNR and time. For quantitative comparison, we show the PSNR values and running times in Figures 6 and 7 for settings of $(\rho_{\mathrm{obs}}, \rho_s) = (0.9, 0.1)$ and $(\rho_{\mathrm{obs}}, \rho_s) = (0.8, 0.2)$, respectively. Several visual examples are shown in Figure 8 for qualitative comparison for the setting of $(\rho_{\mathrm{obs}}, \rho_s) = (0.8, 0.2)$. We can see from Figures 6–8 that the proposed Algorithm 1 has the highest recovery quality and the proposed Algorithm 2 has the second highest quality but the fastest running time.

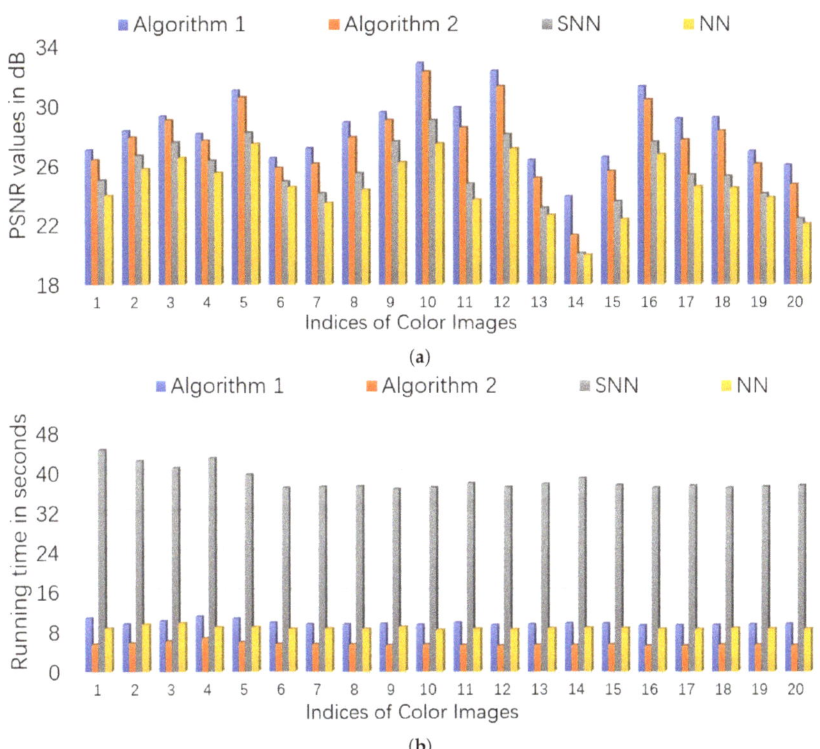

Figure 6. The quantitative comparison in PSNR and time on color images. First, 10% entries of each image is corrupted by *i.i.d.* symmetric Bernoulli variable, then polluted by Gaussian noise of noise level $c = 0.05$, and finally 10% of the corrupted entries are missing uniformly at random. (**a**): the PSNR values of each algorithm; (**b**): the running time of each algorithm.

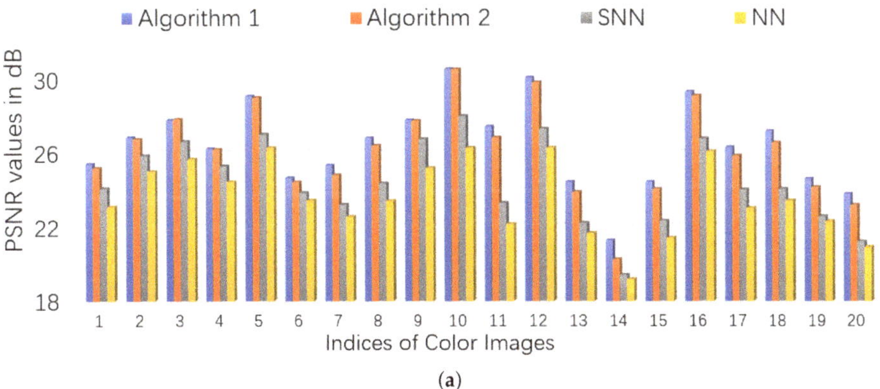

(**a**)

Figure 7. *Cont.*

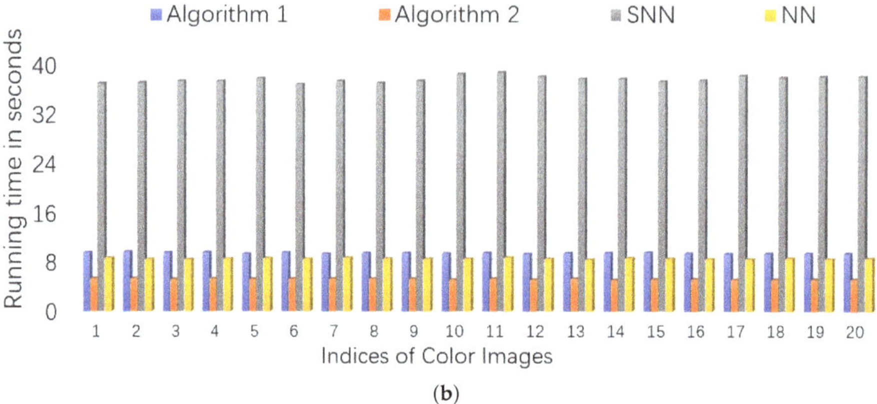

(**b**)

Figure 7. The quantitative comparison in PSNR and time on color images. First, 20% entries of each image is corrupted by *i.i.d.* symmetric Bernoulli variable, then polluted by Gaussian noise of noise level $c = 0.05$, and finally 20% of the corrupted entries are missing uniformly at random. (**a**): the PSNR values of each algorithm; (**b**): the running time of each algorithm.

Figure 8. The visual results for image recovery of different algorithms. First, 20% entries of each image is corrupted by *i.i.d.* symmetric Bernoulli variable, then polluted by Gaussian noise of noise level $c = 0.05$, and finally 20% of the corrupted entries are missing uniformly at random. (**a**): the original image; (**b**): the corrupted image; (**c**) image recovered by Algorithm 1; (**d**) image recovered by Algorithm 2; (**e**) image recovered by the matrix nuclear norm (NN)-based Model (53); (**f**) image recovered by the SNN-based Model (55).

5.2.2. Videos

In this subsection, video restoration experiments are conducted on four broadly used YUV videos (They can be downloaded from https://sites.google.com/site/subudhibadri/fewhelpfuldownloads: Akiyo_qcif, Scilent_qcif, Carphone_qcif, and Claire_qcif.) Owing to computational limitation, we simply use the first 32 frames of the Y components of all the videos which results in four 144-by-176-by-30 tensors. For a 3-way data tensor $\underline{L}_0 \in \mathbb{R}^{n_1 \times n_2 \times n_3}$, To generate corruption $\underline{S}_0$, the support is chosen uniformly at random with ratio ρ_s and then elements in the support are filled in with *i.i.d.* symmetric Bernoulli variables. The noise tensor $\underline{E}_0$ is also generated with *i.i.d.* zero-mean Gaussian entries whose standard deviation is given by $\sigma = 0.05\|\underline{L}_0\|_F/\sqrt{n_1 n_2 n_3}$. Then, the binary observation mask $\underline{B}$ is formed thorough choosing its support uniformly at random with ratio ρ_{obs}. Finally, the partially observed corruption $\underline{M} = \underline{B} \odot (\underline{L}_0 + \underline{S}_0 + \underline{E}_0)$ are formed.

We also consider two scenarios by setting $(\rho_{obs}, \rho_s) \in \{(0.9, 0.1), (0.8, 0.2)\}$. NN-II Model (54) is used in video restoration. For NN-II, we set the regularization parameters $\lambda = 1/\sqrt{n_1 n_2 \rho_{obs}}$ (suggested by [46]), and set parameter $\gamma = \|\underline{E}_0\|_{sp}$ where $\|\underline{E}_0\|_{sp}$ is estimated as $6.5\sigma\sqrt{\rho_{obs} n_1 n_3 \log((n_1 + n_2)n_3)}$ (suggested by [5]). For SNN, the parameters are chosen to satisfy $\gamma = 0.05$, $\alpha_1 = \alpha_2 = \sqrt{n_1 n_3 \rho_{obs}}$, $\alpha_3 = 5\sqrt{n_1 n_3 \rho_{obs}}$. For Algorithm 1 and Algorithm 2, we set $\gamma = 0.3\|\underline{E}_0\|_{sp}$, and $\lambda = 1/\sqrt{\max\{n_1, n_2\} n_3 \rho_{obs}}$ after careful parameter tuning. The initialized rank r in Algorithm 2 is set as 60. In each setting, we test each video for 10 times and report the averaged PSNR and time. For quantitative comparison, we show the PSNR values and running times in Table 4. It can be seen that Algorithm 1 has the highest recovery quality and the proposed Algorithm 2 has the second highest quality but the fastest running time.

Table 4. PSNR values and running time (in seconds) of different algorithms on video data. First, $\rho_s n_1 n_2 n_3$ entries of each image is corrupted by *i.i.d.* symmetric Bernoulli variable, then polluted by Gaussian noise of noise level $c = 0.05$, and finally $(1 - \rho_{obs}) n_1 n_2 n_3$ of the corrupted entries are missing uniformly at random. The items with **highest PSNR values** are highlighted with bold face, and the items with shortest running time are highlighted with underline.

Data Set	(ρ_{obs}, ρ_s)	Index	NN, Model (54)	SNN, Model (55)	Algorithm 1	Algorithm 2
Akiyo	(0.9,0.1)	PSNR	31.74	32.09	**33.94**	33.36
		time/s	29.48	51.13	20.10	12.39
	(0.8,0.2)	PSNR	30.59	30.70	**32.44**	32.07
		time/s	30.65	51.17	19.53	14.92
Silent	(0.9,0.1)	PSNR	28.26	30.39	**31.74**	31.23
		time/s	28.91	49.79	21.21	14.76
	(0.8,0.2)	PSNR	26.95	27.60	**30.42**	30.07
		time/s	36.51	60.81	22.43	15.62
Carphone	(0.9,0.1)	PSNR	26.87	28.79	**29.15**	28.94
		time/s	28.55	47.17	22.12	14.41
	(0.8,0.2))	PSNR	26.12	26.43	**28.17**	27.99
		time/s	26.72	49.21	20.55	14.74
Claire	(0.9,0.1)	PSNR	30.56	32.20	**34.27**	34.02
		time/s	29.75	47.32	21.43	13.52
	(0.8,0.2)	PSNR	29.94	30.43	**32.96**	32.78
		time/s	29.43	50.46	19.47	13.04

6. Conclusions

This paper studied the problem of stable tensor principal component pursuit which aims to recover a tensor from noises and sparse corruptions. We proposed a constrained tubal nuclear norm-based model and established upper bounds on the estimation error. In contrast to prior work [37], our theory can guarantee exact recovery in the noiseless setting. We also designed two algorithms, the first

ADMM algorithm can be accelerated by the second Algorithm which adopts a factorization strategy. We validated the correctness of our theory by simulations on synthetic data, and evaluated the effectiveness and efficiency of the proposed algorithms via experiments on color images and videos.

For future directions, it is a natural and interesting extension to consider recovery of 4-way tensors [35] with arbitrary linear transformation [53,54]. It is also interesting to use tensor factorization-based methods [55,56] for STPCP. Another challenging future direction is developing tools to verify whether the unknown tensor satisfies the tensor incoherence condition from its incomplete or corrupted observations.

For extensions of the proposed approach to higher-way tensors, we produce the following two ideas:

1. By recursively applying DFT over successive modes higher than 3 and then unfolding the obtained tensor into 3-way [57], the proposed algorithms and theoretical analysis can be extended to higher-way tensors.
2. By using the overlapped orientation invariant tubal nuclear norm [58], we can extend the proposed algorithm to higher-order cases and obtain orientation invariance.

Author Contributions: Conceptualization, W.F. Data curation, D.W. and R.Z. Formal analysis, W.F. Methodology, W.F., D.W and R.Z. Software, D.W. and R.Z. Writing, original draft, W.F., D.W. and R.Z.

Funding: This research was funded by the Key Projects of Natural Science Research in Universities in Anhui Province under grant number KJ2019A0994.

Acknowledgments: We sincerely thank Andong Wang who shared the codes of [37] and gave us some suggestions of the proof.

Conflicts of Interest: The authors declare no conflict of interest.

Appendix A. Proofs of Lemmas and Theorems

Appendix A.1. The Proof of Theorem 1

Appendix A.1.1. Key Lemmas for the Proof of Theorem 1

Before Proving Theorem 1, we should define some notations and operators first.

Suppose $\underline{L}_0 \in \mathbb{R}^{n_1 \times n_2 \times n_3}$ with tubal rank r has the skinny t-SVD $\underline{L}_0 = \underline{U} * \underline{\Lambda} * \underline{V}^\top$, where $\underline{U} \in \mathbb{R}^{n_1 \times r \times n_3}, \underline{V} \in \mathbb{R}^{r \times n_2 \times n_3}$ are orthogonal tensors, and $\underline{\Lambda} \in \mathbb{R}^{r \times r \times n_3}$ is an f-diagonal tensor. Define the following set:

$$\mathsf{T} := \left\{ \underline{U} * \underline{A} + \underline{B} * \underline{V}^\top \mid \underline{A} \in \mathbb{R}^{r \times n_2 \times n_3}, \underline{B} \in \mathbb{R}^{n_1 \times r \times n_3} \right\} \subset \mathbb{R}^{n_1 \times n_2 \times n_3}. \tag{A1}$$

Then, define the projector onto T for any tensor $\underline{T} \in \mathbb{R}^{n_1 \times n_2 \times n_3}$ as follows:

$$\begin{aligned} \mathcal{P}_{\mathsf{T}}(\underline{T}) &:= \underline{U} * \underline{U}^\top * \underline{T} + \underline{T} * \underline{V} * \underline{V}^\top - \underline{U} * \underline{U}^\top * \underline{T} * \underline{V} * \underline{V}^\top, \\ \mathcal{P}_{\mathsf{T}^\perp}(\underline{T}) &:= (\underline{I} - \underline{U} * \underline{U}^\top) * \underline{T} * (\underline{I} - \underline{V} * \underline{V}^\top). \end{aligned} \tag{A2}$$

Let $\Omega^\perp$ be the complement of $\Omega \subset [n_1] \times [n_2] \times [n_3]$ which is the support of $\underline{S}_0$. Then, define two operators $\mathcal{P}_\Omega, \mathcal{P}_{\Omega^\perp}$ as follows:

$$\mathcal{P}_\Omega(\underline{T}) := \sum_{(i,j,k) \in \Omega} \left\langle \underline{T}, \mathring{\mathfrak{e}}_i * \dot{\mathfrak{e}}_k * \mathring{\mathfrak{e}}_j^\top \right\rangle, \quad \mathcal{P}_{\Omega^\perp}(\underline{T}) := \sum_{(i,j,k) \in \Omega^\perp} \left\langle \underline{T}, \mathring{\mathfrak{e}}_i * \dot{\mathfrak{e}}_k * \mathring{\mathfrak{e}}_j^\top \right\rangle, \tag{A3}$$

for any $\underline{T} \in \mathbb{R}^{n_1 \times n_2 \times n_3}$.

Define two sets Γ and $\Gamma^\perp$ as follows:

$$\Gamma = \{(\underline{A}, \underline{A}) \mid \underline{A} \in \mathbb{R}^{n_1 \times n_2 \times n_3}\}, \quad \Gamma^\perp = \{(\underline{A}, -\underline{A}) \mid \underline{A} \in \mathbb{R}^{n_1 \times n_2 \times n_3}\}. \tag{A4}$$

Then, for any tensors $\underline{X}_\iota, \underline{X}_s \in \mathbb{R}^{n_1 \times n_2 \times n_3}$, the projectors of the tensor $\underline{\mathbf{X}} = (\underline{X}_\iota, \underline{X}_s)$ into the sets Γ and $\Gamma^\perp$ are given as follows, respectively:

$$\mathcal{P}_\Gamma(\underline{\mathbf{X}}) = \left(\frac{\underline{X}_\iota + \underline{X}_s}{2}, \frac{\underline{X}_\iota + \underline{X}_s}{2}\right), \quad \mathcal{P}_{\Gamma^\perp}(\underline{\mathbf{X}}) = \left(\frac{\underline{X}_\iota - \underline{X}_s}{2}, \frac{\underline{X}_s - \underline{X}_\iota}{2}\right). \tag{A5}$$

For any tensors $\underline{X}_\iota, \underline{X}_s \in \mathbb{R}^{n_1 \times n_2 \times n_3}$, define two operators on $\underline{\mathbf{X}} = (\underline{X}_\iota, \underline{X}_s)$ as follows:

$$(\mathcal{P}_\mathrm{T} \times \mathcal{P}_\Omega)(\underline{\mathbf{X}}) = (\mathcal{P}_\mathrm{T}(\underline{X}_\iota), \mathcal{P}_\Omega(\underline{X}_s)), \quad (\mathcal{P}_{\mathrm{T}^\perp} \times \mathcal{P}_{\Omega^\perp})(\underline{\mathbf{X}}) = (\mathcal{P}_{\mathrm{T}^\perp}(\underline{X}_\iota), \mathcal{P}_{\Omega^\perp}(\underline{X}_s)). \tag{A6}$$

Also define two norms as follows:

$$\|\underline{\mathbf{X}}\|_\mathrm{F} = \sqrt{\|\underline{X}_\iota\|_\mathrm{F}^2 + \|\underline{X}_s\|_\mathrm{F}^2}, \quad \|\underline{\mathbf{X}}\|_{\mathrm{F},\mu} = \sqrt{\|\underline{X}_\iota\|_\mathrm{F}^2 + \mu^2\|\underline{X}_s\|_\mathrm{F}^2}. \tag{A7}$$

where μ is a constant that will be determined afterwards.

We first give Lemma A1 which can be seen as a modified version of Lemma C.1 in [2].

Lemma A1. *Assume that* $\|\mathcal{P}_\Omega\mathcal{P}_\mathrm{T}\| \le \frac{1}{2}$, *and* $\lambda \le \frac{1}{2\sqrt{n_3}}$. *Suppose there exists a tensor* $\underline{G}^*$ *satisfying the following conditions:*

$$\begin{cases} \mathcal{P}_\mathrm{T}(\underline{G}^*) = \underline{U} * \underline{V}^\top, \\ \|\mathcal{P}_{\mathrm{T}^\perp}(\underline{G}^*)\|_{\mathrm{sp}} \le \dfrac{1}{2}, \\ \|\mathcal{P}_\Omega(\underline{G}^* - \lambda \mathrm{sign}(\underline{S}_0))\|_\mathrm{F} \le \dfrac{\lambda}{4}, \\ \|\mathcal{P}_{\Omega^\perp}(\underline{G}^*)\|_\infty \le \dfrac{\lambda}{2}. \end{cases} \tag{A8}$$

Then for any perturbation $\underline{\Delta} \in \mathbb{R}^{n_1 \times n_2 \times n_3}$, *one has:*

$$\begin{aligned} &\|\underline{L}_0 + \underline{\Delta}\|_\mathrm{TNN} + \lambda\|\underline{S}_0 - \underline{\Delta}\|_1 \\ &\ge \|\underline{L}_0\|_\mathrm{TNN} + \lambda\|\underline{S}_0\|_1 + (\tfrac{3}{4} - \|\mathcal{P}_{\mathrm{T}^\perp}(\underline{G})\|_{\mathrm{sp}})\,\|\mathcal{P}_{\mathrm{T}^\perp}(\underline{\Delta})\|_\mathrm{TNN} + (\tfrac{3}{4}\lambda - \mathcal{P}_{\Omega^\perp}(\|\underline{G}\|_\infty))\,\|\mathcal{P}_{\Omega^\perp}(\Delta)\|_1. \end{aligned} \tag{A9}$$

Proof. Let $\underline{G}_\iota \in \partial\|\underline{L}_0\|_\mathrm{TNN}$, i.e., any sub-gradient of $\|\cdot\|_\mathrm{TNN}$ at $\underline{L}_0$, then it satisfies:

$$\mathcal{P}_\mathrm{T}(\underline{G}_\iota) = \underline{U} * \underline{V}^\top, \quad \|\mathcal{P}_{\mathrm{T}^\perp}(\underline{G}_\iota)\|_{\mathrm{sp}} \le 1. \tag{A10}$$

$\underline{G}_\iota \in \partial\|\underline{L}_0\|_\mathrm{TNN}$ and $\underline{G}_s \in \partial(\lambda\|\underline{S}_0\|_1)$. According to the convexity of $\|\cdot\|_\mathrm{TNN}$ and $\|\cdot\|_1$, we have:

$$\|\underline{L}_0 + \underline{\Delta}\|_\mathrm{TNN} \ge \|\underline{L}_0\|_\mathrm{TNN} + \langle \underline{G}_\iota, \underline{\Delta}\rangle, \quad \lambda\|\underline{S}_0 - \underline{\Delta}\|_1 \ge \lambda\|\underline{S}_0\|_1 - \langle \underline{G}_s, \underline{\Delta}\rangle. \tag{A11}$$

By choosing $\underline{G}_\iota = \underline{U} * \underline{V}^\top + \underline{P} * \underline{Q}^\top$, where $\underline{P}$ and $\underline{Q}$ comes from the skinny t-SVD of $\mathcal{P}_{\mathrm{T}^\perp}(\underline{\Delta}) = \underline{P} * \underline{\Sigma} * \underline{Q}^\top$, one has:

$$\begin{aligned} \langle \underline{G}_\iota, \underline{\Delta}\rangle &= \langle \underline{G}, \underline{\Delta}\rangle + \langle \underline{G}_\iota - \underline{G}, \underline{\Delta}\rangle = \langle \underline{G}, \underline{\Delta}\rangle + \langle \mathcal{P}_{\mathrm{T}^\perp}(\underline{G}_\iota), \mathcal{P}_{\mathrm{T}^\perp}(\underline{\Delta})\rangle - \langle \mathcal{P}_{\mathrm{T}^\perp}(\underline{G}), \mathcal{P}_{\mathrm{T}^\perp}(\underline{\Delta})\rangle \\ &= \langle \underline{G}, \underline{\Delta}\rangle - (1 - \|\mathcal{P}_{\mathrm{T}^\perp}(\underline{G})\|_{\mathrm{sp}})\|\mathcal{P}_{\mathrm{T}^\perp}(\underline{\Delta})\|_\mathrm{TNN}. \end{aligned} \tag{A12}$$

Also, by choosing $\underline{G}_s = \lambda\mathrm{sign}(\underline{S}_0) - \mathrm{sign}(\mathcal{P}_{\Omega^\perp}(\underline{\Delta}))$, one has:

$$\begin{aligned} -\langle \underline{G}_s, \underline{\Delta}\rangle &= -\langle \underline{G}, \underline{\Delta}\rangle - \langle \underline{G}_s - \underline{G}, \underline{\Delta}\rangle \\ &= -\langle \underline{G}, \underline{\Delta}\rangle - \langle \mathcal{P}_\Omega(\lambda\mathrm{sign}(\underline{S}_0) - \underline{G}), \mathcal{P}_\Omega(\underline{\Delta})\rangle - \langle \mathcal{P}_{\Omega^\perp}(\underline{G}_s), \mathcal{P}_{\Omega^\perp}(\underline{\Delta})\rangle + \langle \mathcal{P}_{\Omega^\perp}(\underline{G}), \mathcal{P}_{\Omega^\perp}(\underline{\Delta})\rangle \\ &\ge -\langle \underline{G}, \underline{\Delta}\rangle - \|\mathcal{P}_\Omega(\lambda\mathrm{sign}(\underline{S}_0) - \underline{G})\|_\mathrm{F}\|\mathcal{P}_\Omega(\underline{\Delta})\|_\mathrm{F} + \|\mathcal{P}_{\Omega^\perp}(\underline{\Delta})\|_1 - \|\mathcal{P}_{\Omega^\perp}(\underline{G})\|_\infty\|\mathcal{P}_{\Omega^\perp}(\underline{\Delta})\|_1 \\ &\ge -\langle \underline{G}, \underline{\Delta}\rangle - \frac{\lambda}{4}\|\mathcal{P}_\Omega(\underline{\Delta})\|_\mathrm{F} + (1 - \|\mathcal{P}_{\Omega^\perp}(\underline{G})\|_\infty)\|\mathcal{P}_{\Omega^\perp}(\underline{\Delta})\|_1 \end{aligned} \tag{A13}$$

Also note that:

$$\begin{aligned}\|\mathcal{P}_{\Omega}(\underline{\Delta})\|_{\mathrm{F}} &\leq \|\mathcal{P}_{\Omega}\mathcal{P}_{\mathrm{T}}(\underline{\Delta})\|_{\mathrm{F}} + \|\mathcal{P}_{\Omega}\mathcal{P}_{\mathrm{T}^{\perp}}(\underline{\Delta})\|_{\mathrm{F}} \\ &\leq \|\mathcal{P}_{\Omega}\mathcal{P}_{\mathrm{T}}(\underline{\Delta})\|_{\mathrm{F}} + \|\mathcal{P}_{\Omega}\mathcal{P}_{\mathrm{T}^{\perp}}(\underline{\Delta})\|_{\mathrm{F}} \\ &\leq \frac{1}{2}\|\underline{\Delta}\|_{\mathrm{F}} + \|\mathcal{P}_{\Omega}\mathcal{P}_{\mathrm{T}^{\perp}}(\underline{\Delta})\|_{\mathrm{F}} \\ &\leq \frac{1}{2}\|\mathcal{P}_{\Omega}(\underline{\Delta})\|_{\mathrm{F}} + \frac{1}{2}\|\mathcal{P}_{\Omega^{\perp}}(\underline{\Delta})\|_{\mathrm{F}} + \|\mathcal{P}_{\Omega}\mathcal{P}_{\mathrm{T}^{\perp}}(\underline{\Delta})\|_{\mathrm{F}}\end{aligned} \tag{A14}$$

which leads to:

$$\|\mathcal{P}_{\Omega}(\underline{\Delta})\|_{\mathrm{F}} \leq \|\mathcal{P}_{\Omega^{\perp}}(\underline{\Delta})\|_{\mathrm{F}} + 2\|\mathcal{P}_{\Omega}\mathcal{P}_{\mathrm{T}^{\perp}}(\underline{\Delta})\|_{\mathrm{F}} \leq \|\mathcal{P}_{\Omega^{\perp}}(\underline{\Delta})\|_{1} + 2\sqrt{n_3}\|\mathcal{P}_{\mathrm{T}^{\perp}}(\underline{\Delta})\|_{\mathrm{TNN}}. \tag{A15}$$

Putting things together, we have:

$$\begin{aligned}&\|\underline{\mathrm{L}}_0 + \underline{\Delta}\|_{\mathrm{TNN}} + \lambda\|\underline{\mathrm{S}}_0 - \underline{\Delta}\|_1 - (\|\underline{\mathrm{L}}_0\|_{\mathrm{TNN}} + \lambda\|\underline{\mathrm{S}}_0\|_1) \\ &\geq \left(1 - \frac{\lambda\sqrt{n_3}}{2} - \|\mathcal{P}_{\mathrm{T}^{\perp}}(\underline{\mathrm{G}})\|_{\mathrm{sp}}\right)\|\mathcal{P}_{\mathrm{T}^{\perp}}(\underline{\Delta})\|_{\mathrm{TNN}} + \left(\frac{3}{4}\lambda - \mathcal{P}_{\Omega^{\perp}}(\|\underline{\mathrm{G}}\|_{\infty})\right)\|\mathcal{P}_{\Omega^{\perp}}(\underline{\Delta})\|_1.\end{aligned} \tag{A16}$$

Since $\lambda \leq \frac{1}{2\sqrt{n_3}}$, it holds that:

$$\begin{aligned}&\|\underline{\mathrm{L}}_0 + \underline{\Delta}\|_{\mathrm{TNN}} + \lambda\|\underline{\mathrm{S}}_0 - \underline{\Delta}\|_1 \\ &\geq \|\underline{\mathrm{L}}_0\|_{\mathrm{TNN}} + \lambda\|\underline{\mathrm{S}}_0\|_1 + \left(\frac{3}{4} - \|\mathcal{P}_{\mathrm{T}^{\perp}}(\underline{\mathrm{G}})\|_{\mathrm{sp}}\right)\|\mathcal{P}_{\mathrm{T}^{\perp}}(\underline{\Delta})\|_{\mathrm{TNN}} + \left(\frac{3}{4}\lambda - \mathcal{P}_{\Omega^{\perp}}(\|\underline{\mathrm{G}}\|_{\infty})\right)\|\mathcal{P}_{\Omega^{\perp}}(\Delta)\|_1,\end{aligned}$$

for any perturbation $\underline{\Delta} \in \mathbb{R}^{n_1 \times n_2 \times n_3}$. □

Lemma A2. *Suppose that $\|\mathcal{P}_{\Omega}\mathcal{P}_{\mathrm{T}}\| \leq 1/2$, then for any $\underline{\mathbf{X}} = (\underline{\mathrm{X}}_i, \underline{\mathrm{X}}_s)$, we have:*

$$\|\mathcal{P}_{\Gamma}(\mathcal{P}_{\mathrm{T}} \times \mathcal{P}_{\Omega})(\underline{\mathbf{X}})\|_{\mathrm{F},\mu}^2 \geq \frac{1+\mu^2}{8}\|\mathcal{P}_{\mathrm{T}} \times \mathcal{P}_{\Omega}(\underline{\mathbf{X}})\|_{\mathrm{F}}^2. \tag{A17}$$

Proof. According to the definitions of $\mathcal{P}_{\Gamma}$ and $\mathcal{P}_{\mathrm{T}} \times \mathcal{P}_{\Omega}$, we have:

$$\mathcal{P}_{\Gamma}(\mathcal{P}_{\mathrm{T}} \times \mathcal{P}_{\Omega})(\underline{\mathbf{X}}) = \left(\frac{\mathcal{P}_{\mathrm{T}}(\underline{\mathrm{X}}_i) + \mathcal{P}_{\Omega}(\underline{\mathrm{X}}_s)}{2}, \frac{\mathcal{P}_{\mathrm{T}}(\underline{\mathrm{X}}_i) + \mathcal{P}_{\Omega}(\underline{\mathrm{X}}_s)}{2}\right). \tag{A18}$$

Then, we have:

$$\begin{aligned}\|\mathcal{P}_{\Gamma}(\mathcal{P}_{\mathrm{T}} \times \mathcal{P}_{\Omega})(\underline{\mathbf{X}})\|_{\mathrm{F},\mu}^2 &= (1+\mu^2) \cdot \frac{1}{4} \cdot \left(\|\mathcal{P}_{\mathrm{T}}(\underline{\mathrm{X}}_i)\|_{\mathrm{F}}^2 + \|\mathcal{P}_{\Omega}(\underline{\mathrm{X}}_s)\|_{\mathrm{F}}^2 + 2\langle\mathcal{P}_{\mathrm{T}}(\underline{\mathrm{X}}_i), \mathcal{P}_{\Omega}(\underline{\mathrm{X}}_s)\rangle\right) \\ &= \frac{(1+\mu^2)}{4}\left(\|\mathcal{P}_{\mathrm{T}}(\underline{\mathrm{X}}_i)\|_{\mathrm{F}}^2 + \|\mathcal{P}_{\Omega}(\underline{\mathrm{X}}_s)\|_{\mathrm{F}}^2 + 2\langle\mathcal{P}_{\Omega}\mathcal{P}_{\mathrm{T}}\mathcal{P}_{\mathrm{T}}(\underline{\mathrm{X}}_i), \mathcal{P}_{\Omega}(\underline{\mathrm{X}}_s)\rangle\right) \\ &\geq \frac{(1+\mu^2)}{4}\left(\|\mathcal{P}_{\mathrm{T}}(\underline{\mathrm{X}}_i)\|_{\mathrm{F}}^2 + \|\mathcal{P}_{\Omega}(\underline{\mathrm{X}}_s)\|_{\mathrm{F}}^2 - 2\|\mathcal{P}_{\Omega}\mathcal{P}_{\mathrm{T}}\|\|\mathcal{P}_{\mathrm{T}}(\underline{\mathrm{X}}_i)\|_{\mathrm{F}}\|\mathcal{P}_{\Omega}(\underline{\mathrm{X}}_s)\|_{\mathrm{F}}\right) \\ &\geq \frac{(1+\mu^2)}{4}\left(\|\mathcal{P}_{\mathrm{T}}(\underline{\mathrm{X}}_i)\|_{\mathrm{F}}^2 + \|\mathcal{P}_{\Omega}(\underline{\mathrm{X}}_s)\|_{\mathrm{F}}^2 - 2 \cdot \frac{1}{2}\|\mathcal{P}_{\mathrm{T}}(\underline{\mathrm{X}}_i)\|_{\mathrm{F}}\|\mathcal{P}_{\Omega}(\underline{\mathrm{X}}_s)\|_{\mathrm{F}}\right) \\ &\geq \frac{(1+\mu^2)}{4}\left(\|\mathcal{P}_{\mathrm{T}}(\underline{\mathrm{X}}_i)\|_{\mathrm{F}}^2 + \|\mathcal{P}_{\Omega}(\underline{\mathrm{X}}_s)\|_{\mathrm{F}}^2 - \frac{\|\mathcal{P}_{\mathrm{T}}(\underline{\mathrm{X}}_i)\|_{\mathrm{F}}^2 + \|\mathcal{P}_{\Omega}(\underline{\mathrm{X}}_s)\|_{\mathrm{F}}^2}{2}\right) \\ &= \frac{(1+\mu^2)}{8}\|\mathcal{P}_{\mathrm{T}} \times \mathcal{P}_{\Omega}(\underline{\mathbf{X}})\|_{\mathrm{F}}^2.\end{aligned} \tag{A19}$$

Hence completes the proof. □

Appendix A.1.2. Proof of Theorem 1

Proof. For $\underline{\mathbf{X}} = (\underline{\mathrm{L}}, \underline{\mathrm{S}})$, define $\|\underline{\mathbf{X}}\|_{\diamond} = \|\underline{\mathrm{L}}\|_{\mathrm{TNN}} + \lambda\|\underline{\mathrm{S}}\|_1$. Let $\hat{\underline{\mathbf{X}}} = (\hat{\underline{\mathrm{L}}}, \hat{\underline{\mathrm{S}}}), \underline{\mathbf{X}}^* = (\underline{\mathrm{L}}_0, \underline{\mathrm{S}}_0)$. According to the optimality of $(\hat{\underline{\mathrm{L}}}, \hat{\underline{\mathrm{S}}})$ and the feasibility of $(\underline{\mathrm{L}}_0, \underline{\mathrm{S}}_0)$, we directly have:

$$\|\hat{\underline{\mathbf{X}}}\|_{\diamond} \le \|\underline{\mathbf{X}}^*\|_{\diamond}, \tag{A20}$$

$$\|\hat{\underline{\mathrm{L}}} + \hat{\underline{\mathrm{S}}} - \underline{\mathrm{M}}\|_{\mathrm{F}} \le \delta, \tag{A21}$$

$$\|\underline{\mathrm{L}}_0 + \underline{\mathrm{S}}_0 - \underline{\mathrm{M}}\|_{\mathrm{F}} \le \delta. \tag{A22}$$

Let $\underline{\Delta}_l = \hat{\underline{\mathrm{L}}} - \underline{\mathrm{L}}_0, \underline{\Delta}_s = \hat{\underline{\mathrm{S}}} - \underline{\mathrm{S}}_0$. Then, we have:

$$\|\underline{\Delta}_l + \underline{\Delta}_s\|_{\mathrm{F}} = \|\hat{\underline{\mathrm{L}}} + \hat{\underline{\mathrm{S}}} - \underline{\mathrm{M}} - (\underline{\mathrm{L}}_0 + \underline{\mathrm{S}}_0 - \underline{\mathrm{M}})\|_{\mathrm{F}} \le \|\hat{\underline{\mathrm{L}}} + \hat{\underline{\mathrm{S}}} - \underline{\mathrm{M}}\|_{\mathrm{F}} + \|\underline{\mathrm{L}}_0 + \underline{\mathrm{S}}_0 - \underline{\mathrm{M}}\|_{\mathrm{F}} \le 2\delta. \tag{A23}$$

Define the pair of error tensors $\underline{\mathbf{\Delta}} = \hat{\underline{\mathbf{X}}} - \underline{\mathbf{X}}^* = (\underline{\Delta}_l, \underline{\Delta}_s)$. The goal is to bound $\|\underline{\mathbf{\Delta}}\|_{\mathrm{F},\mu}$.

First, we use the decomposition $\underline{\mathbf{\Delta}} = \mathcal{P}_{\Gamma}(\underline{\mathbf{\Delta}}) + \mathcal{P}_{\Gamma^\perp}(\underline{\mathbf{\Delta}})$, and let $\underline{\mathbf{\Delta}}^{\Gamma} = \mathcal{P}_{\Gamma}(\underline{\mathbf{\Delta}}) = (\underline{\Delta}_l^{\Gamma}, \underline{\Delta}_s^{\Gamma}) = (\frac{\underline{\Delta}_l + \underline{\Delta}_s}{2}, \frac{\underline{\Delta}_l + \underline{\Delta}_s}{2}), \underline{\mathbf{\Delta}}^{\Gamma^\perp} = \mathcal{P}_{\Gamma^\perp}(\underline{\mathbf{\Delta}}) = (\underline{\Delta}_l^{\Gamma^\perp}, \underline{\Delta}_s^{\Gamma^\perp}) = (\frac{\underline{\Delta}_l - \underline{\Delta}_s}{2}, \frac{\underline{\Delta}_s - \underline{\Delta}_l}{2})$ for simplicity. Then, we have:

$$\|\underline{\mathbf{\Delta}}\|_{\mathrm{F},\mu} = \|\underline{\mathbf{\Delta}}^{\Gamma} + \underline{\mathbf{\Delta}}^{\Gamma^\perp}\|_{\mathrm{F},\mu} \le \|\underline{\mathbf{\Delta}}^{\Gamma}\|_{\mathrm{F},\mu} + \|\underline{\mathbf{\Delta}}^{\Gamma^\perp}\|_{\mathrm{F},\mu}. \tag{A24}$$

Please note that $\underline{\Delta}_l^{\Gamma} = \underline{\Delta}_s^{\Gamma} = \frac{\underline{\Delta}_l + \underline{\Delta}_s}{2}$, thus $\|\underline{\mathbf{\Delta}}^{\Gamma}\|_{\mathrm{F},\mu}$ can be bounded easily as follows:

$$\|\underline{\mathbf{\Delta}}^{\Gamma}\|_{\mathrm{F},\mu} = \sqrt{\|\underline{\Delta}_l^{\Gamma}\|_{\mathrm{F}}^2 + \mu^2\|\underline{\Delta}_s^{\Gamma}\|_{\mathrm{F}}^2} = \frac{\sqrt{1+\mu^2}}{2}\|\underline{\Delta}_l + \underline{\Delta}_s\|_{\mathrm{F}} \le \delta\sqrt{1+\mu^2}. \tag{A25}$$

Then, it remains to bound $\|\underline{\mathbf{\Delta}}^{\Gamma^\perp}\|_{\mathrm{F},\mu}$. Due to the triangular inequality:

$$\|\underline{\mathbf{\Delta}}^{\Gamma^\perp}\|_{\mathrm{F},\mu} \le \|(\mathcal{P}_{\mathrm{T}} \times \mathcal{P}_{\Omega})\underline{\mathbf{\Delta}}^{\Gamma^\perp}\|_{\mathrm{F},\mu} + \|(\mathcal{P}_{\mathrm{T}^\perp} \times \mathcal{P}_{\Omega^\perp})\underline{\mathbf{\Delta}}^{\Gamma^\perp}\|_{\mathrm{F},\mu}, \tag{A26}$$

(A) bound $\|(\mathcal{P}_{\mathrm{T}^\perp} \times \mathcal{P}_{\Omega^\perp})\underline{\mathbf{\Delta}}^{\Gamma^\perp}\|_{\mathrm{F},\mu}$. According to the convexity of $\|\cdot\|_{\diamond}$ we have

$$\|\underline{\mathbf{X}}^* + \underline{\mathbf{\Delta}}\|_{\diamond} = \|\underline{\mathbf{X}}^* + \underline{\mathbf{\Delta}}^{\Gamma} + \underline{\mathbf{\Delta}}^{\Gamma^\perp}\|_{\diamond} \ge \|\underline{\mathbf{X}}^* + \underline{\mathbf{\Delta}}^{\Gamma^\perp}\|_{\diamond} - \|\underline{\mathbf{\Delta}}^{\Gamma}\|_{\diamond}. \tag{A27}$$

Using Lemma A1, we have:

$$\begin{aligned}\|\underline{\mathbf{X}}^* + \underline{\mathbf{\Delta}}^{\Gamma^\perp}\|_{\diamond} &\ge \|\underline{\mathbf{X}}^*\|_{\diamond} + (\tfrac{3}{4} - \|\mathcal{P}_{\mathrm{T}^\perp}(\underline{\mathrm{G}})\|_{\mathrm{sp}})\,\|\mathcal{P}_{\mathrm{T}^\perp}(\underline{\Delta}_l^{\Gamma^\perp})\|_{\mathrm{TNN}} + (\tfrac{3}{4}\lambda - \mathcal{P}_{\Omega^\perp}(\|\underline{\mathrm{G}}\|_{\infty}))\,\|\mathcal{P}_{\Omega^\perp}(\underline{\Delta}_s^{\Gamma^\perp})\|_1 \\ &\ge \|\underline{\mathbf{X}}^*\|_{\diamond} + \tfrac{1}{4}\|(\mathcal{P}_{\mathrm{T}^\perp} \times \mathcal{P}_{\Omega^\perp})\underline{\mathbf{\Delta}}^{\Gamma^\perp}\|_{\diamond}.\end{aligned} \tag{A28}$$

Combining Equations (A20), (A27) and (A28), we have:

$$\|\underline{\mathbf{\Delta}}^{\Gamma}\|_{\diamond} \ge \frac{1}{4}\|(\mathcal{P}_{\mathrm{T}^\perp} \times \mathcal{P}_{\Omega^\perp})\underline{\mathbf{\Delta}}^{\Gamma^\perp}\|_{\diamond} \tag{A29}$$

Then, with $\mu = \sqrt{n_3}\lambda$, we reach a bound on $\|(\mathcal{P}_{\mathsf{T}^\perp} \times \mathcal{P}_{\Omega^\perp})\underline{\mathbf{\Delta}}^{\Gamma^\perp}\|_{\mathrm{F},\mu}$ as follows:

$$\begin{aligned}
\|(\mathcal{P}_{\mathsf{T}^\perp} \times \mathcal{P}_{\Omega^\perp})\underline{\mathbf{\Delta}}^{\Gamma^\perp}\|_{\mathrm{F},\mu} &\leq \|\mathcal{P}_{\mathsf{T}^\perp}(\underline{\Delta}_t^{\Gamma^\perp})\|_{\mathrm{F}} + \mu\|\mathcal{P}_{\Omega^\perp}(\underline{\Delta}_s^{\Gamma^\perp})\|_{\mathrm{F}} \\
&\leq \sqrt{n_3}\|\mathcal{P}_{\mathsf{T}^\perp}(\underline{\Delta}_t^{\Gamma^\perp})\|_{\mathrm{TNN}} + \mu\|\mathcal{P}_{\Omega^\perp}(\underline{\Delta}_s^{\Gamma^\perp})\|_1 \\
&\leq \sqrt{n_3}\left(\|\mathcal{P}_{\mathsf{T}^\perp}(\underline{\Delta}_t^{\Gamma^\perp})\|_{\mathrm{TNN}} + \lambda\|\mathcal{P}_{\Omega^\perp}(\underline{\Delta}_s^{\Gamma^\perp})\|_1\right) \\
&\leq \sqrt{n_3}\|(\mathcal{P}_{\mathsf{T}^\perp} \times \mathcal{P}_{\Omega^\perp})\underline{\mathbf{\Delta}}^{\Gamma^\perp}\|_\diamond \\
&\leq 4\sqrt{n_3}\|\underline{\mathbf{\Delta}}^{\Gamma}\|_\diamond \\
&\leq 4\sqrt{n_3}\left(\|\underline{\Delta}_t^{\Gamma}\|_{\mathrm{TNN}} + \lambda\|\underline{\Delta}_s^{\Gamma}\|_1\right) \\
&\leq 4\sqrt{n_3}\left(\sqrt{\min\{n_1,n_2\}}\|\underline{\Delta}_t^{\Gamma}\|_{\mathrm{F}} + \lambda\sqrt{n_1n_2n_3}\|\underline{\Delta}_s^{\Gamma}\|_{\mathrm{F}}\right) \\
&= 4\sqrt{n_3}\left(\sqrt{\min\{n_1,n_2\}} + \lambda\sqrt{n_1n_2n_3}\right)\|\underline{\Delta}_t^{\Gamma}\|_{\mathrm{F}} \\
&\leq 4\sqrt{n_3}\left(\sqrt{\min\{n_1,n_2\}} + \lambda\sqrt{n_1n_2n_3}\right)\delta.
\end{aligned} \tag{A30}$$

(B) bound $\|(\mathcal{P}_{\mathsf{T}} \times \mathcal{P}_{\Omega})\underline{\mathbf{\Delta}}^{\Gamma^\perp}\|_{\mathrm{F},\mu}$. Please note that:

$$\mathcal{P}_\Gamma(\underline{\mathbf{\Delta}}^{\Gamma^\perp}) = \underline{\mathbf{0}} = \mathcal{P}_\Gamma(\mathcal{P}_{\mathsf{T}} \times \mathcal{P}_{\Omega})(\underline{\mathbf{\Delta}}^{\Gamma^\perp}) + \mathcal{P}_\Gamma(\mathcal{P}_{\mathsf{T}^\perp} \times \mathcal{P}_{\Omega^\perp})(\underline{\mathbf{\Delta}}^{\Gamma^\perp}), \tag{A31}$$

which means:

$$\|\mathcal{P}_\Gamma(\mathcal{P}_{\mathsf{T}} \times \mathcal{P}_{\Omega})(\underline{\mathbf{\Delta}}^{\Gamma^\perp})\|_{\mathrm{F},\mu} = \|\mathcal{P}_\Gamma(\mathcal{P}_{\mathsf{T}^\perp} \times \mathcal{P}_{\Omega^\perp}(\underline{\mathbf{\Delta}}^{\Gamma^\perp}))\|_{\mathrm{F},\mu} \leq \|\mathcal{P}_{\mathsf{T}^\perp} \times \mathcal{P}_{\Omega^\perp}(\underline{\mathbf{\Delta}}^{\Gamma^\perp})\|_{\mathrm{F},\mu}. \tag{A32}$$

According to Lemma A2, we have:

$$\begin{aligned}
\|(\mathcal{P}_{\mathsf{T}} \times \mathcal{P}_{\Omega})(\underline{\mathbf{\Delta}}^{\Gamma^\perp})\|_{\mathrm{F},\mu} &\leq \|\mathcal{P}_{\mathsf{T}}(\underline{\Delta}_t^{\Gamma^\perp})\|_{\mathrm{F}} + \mu\|\mathcal{P}_{\Omega}(\underline{\Delta}_t^{\Gamma^\perp})\|_{\mathrm{F}} \\
&\leq \sqrt{1+\mu^2}\sqrt{\|\mathcal{P}_{\mathsf{T}}(\underline{\Delta}_t^{\Gamma^\perp})\|_{\mathrm{F}}^2 + \|\mathcal{P}_{\Omega}(\underline{\Delta}_s^{\Gamma^\perp})\|_{\mathrm{F}}^2} \\
&= \sqrt{1+\mu^2}\|\mathcal{P}_{\mathsf{T}} \times \mathcal{P}_{\Omega}(\underline{\mathbf{\Delta}}^{\Gamma^\perp})\|_{\mathrm{F}} \\
&\leq \sqrt{1+\mu^2}\cdot\sqrt{\frac{8}{\sqrt{1+\mu^2}}}\cdot\|\mathcal{P}_\Gamma(\mathcal{P}_{\mathsf{T}} \times \mathcal{P}_{\Omega})(\underline{\mathbf{\Delta}}^{\Gamma^\perp})\|_{\mathrm{F}}.
\end{aligned} \tag{A33}$$

According to Equations (A32) and (A33), we obtain:

$$\|(\mathcal{P}_{\mathsf{T}} \times \mathcal{P}_{\Omega})(\underline{\mathbf{\Delta}}^{\Gamma^\perp})\|_{\mathrm{F},\mu} \leq 2\sqrt{2}\|(\mathcal{P}_{\mathsf{T}^\perp} \times \mathcal{P}_{\Omega^\perp})(\underline{\mathbf{\Delta}}^{\Gamma^\perp})\|_{\mathrm{F},\mu}. \tag{A34}$$

Thus, combing Equations (A24), (A25), (A30) and (A34), and setting $\mu = \sqrt{n_3}\lambda$, we obtain:

$$\|\underline{\mathbf{\Delta}}\|_{\mathrm{F},\mu} \leq \left(\sqrt{1+n_3\lambda^2} + 4(1+2\sqrt{2})\left(\sqrt{\min\{n_1,n_2\}n_3} + n_3\lambda\sqrt{n_1n_2}\right)\right)\delta. \tag{A35}$$

Since $\lambda = \frac{1}{\sqrt{\max\{n_1,n_2\}n_3}}$, we have:

$$\|\underline{\mathbf{\Delta}}\|_{\mathrm{F},\mu} \leq \left(\sqrt{1+\frac{1}{\max\{n_1,n_2\}}} + 8(1+2\sqrt{2})\sqrt{\min\{n_1,n_2\}n_3}\right)\delta, \tag{A36}$$

which indicates that:

$$\begin{aligned} \|\hat{\underline{\mathrm{L}}} - \underline{\mathrm{L}}_0\|_{\mathrm{F}} &\leq \left(\sqrt{1 + \frac{1}{\max\{n_1, n_2\}}} + 8(1+2\sqrt{2})\sqrt{\min\{n_1, n_2\} n_3} \right) \delta \\ \|\hat{\underline{\mathrm{S}}} - \underline{\mathrm{S}}_0\|_{\mathrm{F}} &\leq \left(\sqrt{1 + \max\{n_1, n_2\}} + 8(1+2\sqrt{2})\sqrt{n_1 n_2 n_3} \right) \delta. \end{aligned} \tag{A37}$$

Moreover, according to the analysis in [2], the conditions $\|\mathcal{P}_\Omega \mathcal{P}_{\mathrm{T}}\| \leq \frac{1}{2}$ and Equation (A8) in Lemma A1 hold with probability at least $1 - c_1(n_3 \max\{n_1, n_2\})^{-c_2}$, where c_1 and c_2 are positive constants.

In this way, the proof of Theorem 1 is completed. □

Appendix A.2. Proof of Theorem 2

Proof. The key idea is to rewrite Problem (29) into a standard two-block ADMM problem. For notational simplicity, let:

$$f(\mathbf{x}) = \frac{1}{2}\|\underline{\mathrm{L}} + \underline{\mathrm{S}} - \underline{\mathrm{Y}}\|_{\mathrm{F}}^2, \qquad g(\mathbf{z}) = \gamma \|\underline{\mathrm{K}}\|_{\mathrm{TNN}} + \gamma\lambda R(\underline{\mathrm{S}}),$$

where $\mathbf{x}, \mathbf{y}, \mathbf{z}$ and $\mathbf{A}$ are defined as follows:

$$\mathbf{x} = \begin{bmatrix} \mathrm{vec}(\underline{\mathrm{L}}) \\ \mathrm{vec}(\underline{\mathrm{S}}) \end{bmatrix}, \ \mathbf{y} = \begin{bmatrix} \mathrm{vec}(\underline{\mathrm{Y}}_1) \\ \mathrm{vec}(\underline{\mathrm{Y}}_2) \end{bmatrix}, \ \mathbf{z} = \begin{bmatrix} \mathrm{vec}(\underline{\mathrm{K}}) \\ \mathrm{vec}(\underline{\mathrm{R}}) \end{bmatrix}, \ \mathbf{A} = \begin{bmatrix} \mathrm{diag}(\mathrm{vec}(\underline{\mathrm{B}})) & \mathbf{0} \\ \mathbf{0} & \mathrm{diag}(\mathrm{vec}(\underline{\mathrm{B}})) \end{bmatrix},$$

and $\mathrm{vec}(\cdot)$ denotes an operation of tensor vectorization (see [40]).

It can be verified that $f(\cdot)$ and $g(\cdot)$ are closed, proper convex functions. Then, Problem (29) can be re-written as follows:

$$\begin{aligned} \min_{\mathbf{x},\mathbf{z}} \ & f(\mathbf{x}) + g(\mathbf{z}) \\ \text{s.t.} \ & \mathbf{A}\mathbf{x} - \mathbf{z} = \mathbf{0}. \end{aligned}$$

According to the convergence analysis in [48], we have:

$$\begin{aligned} &\text{objective convergence:} && \lim_{t\to\infty} f(\mathbf{x}^t) + g(\mathbf{z}^t) = f^\star + g^\star, \\ &\text{dual variable convergence:} && \lim_{t\to\infty} \mathbf{y}^t = \mathbf{y}^\star, \\ &\text{constraint convergence:} && \lim_{t\to\infty} \mathbf{A}\mathbf{x}^t - \mathbf{z}^t = \mathbf{0}, \end{aligned}$$

where $f^\star, g^\star$ are the optimal values of $f(\mathbf{x})$, $g(\mathbf{z})$, respectively. Variable $\mathbf{y}^\star$ is a dual optimal point defined as:

$$\mathbf{y}^\star = \begin{bmatrix} \mathrm{vec}(\underline{\mathrm{Y}}_1^\star) \\ \mathrm{vec}(\underline{\mathrm{Y}}_2^\star) \end{bmatrix},$$

where $(\underline{\mathrm{Y}}_1^\star, \underline{\mathrm{Y}}_2^\star)$ is the dual component of a saddle point $(\underline{\mathrm{L}}^\star, \underline{\mathrm{S}}^\star, \underline{\mathrm{K}}^\star, \underline{\mathrm{R}}^\star, \underline{\mathrm{Y}}_1^\star, \underline{\mathrm{Y}}_2^\star)$ of the unaugmented Lagrangian $L(\underline{\mathrm{L}}, \underline{\mathrm{S}}, \underline{\mathrm{K}}, \underline{\mathrm{R}}, \underline{\mathrm{Y}}_1, \underline{\mathrm{Y}}_2)$. □

Appendix A.3. Proof of Lemma 4

Proof. Let the full t-SVD of $\underline{\mathrm{X}}$ be $\underline{\mathrm{X}} = \underline{\mathrm{U}} * \underline{\Lambda} * \underline{\mathrm{V}}^\top$, where $\underline{\mathrm{U}}, \underline{\mathrm{V}} \in \mathbb{R}^{r\times r\times n_3}$ are orthogonal tensors and $\underline{\Lambda} \in \mathbb{R}^{r\times r\times n_3}$ is f-diagonal. Then:

$$\|\underline{\mathrm{X}}\|_{\mathrm{TNN}} = \left\|\overline{\underline{\mathrm{U}} * \underline{\Lambda} * \underline{\mathrm{V}}^\top}\right\|_* = \left\|\overline{\underline{\mathrm{U}}} \cdot \overline{\underline{\Lambda}} \cdot \overline{\underline{\mathrm{V}}^\top}\right\|_* = \left\|\overline{\underline{\Lambda}}\right\|_*. \tag{A38}$$

Then $\underline{Q} * \underline{X} = (\underline{Q} * \underline{U}) * \underline{\Lambda} * \underline{V}^{\top}$. Since

$$(\underline{Q} * \underline{U})^{\top} * (\underline{Q} * \underline{U}) = \underline{U}^{\top} * \underline{Q}^{\top} * \underline{Q} * \underline{U} = \underline{I}, \tag{A39}$$

we obtain that:

$$\begin{aligned} \|\underline{Q} * \underline{X}\|_{\mathrm{TNN}} &= \|\overline{\underline{Q} * \underline{X}}\|_* \\ &= \|\overline{(\underline{Q} * \underline{U}) * \underline{\Lambda} \underline{V}^{\top}}\|_* \\ &= \|\overline{(\underline{Q} * \underline{U})} \cdot \overline{\underline{\Lambda}} \cdot \overline{\underline{V}^{\top}}\|_* \\ &= \|\overline{\underline{\Lambda}}\|_*. \end{aligned} \tag{A40}$$

Thus, $\|\underline{Q} * \underline{X}\|_{\mathrm{TNN}} = \|\underline{X}\|_{\mathrm{TNN}}$. □

Appendix A.4. Proof of Theorem 3

Proof. Please note that $(\underline{Q}_* * \underline{X}_*, \underline{S}_*)$ is a feasible point of Problem (28), then we have:

$$\begin{aligned} &\frac{1}{2}\|\underline{B} \odot (\underline{L}^\star + \underline{S}^\star - \underline{M})\|_F^2 + \gamma(\|\underline{L}^\star\|_{\mathrm{TNN}} + \lambda\|\underline{S}^\star\|_1) \\ \leq &\frac{1}{2}\|\underline{B} \odot (\underline{Q}_* * \underline{X}_* + \underline{S}_* - \underline{M})\|_F^2 + \gamma(\|\underline{Q}_* * \underline{X}_*\|_{\mathrm{TNN}} + \lambda\|\underline{S}_*\|_1) \\ = &\frac{1}{2}\|\underline{B} \odot (\underline{Q}_* * \underline{X}_* + \underline{S}_* - \underline{M})\|_F^2 + \gamma(\|\underline{X}_*\|_{\mathrm{TNN}} + \lambda\|\underline{S}_*\|_1) \end{aligned} \tag{A41}$$

By the assumption that $r_{\mathrm{tubal}}(\underline{L}^\star) \leq r$, there exists a decomposition $\underline{L}^\star = \underline{Q}^\star * \underline{X}^\star$, such that $(\underline{Q}^\star, \underline{X}^\star, \underline{S}^\star)$ is also a feasible point of Problem (39).

Moreover, since $(\underline{Q}_*, \underline{X}_*, \underline{S}_*)$ is a global optimal solution to Problem (39), then we have that

$$\begin{aligned} &\frac{1}{2}\|\underline{B} \odot (\underline{Q}_* * \underline{X}_* + \underline{S}_* - \underline{M})\|_F^2 + \gamma(\|\underline{X}_*\|_{\mathrm{TNN}} + \lambda\|\underline{S}_*\|_1) \\ \leq &\frac{1}{2}\|\underline{B} \odot (\underline{Q}^\star * \underline{X}^\star + \underline{S}^\star - \underline{M})\|_F^2 + \gamma(\|\underline{X}^\star\|_{\mathrm{TNN}} + \lambda\|\underline{S}^\star\|_1). \end{aligned}$$

By $\underline{L}^\star = \underline{Q}^\star * \underline{X}^\star$, we have:

$$\|\underline{L}^\star\|_{\mathrm{TNN}} = \|\underline{Q}^\star * \underline{X}^\star\|_{\mathrm{TNN}} = \|\underline{X}^\star\|_{\mathrm{TNN}}. \tag{A42}$$

Thus, we deduce:

$$\begin{aligned} &\frac{1}{2}\|\underline{B} \odot (\underline{Q}_* * \underline{X}_* + \underline{S}_* - \underline{M})\|_F^2 + \gamma(\|\underline{X}_*\|_{\mathrm{TNN}} + \lambda\|\underline{S}_*\|_1) \\ \leq &\frac{1}{2}\|\underline{B} \odot (\underline{L}^\star + \underline{S}^\star - \underline{M})\|_F^2 + \gamma(\|\underline{L}^\star\|_{\mathrm{TNN}} + \lambda\|\underline{S}^\star\|_1). \end{aligned} \tag{A43}$$

According to Equations (A41) and (A43), we further have:

$$\begin{aligned} &\frac{1}{2}\|\underline{B} \odot (\underline{Q}_* * \underline{X}_* + \underline{S}_* - \underline{M})\|_F^2 + \gamma(\|\underline{X}_*\|_{\mathrm{TNN}} + \lambda\|\underline{S}_*\|_1) \\ \leq &\frac{1}{2}\|\underline{B} \odot (\underline{L}^\star + \underline{S}^\star - \underline{M})\|_F^2 + \gamma(\|\underline{L}^\star\|_{\mathrm{TNN}} + \lambda\|\underline{S}^\star\|_1). \end{aligned} \tag{A44}$$

In this way, $(\underline{Q}_* * \underline{X}_*, \underline{S}_*)$ is also the optimal solution to Problem (28). □

References

1. Liu, J.; Musialski, P.; Wonka, P.; Ye, J. Tensor completion for estimating missing values in visual data. *IEEE Trans. Pattern Anal. Mach. Intell.* **2013**, *35*, 208–220. [CrossRef] [PubMed]
2. Lu, C.; Feng, J.; Chen, Y.; Liu, W.; Lin, Z.; Yan, S. Tensor robust principal component analysis with a new tensor nuclear norm. *IEEE Trans. Pattern Anal. Mach. Intell.* **2019**. [CrossRef]
3. Xu, Y.; Hao, R.; Yin, W.; Su, Z. Parallel matrix factorization for low-rank tensor completion. *Inverse Probl. Imaging* **2015**, *9*, 601–624. [CrossRef]
4. Liu, Y.; Shang, F. An Efficient Matrix Factorization Method for Tensor Completion. *IEEE Signal Process. Lett.* **2013**, *20*, 307–310. [CrossRef]
5. Wang, A.; Wei, D.; Wang, B.; Jin, Z. Noisy Low-Tubal-Rank Tensor Completion Through Iterative Singular Tube Thresholding. *IEEE Access* **2018**, *6*, 35112–35128. [CrossRef]
6. Tan, H.; Feng, G.; Feng, J.; Wang, W.; Zhang, Y.J.; Li, F. A tensor-based method for missing traffic data completion. *Transp. Res. Part C* **2013**, *28*, 15–27. [CrossRef]
7. Peng, Y.; Lu, B.L. Discriminative extreme learning machine with supervised sparsity preserving for image classification. *Neurocomputing* **2017**, *261*, 242–252. [CrossRef]
8. Cichocki, A.; Mandic, D.; De Lathauwer, L.; Zhou, G.; Zhao, Q.; Caiafa, C.; Phan, H.A. Tensor decompositions for signal processing applications: From two-way to multiway component analysis. *IEEE Signal Process. Mag.* **2015**, *32*, 145–163. [CrossRef]
9. Vaswani, N.; Bouwmans, T.; Javed, S.; Narayanamurthy, P. Robust subspace learning: Robust PCA, robust subspace tracking, and robust subspace recovery. *IEEE Signal Process. Mag.* **2018**, *35*, 32–55. [CrossRef]
10. Cichocki, A.; Lee, N.; Oseledets, I.; Phan, A.H.; Zhao, Q.; Mandic, D.P. Tensor Networks for Dimensionality Reduction and Large-scale Optimization: Part 1 Low-Rank Tensor Decompositions. *Found. Trends® Mach. Learn.* **2016**, *9*, 249–429. [CrossRef]
11. Yuan, M.; Zhang, C.H. On Tensor Completion via Nuclear Norm Minimization. *Found. Comput. Math.* **2016**, *16*, 1–38. [CrossRef]
12. Candès, E.J.; Tao, T. The power of convex relaxation: Near-optimal matrix completion. *IEEE Trans. Inf. Theory* **2010**, *56*, 2053–2080. [CrossRef]
13. Hillar, C.J.; Lim, L. Most Tensor Problems Are NP-Hard. *J. ACM* **2009**, *60*, 45. [CrossRef]
14. Yuan, M.; Zhang, C.H. Incoherent Tensor Norms and Their Applications in Higher Order Tensor Completion. *IEEE Trans. Inf. Theory* **2017**, *63*, 6753–6766. [CrossRef]
15. Tomioka, R.; Suzuki, T. Convex tensor decomposition via structured schatten norm regularization. In Proceedings of the Advances in Neural Information Processing Systems, Lake Tahoe, NV, USA, 5–10 December 2013; pp. 1331–1339.
16. Semerci, O.; Hao, N.; Kilmer, M.E.; Miller, E.L. Tensor-Based Formulation and Nuclear Norm Regularization for Multienergy Computed Tomography. *IEEE Trans. Image Process.* **2014**, *23*, 1678–1693. [CrossRef]
17. Mu, C.; Huang, B.; Wright, J.; Goldfarb, D. Square Deal: Lower Bounds and Improved Relaxations for Tensor Recovery. In Proceedings of the International Conference on Machine Learning, Beijing, China, 21–26 June 2014; pp. 73–81.
18. Zhao, Q.; Meng, D.; Kong, X.; Xie, Q.; Cao, W.; Wang, Y.; Xu, Z. A Novel Sparsity Measure for Tensor Recovery. In Proceedings of the IEEE International Conference on Computer Vision, Santiago, Chile, 7–13 December 2015; pp. 271–279.
19. Wei, D.; Wang, A.; Wang, B.; Feng, X. Tensor Completion Using Spectral (k,p) -Support Norm. *IEEE Access* **2018**, *6*, 11559–11572. [CrossRef]
20. Tomioka, R.; Hayashi, K.; Kashima, H. Estimation of low-rank tensors via convex optimization. *arXiv* **2010**, arXiv:1010.0789.
21. Chretien, S.; Wei, T. Sensing tensors with Gaussian filters. *IEEE Trans. Inf. Theory* **2016**, *63*, 843–852. [CrossRef]
22. Ghadermarzy, N.; Plan, Y.; Yılmaz, Ö. Near-optimal sample complexity for convex tensor completion. *arXiv* **2017**, arXiv:1711.04965.
23. Ghadermarzy, N.; Plan, Y.; Yılmaz, Ö. Learning tensors from partial binary measurements. *arXiv* **2018**, arXiv:1804.00108.

24. Liu, Y.; Shang, F.; Fan, W.; Cheng, J.; Cheng, H. Generalized Higher-Order Orthogonal Iteration for Tensor Decomposition and Completion. In Proceedings of the Advances in Neural Information Processing Systems, Montreal, QC, Canada, 8–13 December 2014; pp. 1763–1771.
25. Zhang, Z.; Ely, G.; Aeron, S.; Hao, N.; Kilmer, M. Novel methods for multilinear data completion and de-noising based on tensor-SVD. In Proceedings of the IEEE Conference on Computer Vision and Pattern Recognition, Columbus, OH, USA, 23–28 June 2014; pp. 3842–3849.
26. Lu, C.; Feng, J.; Lin, Z.; Yan, S. Exact Low Tubal Rank Tensor Recovery from Gaussian Measurements. In Proceedings of the 28th International Joint Conference on Artificial Intelligence, Stockholm, Sweden, 13–19 July 2018; pp. 1948–1954.
27. Jiang, J.Q.; Ng, M.K. Exact Tensor Completion from Sparsely Corrupted Observations via Convex Optimization. *arXiv* **2017**, arXiv:1708.00601.
28. Xie, Y.; Tao, D.; Zhang, W.; Liu, Y.; Zhang, L.; Qu, Y. On Unifying Multi-view Self-Representations for Clustering by Tensor Multi-rank Minimization. *Int. J. Comput. Vis.* **2018**, *126*, 1157–1179. [CrossRef]
29. Ely, G.T.; Aeron, S.; Hao, N.; Kilmer, M.E. 5D seismic data completion and denoising using a novel class of tensor decompositions. *Geophysics* **2015**, *80*, V83–V95. [CrossRef]
30. Liu, X.; Aeron, S.; Aggarwal, V.; Wang, X.; Wu, M. Adaptive Sampling of RF Fingerprints for Fine-grained Indoor Localization. *IEEE Trans. Mob. Comput.* **2016**, *15*, 2411–2423. [CrossRef]
31. Wang, A.; Lai, Z.; Jin, Z. Noisy low-tubal-rank tensor completion. *Neurocomputing* **2019**, *330*, 267–279. [CrossRef]
32. Sun, W.; Chen, Y.; Huang, L.; So, H.C. Tensor Completion via Generalized Tensor Tubal Rank Minimization using General Unfolding. *IEEE Signal Process. Lett.* **2018**, *25*, 868–872. [CrossRef]
33. Kilmer, M.E.; Braman, K.; Hao, N.; Hoover, R.C. Third-order tensors as operators on matrices: A theoretical and computational framework with applications in imaging. *SIAM J. Matrix Anal. Appl.* **2013**, *34*, 148–172. [CrossRef]
34. Liu, X.Y.; Aeron, S.; Aggarwal, V.; Wang, X. Low-tubal-rank tensor completion using alternating minimization. *arXiv* **2016**, arXiv:1610.01690.
35. Liu, X.Y.; Wang, X. Fourth-order tensors with multidimensional discrete transforms. *arXiv* **2017**, arXiv:1705.01576.
36. Gu, Q.; Gui, H.; Han, J. Robust tensor decomposition with gross corruption. In Proceedings of the Advances in Neural Information Processing Systems, Montreal, QC, Canada, 8–13 December 2014; pp. 1422–1430.
37. Wang, A.; Jin, Z.; Tang, G. Robust tensor decomposition via t-SVD: Near-optimal statistical guarantee and scalable algorithms. *Signal Process.* **2020**, *167*, 107319. [CrossRef]
38. Zhang, Z.; Aeron, S. Exact Tensor Completion Using t-SVD. *IEEE Trans. Signal Process.* **2017**, *65*, 1511–1526. [CrossRef]
39. Goldfarb, D.; Qin, Z. Robust low-rank tensor recovery: Models and algorithms. *SIAM J. Matrix Anal. Appl.* **2014**, *35*, 225–253. [CrossRef]
40. Kolda, T.G.; Bader, B.W. Tensor decompositions and applications. *SIAM Rev.* **2009**, *51*, 455–500. [CrossRef]
41. Cheng, L.; Wu, Y.C.; Zhang, J.; Liu, L. Subspace identification for DOA estimation in massive/full-dimension MIMO systems: Bad data mitigation and automatic source enumeration. *IEEE Trans. Signal Process.* **2015**, *63*, 5897–5909. [CrossRef]
42. Cheng, L.; Xing, C.; Wu, Y.C. Irregular Array Manifold Aided Channel Estimation in Massive MIMO Communications. *IEEE J. Sel. Top. Signal Process.* **2019**, *13*, 974–988. [CrossRef]
43. Zhao, Q.; Zhou, G.; Zhang, L.; Cichocki, A.; Amari, S.I. Bayesian robust tensor factorization for incomplete multiway data. *IEEE Trans. Neural Networks Learn. Syst.* **2016**, *27*, 736–748. [CrossRef]
44. Zhou, Y.; Cheung, Y. Bayesian Low-Tubal-Rank Robust Tensor Factorization with Multi-Rank Determination. *IEEE Trans. Pattern Anal. Mach. Intell.* **2019**. [CrossRef]
45. Zhou, Z.; Li, X.; Wright, J.; Candes, E.; Ma, Y. Stable principal component pursuit. In Proceedings of the 2010 IEEE International Symposium on Information Theory, Austin, TX, USA, 12–18 June 2010; pp. 1518–1522.
46. Candès, E.J.; Li, X.; Ma, Y.; Wright, J. Robust principal component analysis? *J. ACM* **2011**, *58*, 11. [CrossRef]
47. Lu, C.; Feng, J.; Chen, Y.; Liu, W.; Lin, Z.; Yan, S. Tensor Robust Principal Component Analysis: Exact Recovery of Corrupted Low-Rank Tensors via Convex Optimization. In Proceedings of the IEEE Conference on Computer Vision and Pattern Recognition, Las Vegas, NV, USA, 27–30 June 2016; pp. 5249–5257.
48. Boyd, S.; Parikh, N.; Chu, E.; Peleato, B.; Eckstein, J. Distributed optimization and statistical learning via the alternating direction method of multipliers. *Found. Trends® Mach. Learn.* **2011**, *3*, 1–122. [CrossRef]

49. Peng, Y.; Lu, B.L. Robust structured sparse representation via half-quadratic optimization for face recognition. *Multimed. Tools Appl.* **2017**, *76*, 8859–8880. [CrossRef]
50. Liu, G.; Yan, S. Active subspace: Toward scalable low-rank learning. *Neural Comput.* **2012**, *24*, 3371–3394. [CrossRef] [PubMed]
51. Wang, A.; Jin, Z.; Yang, J. *A Factorization Strategy for Tensor Robust PCA*; ResearchGate: Berlin, Germany, 2019.
52. Jiang, Q.; Ng, M. Robust Low-Tubal-Rank Tensor Completion via Convex Optimization. In Proceedings of the 28th International Joint Conference on Artificial Intelligence, Macao, China, 10–16 August 2019; pp. 2649–2655.
53. Kernfeld, E.; Kilmer, M.; Aeron, S. Tensor–tensor products with invertible linear transforms. *Linear Algebra Its Appl.* **2015**, *485*, 545–570. [CrossRef]
54. Lu, C.; Peng, X.; Wei, Y. Low-Rank Tensor Completion With a New Tensor Nuclear Norm Induced by Invertible Linear Transforms. In Proceedings of the IEEE Conference on Computer Vision and Pattern Recognition, Long Beach, CA, USA, 16–20 June 2019; pp. 5996–6004.
55. Liu, X.Y.; Aeron, S.; Aggarwal, V.; Wang, X. Low-tubal-rank tensor completion using alternating minimization. In Proceedings of the SPIE Defense+ Security, Baltimore, MD, USA, 17–21 April 2016; International Society for Optics and Photonics: Bellingham, DC, USA, 2016; p. 984809.
56. Zhou, P.; Lu, C.; Lin, Z.; Zhang, C. Tensor Factorization for Low-Rank Tensor Completion. *IEEE Trans. Image Process.* **2018**, *27*, 1152–1163. [CrossRef] [PubMed]
57. Martin, C.D.; Shafer, R.; Larue, B. An Order-p Tensor Factorization with Applications in Imaging. *SIAM J. Sci. Comput.* **2013**, *35*, A474–A490. [CrossRef]
58. Wang, A.; Jin, Z. Orientation Invariant Tubal Nuclear Norms Applied to Robust Tensor Decomposition. Available online: https://www.researchgate.net/publication/329116872_Orientation_Invariant_Tubal_Nuclear_Norms_Applied_to_Robust_Tensor_Decomposition (accessed on 3 December 2019).

Article

Fingerprinting-Based Indoor Localization Using Interpolated Preprocessed CSI Phases and Bayesian Tracking

Wenxu Wang [1], Damián Marelli [1,2,*] and Minyue Fu [1,3]

1 School of Automation, Guangdong University of Technology, Guangzhou 510006, China; wangwenxu0909@126.com (W.W.); minyue.fu@newcastle.edu.au (M.F.)
2 French Argentine International Center for Information and Systems Sciences, National Scientific and Technical Research Council, 2000 Rosario, Argentina
3 School of Electrical Engineering and Computer Science, University of Newcastle, Callaghan NSW 2308, Australia
* Correspondence: damian.marelli@newcastle.edu.au

Received: 7 April 2020; Accepted: 7 May 2020; Published: 18 May 2020

Abstract: Indoor positioning using Wi-Fi signals is an economic technique. Its drawback is that multipath propagation distorts these signals, leading to an inaccurate localization. An approach to improve the positioning accuracy consists of using fingerprints based on channel state information (CSI). Following this line, we propose a new positioning method which consists of three stages. In the first stage, which is run during initialization, we build a model for the fingerprints of the environment in which we do localization. This model permits obtaining a precise interpolation of fingerprints at positions where a fingerprint measurement is not available. In the second stage, we use this model to obtain a preliminary position estimate based only on the fingerprint measured at the receiver's location. Finally, in the third stage, we combine this preliminary estimation with the dynamical model of the receiver's motion to obtain the final estimation. We compare the localization accuracy of the proposed method with other rival methods in two scenarios, namely, when fingerprints used for localization are similar to those used for initialization, and when they differ due to alterations in the environment. Our experiments show that the proposed method outperforms its rivals in both scenarios.

Keywords: indoor localization; CSI; fingerprinting; Bayesian tracking

---

1. Introduction

The global positioning system (GPS) permits solving the positioning problem in a reliable manner. However, this approach is limited to outdoor environments, since GPS signals do not reach indoor receivers. A number of techniques are available for indoor positioning systems [1,2]. These techniques include the use of ultra-wideband signals [3–5], Wi-Fi signals [6–8], bluetooth signals [9,10], radio-frequency identification [11–13], odometry measurements [14], etc. However, the accuracy of positioning systems based on these techniques is severely affected by the multipath propagation of signals. Therefore, obtaining a high-precision and reliable indoor positioning method has become the main research problem in various location-based services.

A popular technique consists of using Wi-Fi signals for indoor positioning. The main reason for doing so is the widespread and low cost of these signals in comparison with other alternatives, e.g., Zigbee, Bluetooth, ultrasound, ultra-wideband, etc. There are two main approaches for indoor positioning using Wi-Fi signals, namely, geometric mapping and fingerprinting [6]. Geometric mapping is based on the estimation of geometric parameters, such as distance or angles with respect to certain reference points [15,16]. The problem with this approach is, as we mentioned earlier, the irregularities in the signals caused by multipath effects. To go around this, the fingerprint approach builds a database of certain features obtained from the received Wi-Fi signals, depicting a unique pattern based on its location. The positioning problem then becomes one of finding the best match for the signal received at an unknown location from the information available within the database.

For an indoor positioning system, which kind of signal feature is selected as a fingerprint is critical, as it must faithfully represent its location. A popular choice for this feature is a vector of received signal strength indicators (RSSIs) from multiple Wi-Fi access points (APs) [17,18]. Then, the position is estimated as the one from the database with the highest similarity score with the fingerprint of the received signal. The disadvantage of this method is that each location is only matched to one within the database. To solve this, Youssef [19] estimates the position by building its conditional probability given the available measurements. A theoretical backing for this approach is given by the fact that RSSI measurements can be approximately considered normally distributed with negligible accuracy loss [20].

The advantages of using RSSI as fingerprints are its simplicity and low computational requirements. However, since the database only stores signal features at a finite number of points, the accuracy drops significantly at other points. This is because RSSI does not contain all the information available in the Wi-Fi signal, which could be used to improve the positioning accuracy. The whole available information appears in the channel state information (CSI). This is a complex vector whose entries represent the gain (amplitude and phase) of each subcarrier of the OFDM channel [6]. In [21], CSI was used to propose the fine-grained indoor localization (FILA) method. In this method, the fingerprint is chosen to be the square sum of CSI amplitudes. A popular choice of fingerprint consists of using the real vector formed by the amplitudes of each subcarrier [22–24]. However, the instability of CSI amplitudes limit the accuracy achievable with this kind of fingerprints. In an attempt to go around this, inter-subcarrier phase differences are used as fingerprints in [25]. A drawback of this approach is that CSI phases are highly sensitive to hardware imperfections. To cope with this, a phase calibration method was proposed in [26] to compensate for hardware effects.

Regardless of the choice of fingerprint, two approaches are available for estimating the position. The first one consists in dividing the area into regions, and using a classifier, based on machine learning techniques, to decide the region at which the receiver is located. In [27], this is done using a neural network, and in [28], a random forest model. A drawback of these methods is that the positioning accuracy is determined by the regions' size.

The second positioning approach aims at solving the above limitation by interpolating a position from those available within the fingerprint database. In [21], this is done using Pearson correlation coefficients between the measured fingerprints at the unknown location and the fingerprints from the database. In [26], this is done using a three-hidden-layer deep network, a deep autoencoder is used in [29,30], a deep convolutional neural networks in [31] and a k-nearest neighbor average in [25]. Nevertheless, a common problem with these methods is that their inaccuracy is still significant at locations outside the database.

For moving targets, the above positioning techniques are used for carrying out a Bayesian tracking task, which uses, at each sample time, all available measurements to provide a position estimate [32,33]. The case in which positioning is based on Wi-Fi signals, the information provided by these signals are often combined (fused) with certain extra information, such as infra-red motion sensors [34], inertial sensors [35], orientation sensors and landmarks [36].

In this work, a new fingerprint method for indoor positioning and Bayesian tracking is proposed, based on CSI. Following [26], we also calibrate phases. However, instead of using them as fingerprints, we use their differences with respect to the first subcarrier. We do so because these phase differences change more smoothly than absolute phases as we move along a straight line. This largely simplifies the task of interpolating fingerprints at locations outside the database. To do the interpolation, we construct a model of the room's fingerprints, i.e., a function mapping positions within the room to fingerprints. We do so using a Gaussian kernel expansion. Using this model, the positioning task is split into two steps. In the first one, we obtain a preliminary estimate of the position using only the fingerprint measured by the receiver. To this end, we use the maximum likelihood criterion. In the second stage, we combine this preliminary estimation with a dynamic model of the receiver's motion, to obtain the final estimation. We do so using a Bayesian tracking technique. The most popular among these techniques are extended Kalman filtering and particle filtering. The advantage of the former is its computational efficiency. However, since it is based on linearizing a non-linear model, it is often inaccurate to the extent that it can lead to the instability of the estimate. Particle filtering, on the other hand, can be made arbitrarily accurate. However, its computational complexity makes it often unfeasible for real-time applications. In order to obtain a reliable and numerically efficient Bayesian tracking implementation, we use a recently proposed technique called maximum likelihood Kalman filtering [37]. It uses the maximum likelihood estimate obtained in the first step, guarantees the stability of the estimate and is asymptotically optimal as the dimension of the fingerprint vector tends to infinity [37]. To summarize, the contribution of our paper is to effectively combine the use of phase information, its calibration and phase differences for fingerprints, and to apply Gaussian kernel modeling and maximum likelihood Kalman filtering for positioning. The resulting algorithm is shown, via experiments, to have more accurate position estimates, in comparison with other available algorithms, at locations outside the database.

The rest of this paper is organized as follows. In Section 2 we describe how we build fingerprints. In Section 3 we formulate the positioning problem. The proposed localization method is introduced in Section 4. Experimental results are given in Section 5 and concluding remarks in Section 6.

2. Choice of Fingerprints

2.1. Channel State Information

In an OFDM system, the signal received from a multipath channel can be described as

$$r = Cs + n, \tag{1}$$

where $r = [r_1, \cdots, r_L] \in \mathbb{C}^L$ and $s = [s_1, \cdots, s_L] \in \mathbb{C}^L$ represent the received and transmitted signal vectors, respectively, L is the number of subcarriers, n is additive noise, $C = diag(c) \in \mathbb{C}^{L\times L}$ is the channel frequency response (CFR) matrix, with $c = [c_1, \cdots, c_L]$ and c_m denoting the channel's gain at the m-th subcarrier.

We call vector c the CSI. According to the 802.11n protocol, this vector is used to recover s from r by equalizing the channel distortion caused by multipaths. Hence, the CSI c is readily available in every OFDM receiver.

2.2. Phase Information

The CSI value c_m, at each subcarrier $m = 1, \cdots, L$, is a complex number. Most CSI-based fingerprint methods choose amplitudes $|c_m|$ as their fingerprints. However, Sen et al. [38] proposed a linear calibration method for the (raw) phase information $\angle c_m$. As we show below, the calibrated phases so obtained are more stable than amplitudes, in the sense that they incur smaller changes between consecutive measurements. We summarize this calibration method below.

We assume that raw CSI measurements c have certain packet-by-packet random phase shifts due to synchronization errors at the receiver. More precisely, let $\theta_m = \angle c_m$ denote the raw phase of the m-th subcarrier and ϕ_m denote the ideal phase that we would have received if there were no random phase shifts. We then have

$$\theta_m = \phi_m + \frac{2\pi}{L\Delta t} l_m + q, \tag{2}$$

where q denotes the unknown center frequency shift and Δt denotes the delay due to both, packet detection delay (PDD) and sampling frequency offset (SFO). By l_m, we denote the m-th entry of the subcarrier index vector l, and, as above, L is the number of subcarriers. For example, in the 20 MHz bandwidth 802.11n protocol, $L = 56$ and

$$l = [-28, -27, ..., -2, -1, 1, 2, ..., 27, 28]^{\top}.$$

In order to compensate for the effects of random phase shifts, we use

$$\phi_m = \theta_m - o l_m - q, \tag{3}$$

with

$$o = \frac{\theta_L - \theta_1}{l_L - l_1},$$
$$q = \frac{1}{L} \sum_{m=1}^{L} \theta_m.$$

Figure 1 illustrates the effect of phase calibration. On the left, we see 100 raw measurements of the CSI at subcarrier $m = 1$, and on the right, the same CSI measurements with calibrated phases. We see how, after calibration, phases are more stable than amplitudes.

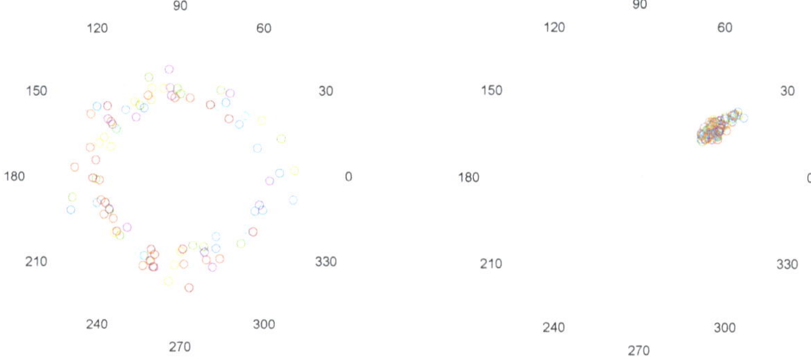

Figure 1. (**Left**) Complex values, in polar coordinates, corresponding to subcarrier $m = 1$ of 100 raw channel state information (CSI) measurements. (**Right**) The same values after phase calibration.

2.3. Phase Differences

The calibrated phase is stable at a fixed position. However, it changes drastically from one position to another. This occurs because, theoretically, for a subcarrier at 2.4 GHz, the phase goes through an entire cycle in a wavelength distance (about 12.5 cm). This is much shorter than the distance between positions used to build the fingerprint database. To go around this, instead of using the compensated phases ϕ_m, $m = 1, \cdots, L$, we use their differences $\psi_m = \phi_{m+1} - \phi_1$, $m = 1, \cdots, L-1$, with respect to the first component, and unwrap them so as to avoid that jumps $|\psi_{m+1} - \psi_m|$, $m = 1, \cdots, L-2$, are bigger than or equal to π. This is equivalent to consider that the frequency of the first subcarrier is zero, and the frequencies of each other subcarriers are 312.5 KHz, 625 KHz,..., 20 MHz. Then, the smallest wavelength among these subcarriers becomes $(300 \times 10^6 \text{ m/s})/(20 \times 10^6 \text{ Hz}) = 15$ m. Figure 2, shows how phase differences change more consistently than plain phases in a sequence of three points, arranged in a straight line, 0.5 m away from each other.

We use y_m, $m = 1, \cdots, L-1$, to denote the resulting unwrapped calibrated phase differences (UCPDs), and use them as fingerprints for indoor localization.

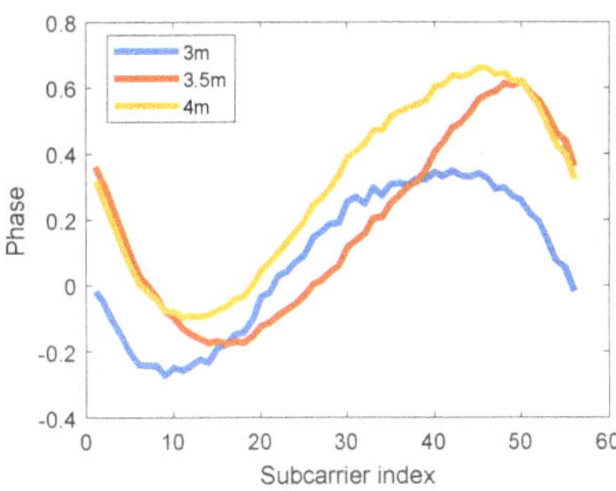

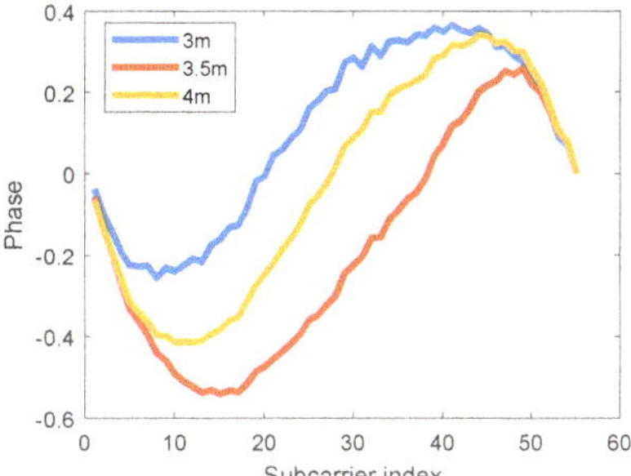

Figure 2. (**Left**) Compensated CSI phase at three adjacent positions. (**Right**) Phase differences at the same positions.

3. Problem Description

We consider a room ($\mathbb{R}^2$) having A Wi-Fi access points (APs) with B antennas. At time $t \in \mathbb{N}$, a receiver, having D antennas, is located at $x(t) \in \mathbb{R}^2$, within the room. The OFDM channel establishing the communication between each AP and each receiver's antenna, has L, subcarriers, whose frequencies are $f_m \in \mathbb{R}$, $m = 1, \cdots, L$. The UCPD of the m-th subcarrier of the channel from AP a antenna b to receiver antenna d is denoted by $y_m^{(a,b,d)}(t)$, $m = 1, \cdots, L-1$. Let $y^{(a,b,d)}(t) = \left[y_1^{(a,b,d)}(t), \cdots y_{L-1}^{(a,b,d)}(t)\right] \in \mathbb{R}^{L-1}$ be the set of all UCPDs of the same channel, $y^{(a,b)\top}(t) = \left[y^{(a,b,1)\top}(t), \cdots, y^{(a,b,D)\top}(t)\right]$ be the set of all UCPDs from AP a antenna b and $y^\top(t) = \left[y^{(1,1)\top}(t), \cdots, y^{(A,1)\top}(t), ..., y^{(1,B)\top}(t), \cdots, y^{(A,B)\top}(t)\right]$ be the set of all UCPDs. The fingerprint at $x(t)$ is given by $y(t) \in \mathbb{R}^N$, with $N = (L-1)ABD$.

We assume that $x(t)$ and $y(t)$ are generated by the following model:

$$\xi(t+1) = F\xi(t) + w(t), \tag{4}$$
$$y(t) = h\left(J\xi(t)\right) + \iota(t), \tag{5}$$

where the state $\xi(t) = \left[x^\top(t), \dot{x}^\top(t)\right]^\top$ contains the position and velocity of the receiver at time t, $F \in \mathbb{R}^{4\times 4}$ is the state-transition matrix, $w(t) \sim \mathcal{N}(0, Q)$ is the process noise, $\iota(t) \sim \mathcal{N}(0, R)$ is the measurement noise, $w(t)$ and $\iota(t)$ are white and jointly independent, and $J = [\mathcal{I}, 0]^\top$, with $\mathcal{I}$ being the identity matrix, is a selection matrix such that $J\xi(t) = x(t)$.

We assume that matrices F and Q, as well as the fingerprints $y(t)$ are known. On the other hand, matrix R and the map $\jmath : \mathbb{R}^2 \to \mathbb{R}^N$ are not known.

Problem 1. *Based on the above model and assumptions, our goal is to estimate, at each $t \in \mathbb{N}$, the receiver's location $x(t)$.*

4. Proposed Positioning Method

The proposed method is formed by three stages. The first one is called initialization. In this stage, we estimate the fingerprint model $\jmath : \mathbb{R}^2 \to \mathbb{R}^N$. The second one (static positioning), uses the estimated model and the output Equation (5) to obtain a first estimate $\check{x}(t)$ of the receiver's location $x(t)$, at time $t \in \mathbb{N}$, given the fingerprint $y(t)$. In the third stage (dynamic positioning), the estimated location $\check{x}(t)$ obtained in the second stage is combined with the dynamic motion model (4) to obtain the final estimation $\hat{x}(t)$ of $x(t)$.

4.1. Fingerprint Modeling

The goal of this stage is to obtain estimates of both, the map $h : \mathbb{R}^2 \to \mathbb{R}^N$ and the matrix $R \in \mathbb{R}^{N\times N}$. To do so, we rely on a collection of points $p_i \in \mathbb{R}^2$, $i = 1, \cdots, I$, at each of which K measurements $g_{i,k} \in \mathbb{R}^N$, $k = 1, \cdots, K$ of the fingerprint are available. The ground truth of each p_i is known.

It follows from (5) that

$$g_{i,k} = h(p_i) + \iota_{i,k}, \tag{6}$$

with $\iota_{i,k} \sim \mathcal{N}(0, Q)$ being jointly independent. Let $\bar{g}_i = \frac{1}{K}\sum_{k=1}^{K} g_{i,k}$. We then have

$$\begin{aligned} \bar{g}_i &= h(p_i) + \bar{\iota}_i, \\ \bar{\iota}_i &= \frac{1}{K}\sum_{k=1}^{K} \iota_{i,k}. \end{aligned} \tag{7}$$

Since $\bar{\iota}_i \sim \mathcal{N}\left(0, \frac{1}{K}Q\right)$, for large K we can do the following approximation

$$\bar{g}_i \approx h(p_i). \tag{8}$$

We can then obtain an estimate $\hat{\jmath}$ of $\jmath$ by approxmating $h(p)$ using a Gaussian kernel expansion of the form

$$h(p) \approx \hat{h}(p, \alpha) = \sum_{i=1}^{I} \alpha_i \exp\left(-\gamma \|p - p_i\|^2\right), \tag{9}$$

where $\alpha = [\alpha_1, \cdots, \alpha_I]^\top$. Following [39], we choose $\gamma = \frac{1}{2}I^{1/3}$. Then, using (8), we estimate α as follows

$$\hat{\alpha} = \arg\min_{\alpha} \sum_{i=1}^{I} \left[\hat{h}(p_i, \alpha) - \bar{g}_i\right]^2. \tag{10}$$

Finally, in order to obtain an estimate $\hat{R}$ of R, we do

$$\hat{R} \approx \frac{1}{I}\sum_{i=1}^{I} \frac{1}{K-1}\sum_{k=1}^{K} (g_{i,k} - \bar{g}_i)(g_{i,k} - \bar{g}_i)^\top. \tag{11}$$

4.2. Static Positioning

Knowing the estimate $\hat{\alpha}$, we can replace (5) by

$$y(t) = \hat{h}(J\xi(t), \hat{\alpha}) + \hat{\iota}(t). \tag{12}$$

with $\hat{\iota}(t) \sim \mathcal{N}(0, \hat{R})$. We can then obtain an estimate $\check{x}(t)$ of $x(t)$ using the maximum likelihood criterion, i.e.,

$$\check{x}(t) = \arg\max_{x} L_t(x), \tag{13}$$

where

$$L_t(x) = \log p\left(y(t) | J\xi(t) = x\right), \tag{14}$$

denotes the log-likelihood function and

$$p\left(y(t) | J\xi(t) = x\right) = \mathcal{N}\left(y(t); \hat{h}(x, \hat{\alpha}), \hat{R}\right). \tag{15}$$

4.3. Dynamic Positioning

We aim to use a Bayesian tracking method to estimate $\xi(t)$ based on the model (4), (12), and then obtaining $x(t)$ from the estimated value of $\xi(t)$. In principle, this could be done using the typical approaches for Bayesian tracking, namely, extended Kalman filtering (EKF) and particle filtering (PF). However, since EKF is based on linearizing the non-linear measurement Equation (5), its result is inaccurate and it often causes instability of the estimate. On the other hand, while this problem can be solved using PF, which can be made arbitrarily accurate by using enough particles, the number of required particles makes this method impractical for real-time applications. In order to obtain an accurate and numerically tractable Bayesian tracking implementation, we use a method called maximum likelihood Kalman filter [37]. We summarize it below.

We want to compute

$$\begin{aligned}\hat{\xi}(t) &= \mathcal{E}\left\{\xi(t)|Y(t)\right\} \\ &= \int \xi(t)p\left(\xi(t)|Y(t)\right)d\xi(t),\end{aligned} \tag{16}$$

where $Y(t) = \{y(1), \cdots, y(t)\}$. In order to compute $p\left(\xi(t)|Y(t)\right)$ we alternate two steps, namely, prediction and update, which are described below.

4.3.1. Prediction

In this step we assume that $\xi(t)|Y(t) \sim \mathcal{N}\left(\mu(t|t), \Sigma(t|t)\right)$, with $\mu(t|t)$ and $\Sigma(t|t)$ known. We aim to compute $\mu(t+1|t)$ and $\Sigma(t+1|t)$ such that $\xi(t+1|t)|Y(t) \sim \mathcal{N}\left(\mu(t+1|t), \Sigma(t+1|t)\right)$. We do so using a typical Kalman filtering prediction step, i.e.,

$$\begin{aligned}\mu(t+1|t) &= F\mu(t|t), \\ \Sigma(t+1|t) &= F\Sigma(t|t)F^{\top} + Q.\end{aligned} \tag{17}$$

4.3.2. Update

In this case, we assume that $\xi(t)|Y(t-1) \sim \mathcal{N}\left(\mu(t|t-1), \Sigma(t|t-1)\right)$, with $\mu(t|t-1)$ and $\Sigma(t|t-1)$ known. We aim to compute $\mu(t|t)$ and $\Sigma(t|t)$ such that $\xi(t)|Y(t) \sim \mathcal{N}\left(\mu(t|t), \Sigma(t|t)\right)$. In this case, we cannot do a typical Kalman filtering step, because the measurement equation is nonlinear. To go around this, we use the estimate $\check{x}(t)$ obtained by the static positioning method described in Section 4.2. More precisely, we compute

$$\begin{aligned}\mu(t|t) &= \Sigma(t|t)\left(\Sigma^{-1}(t|t-1)\mu(t|t-1) + \Lambda(t)\lambda(t)\right), \\ \Sigma(t|t) &= \left(\Sigma^{-1}(t|t-1) + \Lambda(t)\right)^{-1},\end{aligned} \tag{18}$$

where

$$\begin{aligned}\lambda(t) &= J^{\top}\check{x}(t), \\ \Lambda(t) &= J^{\top}\check{C}(t)J,\end{aligned}$$

and

$$\check{C}(t) = -\nabla^2 L_t\left(\check{x}(t)\right),$$

with $\nabla^2 L_t(x)$ denoting the Hessian of L_t evaluated at x (recall that $\check{x}(t)$ is computed from (13)).

4.3.3. Final Step

Using the above, at each t, we can compute $\mu(t|t)$ and $\Sigma(t|t)$ such that $\xi(t)|Y(t) \sim \mathcal{N}\left(\mu(t|t), \Sigma(t|t)\right)$. Then, using (16), we obtain

$$\hat{\xi}(t) = \mu(t|t), \tag{19}$$

and finally

$$\hat{x}(t) = J\mu(t|t). \tag{20}$$

5. Experiments

5.1. Static Positioning

In this section, we will experimentally compare our proposed method with three rival methods, namely, the FILA method [21], the DeepFi method [22] (amplitude-based fingerprinting), and the PhaseFi method [26] (Phase-based fingerprinting), described in Section 1. We use TP-Link WDR4310 routers, having the OpenWrt platform installed, and working at a package rate of 100 packets per second. We obtain CSI measurements from the Atheros CSI Tool [40]. For every AP/receiving antenna pair, this tool produces a complex vector whose entries are the amplitude and phase of certain subcarriers from the IEEE 802.11n standard. More precisely, for a 20 MHz channel, values are available for 56 subcarriers; for a 40 MHz channel, values are available for 114 subcarriers. 10 bits are used to represent both, the real and imaginary parts of each subcarrier's gain. The ground truth position of the receiver is obtained using a motion capture camera, which we believe to be sufficiently reliable for our testing purposes.

We use two experimental setups. The first one is an empty room. This setup permits evaluating the positioning performance of the different methods in the ideal situation in which fingerprints used for positioning are as similar as possible to those used during the initialization stage. The setup is depicted in Figure 3, showing the wireless access point in a corner. The figure also shows the $I = 70$ positions (marked as initialization points) used as ground truth positions for constructing the fingerprint model. In addition, there are 26 points for testing the static and dynamic positioning algorithms. Figure 4 shows the first entry of the averaged fingerprints $\bar{g}_i$ at each initialization point p_i, $i = 1, \cdots, I$, together with the value yield by the estimated model $\hat{h}(p, \hat{\alpha})$.

The second setup is an office including people as well as objects like desks, chairs and lab benches. This setup permits evaluating the robustness of the different methods, in the sense of how the performance deteriorates when fingerprints differ from those used to build the model during initialization. The setup is depicted in Figure 5, showing desks represented by rectangles, the wireless access point, the positions of the $I = 56$ initialization points and 24 testing points.

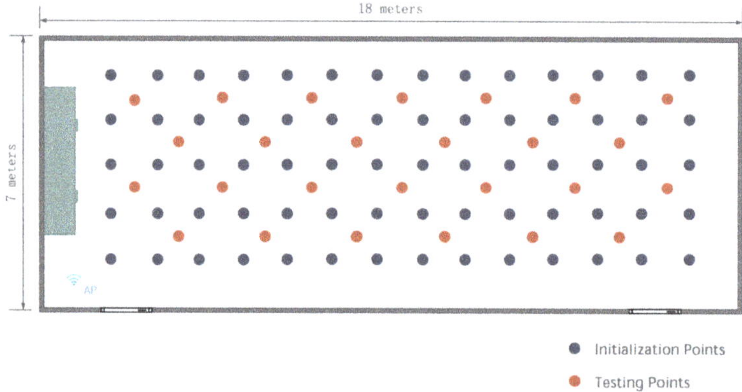

Figure 3. Empty room layout for method comparison. The wireless access point is placed in the lower-left corner.

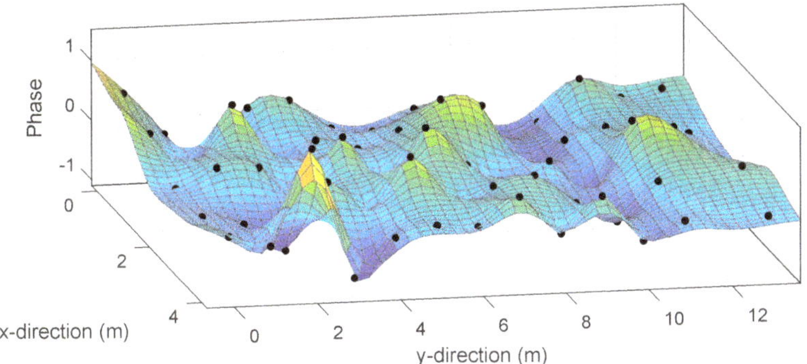

Figure 4. First entry of the averaged fingerprint $\bar{g}_i$ (dots) and the value yield by the model $\hat{h}(p, \hat{\alpha})$.

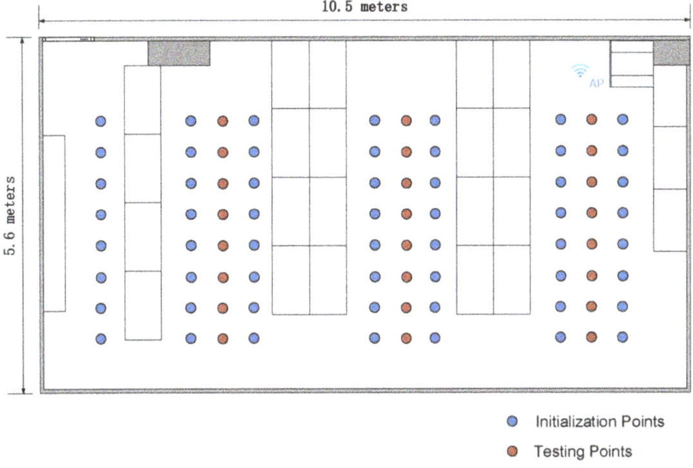

Figure 5. Office layout for method comparison. The wireless access point is placed in the upper-right corner, and rectangles represent desks.

Figures 6 and 7 show the cumulative distribution function (CDF) of the positioning error (in meters), (i.e., the relative number of points at which the localization error is smaller than each distance error in the horizontal axis) for the aforementioned four methods and the two setups. Tables 1 and 2 show the minimum and mean positioning errors also for the four methods and both setups. These results show that our proposed method significantly outperforms the FILA, DeepFi and PhaseFi methods in both scenarios.

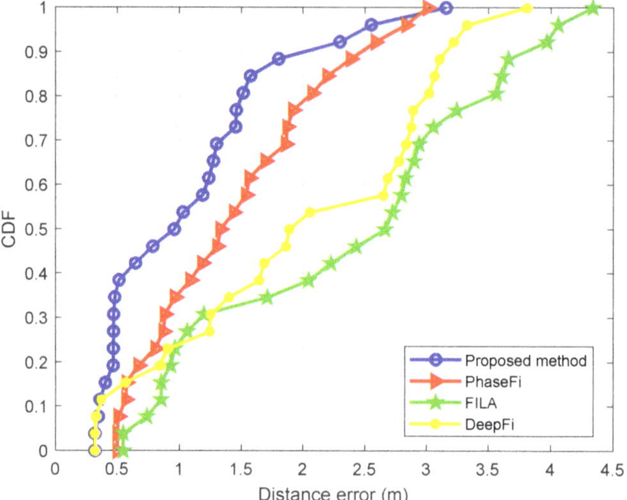

Figure 6. Positioning error CDF (empty room).

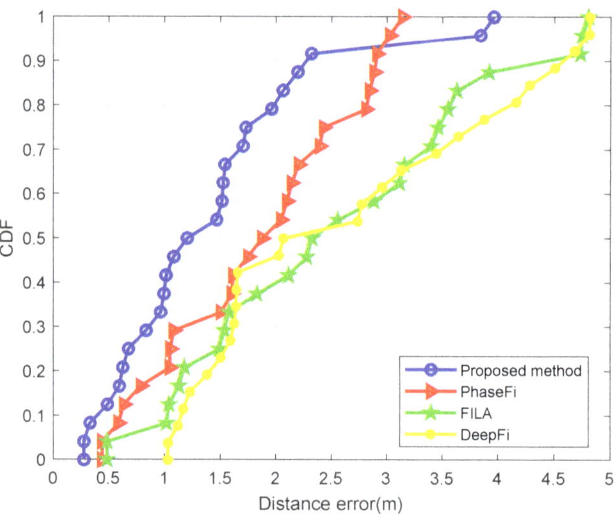

Figure 7. Positioning error CDF (office).

Table 1. Positioning error (empty room).

Methods	Mean Error [meters]	Minimum Error [meters]
Static positioning	1.0970	0.3196
PhaseFi	1.4722	0.5021
FILA	2.3825	0.5511
DeepFi	2.0283	0.3210

Table 2. Positioning error (office).

Methods	Mean Error [meters]	Minimum Error [meters]
Static positioning	1.4551	0.2763
PhaseFi	1.8722	0.4395
FILA	2.5826	0.4863
DeepFi	2.6770	1.0355

5.2. Dynamic Positioning

The dynamic positioning method makes use of the state-transition Equation (4). In order to design this state equation, we use the following model

$$\begin{aligned} p_x(t+1) &= p_x(t) + Tv_x(t) + w_x(t), \\ p_y(t+1) &= p_y(t) + Tv_y(t) + w_y(t), \\ v_x(t+1) &= v_x(t) + \iota_x(t), \\ v_y(t+1) &= v_y(t) + \iota_y(t), \end{aligned} \tag{21}$$

where p_ζ and v_ζ, $\zeta \in \{x, y\}$, denote the position and velocity, respectively, in the ζ axis. We choose $T = 1$ (seconds). We assume that w_x, w_y, ι_x and ι_y are mutually independent, $w_x \sim w_y \sim \mathcal{N}\left(0, \sigma_p^2\right)$, with $\sigma_p^2 = 10^{-4}$, and $\iota_x \sim \iota_y \sim \mathcal{N}\left(0, \sigma_\iota^2\right)$, with $\sigma_\iota^2 = 2.5 \times 10^{-4}$. We therefore have

$$F = \begin{bmatrix} 1 & 0 & T & 0 \\ 0 & 1 & 0 & T \\ 0 & 0 & 1 & 0 \\ 0 & 0 & 0 & 1 \end{bmatrix}, \text{ and } Q = \operatorname{diag}\left(\sigma_p^2, \sigma_p^2, \sigma_\iota^2, \sigma_\iota^2\right).$$

To evaluate this performance, we moved the receiver over a trajectory that is well modeled by this equation. This trajectory is shown in red in Figures 8 and 9. Figure 8 shows the trajectory estimated by our static positioning method. Figure 9 shows the one estimated using a particle filter [41], and the one estimated by our proposed dynamic positioning method based on the maximum likelihood Kalman filter. Table 3 shows the maximum tracking errors, as well as their means, given by the three methods. We see that the proposed dynamic positioning method offers a significant accuracy improvement over its static counterpart, as well as over particle filtering.

Table 3. Tracking Result.

Methods	Mean Error [meters]	Maximum Error [meters]
Static positioning	0.9879	2.4325
Particle filter	1.6475	2.8131
Dynamic positioning	0.4602	1.0706

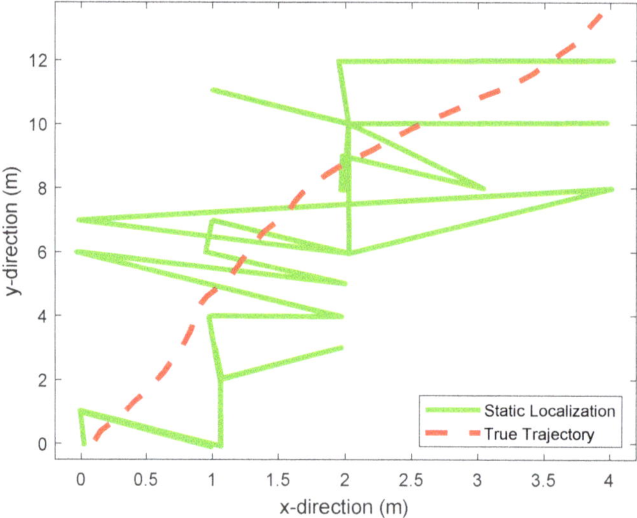

Figure 8. Static positioning result.

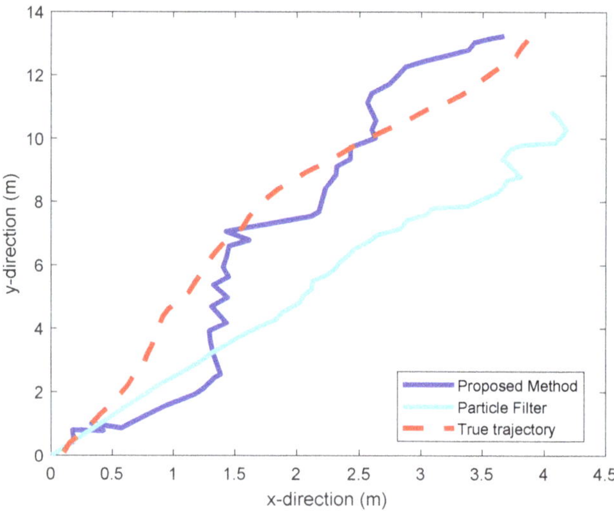

Figure 9. Comparison of the dynamic positioning results yield by particle filtering and our proposed dynamic positioning method.

6. Conclusions

We proposed a new fingerprint-based method for indoor localization based on CSI. The proposed method uses unwrapped calibrated phase differences as fingerprints, and consists of three stages. In the first one (fingerprint modeling), a model of the room's fingerprints is constructed. In the second stage (static positioning) we use this model, together with the maximum likelihood criterion, to obtain a preliminary position estimate based only on the fingerprint measured at the receiver. Finally, in the third stage (dynamic positioning), we use a maximum likelihood Kalman filter to combine this preliminary estimate with a dynamic model for the receiver's motion, to obtain the final estimate. We present experimental results comparing the localization accuracy of the proposed method with that of other rival methods. This is done in two scenarios, namely, in the ideal one in which fingerprints used for localization are similar to those used for initialization, and in a more realistic scenario, in which these fingerprints differ due to alterations in the environment. Our experiments show that our method is more accurate than its rivals in both situations.

The proposed three-stage technique also applies to other types of measurements. A future research direction is to find out how much improvement this technique brings to measurements using Bluetooth and ultra-wideband signals. How to fuse different types of measurements, and how much additional improvement this may bring, will be of interest too. The maximum likelihood Kalman filter method that we use in the dynamic positioning stage is only valid if the motion model of the receiver is linear and Gaussian. However, non-linear motion models are sometimes used, particularly in robotics. An extension of the proposed method consists in modifying the dynamic localization stage so as to handle these kinds of non-linear models. While this is out of the scope of the present work, it is a future research direction as well.

Author Contributions: Conceptualization, D.M. and M.F.; methodology, D.M.; software, W.W.; validation, W.W.; formal analysis, D.M.; investigation, W.W. and D.M.; resources, M.F.; data curation, W.W.; writing—original draft preparation, W.W.; writing—review and editing, D.M. and M.F.; visualization, W.W. and M.F.; supervision, D.M. and M.F.; project administration, M.F.; funding acquisition, M.F. All authors have read and agreed to the published version of the manuscript.

Funding: This research was funded by the National Natural Science Foundation of China grant numbers 61633014, 61803101 and U1701264, Guangdong Basic and Applied Basic Research Foundation grant number 2020A1515011505, and the Argentinean Agency for Scientific and Technological Promotion grant number PICT-201-0985.

Conflicts of Interest: The authors declare no conflict of interest.

References

1. Bensky, A. *Wireless Positioning Technologies and Applications*; Artech House: Norwood, MA, USA, 2016.
2. Liu, H.; Darabi, H.; Banerjee, P.; Liu, J. Survey of wireless indoor positioning techniques and systems. *IEEE Trans. Syst. Man Cybern. Part C Appl. Rev.* **2007**, *37*, 1067–1080. [CrossRef]
3. Sahinoglu, Z.; Gezici, S.; Guvenc, I. *Ultra-Wideband Positioning Systems*; Cambridge: New York, NY, USA, 2008.
4. Shi, G.; Ming, Y. *Survey of Indoor Positioning Systems Based on Ultra-Wideband (UWB) Technology*; Springer: Berlin, Germany, 2016; pp. 1269–1278.
5. Alarifi, A.; Al-Salman, A.; Alsaleh, M.; Alnafessah, A.; Al-Hadhrami, S.; Al-Ammar, M.A.; Al-Khalifa, H.S. Ultra wideband indoor positioning technologies: Analysis and recent advances. *Sensors* **2016**, *16*, 707. [CrossRef] [PubMed]
6. Zheng, Y.; Zhou, Z.; Liu, Y. From RSSI to CSI:Indoor localization via channel response. *ACM Comput. Surv.* **2013**, *46*, 1–32.
7. Ma, R.; Guo, Q.; Hu, C.; Xue, J. An improved WiFi indoor positioning algorithm by weighted fusion. *Sensors* **2015**, *15*, 21824–21843. [CrossRef]
8. Liu, H.H.; Liu, C. Implementation of Wi-Fi signal sampling on an android smartphone for indoor positioning systems. *Sensors* **2018**, *18*, 3. [CrossRef]
9. Feldmann, S.; Kyamakya, K.; Zapater, A.; Lue, Z. An indoor bluetooth-based positioning system: Concept, implementation and experimental evaluation. In Proceedings of the International Conference on Wireless Networks, Las Vegas, NV, USA, 23–26 June 2003; Volume 272.
10. Li, X.; Wang, J.; Liu, C. A Bluetooth/PDR integration algorithm for an indoor positioning system. *Sensors* **2015**, *15*, 24862–24885. [CrossRef]
11. Chon, H.D.; Jun, S.; Jung, H.; An, W. Using RFID for accurate positioning. *Positioning* **2004**, *3*, 32–39. [CrossRef]

12. Jechlitschek, C. A survey paper on Radio Frequency Identification (RFID) trends. *Radio Freq. IDentif.* **2010**, *100*, 765–768.
13. Seco, F.; Jiménez, A.R. Smartphone-based cooperative indoor localization with RFID technology. *Sensors* **2018**, *18*, 266. [CrossRef]
14. Nguyen, T.M.; Qiu, Z.; Cao, M.; Nguyen, T.H.; Xie, L. Single landmark distance-based navigation. *IEEE Trans. Control Syst. Technol.* **2019**, 1–8. [CrossRef]
15. Kotaru, M.; Joshi, K.; Bharadia, D.; Katti, S. SpotFi:Decimeter Level Localization Using WiFi. *ACM Sigcomm Comput. Commun. Rev.* **2015**, *45*, 269–282. [CrossRef]
16. Tian, Z.; Li, Z.; Zhou, M.; Jin, Y.; Wu, Z. PILA: Sub-meter localization using CSI from commodity Wi-Fi devices. *Sensors* **2016**, *16*, 1664. [CrossRef] [PubMed]
17. Bahl, P.; Padmanabhan, V.N. RADAR: An in-building RF-based user location and tracking system. In Proceedings of the IEEE INFOCOM 2000: Conference on Computer Communications, Nineteenth Annual Joint Conference of the IEEE Computer and Communications Societies (Cat. No.00CH37064), Tel Aviv, Israel, 26–30 March 2000.
18. Jan, S.S.; Yeh, S.J.; Liu, Y.W. Received signal strength database interpolation by Kriging for a Wi-Fi indoor positioning system. *Sensors* **2015**, *15*, 21377–21393. [CrossRef]
19. Youssef, M.; Agrawala, A. The Horus WLAN location determination system. In Proceedings of the 3rd International Conference on Mobile Systems, Applications, and Services, Seattle, WA, USA, 6–8 June 2005; pp. 205–218.
20. Marelli, D.; Fu, M. Asymptotic Properties of Statistical Estimators using Multivariate Chi-squared Measurements. *Digit. Signal Process.* **2020**, submitted.
21. Wu, K.; Jiang, X.; Yi, Y.; Chen, D.; Luo, X.; Ni, L.M. CSI-Based Indoor Localization. *IEEE Trans. Parallel Distrib. Syst.* **2013**, *24*, 1300–1309. [CrossRef]
22. Wang, X.; Gao, L.; Mao, S.; Pandey, S. DeepFi: Deep learning for indoor fingerprinting using channel state information. In Proceedings of the IEEE Wireless Communications & Networking Conference (WCNC), New Orleans, LA, USA, 9–12 March 2015.
23. Song, Q.; Guo, S.; Liu, X.; Yang, Y. CSI amplitude fingerprinting-based NB-IoT indoor localization. *IEEE Internet Things J.* **2017**, *5*, 1494–1504. [CrossRef]
24. Wang, W.; Marelli, D.; Fu, M. A statistical CSI model for indoor positioning using fingerprinting. In Proceedings of the 2019 Chinese Control Conference (CCC), Guangzhou, China, 27–30 July 2019; pp. 3630–3633.
25. Chapre, Y.; Ignjatovic, A.; Seneviratne, A.; Jha, S. CSI-MIMO: Indoor Wi-Fi fingerprinting system. In Proceedings of the 39th Annual IEEE Conference on Local Computer Networks, Edmonton, AB, Canada, 8–11 September 2014.
26. Wang, X.; Gao, L.; Mao, S. CSI Phase Fingerprinting for Indoor Localization with a Deep Learning Approach. *IEEE Internet Things J.* **2017**, *3*, 1113–1123. [CrossRef]
27. Bernas, M.; Płaczek, B. Fully connected neural networks ensemble with signal strength clustering for indoor localization in wireless sensor networks. *Int. J. Distrib. Sens. Netw.* **2015**, *11*, 403242. [CrossRef]
28. Wang, Y.; Xiu, C.; Zhang, X.; Yang, D. WiFi indoor localization with CSI fingerprinting-based random forest. *Sensors* **2018**, *18*, 2869. [CrossRef]
29. Wang, X.; Gao, L.; Mao, S. BiLoc: Bi-modal deep learning for indoor localization with commodity 5GHz WiFi. *IEEE Access* **2017**, *5*, 4209–4220. [CrossRef]
30. Khatab, Z.E.; Hajihoseini, A.; Ghorashi, S.A. A fingerprint method for indoor localization using autoencoder based deep extreme learning machine. *IEEE Sens. Lett.* **2017**, *2*, 1–4. [CrossRef]
31. Wang, X.; Wang, X.; Mao, S. CiFi: Deep convolutional neural networks for indoor localization with 5 GHz Wi-Fi. In Proceedings of the 2017 IEEE International Conference on Communications (ICC), Paris, France, 21–25 May 2017; pp. 1–6.
32. Figueiras, J.; Frattasi, S. *Mobile Positioning and Tracking*; Wiley: Hoboken, NJ, USA, 2016.
33. Dardari, D.; Closas, P.; Djuric, P.M. Indoor Tracking: Theory, Methods, and Technologies. *IEEE Trans. Veh. Technol.* **2015**, *64*, 1263–1278. [CrossRef]
34. Van Der Merwe, R.; Wan, E. Sigma-Point Kalman Filters for Probabilistic Inference in Dynamic State-Space Models. Ph.D. Thesis, OGI School of Science & Engineering at OHSU, Portland, OR, USA, 2004.
35. Leppäkoski, H.; Collin, J.; Takala, J. Pedestrian navigation based on inertial sensors, indoor map, and WLAN signals. *J. Signal Process. Syst.* **2013**, *71*, 287–296. [CrossRef]

36. Chen, Z.; Zou, H.; Jiang, H.; Zhu, Q.; Soh, Y.C.; Xie, L. Fusion of WiFi, smartphone sensors and landmarks using the Kalman filter for indoor localization. *Sensors* **2015**, *15*, 715–732. [CrossRef] [PubMed]
37. Marelli, D.; Fu, M.; Ninness, B. Asymptotic Optimality of the Maximum-Likelihood Kalman Filter for Bayesian Tracking With Multiple Nonlinear Sensors. *IEEE Trans. Signal Process.* **2015**, *63*, 4502–4515. [CrossRef]
38. Sen, S.; Radunovic, B.; Choudhury, R.R.; Minka, T. You are facing the Mona Lisa: Spot localization using PHY layer information. In Proceedings of the 10th International Conference on Mobile Systems, Applications, and Services, Low Wood Bay, Lake District, UK, 25–29 June 2012.
39. Hardle, W. *Kernel Density Estimation*; Springer-Verlag New York Inc.: New York, NY, USA, 1991.
40. Xie, Y.; Li, Z.; Li, M. Precise power delay profiling with commodity Wi-Fi. *IEEE Trans. Mob. Comput.* **2018**, *18*, 1342–1355. [CrossRef]
41. Arulampalam, M.S.; Maskell, S.; Gordon, N.; Clapp, T. A tutorial on particle filters for online nonlinear/non-Gaussian Bayesian tracking. *IEEE Trans. Signal Process.* **2002**, *50*, 174–188. [CrossRef]

Article

An Automatic Sleep Stage Classification Algorithm Using Improved Model Based Essence Features

Huaming Shen [1,*], Feng Ran [1], Meihua Xu [1], Allon Guez [2], Ang Li [1] and Aiying Guo [1]

1 School of Mechatronics Engineering and Automation, Shanghai University, Shanghai 200444, China; ranfeng@shu.edu.cn (F.R.); mhxu@shu.edu.cn (M.X.); shulivia@shu.edu.cn (A.L.); gayshh@shu.edu.cn (A.G.)
2 Faculty of Biomedical Engineering, Drexel University, Philadelphia, PA 19104, USA; guezal@drexel.edu
* Correspondence: shm2016@shu.edu.cn

Received: 22 July 2020 ; Accepted: 13 August 2020; Published: 19 August 2020

Abstract: The automatic sleep stage classification technique can facilitate the diagnosis of sleep disorders and release the medical expert from labor-consumption work. In this paper, novel improved model based essence features (IMBEFs) were proposed combining locality energy (LE) and dual state space models (DSSMs) for automatic sleep stage detection on single-channel electroencephalograph (EEG) signals. Firstly, each EEG epoch is decomposed into low-level sub-bands (LSBs) and high-level sub-bands (HSBs) by wavelet packet decomposition (WPD), separately. Then, the DSSMs are estimated by the LSBs and the LE calculation is carried out on HSBs. Thirdly, the IMBEFs extracted from the DSSM and LE are fed into the appropriate classifier for sleep stage classification. The performance of the proposed method was evaluated on three public sleep databases. The experimental results show that under the Rechtschaffen's and Kale's (R&K) standard, the sleep stage classification accuracies of six classes on the Sleep EDF database and the Dreams Subjects database are 92.04% and 78.92%, respectively. Under the American Academy of Sleep Medicine (AASM) standard, the classification accuracies of five classes in the Dreams Subjects database and the ISRUC database reached 79.90% and 81.65%. The proposed method can be used for reliable sleep stage classification with high accuracy compared with state-of-the-art methods.

Keywords: EEG; sleep stage; wavelet packet; state space model

---

1. Introduction

Automatic sleep stage classification is an important research focus due to its importance for the study of sleep related disorders. There are currently two classification criteria for sleep stages. According to Rechtschaffen's and Kale's (R&K) recommendations, sleep stages can be divided into six stages: The Awake stage (Awa), rapid Eye Movement stage (REM), Sleep stage 1 (S1), Sleep stage 2 (S2), Sleep stage 3 (S3), Sleep stage 4 (S4) [1]. Another sleep stage classification standard was provided by the AASM. In this standard, there are five sleep stages: Awa, N1 (S1), N2 (S2), N3 (the merging of stages S3 and S4) and REM [2]. Usually, the detection of each sleep stage requires manual marking by professionals, which requires a lot of work and may produce erroneous markings. Therefore, it is imperative to study the method for automatic sleep stage classification.

According to the characteristics of the adopted features, currently commonly used automatic detection methods can be divided into the following two categories. The first is the method based on statistical features (such as spectral energy) extracted from the one-dimensional EEG signal. The other is the implicit features, which can be obtained by training deep-learning based classifiers. Hassan et al. computed various spectral features by Tunable-Q factor wavelet transform (TQWT) on sleep-EEG signal segments [3]. With the random forest classifier, they achieved accuracies of 90.38%, 91.50%, 92.11%, 94.80%, 97.50% for 6-stage to 2-stage classification of sleep states on the Sleep-EDF database. Diykh et al. adopted different structural and spectral attributes extracted from weighted undirected networks to automatically classify the sleep stages [4]. Kang et al. present a statistical framework to estimate whole-night sleep states in patients with obstructive sleep apnea (OSA)—the most common sleep disorder [5]. In this framework, they extracted 11 spectral features from 60903 epochs to estimate per-night sleep stages with a 5-state hidden Markov model. Abdulla et al. used graph modularity of EEG segments as the features to feed an ensemble classifier which achieved the accuracy of 93.1% with 20265 epochs from Sleep EDF database [6].

In [7], Ghimatgar et al. constructed a features pool by the relevance and redundancy analysis on the sleep EEG epochs. With a random forest classifier and a Hidden Markov Model, this method was evaluated on three public sleep EEG database scored according to R&K and AASM guidelines. They achieved overall accuracies in the range of (79.4–87.4%) and (77.6–80.4%) for six-stage (R&K) and five-stage (AASM) classification, respectively. Taran et al. proposed an optimized flexible analytic wavelet transform (OFAWT) to decompose EEG signals into band-limited basis or sub-bands (SBs) [8]. The experimental results yields classification accuracies for the classification of six to two sleep stages 96.03%, 96.39%, 96.48%, 97.56% and 99.36%, respectively. Sharma et al. computed the discriminatory features namely fuzzy entropy and log energy by the wavelet decomposition coefficients [9]. This approach yielded an accuracy of 91.5% and 88.5% for six-class classification task using small and large datasets, respectively. Hassan et al. extracted various statistical moment based features decomposed by the Empirical Mode Decomposition (EMD) and achieved a good performance on a small database [10]. They also decomposed EEG signal segments using Ensemble Empirical Mode Decomposition (EEMD) to extract various statistical moment based features and achieved 88.07%, 83.49%, 92.66%, 94.23% and 98.15% for 6-state to 2-state classification of sleep stages on Sleep-EDF database [11]. Sharma et al. adopted the Poincare plot descriptors and statistical measures which are calculated by the discrete energy separation algorithm (DESA) as the features [12]. Moreover, the classification accuracy of the two to six categories on 15136 epochs from the Sleep-EDF database was 98.02%, 94.66%, 92.29%, 91.13% and 90.02%, respectively.

Besides the conventional features extraction method, some researchers choose the convolutional neural network (CNN) to classify sleep stages with the time–frequency images which are converted by one-dimensional EEG signals. Zhang et al. converted EEG data to a time–frequency representation via Hilbert–Huang transform and employed an orthogonal convolutional neural network (OCNN) as the classifier [13]. They achieved a total classification accuracy of 88.4% and 87.6% on two public datasets, respectively. Similarly, Xu et al. employed multiple CNN on multi-channel EEG signals to classify the sleep stages [14]. Mousavi [15] directly fed the raw EEG signals to a deep CNN with nine layers followed by two fully connected layers, without involving feature extraction and selection. This method achieved the accuracy of 98.10%, 96.86%, 93.11%, 92.95%, 93.55% for two to six class classification. Long short-term memory (LSTM) is an artificial recurrent neural network (RNN) architecture used in the field of deep learning. It can not only process single data points (such as images), but also entire sequences of data (such as speech or EEG signal). Korkalainen et al. used a combined convolutional and LSTM neural network on the public database and achieved sleep staging accuracy of 83.7% with a single frontal EEG channel [16]. Michielli et al. proposed a novel cascaded RNN architecture based on LSTM for automated scoring of sleep stages on single-channel EEG signals [17]. The network performed four and two classes classification with a classification rate of 90.8% and 83.6%, respectively.

Most of the existing studies only adopted a few epochs or a single database when evaluating the performance of these method and some do not use the k-fold cross-validation, which will cause large fluctuations in the experimental results. Therefore, although the published researches have achieved positive results in automatic sleep stage classification, there is still a need for further validation and improvements to the existing methods. In this paper we proposed a novel IMBEFs extracted from LE and DSSM for automatically detecting the sleep stages with a high degree of accuracy. LE and DSSM are estimated from the two sets coefficients of LSBs and HSBs. The two sets coefficients are coming from the WPD of the sleep EEG epoch based on two wavelet bases separately. After comparing with various kinds of classifiers, the Bagged Trees was finally selected as the suitable classifier for this method to identify the sleep stages. In addition, experiments are conducted on three public sleep databases and the results are compared with state of the art published work in order to fully evaluate and validate the performance of the proposed method.

The paper is organized as follows: In Section 2, the experimental material and methodology of the proposed method are descripted in detail. Section 3 resents the experimental results. In Section 4, the results and findings of this paper are discussed. The conclusions of the paper are drawn in Section 5.

2. Materials and Methods

2.1. Sleep State Classes

According to the AASM and R&K standards, the classes of sleep stages can be divided into two to six classes. Moreover, under the AASM standard, it can be divided into two to five classes. The difference is that the N3 stage of AASM includes the S3 and S4 stages of the R&K standard. The detailed description of classes considered in this work are shown in Tables 1 and 2.

Table 1. The class description considered in this work under the Rechtschaffen's and Kale's (R&K) standard.

Classes	6 Classes	5 Classes	4 Classes	3 Classes	2 Classes
Stages	Awa vs. REM vs. S1 vs. S2 vs. S3 vs. S4	Awa vs. REM vs. S1 vs. S2 vs. S3, S4	Awa vs. REM vs. S1, S2 vs. S3, S4	Awa vs. REM vs. NREM (S1, S2, S3, S4)	Awa vs. Asleep (REM, S1, S2, S3, S4)

Table 2. The class description considered in this work under the American Academy of Sleep Medicine (AASM) standard.

Classes	5 Classes	4 Classes	3 Classes	2 Classes
Stages	Awa vs. REM vs. N1 vs. N2 vs. S3, S4	Awa vs. REM vs. N1, N2 vs. N3	Awa vs. REM vs. NREM (N1, N2, N3)	Awa vs. Asleep (REM, N1, N2, N3)

2.2. Datasets

2.2.1. Sleep EDF (S-EDF) Database

The S-EDF database have 197 whole-night Polysomnography (PSG) sleep recordings, containing EEG, EOG, chin EMG and event markers [18,19]. All the Hypnograms (sleep patterns) were manually scored by well-trained technicians according to the R&K criteria. In this study, 34 EEG recordings from 26 subjects aged 25 to 96 years are randomly selected.

2.2.2. DREAMS Subjects (DRMS) Database

The DRMS Database consists of 20 whole-night PSG recordings coming from healthy subjects, annotated in sleep stages according to both the R&K criteria and the new standard of the AASM [20]. Data collected were acquired in a sleep laboratory of a Belgium hospital using a digital 32-channel polygraph (BrainnetTM System of MEDATEC, Brussels, Belgium). The sampling frequency was 200 Hz.

2.2.3. ISRUC(Subgroup 3, ISRUC3) Database

The ISRUC3 database is the third subgroup of ISRUC database [21]. The data were obtained from human adults, including healthy subjects, subjects with sleep disorders and subjects under the effect of sleep medication. Each recording was randomly selected between PSG recordings that were acquired by the Sleep Medicine Centre of the Hospital of Coimbra University (CHUC).

The S-EDF database was only labeled under the R&K criteria. Moreover, the ISRUC3 database was only labeled by the AASM criteria. The DRMS database was not only labeled by R&K criteria but also the AASM criteria. The annotations of S-EDF database and DRMS database were produced visually by a single expert. The ISRUC3 database was scored by two experts and the label made by the second expert was used in this paper. The Pz-Oz channel of the S-EDF database is used according to the recommendations of various studies [3–7]. At the same time, for the DRMS database, as the researches [9–12] recommended, the Cz-A1 channel was adopted in this work. Moreover, for the ISRUC database, the C3-A2 channel is the best choice [7]. Table 3 lists the detailed information of the above three databases.

Table 3. The specification of the electroencephalograph (EEG) databases included in this study.

Scoring Manual	R&K Criteria		AASM Criteria	
dataset name	S-EDF database	DRMS database	DRMS database	ISRUC3 database
Epoch length(Seconds)	30	20	30	20
Number of subjects	26	20	20	10
Recoding Files	34	20	20	10
Age	25–96	20–65	20–65	30–58
Gender(male-female)	17–17	4–16	4–16	9–1
Sampling frequency (Hz)	100	200	200	200
EEG channel	Pz–Oz	Cz–A1	Cz–A1	C3–A2
Stage	Number of epochs			
Awa	7,3835	5601	3559	1702
REM	6744	4555	3019	1238
S1(N1)	3017	1788	1480	1123
S2(N2)	1,7249	1,3274	8251	2850
S3(N3)	2288	2112	3956	1976
S4	1510	3071	–	–
Total Number of Epochs	10,4643	3,0401	2,0265	8889

2.3. Method

Figure 1 shows the schematic outline of the proposed IMBEFs based sleep statge classification algorithm comprising preprocessing, wavelet package decomposition, locality energy calculation, state space models estimation, features extraction, classifier training and performance evaluation.

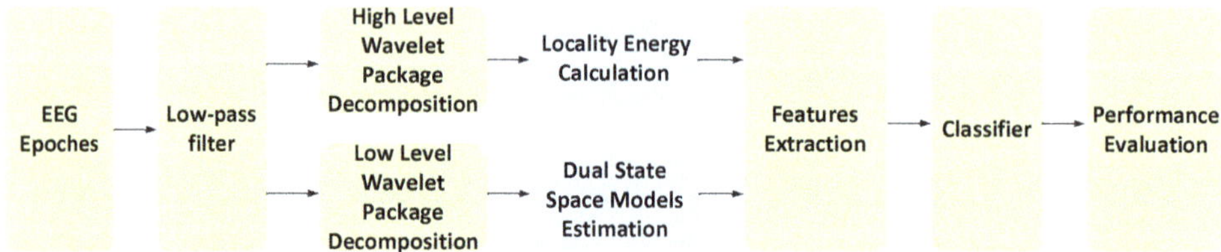

Figure 1. A schematic outline of the proposed improved model based essence features (IMBEFs) based sleep stage classification algorithm.

2.3.1. EEG Data Preprocessing

Firstly, all the single-channel data will be extracted by the Matlab and EEGLAB [22] tools from the three database described previously. According to the prior work [5–11], the 0–35 Hz low pass filter can be used to eject the most of artifact. Once the dataset is filtered, it will be exported as one-dimensional vector without time information and saved as txt file which also can be denoted as the Formula (1).

$$\mathbf{X} = [x_1, x_2, \ldots, x_k, \ldots, x_M], k \in [1, M], x_k \in \mathbb{R} \tag{1}$$

where $\mathbf{X}$ is the vector containing the sampled EEG x_k and where M is the length of vector.

Furthermore, we use a window of length j to divide the full data $\mathbf{X}$ across time without overlap. That is $\mathbf{X}$ is converted into $[\mathbf{X}_1, \mathbf{X}_2, \ldots, \mathbf{X}_i, \ldots, \mathbf{X}_L]^T$ which can be described as (2).

$$\begin{bmatrix} \mathbf{X}_1 \\ \mathbf{X}_2 \\ \vdots \\ \mathbf{X}_i \\ \vdots \\ \mathbf{X}_L \end{bmatrix} = \begin{bmatrix} x_1 & x_2 & \cdots & x_j \\ x_{j+1} & x_{j+2} & \cdots & x_{2j} \\ \vdots & \vdots & \cdots & \vdots \\ x_{(i-1)j+1} & x_{(i-1)j+2} & \cdots & x_{i\times j} \\ \vdots & \vdots & \vdots & \vdots \\ x_{(L-1)j+1} & x_{(L-1)j+2} & \cdots & x_{L\times j} \end{bmatrix}$$
$$L = \left\lfloor \frac{M}{j} \right\rfloor, i \in [1, L] \tag{2}$$

where $j = T_e \times F_s$. The T_e is the length of each epoch. Moreover, the F_s is the sampling frequency of the database. For the S-EDF database, the $T_e = 30$ and the $F_s = 100$, so the j is 3000. Moreover, for the ISRUC3 database, the $T_e = 20$ and the $F_s = 200$, so the j is 4000.

2.3.2. Wavelet Package Decomposition

WPD is a powerful tool to analyze non-stationary EEG signals. In essence, WPD is a wavelet transform where the discrete-time signal is passed through more filters than the discrete wavelet transform, which can provide a multi-level time-frequency decomposition of signals [23]. Compared with discrete wavelet transform, WPD can provide more frequency resolutions. In the discrete wavelet transform, a signal is split into an approximation coefficient and a detail coefficient [24]. The approximation coefficient is then itself split into a second-level approximation coefficients and detail coefficients and the process is repeated. A wavelet packet function $\omega^m_{l,d}(q)$ is defined as (3):

$$\omega^m_{l,d}(q) = 2^{l/2}\omega^m(2^l q - d) \tag{3}$$

where l and d are the scaling (frequency localization) parameter and the translation (time localization) parameter, respectively; $m = 0, 1, 2, \ldots$ is the oscillation parameter.

Wavelet packet (WP) coefficients of the EEG epoch X_i are embedded in the inner product of the signal with every WP function, denoted by $p^{i,m}_l(d), d = ..., -1, 0, 1, ...$ and given below:

$$p^{i,m}_l(d) = \sum x^i(q)\omega^m_{l,d}(q) \tag{4}$$

where $p^{i,m}_l(d)$ denotes the m-th set of WPD coefficients at l-th scale parameter and d is the translation parameter. All frequency components and their occurring times are reflected in $p^{i,m}_l(d)$ through change in m, l, d. Each coefficient $p^m_l(d)$ measures a specific sub-band frequency content, controlled by scaling parameter l and oscillation parameter m. The essential function of WPD is the filtering operation through low-pass filter $h(d)$ and high-pass filter $g(d)$. For the l-th level of decomposition, there are 2^l sets of sub-band coefficients $C^i_{l,m}$, of length $j/2^l$. The wavelet packet coefficients of epoch X_i are given as

$$C^i_{l,m} = \{p^{i,m}_l(d) | d = 1, 2, ..., j/2^l\} \tag{5}$$

It can be seen from the (5) that each node of the WP tree is indexed with a pair of integers (l, m), where l is the corresponding level of decomposition and m is the order of the node position in the specific level. Here, the level l_{LE} and wavelet basis ω_{LE} of WPD on the epoch X_i for LE calculation will be confirmed in the Section 3. Moreover, the wavelet basis ω_{DSSM} for DSSM will be confirmed in the same section.

2.3.3. Locality Energy Calculation

The wavelet package energy $E^i_{l_{LE},m}$ at the m-th node on the level l_{LE} of epoch X_i can be defined as follows [25].

$$E^i_{l_{LE},m} = \sum |p_l^{i,m}(d)|^2 = |C^i_{l_{LE},m}|^2, m = \{1, 2, \ldots, 2^{l_{LE}}\} \tag{6}$$

Then, the locality energy features (LEFs) of each Epoch can be defined as $\{E^i_{l_{LE},m} | m = 1, 2, \ldots, 2^{l_{LE}}\}$.

2.3.4. Dual State Space Models Estimation

As we have described before, after the wavelet packet decomposition, the low-level (the first level) coefficients will be used to estimate the dual state space models which can denoted by the difference Equation (7).

$$\begin{cases} \mathbf{u}_{k+1} = \mathbf{A}\mathbf{u}_k + \mathbf{K}e_k \\ y_k = \mathbf{B}\mathbf{u}_k + e_k \end{cases} \tag{7}$$

The $y_k \in C^i_{1,m}$ is the coefficient at instant $k \in [1, 2, \ldots, j/2]$. Vector $u_k \in \mathbb{R}^{n\times 1}$ is the state vector of process at discrete time instant k and contains the numerical value of n states. Matrix $\mathbf{A} \in \mathbb{R}^{n\times n}$ is the dynamical system matrix. $\mathbf{K} \in \mathbb{R}^{n\times 1}$ is the steady state Kalman gain. $\mathbf{B} \in \mathbb{R}^{1\times n}$ is the output matrix, which describes how the internal state is transferred to the outside world in the observations y_k. The $e_k \in \mathbb{R}$ denotes zero mean white noise.

With the traditional subspace algorithm such as N4SID, the matrix $\widehat{\mathbf{A}}$, $\widehat{\mathbf{B}}$, $\widehat{\mathbf{K}}$ of the state space model of dynamic system can be estimated [26]. In this paper, the order n_{DSSM} of dual state space models will be determined by the experiments in the Section 3. Moreover, the parameter matrixes of state space model estimated by the first level wavelet coefficients $C^i_{1,m}$ can be expressed as

$$\begin{aligned} \widehat{\mathbf{A}}^i_{1,m} &= \begin{bmatrix} a^{i,m}_{1,1} & \cdots & a^{i,m}_{1,n_{DSSM}} \\ \vdots & \cdots & \vdots \\ a^{i,m}_{n_{DSSM},1} & \cdots & a^{i,m}_{n_{DSSM},n_{DSSM}} \end{bmatrix} \\ \widehat{\mathbf{B}}^i_{1,m} &= \begin{bmatrix} b^{i,m}_1 \; b^{i,m}_2 \ldots b^{i,m}_{n_{DSSM}} \end{bmatrix} \\ \widehat{\mathbf{K}}^i_{1,m} &= \begin{bmatrix} k^{i,m}_1 \; b^{i,m}_2 \ldots k^{i,m}_{n_{DSSM}} \end{bmatrix}^T \\ i &\in [1, L], m = \{1, 2\} \end{aligned} \tag{8}$$

Then the DSSM S_i of the $\mathbf{X}_i$ can be defined as:

$$S_i = \begin{bmatrix} s^i_1 \\ s^i_2 \end{bmatrix} = \begin{bmatrix} a^{i,1}_{1,1} & \cdots & a^{i,1}_{n_{DSSM},n_{DSSM}} & b^{i,1}_1 & \cdots & b^{i,1}_{n_{DSSM}} & k^{i,1}_1 & \cdots & k^{i,1}_{n_{DSSM}} \\ a^{i,2}_{1,1} & \cdots & a^{i,2}_{n_{DSSM},n_{DSSM}} & b^{i,2}_1 & \cdots & b^{i,2}_{n_{DSSM}} & k^{i,2}_1 & \cdots & k^{i,2}_{n_{DSSM}} \end{bmatrix} \tag{9}$$

So, the parameters extracted from DSSM here is called DSSM Features (DSSMFs) can be defined as $DSSMFs = \begin{bmatrix} s^i_1 \; s^i_2 \end{bmatrix}$.

2.3.5. IMBEFs Construction

According to the previously calculated LEFs $E^i_{l_{LE},m}$ and the parameters S_i of the DSSM, the features IMBEFs of epoch X_i here are given by

$$F^i_{DSSM} = \left[E^i_{l_{LE},1} \cdots E^i_{l_{LE},2^{l_{LE}}} \; s^i_1 \; s^i_2\right] \tag{10}$$

The feature dimension can be calculated by the Equation (11).

$$Dim_{DSSM} = 2^{l_{LE}} + 2(n^2_{DSSM} + 2 \times n_{DSSM}) \tag{11}$$

Here, the general form of features extracted from LE and multiple state space models (MSSM) which are estimated by the l_{MSSM}-th level WPD coefficients can be depicted as Equation (12).

$$F^i_{MSSM} = \left[E^i_{l_{LE},1} \cdots E^i_{l_{LE},2^{l_{LE}}} \; s^i_1 \cdots s^i_{2^{l_{MSSM}}}\right] \tag{12}$$

The dimension of the F^i_{MSSM} can be calculated by

$$Dim_{MSSM} = 2^{l_{LE}} + 2^{l_{MSSM}}(n^2_{MSSM} + 2 \times n_{MSSM}) \tag{13}$$

where n_{MSSM} is the order of MSSM. Usually, the n_{MSSM} range from 5 to 10. Assume the $n_{MSSM} = 5$, $Dim_{DSSM} = 2^{l_{LE}} + 2^{l_{MSSM}} \times 40$. Then if $l_{MSSM} > 2$, the Dim_{MSSM} will be too large. So the l_{MSSM} is set to 1 in this paper, which means there are two state space models.

3. Experiments and Results

In this section, there are four experimental parts. The first is the experiment for selecting a suitable classifier among several candidate classifiers. Then is the determination of the most suitable wavelet basis ω_{DSSM} and model order n_{DSSM} for DSSM estimation. Next is the selection of the wavelet basis ω_{LE} and the level l_{LE} for the LE calculation. Finally, the test experiment will be conducted on the S-EDF database and ISRUCS3 database with the ω_{DSSM}, ω_{LE}, n_{DSSM} and l_{LE} determined according to the previous experiments.

In the process of selecting these parameters, the DRMS database was adopted for testing under the both R&K and AASM standards. There are several conventional verification strategies, including k-fold cross-validation, leave one-subject-out cross-validation (LOOCV) and corss-dataset validation, etc. In this paper, many commonly-used databases are adopted to verify the performance of the algorithm, in which the S-EDF database and the DRMS database contains lots of subjects. However, some subjects contained in these database possess unevenly distributed samples, which means the incomplete sleep stages. Consequently, the 10-fold cross-validation method would be more suitable for the performance verification in this research. In 10-fold cross-validation, the original sample is randomly partitioned into 10 equal size subsamples. Of the 10 subsamples, a single subsample is retained as the validation data for testing the model and the remaining nine subsamples are used as training data. The cross-validation process is then repeated 10 times, with each of the 10 subsamples used exactly once as the validation data. The 10 results from the folds can then be averaged to produce a single estimation. The advantage of this method is that all observations are used for both training and validation and each observation is used for validation exactly once. The accuracy (ACC) and Cohen's Kappa Coefficient (Kappa) are computed to evaluate the overall classification performance.

$$ACC = \frac{TP + TN}{TP + TN + TN + FN} \times 100\% \tag{14}$$

$$Kappa = \frac{ACC - p_e}{1 - p_e} \tag{15}$$

where TP, TN, FP and FN represent the number of true positive, true negative, false positive and false negative examples respectively. And p_e is the hypothetical probability of agreement by chance.

3.1. Classifier Comparison and Selection

In this section, an algorithm is designed to search the best classifier for the method proposed in this paper. The detailed steps are shown in the Algorithm 1 below. In this algorithm, according to the distribution and characteristic of the samples, the candidate classifiers are including Linear Discriminant, Quadratic Discriminant, Quadratic SVM, Fine KNN, Bagged Trees and RUSBoosted Trees. The candidate wavelet bases include the db1, db2, db3, db4, db5, db6, db8, db16, db32, sym2, sym8, sym16, coif1, coif3 and dmey. The order of DSSM range from 5 to 10. Here only the DSSMFs are used for training and validation.

Table 4 shows the experiment results of Algorithm 1. As can be seen from Table 4, the Bagged Trees is the optimal classifier in the classification of two to six classes. At the meantime, the corresponding order of DSSM is 6. In addition, in the two class classification, the optimal wavelet basis is sym2; the others, however, are db1. Furthermore, the comparison of different classifiers in two classes classification under the condition of $n_{DSSM} = 6$ are listed in Table 5. It can be seen from Table 5 that the accuracy of sym2 is 95.79%, which is a little higher than the 95.71% of db1 and 95.72% of db2. Therefore, considering the results in Tables 4 and 5 , the Bagged Trees will be used as the classifier for subsequent experiments.

Table 4. The outputs of Algorithm 1.

Classes	Optimal Classifier	ω_{DSSM}	n_{DSSM}	Accuracy(%)
2	Bagged Trees	sym2	6	95.79
3	Bagged Trees	db1	6	88.29
4	Bagged Trees	db1	6	83.07
5	Bagged Trees	db1	6	81.45
6	Bagged Trees	db1	6	78.57

Table 5. Comparison of different classifiers in two class classification with different wavelet. The $n_{DSSM} = 6$. Highest values are highlighted in boldface.

Accuracy (%)	db1	db2	db3	db4	db5	db6	db8	db16	db32	sym2	sym8	sym16	coif1	coif3	dmey
Linear Discriminant	94.21	93.94	94.33	93.93	93.52	93.17	92.53	91.72	91.66	93.90	92.35	91.68	93.98	92.59	92.24
Quadratic Discriminant	92.94	93.73	94.42	92.89	91.90	91.07	91.15	87.63	87.13	93.68	90.99	88.37	94.14	91.29	87.27
Quadratic SVM	95.13	95.01	95.16	94.88	94.62	94.43	94.14	93.71	93.47	95.09	94.13	93.59	95.12	94.25	93.68
Fine KNN	91.96	90.96	92.62	91.05	90.06	89.82	88.70	87.60	86.58	90.84	88.43	87.00	90.61	89.52	88.22
Bagged Trees	95.71	95.72	95.70	95.59	95.42	95.43	95.16	94.93	95.00	**95.79**	95.04	94.82	95.71	95.29	95.01
RUSBoosted Trees	94.28	94.08	94.68	94.41	94.07	93.94	93.75	93.02	93.08	94.05	93.78	92.79	94.20	93.76	93.20

Algorithm 1: Search the Optimal Classifier.

```
Input:
WBS:the candidate wavelet base array, including the db1, db2, db3, db4, db5, db6 ,db8 ,db16,
db32, sym2, sym8, sym16, coif1, coif3, dmey;
CCA:the candidate classifier array,including the Linear Discriminant, Quadratic Discriminant,
Quadratic SVM, Fine KNN, Bagged Trees and RUSBoosted Trees;
SSMOV:the state space model order array range from 5 to 10;
Output: the optimal classifier array OCA , the corresponding order array COV and wavelet
        basis array CWBA
Initialize the samples matrix SM, validation Accuracy VA = 0, the matrix VAM to store the
  VA, the current classifier CC, the current wavelet basis CWB, the current model order CMO,
  temp value TMP = 0 ;
for i ∈ [1,5] do
    for j ∈ [1,6] do
        CMO = SSMOV(j);
        for k ∈ [1,15] do
            CWB = WBS(k);
            Select and Construct the SM according to the CMO and CWB
            for m ∈ [1,6] do
                CC = CCA(m);
                Put the SM into CC for training and verification with 10-fold cross-validation
                  method, get the VA
                VAM(m + (j − 1) × 6, k) = VA
            end
        end
    end
    for n ∈ [1,36] do
        for p ∈ [1,15] do
            TMP = 0
            if VAM(n, p) > temp then
                OCA(i)= CCA(n%6);
                COV(i) = SSMOV(ceil(n/6));
                /* the ceil(X) rounds X to the nearest integer greater than or
                   equal to X.                                               */
                CWBA(i) = WBS(p);
            end
        end
    end
end
Output the OCA, COV, CWBA;
```

3.2. Wavelet Basis Comparison and Selection

After the classifier is determined, the model order n_{DSSM} and wavelet basis ω_{DSSM} should be further confirmed through the grid search method. This process can be seen in the step 1 of the Figure 2. The candidate wavelets include db1, db2, db3, db4, db5, db6, db8, db16, db32, sym2, sym8, sym16, coif1, coif3 and dmey. The candidate model order is 5 to 10. The Following Tables 6–14 are the experiments results of the DRMS database without LEFs, in which the highest accuracy values are highlighted in bold.

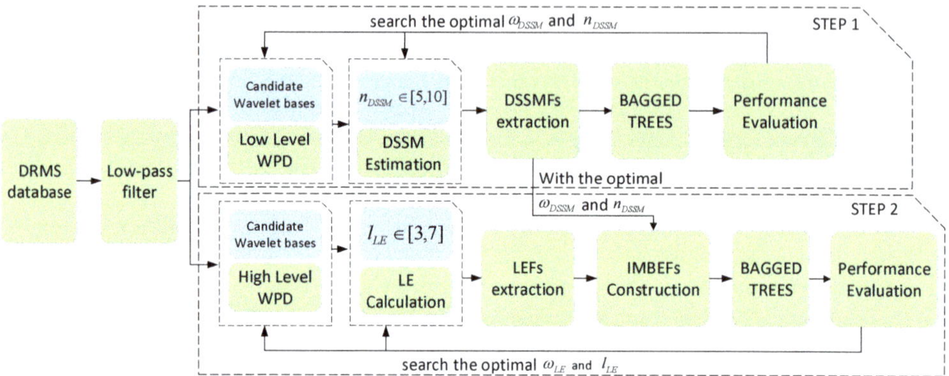

Figure 2. The diagram of the parameter optimization process.

Table 6. The accuracy (%) of the two class sleep stage classification with different wavelet bases and different order of DSSM under R&K standard. Only DSSMFs are used, no LEFs.

	ω_{DSSM}														
n_{DSSM}	db1	db2	db3	db4	db5	db6	db8	db16	db32	sym2	sym8	sym16	coif1	coif3	dmey
5	95.52	95.55	95.69	95.59	95.17	95.16	95.09	94.98	94.96	95.72	94.94	94.92	95.54	95.07	95.24
6	95.71	95.72	95.70	95.59	95.42	95.43	95.16	94.93	95.00	**95.79**	95.04	94.82	95.71	95.29	95.01
7	95.63	95.54	95.57	95.70	95.61	95.38	95.28	94.70	94.80	95.59	95.31	94.58	95.60	95.21	94.94
8	95.47	95.51	95.51	95.64	95.50	95.23	95.15	94.85	94.47	95.45	95.16	94.75	95.44	95.20	94.62
9	95.57	95.52	95.54	95.62	95.71	95.34	95.12	94.93	94.33	95.49	95.26	94.55	95.45	95.21	94.45
10	95.38	95.45	95.46	95.57	95.54	95.28	95.23	94.72	94.12	95.48	95.10	94.76	95.44	95.23	94.32

Table 7. The accuracy (%) of three class sleep stage classification with different wavelet bases and different order of DSSM under R&K standard. Only DSSMFs are used, no LEFs.

	ω_{DSSM}														
n_{DSSM}	db1	db2	db3	db4	db5	db6	db8	db16	db32	sym2	sym8	sym16	coif1	coif3	dmey
5	87.90	87.92	87.72	88.04	86.70	86.73	86.18	85.50	85.25	87.86	86.11	85.55	87.54	86.03	85.43
6	**88.29**	88.03	88.26	88.05	87.85	87.70	86.72	85.38	85.25	87.82	86.54	85.20	87.90	87.09	85.65
7	87.88	87.72	88.13	87.94	87.67	87.76	87.20	84.85	84.27	87.88	87.43	85.00	87.96	87.18	84.82
8	87.67	87.87	88.10	87.96	87.86	87.68	87.24	85.66	84.15	87.88	87.53	85.83	87.71	87.16	84.20
9	87.81	87.79	88.07	87.83	87.96	87.81	87.55	85.99	83.85	87.84	87.55	86.37	87.67	87.24	83.79
10	87.62	87.88	87.84	88.02	87.80	87.78	87.58	86.14	83.93	87.92	87.52	86.62	87.27	87.16	83.53

Table 8. The accuracy (%) of four class sleep stage classification with different wavelet bases and different order of DSSM under R&K standard. Only DSSMFs are used, no LEFs.

	ω_{DSSM}														
n_{DSSM}	db1	db2	db3	db4	db5	db6	db8	db16	db32	sym2	sym8	sym16	coif1	coif3	dmey
5	82.61	82.69	82.51	82.93	81.36	81.37	80.29	79.83	79.53	82.88	80.59	79.59	82.13	80.49	79.76
6	**83.07**	82.71	82.76	82.78	82.29	82.23	81.40	79.63	79.16	82.71	81.07	79.43	82.58	81.75	79.79
7	82.38	82.78	82.81	82.42	82.20	82.35	81.45	78.91	78.37	82.76	81.70	79.03	82.45	81.69	78.62
8	82.29	82.40	82.69	82.32	82.33	82.06	81.51	79.78	78.02	82.39	81.83	79.69	82.36	81.36	78.28
9	82.29	82.56	82.54	82.49	82.24	82.25	81.76	80.00	77.68	82.35	81.99	80.35	82.53	81.58	78.00
10	82.09	82.48	82.71	82.71	82.10	82.38	81.95	79.95	77.70	82.56	81.65	80.50	81.93	81.57	77.48

Table 9. The accuracy (%) of five class sleep stage classification with different wavelet bases and different order of DSSM under R&K standard. Only DSSMFs are used, no LEFs.

		ω_{DSSM}														
		db1	**db2**	**db3**	**db4**	**db5**	**db6**	**db8**	**db16**	**db32**	**sym2**	**sym8**	**sym16**	**coif1**	**coif3**	**dmey**
n_{DSSM}	5	81.14	81.15	80.66	81.17	79.79	79.83	79.04	78.55	78.30	81.20	79.27	78.41	80.51	79.30	78.72
	6	**81.45**	81.38	81.42	81.17	81.00	80.60	79.92	78.38	78.13	81.33	79.66	78.15	81.19	80.23	78.71
	7	80.80	80.87	81.00	80.88	80.63	80.48	79.87	77.75	77.08	80.91	80.27	77.53	80.87	80.07	77.61
	8	80.95	81.07	81.08	80.92	80.62	80.52	80.13	78.13	76.89	81.00	80.15	78.48	80.85	80.16	77.20
	9	80.95	80.90	80.75	80.88	80.67	80.56	80.27	78.73	76.72	80.86	79.99	78.93	80.64	79.97	76.78
	10	80.64	80.86	80.97	80.94	80.71	80.72	80.29	78.42	76.62	80.93	80.22	78.94	80.47	79.90	76.12

Table 10. The accuracy (%) of six class sleep stage classification with different wavelet bases and different order of DSSM under R&K standard. Only DSSMFs are used, no LEFs.

		ω_{DSSM}														
		db1	**db2**	**db3**	**db4**	**db5**	**db6**	**db8**	**db16**	**db32**	**sym2**	**sym8**	**sym16**	**coif1**	**coif3**	**dmey**
n_{DSSM}	5	78.01	77.96	77.56	77.94	76.47	76.46	75.44	74.99	74.58	77.67	75.80	75.02	77.17	75.58	74.99
	6	**78.57**	78.20	78.26	78.16	77.75	77.56	76.88	75.41	75.07	78.40	76.43	75.02	78.38	77.02	75.34
	7	77.31	77.39	77.52	77.38	76.89	77.00	76.32	74.11	73.63	77.39	76.54	74.22	77.03	76.77	74.09
	8	77.36	77.43	77.47	77.40	76.93	76.84	76.15	74.76	73.07	77.58	76.55	74.75	77.18	76.21	73.54
	9	77.25	77.38	77.56	77.44	76.94	76.89	76.39	74.91	73.01	77.22	76.79	75.21	77.07	76.34	73.10
	10	76.76	77.31	77.66	77.59	76.99	76.89	76.64	74.64	72.88	77.24	76.34	75.22	77.08	76.45	72.47

From Tables 6–10, we can see that under the R&K standard, when the order of the DSSM is 6 and the wavelet basis is selected as db1, the classification accuracy for three to six classes can reach the highest. When the wavelet basis is selected as sym2, the accuracy of the two classes is the highest. Through further analysis, it can be seen that in the results of two class classification, the difference between the accuracy of the db1 and the highest is very small.

Table 11. The accuracy (%) of two class sleep stage classification with different wavelet bases and different order of DSSM under AASM standard. Only DSSMFs are used, no LEFs.

		ω_{DSSM}														
		db1	**db2**	**db3**	**db4**	**db5**	**db6**	**db8**	**db16**	**db32**	**sym2**	**sym8**	**sym16**	**coif1**	**coif3**	**dmey**
n_{DSSM}	5	95.09	95.01	95.20	94.99	94.63	94.53	94.63	94.67	94.44	95.10	94.40	94.54	94.91	94.54	94.56
	6	95.59	95.63	95.68	95.55	95.42	95.29	95.10	94.75	94.79	**95.73**	95.02	94.79	95.72	95.12	94.90
	7	95.19	95.21	95.50	95.30	95.11	94.89	94.91	94.33	94.26	95.13	94.99	94.35	95.18	95.00	94.30
	8	94.82	95.14	95.09	95.31	94.90	94.73	94.71	94.25	93.83	95.17	94.62	94.16	94.99	94.74	94.03
	9	95.16	95.28	95.28	95.10	95.26	95.04	94.96	94.16	93.79	95.22	94.68	94.25	95.11	94.81	93.97
	10	95.07	95.20	95.18	95.40	95.09	94.98	94.95	94.24	93.41	95.10	94.72	94.20	95.17	95.09	93.46

Table 12. The accuracy (%) of three class sleep stage classification with different wavelet bases and different order of DSSM under AASM standard. Only DSSMFs are used, no LEFs.

		ω_{DSSM}														
		db1	**db2**	**db3**	**db4**	**db5**	**db6**	**db8**	**db16**	**db32**	**sym2**	**sym8**	**sym16**	**coif1**	**coif3**	**dmey**
n_{DSSM}	5	87.12	87.04	86.90	87.17	85.81	85.84	85.13	84.83	84.69	86.97	84.81	84.82	86.45	85.45	84.85
	6	**87.52**	87.35	87.55	87.63	87.12	86.88	86.16	84.49	84.61	87.26	85.87	84.48	87.21	86.46	84.85
	7	87.62	87.32	87.83	87.49	87.15	87.30	86.58	84.55	83.95	87.49	87.00	84.38	87.44	86.85	84.18
	8	86.92	87.56	87.59	87.44	87.15	87.26	86.73	85.55	83.23	87.20	86.77	85.10	87.14	86.67	83.53
	9	87.52	87.28	87.68	87.43	87.34	87.43	87.16	85.56	83.36	87.55	86.95	85.74	87.20	86.86	83.07
	10	87.38	87.51	87.52	87.62	87.14	87.46	87.32	85.64	83.19	87.37	87.02	86.09	87.25	86.87	82.79

Table 13. The accuracy (%) of four class sleep stage classification with different wavelet bases and different order of DSSM under AASM standard. Only DSSMFs are used, no LEFs.

		ω_{DSSM}														
		db1	**db2**	**db3**	**db4**	**db5**	**db6**	**db8**	**db16**	**db32**	**sym2**	**sym8**	**sym16**	**coif1**	**coif3**	**dmey**
n_{DSSM}	5	81.26	80.82	80.65	80.97	79.27	78.90	78.14	77.80	77.40	80.52	78.48	77.86	80.22	78.13	77.87
	6	**81.51**	80.87	81.18	81.12	80.44	79.86	79.35	77.16	77.01	80.89	79.27	77.35	80.81	79.40	77.56
	7	81.13	80.91	81.40	80.86	80.29	80.21	79.61	76.98	75.92	80.74	80.05	77.19	80.48	79.60	76.52
	8	80.22	80.60	80.34	80.56	80.10	79.67	79.37	77.82	75.32	80.66	79.42	77.89	80.15	79.41	76.05
	9	80.66	80.75	80.77	80.47	80.25	80.38	79.72	77.46	75.54	81.07	79.89	78.11	80.89	79.82	75.01
	10	80.50	80.83	81.06	81.29	80.33	80.43	80.21	77.78	75.52	80.69	79.53	78.34	80.15	79.87	74.99

Table 14. The accuracy (%) of five class sleep stage classification with different wavelet bases and different order of DSSM under AASM standard. Only DSSMFs are used, no LEFs.

		ω_{DSSM}														
		db1	**db2**	**db3**	**db4**	**db5**	**db6**	**db8**	**db16**	**db32**	**sym2**	**sym8**	**sym16**	**coif1**	**coif3**	**dmey**
n_{DSSM}	5	78.70	78.49	78.22	78.83	77.07	76.94	76.10	75.16	75.45	78.70	76.34	75.41	77.90	76.14	75.82
	6	**78.72**	78.55	78.54	78.67	77.74	77.73	77.08	75.36	74.99	78.58	77.22	75.15	78.29	77.21	75.72
	7	78.57	78.49	78.38	78.15	77.95	77.77	77.42	74.81	73.54	78.41	77.19	74.64	77.93	77.02	74.52
	8	78.06	78.11	77.94	77.83	77.43	77.56	77.09	75.55	73.45	78.23	77.35	75.33	77.87	77.18	73.64
	9	78.28	78.30	78.30	77.74	77.77	78.03	77.64	75.49	73.27	78.60	77.41	75.88	78.14	77.11	73.29
	10	77.91	78.29	78.26	78.29	77.87	77.83	77.63	75.12	73.25	78.20	77.49	75.55	78.26	77.37	72.38

As can be seen from Tables 11–14, when the order n_{DSSM} is 6, the highest classification accuracy can be obtained in two to five classes sleep state classification. Moreover, in the three to five classes classifications, when the wavelet basis is db1, the highest classification accuracy can be achieved. In the two classes of sleep classification, when the wavelet base is db1, the accuracy is 0.14% lower than the highest accuracy. Combining the classification results of the above tables, in order to facilitate subsequent calculations, the db1 was uniformly used as the wavelet basis for DSSM estimation and the model order of DSSM adopts 6.

Then, the wavelet basis ω_{LE} and level l_{LE} which are required to calculate LE should be further determined according to the experimental results in the next step. That is, on the basis of the features previously extracted from the DSSM, LEFs will be added which have been shown in the Step 2 of the Figure 2. Tables 15–19 are the classification accuracies of 2–6 classes under the R&K standard, in which the highest accuracy values are highlighted in bold.

Table 15. The accuracy (%) of two class sleep stage classification with different ω_{LE} and l_{LE} under R&K standard.

		ω_{LE}														
		db1	**db2**	**db3**	**db4**	**db5**	**db6**	**db8**	**db16**	**db32**	**sym2**	**sym8**	**sym16**	**coif1**	**coif3**	**dmey**
l_{LE}	3	95.71	95.69	95.70	95.84	95.68	95.64	95.68	95.71	95.64	95.67	95.66	95.77	95.69	95.70	95.69
	4	95.74	95.69	95.74	95.96	95.69	95.75	95.71	95.63	95.71	95.71	95.58	95.70	95.73	95.56	95.63
	5	95.82	95.89	95.90	**96.17**	95.88	95.87	95.81	95.72	95.85	95.80	95.74	95.74	95.89	95.74	95.72
	6	95.80	95.71	95.93	96.06	95.70	95.73	95.69	95.67	95.58	95.84	95.81	95.63	95.82	95.65	95.49
	7	95.76	95.74	95.64	95.70	95.71	95.57	95.61	95.57	95.60	95.58	95.53	95.72	95.64	95.54	95.51

Table 16. The accuracy (%) of three class sleep stage classification with different ω_{LE} and l_{LE} under R&K standard.

		ω_{LE}														
		db1	**db2**	**db3**	**db4**	**db5**	**db6**	**db8**	**db16**	**db32**	**sym2**	**sym8**	**sym16**	**coif1**	**coif3**	**dmey**
l_{LE}	3	88.12	88.07	87.91	88.48	88.12	88.11	88.21	87.86	87.97	88.09	88.02	87.96	88.15	88.10	88.10
	4	88.22	88.22	88.22	88.59	88.81	88.12	87.86	88.08	87.91	88.02	88.05	87.92	88.02	87.85	87.94
	5	88.51	88.56	88.61	88.72	**88.89**	88.75	88.14	88.18	88.09	88.47	88.37	88.14	88.64	88.18	87.98
	6	88.11	88.09	88.00	88.66	88.73	88.00	87.88	87.63	87.54	88.17	87.70	87.84	87.82	87.76	87.43
	7	87.92	87.76	87.58	88.62	88.62	87.65	87.27	87.34	87.35	87.83	87.49	87.39	87.61	87.46	87.27

Table 17. The accuracy (%) of four class sleep stage classification with different ω_{LE} and l_{LE} under R&K standard.

l_{LE}	ω_{LE}														
	db1	**db2**	**db3**	**db4**	**db5**	**db6**	**db8**	**db16**	**db32**	**sym2**	**sym8**	**sym16**	**coif1**	**coif3**	**dmey**
3	82.87	82.76	82.98	83.06	82.99	82.80	82.87	82.98	82.75	82.85	82.82	82.91	82.87	82.97	83.02
4	82.97	82.91	83.23	83.54	83.02	82.89	82.85	82.86	82.83	82.93	82.92	82.68	82.89	82.79	82.75
5	83.53	83.52	83.53	**83.97**	83.71	83.25	83.04	82.99	83.18	83.57	83.07	83.07	83.38	82.98	82.84
6	83.25	82.91	83.67	83.80	82.55	82.76	82.25	82.38	82.40	82.91	82.42	82.11	82.87	82.37	82.17
7	82.57	82.57	82.52	82.60	82.40	82.29	81.93	81.90	82.03	82.56	82.13	81.92	82.43	81.85	81.89

Table 18. The accuracy (%) of five class sleep stage classification with different ω_{LE} and l_{LE} under R&K standard.

l_{LE}	ω_{LE}														
	db1	**db2**	**db3**	**db4**	**db5**	**db6**	**db8**	**db16**	**db32**	**sym2**	**sym8**	**sym16**	**coif1**	**coif3**	**dmey**
3	81.24	81.42	81.52	81.66	81.45	81.29	81.31	81.28	81.35	81.37	81.19	81.13	81.27	81.32	81.30
4	81.49	81.55	81.67	81.93	81.70	81.09	81.18	81.10	81.21	81.30	81.45	81.33	81.43	81.28	80.97
5	81.87	81.66	**82.32**	82.28	81.61	81.42	81.09	80.95	81.01	81.34	80.93	80.92	81.19	81.01	80.79
6	81.59	81.48	81.29	81.82	81.07	80.98	80.67	80.85	80.75	81.52	80.90	80.89	81.36	80.68	80.44
7	81.07	81.04	80.89	81.47	80.70	80.38	80.58	80.59	80.56	80.80	80.69	80.44	80.76	80.47	80.55

Table 19. The accuracy (%) of six class sleep stage classification with different ω_{LE} and l_{LE} under R&K standard.

l_{LE}	ω_{LE}														
	db1	**db2**	**db3**	**db4**	**db5**	**db6**	**db8**	**db16**	**db32**	**sym2**	**sym8**	**sym16**	**coif1**	**coif3**	**dmey**
3	78.25	78.39	78.44	78.52	78.18	78.35	77.99	78.08	78.15	78.29	78.06	78.03	78.21	78.14	78.20
4	78.63	78.64	78.67	78.80	78.67	78.65	78.56	78.54	78.54	78.62	78.44	78.45	78.63	78.48	78.28
5	78.91	78.91	78.88	**78.92**	78.23	78.42	78.38	78.22	78.38	78.77	78.24	78.20	78.81	78.44	77.98
6	78.45	78.35	78.26	78.64	77.97	77.90	77.88	77.68	77.78	78.36	77.74	77.57	78.22	77.64	77.36
7	77.92	77.66	77.48	77.55	77.31	77.22	77.25	77.08	77.63	77.60	77.45	77.18	77.81	77.22	77.02

As can be seen from Tables 15–19, when $l_{LE} = 5$, the ω_{LE} is db4, the accuracy of two, four and six classes is the highest. Moreover, when the ω_{LE} is set to the db5 and db3, the classification accuracy of three and five classes can reach the highest respectively. The Table 20 is the confusion matrix of six classes sleep state classification on DRMS database with IMBEFs under the R&K standard. As shown in the Table 20, the sensitivity of Awa, REM, S1, S2, S3 and S4 are 93.68%, 81.16%, 14.37%, 89.29%, 25.71% and 77.99%, respectively. Moreover, the overall accuracy of the six classes classification is 78.92%.

Table 20. The confusion matrix of six classes sleep state classification on DRMS database under the R&K standard. The $l_{LE} = 5$, $\omega_{LE} = db4$, $n_{DSSM} = 6$, $\omega_{DSSM} = db1$.

		Automatic Classification							
		Awa	**REM**	**S1**	**S2**	**S3**	**S4**	**Sen (%)**	**Overall Accuracy (%)**
Expert	Awa	5247	96	85	155	3	15	93.68	78.92
	REM	203	3697	86	566	1	2	81.16	
	S1	418	784	257	328	1	0	14.37	
	S2	262	731	62	11852	248	119	89.29	
	S3	20	0	0	1022	543	527	25.71	
	S4	174	0	0	231	271	2395	77.99	

Tables 21–24 show the classification accuracy of 2–5 classes with LEFs on the DRMS database under the AASM, in which the highest accuracy values are highlighted in boldface. As can be seen from these tables, after adding LEFs, the accuracy of each classification has been greatly improved. Among them, the highest accuracy can be obtained when using the LEFs extracted from the 5 level

WPD and there are three corresponding wavelet bases, which are db1, db2 and db4. When the wavelet basis is selected as db4, the accuracy of two classes and four classes can reach the highest. In addition, the accuracy of three and five classes are 88.22% and 79.90% respectively, which is not much different from 88.26% and 79.97% of the corresponding highest classification accuracy. Therefore, the parameter of l_{LE} will be set as 5 and ω_{LE} will be set as db4 in this paper.

Table 21. The accuracy (%) of two class sleep stage classification with different ω_{LE} and l_{LE} under AASM standard.

		ω_{LE}														
		db1	**db2**	**db3**	**db4**	**db5**	**db6**	**db8**	**db16**	**db32**	**sym2**	**sym8**	**sym16**	**coif1**	**coif3**	**dmey**
l_{LE}	3	95.39	95.21	95.34	96.18	95.25	95.25	95.33	95.21	95.26	95.99	95.28	95.21	95.35	95.24	95.30
	4	96.00	95.70	95.84	96.24	95.75	95.37	95.34	95.25	95.34	96.24	95.31	95.24	95.85	95.39	95.43
	5	95.87	95.91	95.94	**96.48**	95.95	95.36	95.41	95.46	95.45	96.41	95.45	95.52	95.53	95.40	95.45
	6	95.61	95.37	95.79	96.16	95.33	95.36	95.37	95.28	95.25	96.34	95.38	95.42	95.32	95.30	95.27
	7	95.42	95.31	95.26	95.90	95.38	95.26	95.24	95.34	95.34	95.35	95.21	95.30	95.38	95.22	95.11

Table 22. The accuracy (%) of three class sleep stage classification with different ω_{LE} and l_{LE} under AASM standard.

		ω_{LE}														
		db1	**db2**	**db3**	**db4**	**db5**	**db6**	**db8**	**db16**	**db32**	**sym2**	**sym8**	**sym16**	**coif1**	**coif3**	**dmey**
l_{LE}	3	87.99	87.92	87.62	87.56	87.75	87.67	87.52	87.85	87.55	87.63	87.64	87.77	87.73	87.51	87.59
	4	87.80	87.65	87.67	87.71	87.68	87.64	87.51	87.69	87.77	87.76	87.78	87.70	87.79	87.58	87.88
	5	88.00	**88.26**	88.04	88.22	87.96	87.99	87.85	87.92	87.84	88.17	88.04	88.04	88.04	87.99	87.87
	6	87.89	87.78	87.75	87.71	87.42	87.64	87.59	87.34	87.55	87.64	87.58	87.34	87.82	87.44	87.26
	7	87.13	87.71	87.43	87.92	87.23	87.62	87.38	87.39	87.52	87.75	87.55	87.34	87.51	87.58	87.11

Table 23. The accuracy (%) of four class sleep stage classification with different ω_{LE} and l_{LE} under AASM standard.

		ω_{LE}														
		db1	**db2**	**db3**	**db4**	**db5**	**db6**	**db8**	**db16**	**db32**	**sym2**	**sym8**	**sym16**	**coif1**	**coif3**	**dmey**
l_{LE}	3	81.14	81.43	81.45	81.42	81.23	81.40	81.44	81.53	81.33	81.57	81.07	81.55	81.65	81.29	81.15
	4	81.64	81.47	81.36	81.36	81.46	81.33	81.25	81.47	81.37	81.75	81.09	81.31	81.61	81.26	81.37
	5	81.79	82.03	81.91	**82.08**	81.84	81.93	81.59	81.82	81.73	82.23	81.51	81.65	82.12	81.64	81.40
	6	81.83	81.66	81.66	81.76	81.39	81.40	81.37	81.32	80.99	81.51	81.16	81.01	81.85	80.99	80.65
	7	81.35	81.28	81.19	81.47	81.06	80.96	80.93	80.66	80.75	81.37	80.85	80.68	81.40	80.50	80.40

Table 24. The accuracy (%) of five class sleep stage classification with different ω_{LE} and l_{LE} under AASM standard.

		ω_{LE}														
		db1	**db2**	**db3**	**db4**	**db5**	**db6**	**db8**	**db16**	**db32**	**sym2**	**sym8**	**sym16**	**coif1**	**coif3**	**dmey**
l_{LE}	3	79.24	79.02	78.87	79.20	79.08	78.99	78.86	78.96	78.77	78.89	79.16	79.09	79.03	78.66	79.09
	4	79.41	79.26	79.33	79.25	79.14	78.80	79.08	79.03	78.91	78.93	78.82	79.16	79.05	79.03	79.01
	5	**79.97**	79.54	79.48	79.90	79.45	79.34	79.36	79.38	79.62	79.77	79.36	79.15	79.53	79.43	79.05
	6	79.10	79.29	79.20	79.23	79.13	78.94	78.89	78.68	78.64	79.20	78.86	78.52	79.39	78.43	78.01
	7	79.04	78.69	78.89	78.85	78.67	78.39	78.48	78.33	78.16	78.98	78.71	78.22	78.66	78.30	77.67

The confusion matrix of five classes sleep state classification is listed in the Table 25. As can be seen in this table, the overall accuracy is 79.90%. The sensitivity of Awa, REM, N1, N2, N3 are 92.89%, 81.22%, 17.57%, 85.52% and 78.79%. Furthermore, the receiver operating characteristic (ROC) curve of the classifier trained by this dataset with the confirmed parameter is shown in Figure 3.

As can be seen in the Figure 3, when the positive samples is Awa, the true positive rate is 0.93 and the false positive rate is 0.05. In addition, when the positive samples are REM, N2 and N3, the corresponding positive sample rates are 0.81, 0.86 and 0.79. When the positive samples are N1, the area under the curve (AUC) area is only 0.18. Moreover, the issue of low classification accuracy of S1(N1) will be discussed in the Section 4.

Table 25. The confusion matrix of five classes sleep state classification on DRMS database under the AASM standard. The $l_LE = 5$, $\omega_{LE} = db4$, $n_{DSSM} = 6$, $\omega_{DSSM} = db1$.

		Automatic Classification						
		Awa	REM	N1	N2	N3	Sen(%)	Overall Accuracy (%)
Expert	Awa	3306	53	68	111	21	92.89	79.90
	REM	131	2452	93	330	13	81.22	
	N1	341	480	260	389	10	17.57	
	N2	229	499	50	7056	417	85.52	
	N3	77	1	0	761	3117	78.79	

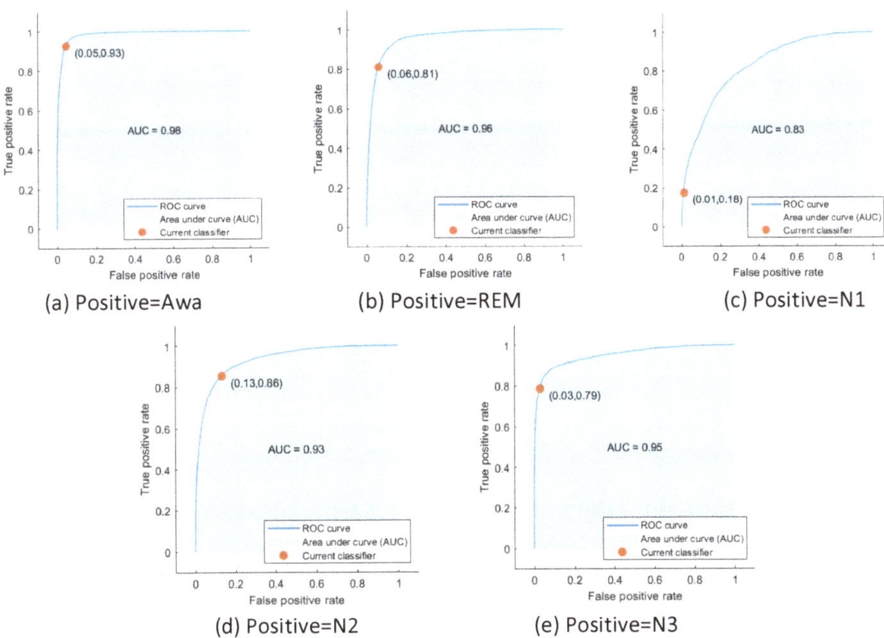

Figure 3. The ROC curve of the classifier to classify the five classes of DRMS database under the AASM standard.

3.3. *Experiments on S-EDF and ISRUC3 Database*

After experiments on the DRMS database, through the comprehensive comparison and selection, the classifier is selected as the Bagged Tress, n_{DSSM} is set to 6, l_{LE} is set to 5, ω_{DSSM} is set to db1 and ω_{LE} is set to db4. In order to further evaluate the performance of the method proposed in this paper, we will use these parameters to conduct experiments on the S-EDF database and the ISRUC3 database.

The classification accuracy and Cohen's Kappa Coefficients of the 2–6 classes on the S-EDF database are shown in Table 26. Furthermore, the confusion matrix of six class classification is listed for further analysis in Table 27.

Table 26. The classification accuracy and Cohen's Kappa Coefficient of 2–6 class sleep classification on S-EDF database.

	6 Classes	5 Classes	4 Classes	3 Classes	2 Classes
Accuracy	92.04%	92.50%	93.87%	94.90%	98.74%
Cohen's Kappa Coefficient	0.8266	0.8364	0.8646	0.8834	0.9697

Table 27. the Confusion matrix of six classes sleep state classification on S-EDF database.

		Automatic Classification							
		Awa	**REM**	**S1**	**S2**	**S3**	**S4**	**Sen (%)**	**Overall Accuracy (%)**
Expert	Awa	73165	483	46	141	0	0	99.09	92.04
	REM	876	4819	61	988	0	0	71.46	
	S1	578	1174	583	682	0	0	19.32	
	S2	375	863	76	15631	254	50	90.62	
	S3	71	0	0	1003	1055	159	46.11	
	S4	23	0	0	171	256	1060	70.20	

Similarly, the method proposed in this paper was also tested on the ISRUC3 database. The experimental results are shown in the following Tables 28 and 29.

Table 28. The classification accuracy and Cohen's Kappa Coefficient of 2–5 class sleep classification on ISRUC3 database.

Classes	5 Classes	4 Classes	3 Classes	2 Classes
accuracy	81.65%	84.68%	90.54%	96.18%
Cohen's Kappa Coefficient	0.7629	0.7729	0.8112	0.878

Table 29. The confusion matrix for five classes case on ISRUC3 database.

		Automatic Classification						
		Awa	**REM**	**N1**	**N2**	**N3**	**Sen(%)**	**Overall Accuracy (%)**
Expert	Awa	1537	13	84	53	15	90.31	81.65
	REM	36	1032	89	68	13	83.36	
	N1	91	135	648	242	7	57.70	
	N2	68	85	128	2312	257	81.12	
	N3	15	1	2	229	1729	87.50	

As can be seen from Table 28, the classification accuracies of two to five classes are 96.18%, 90.54%, 84.68% and 81.65%, respectively. In the five class classification, the sensitivity of Awa, REM, N1, N2, N3 are 90.31%, 83.36%, 57.70%, 81.12% and 87.50%, respectively.

4. Discussion

Table 30 shows the comparison of the classification accuracy from two to six classes of the various published method and the method proposed in this paper on the DRMS database under the R&K standard.

Table 30. The accuracy comparison of various published methods on DRMS database under the R&K standard. Highest accuracy in each case is highlighted in bold.

	Epoch Mumber	6 Classes (%)	5 Classes (%)	4 Classes (%)	3 Classes (%)	2 Classes (%)	Cross-Validation
Hassan et al. [3]	30401	70.73	73.50	79.12	84.4	93.3	10-fold
Hassan et al. [11]	30401	68.74	73.05	78.8	82.96	94.02	0.5/0.5
Shen et al. [27]	30401	78.2	80.9	82.7	87.7	94.9	10-fold
Proposed method Without LEFs	30401	78.52	81.26	82.81	87.95	95.59	10-fold
Proposed method with IMBEFs	30401	**78.92**	**82.28**	**83.97**	**88.72**	**96.17**	10-fold

As can be seen from the Table 30 above, when the only DSSMFs is used, the method proposed in this paper has a certain improvement in accuracy compared with the others. After adding LEFs on the basis of DSSMFs, the classification accuracies of two to six classes are improved by 1.27%, 1.02%, 1.27%, 1.38% and 0.72% compared with our previous study [27].

It can be seen from Table 31 that the method proposed in this paper has a certain improvement in the sleep stage classification of 3–5 classes on the DRMS database compared with the current existing methods. The N1 sensitivity of this method on the DRMS database is 17.57%, which is higher than 14.3% of Ghimatgar [7]. Moreover, Table 32 is the accuracy comparison of various published methods on S-EDF database.

Table 31. The accuracy comparison of various published methods on the Dreams Subjects database under the AASM standard. Highest accuracy in each case is highlighted in bold.

	Epoch number	5 Classes (%)	4 Classes (%)	3 Classes (%)	2 Classes (%)	Cross-validation
Hassan et al.[3]	20265	72.28	79.44	83.75	95.2	10-fold
Hassan et al.[11]	20265	74.59	80.0	85.42	**97.2**	10-fold
Ghimatgar et al.[7]	20265	78.08	80.38	86.88	94.8	20-fold
Proposed Method Without LEFs	20265	78.72	80.9	87.52	95.7	10-fold
Proposed Method with IMBEFs	20265	**79.90**	**82.08**	**88.22**	96.48	10-fold

Table 32. The accuracy comparison of various published methods on the S-EDF database under the R&K standard. Highest accuracy in each case is highlighted in bold.

	Epoch Number	6 Classes (%)	5 Classes (%)	4 Classes (%)	3 Classes (%)	2 Classes (%)	Cross-Validation
Hassan et al. [3]	15188	90.38	91.50	92.11	94.8	97.5	0.5/0.5
Abdulla et al. [6]	23806	**93**	–	–	–	–	-
Ghimatgar et al. [7]	15188	89.91	91.11	92.19	94.65	98.19	0.5/0.5
Ghimatgar et al. [7]	40100	79.13	81.86	83.71	88.39	95.98	0.5/0.5
Hassan et al. [10]	15188	88.62	90.11	91.2	93.55	97.73	0.5/0.5
Hassan et al. [11]	15188	88.07	83.49	92.66	94.23	98.15	0.5/0.5
Sharma et al. [12]	15139	90.03	91.13	92.29	94.66	98.02	10-fold CV
Michielli et al. [17]	10280	–	86.7	–	–	–	10-fold CV
Shen et al. [27]	103505	91.9	92.3	93.0	93.9	98.6	10-fold CV
Sharma et al. [28]	85900	91.5	91.7	92.1	93.9	98.3	10-fold CV
Liang et al. [29]	3708	–	83.6	–	–	–	0.5/0.5
Hsu et al. [30]	2880	–	87.2	–	–	–	10-fold CV
Hassan et al. [31]	15188	89.6	90.8	91.6	93.9	97.2	0.5/0.5
Zhu et al. [32]	14963	87.5	88.9	89.3	92.6	97.9	10-fold CV
Jiang et al. [33]	36972	–	91.5	–	–	–	2-fold CV
Rahman et al. [34]	15188	90.26	91.02	92.89	94.1	98.24	0.5/0.5
Supratak et al. [35]	41950	–	79.8	–	–	–	20-fold CV
Proposed Method	104368	92.04	**92.50**	**93.87**	**94.90**	**98.74**	10-fold CV

It can be seen from Table 32 that when a large number of samples are used, the accuracy is also improved compared with other published methods. Among them, the accuracy for the classification of four classes is 93.87%, while the Sharma [28] is 92.1% and the Shen [27] is 93.0%. In the classification of two classes, Abdulla et al. [6] has the highest accuracy of 93%; however, the number of epoch they used is only 23806. The sensitivity of S1 in this paper is 19.32%, which is higher than 18.3% of Ghimatgar [7] and 15.9% of Shen [27].

The experiments results of the proposed method on ISRUC3 database are also compared with other methods, which can be seen in the following Table 33.

Table 33. The accuracy comparison of the ISRUC3 database with the AASM standard. Highest accuracy in each case is highlighted in bold.

		Epoch number	5 Classes	4 Classes	3 Classes	2 Classes
Overall Accuracy	Ghimatgar et al.[7]	8889	77.56	82.74	88.26	93.76
	Proposed Method	8889	**81.65**	**84.68**	**90.54**	**96.18**
Cohen's kappa Coefficient	Ghimatgar et al.[7]	8889	0.71	0.75	0.77	0.79
	Proposed Method	8889	**0.7629**	**0.7729**	**0.8112**	**0.878**

As can be seen from the Table 33, compared with Ghimatgar [7], the detection accuracy of two and three classes is improved by more than 2 points. The sensitivity of S1 in Table 29 is 57.70%, which is higher than 33% of Ghimatgar [7]. Furthermore, the Cohen's kappa Coefficient is also much higher than Ghimatgar [7].

It should be noted that the classification of S1 which is an enormous challenge to all of the published method. From neurophysiological standpoint, S1(N1) is a transition phase and is a mixture of wakefulness and sleep resulting in similarity with the neural oscillations of S1 and Awa. In REM state, the cortex shows 40–60 Hz gamma waves as it does in waking. So the S1 state is often misclassified as REM or Awa state during the visual inspection by experts [3,11]. This is why many of the S1 epochs are misclassified as REM, Awa or S2 stages in this work. In addition, with different databases, the classification accuracy of S1 (N1) are also different. The detection accuracy of N1 on the ISRUC3 database reached 57.7%; on the DRMS database and the S-EDF database, however, it is less than 20%. This is also related to the different proportions of S1 stages in each database. Under the same AASM standard, on the ISRUC3 database, the S1 accounted for 12.65%; however, on the DRMS database, the S1 accounted for only 7.3%. Furthermore, under the R&K standard, the sensitivity of S3 on the S-EDF and DRMS databases is low, only 46.11% and 25.71%, respectively. The reason relate to this phenomenon rely mainly on that the S3 is a transition phase of S2 and S4. Thus the further research should be conducted to improve the S3 detection accuracy. Moreover, as can be seen in Table 20, a large number of S3 is misclassified as S2 and the other large part is misclassified as S4. Similarly, in Table 27, almost half of S3 epochs are misclassified as S2 and a small part are misclassified as S4. In addition, when under the AASM standard, after combining the S3 and S4 into N3, the sensitivity of N3 has been improved. As shown in Table 25, only 761 epochs of N3 were misclassified as N2; however, in Table 20, 1022 epochs of S3 were misclassified as S2 and 231 epochs of S4 were misclassified as S2. Therefore, the AASM standard is more suitable for guiding the researchers to annotate the sleep stages than the R&K standard.

5. Conclusions

In this paper, a novel IMBEF based automatic sleep stage classification method is proposed. Moreover, a grid search strategy was presented to determine a suitable model order n_{DSSM} and a wavelet basis ω_{DSSM} for estimating the DSSM among 15 candidate wavelets and 6 candidate model orders. With the same search strategy, a proper wavelet basis ω_{LE} and the WPD level l_{LE} for LE calculation are determined under 15 candidate wavelets and multilevel decomposition. The fused IMBEFs extracted from the DSSM and LE would be used as the input features of the suitable classifier which can be selected by comparing a variety of classifiers' experiment results. In order to precisely verify the performance of the proposed IMBEF based automatic sleep stage classification method, experiments were carried out on three public databases. The comparison results with other state-of-the-art methods show that the proposed algorithm can achieve higher accuracy.

We demonstrated in this paper measurable improvements in automatic sleep stage classification, providing better understanding and diagnostic of the sleep phenomenon, clearly essential in medical, wellness and other fields.

Author Contributions: Conceptualization, H.S. and A.L.; methodology, H.S., A.G. (Allon Guez) and A.L.; software, H.S.; validation, H.S., A.G. (Allon Guez) and A.G. (Aiying Guo); formal analysis, H.S., A.L.; investigation, A.G. (Aiying Guo); resources, F.R.; data curation, A.L.; writing–original draft preparation, H.S. and A.L.; writing–review and editing, H.S., A.G. (Allon Guez) and A.L.; visualization, M.X.; supervision, M.X.; project administration, M.X.; funding acquisition, F.R. All authors have read and agreed to the published version of the manuscript.

Funding: This research was funded by the National Natural Science Foundation of China grant number 61674100.

Acknowledgments: We acknowledge the support provided by the Microelectronics Research and Development Center of Shanghai University.

Conflicts of Interest: The authors declare no conflict of interest.

References

1. Rectschaffen, A.; Kales, A. *A Manual of Standardized Terminology, Techniques and Scoring Systems for Sleep Stages of Human Subjects*; National Government Publication: Los Angeles, CA, USA, 1968.
2. Iber, C.; Ancoliisrael, S.; Chesson, A.; Quan, S.F. *The AASM Manual for The Scoring of Sleep and Associated Events: Rules, Terminology and Technical Specifications*; American Academy of Sleep Medicine: Darien, IL, USA, 2007.
3. Hassan, A.R.; Bhuiyan, M.I.H. A decision support system for automatic sleep staging from EEG signals using tunable Q-factor wavelet transform and spectral features. *J. Neurosci. Methods* **2016**, *271*, 107–118. [PubMed]
4. Diykh, M.; Li, Y.; Abdulla, S. EEG sleep stages identification based on weighted undirected complex networks. *Comput Methods Programs Biomed.* **2020**, *184*, 105–116. [CrossRef] [PubMed]
5. Kang, D.Y.; Deyoung, P.N.; Malhotra, A.; Owens, R.L.; Coleman, T.P. A State Space and Density Estimation Framework for Sleep Staging in Obstructive Sleep Apnea. *IEEE Trans. Biomed. Eng.* **2017**, *65*, 1201-1-212. [PubMed]
6. Abdulla, S.; Diykh, M.; Laft, R.L.; Saleh, K.; Deo, R.C. Sleep EEG Signal Analysis Based on Correlation Graph Similarity Coupled with an Ensemble Extreme Machine Learning Algorithm. *Expert Syst. Appl.* **2019**, *138*, 112790–112804. [CrossRef]
7. Ghimatgar, H.; Kazemi, K.; Helfroush, M.S.; Aarabi, A. An automatic single-channel EEG-based sleep stage scoring method based on hidden Markov model. *J. Neurosci. Methods* **2019**, *324*, 180320–180336. [CrossRef]
8. Taran, S.; Sharma, P.C.; Bajaj, V. Automatic sleep stage classification using optimize flexible analytic wavelet transform. *Knowl. Based Syst.* **2020**, *192*, 105367–105374. [CrossRef]
9. Sharma, M.; Patel, S.; Choudhary, S.; Acharya, U.R. Automated Detection of Sleep Stages Using Energy-Localized Orthogonal Wavelet Filter Banks. *Arab. J. Sci. Eng.* **2020**, *45*, 2531–2544. [CrossRef]
10. Hassan, A.R.; Bhuiyan, M.I.H. Automatic sleep scoring using statistical features in the EMD domain and ensemble methods. *Biocybern Biomed. Eng.* **2016**, *36*, 248–255. [CrossRef]
11. Hassan, A.R.; Bhuiyan, M.I.H. Automated identification of sleep states from EEG signals by means of ensemble empirical mode decomposition and random under sampling boosting. *Comput. Methods Programs Biomed.* **2017**, *140*, 201–210. [CrossRef]
12. Sharma, R.; Pachori, R.B.; Upadhyay, A. Automatic sleep stage classification based on iterative filtering of electroencephalogram signals. *Neural. Comput. Appl.* **2017**, *28*, 2959–2978. [CrossRef]
13. Zhang, J.; Yao, R.; Ge, W.; Gao, J. Orthogonal convolutional neural networks for automatic sleep stage classification based on single-channel EEG. *Comput. Methods Programs Biomed.* **2020**, *183*, 105089–105100. [CrossRef] [PubMed]
14. Zhang, X.; Xu, M.; Li, Y.; Su, M.; Xu, Z.; Wang, C.; Kang, D. Automated multi-model deep neural network for sleep stage scoring with unfiltered clinical data. *Sleep Breath* **2020**, *24*, 581–590. [CrossRef]
15. Mousavi, Z.; Rezaii, T.Y.; Sheykhivand, S.; Farzamnia, A.; Razavi, S. N. Deep convolutional neural network for classification of sleep stages from single-channel EEG signals. *J. Neurosci. Methods* **2019**, *324*, 108312–108320. [CrossRef] [PubMed]
16. Korkalainen, H.; Aakko, J.; Nikkonen, S.; Kainulainen, S.; Leino, A.; Duce, B.; Leppänen, T. Accurate Deep Learning-Based Sleep Staging in a Clinical Population with Suspected Obstructive Sleep Apnea. *IEEE J. Biomed. Health Inform.* **2019**, *27*, 2073–2081. [CrossRef] [PubMed]
17. Michielli, N.; Acharya, U.R.; Molinari, F. Cascaded LSTM recurrent neural network for automated sleep stage classification using single-channel EEG signals. *Comput. Biol. Med.* **2019**, *106*, 71–81. [CrossRef]

18. Kemp, B.; Zwinderman, A.H.; Tuk, B.; Kamphuisen, H.A.; Oberye, J.J. Analysis of a sleep-dependent neuronal feedback loop: the slow-wave microcontinuity of the EEG. *IEEE. Trans. Biomed. Eng.* **2000**, *47*, 1185–1194. [CrossRef] [PubMed]
19. Goldberger, A.L.; Amaral, L.A.; Glass, L.; Hausdorff, J.M.; Ivanov, P.C.; Mark, R.G.; Mietus, J.E.; Moody, G.B.; Peng, C.K.; Stanley, H.E. PhysioBank, PhysioToolkit and PhysioNet: components of a new research resource for complex physiologic signals. *Circ. Res.* **2000**, *101*, e215–e220. [CrossRef]
20. Stephanie, D.; Myriam, K.; Stenuit, P.; Kerkhofs, M.; Stanus, E. Cancelling ECG artifacts in EEG using a modified independent component analysis approach. *EURASIP J. Adv. Signal Process.* **2008**, *1*, 747325. *Zenodo.* Available online: https://zenodo.org/record/2650142#.Xztj3igzabh (accessed on 1 January 2009)
21. Khalighi, S.; Sousa, T.; Santos, J.M.; Nunes, U. ISRUC-Sleep: A comprehensive public dataset for sleep researchers. *Comput. Methods Programs Biomed.* **2016**, *124*, 180–192. [CrossRef]
22. Delorme, A.; Makeig, S. EEGLAB: An open source toolbox for analysis of single-trial EEG dynamics including independent component analysis. *J. Neurosci. Methods* **2004**, *134*, 9–21. [CrossRef]
23. Zhang, Y.; Liu, B.; Ji, X.; Huang, D. Classification of EEG Signals Based on Autoregressive Model and Wavelet Packet Decomposition. *Neural. Process Lett.* **2017**, *45*, 365–378. [CrossRef]
24. Law, L.S.; Kim, J.H.; Liew, W.Y.; Lee, S.K. An approach based on wavelet packet decomposition and Hilbert–Huang transform (WPD–HHT) for spindle bearings condition monitoring. *Mech. Syst. Signal Process* **2012**, *33*, 197–211. [CrossRef]
25. Cao, Y.; Sun, Y.; Xie, G.; Wen, T. Fault Diagnosis of Train Plug Door Based on a Hybrid Criterion for IMFs Selection and Fractional Wavelet Package Energy Entropy. *IEEE Trans. Veh. Technol.* **2019**, *68*, 7544–7551. [CrossRef]
26. Van, O.P.; De, M.B. N4SID: Numerical Algorithms for State Space Subspace System Identification. In *Associated Technologies and Recent Developments, Proceedings of the 12th Triennal World Congress of the International Federation of Automatic Control, Sydney, Australia, 18–23 July 1993*; Elsevier: London, UK, 1993; Volume 26, pp. 55–58.
27. Shen, H.; Xu, M.; Guez, A.; Li, A.; Ran, F. An accurate sleep stage classification method based on state space model. *IEEE Access* **2019**, *7*, 125268–125279. [CrossRef]
28. Sharma, M.; Goyal, D.; Achuth, P.V.; Acharya, U.R. An accurate sleep stage classification system using a new class of optimally time-frequency localized three-band wavelet filter bank. *Comput. Biol. Med.* **2018**, *98*, 58–75. [CrossRef]
29. Liang, S.F.; Kuo, C.E.; Hu, Y.H.; Pan, Y.H.; Wang, Y.H. Automatic stage scoring of single-channel sleep EEG by using multiscale entropy and autoregressive models. *IEEE Trans. Instrum. Meas.* **2012**, *61*, 1649–1657. [CrossRef]
30. Hsu, Y.L.; Yang, Y.T.; Wang, J.S.; Hsu, C.Y. Automatic sleep stage recurrent neural classifier using energy features of EEG signals. *Neurocomputing* **2013**, *104*, 105–114. [CrossRef]
31. Hassan, A.R.; Subasi, A. A decision support system for automated identification of sleep stages from single-channel EEG signals. *Knowl. Based Syst.* **2017**, *128*, 115–124. [CrossRef]
32. Zhu, G.; Li, Y.; Wen, P. Analysis and classification of sleep stages based on difference visibility graphs from a single-channel EEG signal. *IEEE J. Biomed. Health Inform.* **2014**, *18*, 1813–1821. [CrossRef]
33. Jiang, D.; Lu, Y.N.; Yu, M.A.; Wang, Y. Robust sleep stage classification with single-channel EEG signals using multimodal decomposition and HMM-based refinement. *Expert Syst. Appl.* **2019**, *121*, 188–203. [CrossRef]
34. Rahman, M.M.; Bhuiyan, M.I.H.; Hassan, A.R. Sleep stage classification using single-channel EOG. *Comput. Biol. Med.* **2018**, *10*, 211–220. [CrossRef] [PubMed]
35. Supratak, A.; Dong, H.; Wu, C.; Guo, Y. A model for automatic sleep stage scoring based on raw single-channel EEG. *IEEE T. Neur. Sys. Reh.* **2017**, *25*, 1998–2008. [CrossRef] [PubMed]

Article

On the Better Performance of Pianists with Motor Imagery-Based Brain-Computer Interface Systems

José-Vicente Riquelme-Ros [1], Germán Rodríguez-Bermúdez [2], Ignacio Rodríguez-Rodríguez [3], José-Víctor Rodríguez [4,*] and José-María Molina-García-Pardo [4]

1 Consejería de Educación y Cultura de la Región de Murcia, E30003 Murcia, Spain; josevicente.riquelme@murciaeduca.es
2 University Center of Defense, San Javier Air Force Base, Ministerio de Defensa-Universidad Politécnica de Cartagena, E30720 Santiago de la Ribera, Spain; german.rodriguez@cud.upct.es
3 Departamento de Ingeniería de Comunicaciones, ATIC Research Group, Universidad de Málaga, E29071 Málaga, Spain; ignacio.rodriguez@ic.uma.es
4 Departamento de Tecnologías de la Información y las Comunicaciones, Universidad Politécnica de Cartagena, E30202 Cartagena, Spain; josemaria.molina@upct.es
* Correspondence: jvictor.rodriguez@upct.es

Received: 2 June 2020; Accepted: 7 August 2020; Published: 10 August 2020

Abstract: Motor imagery (MI)-based brain-computer interface (BCI) systems detect electrical brain activity patterns through electroencephalogram (EEG) signals to forecast user intention while performing movement imagination tasks. As the microscopic details of individuals' brains are directly shaped by their rich experiences, musicians can develop certain neurological characteristics, such as improved brain plasticity, following extensive musical training. Specifically, the advanced bimanual motor coordination that pianists exhibit means that they may interact more effectively with BCI systems than their non-musically trained counterparts; this could lead to personalized BCI strategies according to the users' previously detected skills. This work assessed the performance of pianists as they interacted with an MI-based BCI system and compared it with that of a control group. The Common Spatial Patterns (CSP) and Linear Discriminant Analysis (LDA) machine learning algorithms were applied to the EEG signals for feature extraction and classification, respectively. The results revealed that the pianists achieved a higher level of BCI control by means of MI during the final trial (74.69%) compared to the control group (63.13%). The outcome indicates that musical training could enhance the performance of individuals using BCI systems.

Keywords: brain-computer interface; motor imagery; machine learning; internet of things; pianists

1. Introduction

A brain-computer interface (BCI) system uses various techniques to recognize brain activity and transform this biological signal into a command that can be used by computer systems to complete certain tasks [1]. At the same time, it provides feedback to the user on how the intentions are being transformed into actions. In short, a BCI system transforms mental activity into a command that can affect the surroundings without the user making a physical effort. This command can be used for various applications, such as moving orthopedic prostheses through imagery [2]. Using a similar strategy, a recent work explored the use of noninvasive neuroimaging to enhance the control of a robotic device to complete daily tasks [3]. Similarly, BCIs are also used to synthesize speech by people who are unable to communicate due to neurological impairments [4]. BCIs have also been used to improve rehabilitation after a stroke, translating brain signals into the intended movements of a paralyzed limb [5] and, more recently, for the control through thought of applications for smart homes or robots within the Internet of Things (IoT) context [6,7].

It has been established that musical practice involves the activation of numerous areas of the brain that carry out different yet complementary functionalities to perform the complex task of playing a musical instrument, including reading a score, performing complex, highly specific movements, performing from memory, increased attention and concentration levels during performance, controlling the tuning of the instrument, and even improvisation. It is not unreasonable to conclude that musicians' brains have anatomical and functional differences compared to non-practitioners [8]. Therefore, the musician's brain offers a unique example with which to investigate what influence musical exercise can have on brain structures.

Of all the abilities musicians possess, their high motor coordination capacity is especially interesting. Engaging in a musical profession requires years of practice, with a great deal of time each day devoted to rehearsing specific and concrete movements involving the fine musculature of the hands or body. This undoubtedly effects a greater skill in handling the instrument, which is reflected in the individual's brain structures. The hand movements eventually display greater precision and coordination after the years of practice necessary to become a good musician.

Motor imagery, as a strategy to control a BCI system, has been investigated in groups of people with different characteristics, such as airplane pilots carrying out motor imagery tasks [9]. The present work aims to explore the handling of a BCI system through motor imagery by musicians, specifically pianists, for potential applications. It is hereby assumed that motor imagery control would be more intuitive for pianists and they would, therefore, achieve better performance, building on their previous muscle memory experience. Hence, this research aims to examine motor imagery in pianists to instantiate the benefits of musical education as well as show whether it is possible to consider a personalized BCI strategy for each subject according to their previously identified skills.

Following these introductory aspects, Section 2 outlines the prior work on the subject. Section 3 explains the experimental phase, detailing the characteristics of the test subjects, the resources used, and the method of data collection. The results are presented in Section 4. The work ends with the conclusions and future avenues for research in Section 5.

2. Music Training and Motor Imagery

Musicians' brains have been studied extensively over the last few years as a textbook case of neuroscience research. Given that the main theme of this paper is to explore whether musical training (specifically in pianists) has an influence on the control of a BCI system through motor imagery, this section will present the previous work supporting this hypothesis.

Numerous examples in the literature have analyzed the neuroscientific foundations of music. The work of [10] presents an exhaustive review of how musical production and perception influence cognitive abilities, involving the areas of the auditory cortex and the motor cortex. Münte et al. [8] already examined the neuroanatomical peculiarities of musicians' brains, highlighting their greater neuroplasticity. Later works, such as [11], delved further into the preceding ideas. Indeed, learning to play an instrument is a highly complex task that involves the interaction of higher-order cognitive functions and leads to behavioral, structural, and functional changes in the brain. Consequently, due to the need to constantly engage in musical practice, multiple differences have been shown to appear in the following areas of musicians' brains:

- Corpus callosum. This connects both cerebral hemispheres. It has been observed to be larger in professional musicians, especially those who began musical studies at an early age [12]. This larger size implies a higher interhemispheric transfer rate.
- Motor cortical regions. Numerous studies have shown that professional musicians have a much greater symmetry between the two hemispheres and that the representation of the hand in the motor cortex is much larger in musicians [13].
- Cerebellum. This structure, among other cognitive functions, is involved in the temporal sequencing and coordination of movements, which is undoubtedly fundamental in musical praxis. It has been demonstrated that musicians possess a greater cerebellum volume [14]. Furthermore,

it has been proven that this greater size is related to the intensity of musical training (number of hours practiced per day throughout life) as well as the fact of having initiated this training at an early age.

- Brain stem. This structure deals with basic sensory mechanisms. It has been possible to register faster reactions in musicians responding to certain musical and linguistic stimuli [15].

Taking the above into account, it can be stated that there is a positive correlation between the intensity and frequency of musical practice and the anatomical changes in the brain structures.

Musical performance is the complex activity responsible for these structural changes. It is considered extremely complex as it requires three special skills, i.e., the basic motor controls of coordination, sequencing, and spatial organization of movement [16]. Coordination refers to a good arrangement of the rhythmic aspect of music, while sequencing and spatial organization of movement imply the musician playing the notes on the instrument. It has been observed that more complex note sequences require the activity of structures such as the basal ganglia, dorsal premotor cortex, and cerebellum. The spatial organization of the different movements required to play an instrument involves the integration of different channels of spatial, sensory, and motor information, whereby the activation of the parietal, sensorimotor, and premotor cortex is observed.

In addition, audiomotor interactions occur in the brain when performing music as well as in the passive activity of listening to music. The premotor cortex is the link between the auditory system and the motor control, and for this activation to occur, the person must have an identified sound/action relationship [16].

It is evident that musical practice involves an increase in motor skills. In concrete studies focused on secondary motor areas, musicians show a much smaller activation area in these zones than non-musicians, demonstrating that pianists require smaller neural networks than non-musicians when it comes to motor skills, which in turn indicates that they are more efficient in controlling movements [17,18]. For example, practicing a complex fingering task for several months leads to an increase of approximately 25% in the primary motor area activation (M1). Furthermore, while musicians repeating the same sequence show a small area of activation (habituation) when a new music piece is trained for the first time, there is a larger area of activation (enhancement) [19]. During a fingering task performed by musicians and non-musicians, the former showed a rapid increase in the primary motor cortex (M1), while this was not seen in the latter [20].

Another study has shown that there are functional changes in the brains of children after 15 months of music training [10]. Two groups were studied, one receiving musical training and the other not. In the initial phase, the authors of such study found no differences between the groups. However, after the indicated period, it was found that the children who had been trained improved in motor control and melodic-rhythmic tasks, which supports the fact that the changes seen in adults (musicians) are due to musical practice.

As can be seen, intensive musical training leads not only to structural but also functional modifications in the youth's brain. However, these changes can also be induced in adults, thus preserving areas of gray and white substance [21].

With this, functional changes in the motor cortex occur. The motor cortex changes when performing simple piano exercises with five fingers and also increases the activity of the basal ganglia and the cerebellum. Most significantly, these changes take place either if the practice is performed physically or mentally [15]. In fact, the musician's brain is a paradigm of neuroplasticity [8].

Considering the above, it seems clear that musicians have greater motor coordination capacity than non-musicians, derived from intensive musical practice. However, the present study focuses on the performance of motor imagery. The question is whether undertaking music training in a mental (imagined) way can improve both motor coordination and the actual exercise. Indeed, using transcranial magnetic stimulation (TMS), Pascual-Leone et al. [22] demonstrated that the mere mental practice of an exercise in fingering a specific sequence for two hours a day over five days was sufficient to produce a certain reorganization of the motor cortex.

Subsequently, it was shown that the areas involved in motor imagery are approximately the same as those activated in real musical perception [23]. The same conclusion was drawn in another study [24] examining the MRI activity of seven pianists and seven participants with no musical experience. A few years earlier, [25] had indicated that many professional athletes and musicians can use movement imagery to improve their motor skills.

In the specific case of pianists, the most recent study by Zabielska-Mendyk et al. [26] compared the EEG patterns of pianists and non-pianists while executing both real and imagined fingering of different complexities. The power of the alpha and beta bands (mu rhythm modulation) decreased with decreasing fingering complexity (in both real and imagined cases), and this only occurred with the pianists; the non-musicians did not exhibit this attenuation. This capacity varies according to the experience in years that the musician accumulates, which is acquired progressively, as per [27]. This result already suggests a different behavior in terms of BCI performance.

Throughout these reviewed studies, it seems clear that circumstances may exist that improve the motor imagery skills of some individuals over those of others. However, how this advantage can lead to better performance when using a BCI system has thus far not been described. The presence of significant differences in this performance was investigated by Dobrea et al. [28], while it has been suggested that certain individual traits act as precursors in predicting performance in using a BCI system [29]; other predictors include spatial (motor) skills, which encompass the practice of a musical instrument.

Notably, some people appear to have no capacity to control a BCI, a phenomenon that numerous analysts in the literature have termed "BCI illiteracy" [30]. Hereby, it should be noted that BCIs are generally not easy to control, and even with proper instruction, some users cannot control their systems as desired. Nevertheless, BCI illiteracy is an inadequate concept for clarifying the trouble that users can have when working with BCI frameworks. First, it is a methodologically frail idea that depends on the imperfect assumption that BCI users have physiological or useful qualities that forestall capable performance during BCI use [31]. Second, the term BCI illiteracy invites a comparison between learning to use BCIs and spoken or written language acquisition. Hence, to avoid conceptual snares in terms of how BCI use may or may not relate to language learning, some researchers have chosen to use the term "BCI inefficiency".

Various aspects associated with music are widely used to control BCI systems. For example, Makeig et al. [32] set out to control a BCI system by recreating the emotions produced by different pieces of music. In other words, they sought to recognize the emotions generated by a melody. Up to 84% success in certain experiments was achieved with this method.

Next, as this article uses motor imagery to control a BCI system, it reviews the literature evaluating the performance achieved through this control strategy. In [30], the authors explored the different outcomes in motor imagery achieved with different participants, drawing a distinction between the variability among different participants and that among different states of the same participant. They covered previous works, comparing personal characteristics, psychological mood, and anatomical and physiological aspects, concluding that all these components are essential when discussing future BCI performance.

Randolph et al. [33] developed what has become one of the main works through which we delimit our study area and define our research hypothesis. The authors considered factors such as age, sex, playing sports, playing video games, taking psychiatric medication, and playing a musical instrument. They concluded that a series of personal characteristics influence the modulation of the mu rhythm, leading to better outcomes in the control of a BCI. Specifically, having motor dexterity of the hands leads to better control over a BCI device. Furthermore, they examined the characteristics of age, time spent typing per day, the performance of hand-arm movements, and whole-body movements. They found that both age and hand-arm movements correlate positively with the ability to modulate rhythm induced by both real and imagined movements. The possibility of doing sports or playing a musical instrument is implied within these movements.

A large proportion of motor training occurs when the brain anticipates a movement being executed, i.e., if substantial repetition occurs prior to a movement, cerebral training results in a wave anticipating the movement [34]. Overall, some previous studies have pointed to the importance of the chosen movement used to control the BCI system. For instance, [28] discussed different types of tasks, e.g., motor, mathematics, and linguistics skills, whereby the motor tasks included movements of the fingers of the hand (left/right) and arm (left/right), while Soriano et al. [35] reviewed the different imagined movements that have been used in BCI. In terms of specific results, [36] showed that the movements of the right hand generate a differentiated signal on the EEG and cause hemodynamic activity in the motor cortex of the left hemisphere, detected by fMRI. When such movements are imagined, they generate similar, but less stable, patterns [37].

Other movements that have been examined in this context include grasping with the hand [38], the general use of the fingers (fingering) [39], the use of the index finger [40], the use of the big toe [41], and the maximum contraction of the hand [37]. Soriano et al. [35] undertook a comparison of these movements. However, despite the wealth of previous research, there is no specific study analyzing the performance of BCI system control by professional pianists using their high-level skills with fingers and hands.

3. Methodology

3.1. Sample Characteristics

This work aims to investigate whether pianists can control a BCI system by means of motor imagination more efficiently than non-musicians. To this end, we conducted an experiment analyzing the BCI performances of a group of pianists and a group of non-pianists (generally non-musicians), which functioned as a control group. The characteristics of the participants in both groups were collected, such as their sex and age as well as other more specific features, such as musical practice, the practice of other activities that involve finger movement (typing, video games), and playing sports.

The sample size in this experiment was 8 individuals, with 4 in each group. During the experiment, all participants were duly informed of how it would be conducted (passive and non-invasive measurement) as well as how the collected data would be handled. Furthermore, the anonymity of their personal and EEG data was guaranteed at all times. All experimentation was conducted in accordance with the Declaration of Helsinki and the ethics committees of the involved institutions were asked for approval before the sessions began.

In the case of the pianist group, participants were sought who had at least 10 years of musical training. Both men and women were included, some of whom were still in the process of musical training. The group rehearsed for an average of 5.75 h a day, ensuring significant skill in motor coordination. The characteristics of this group are summarized in Table 1.

Table 1. Description of the group of pianists.

Characteristic	Values	
Subjects	4	
Sex	2 men and 2 women	
Level of education	2 students and 2 professionals	
	Mean	**Standard Deviation**
Age (years)	24.50	±1.50
Time playing (years)	12.75	±1.78
Musical practice (hours/day)	5.75	±1.47

Various factors were considered in the creation of the control group (non-pianists). First, the study avoided including people in this group who had some musical ability, either with the piano or another instrument. Second, as revealed in the previous literature, some factors can increase the performance in using BCI management systems, such as the practice of tasks that involve digital motor coordination (video games, typing, other professions that require precision motor skills, etc.) and the practice of sports that result in a substantial improvement in motor coordination. Therefore, an attempt was made to choose participants in such a way as to minimize these aspects. Table 2 summarizes the characteristics of the control group.

Table 2. Description of the control group.

Characteristic	Values	
Subjects	4	
Sex	2 men and 2 women	
	Mean	**Standard Deviation**
Age (years)	32.75	±5.44
Sport (hours/week)	3.50	±1.11
Digital practice (hours/day)	1.63	±0.96

The non-pianists in the control group were asked about their musical knowledge and musical practice. Those who volunteered for the control group and showed some musical knowledge and/or practiced with musical instruments of any kind (piano or other) were not selected. Likewise, those who practiced a sport at an almost professional level were not selected. Furthermore, all members of the control group had a normal level of digital activity that did not go beyond one or two hours a day spent on work-related typing on a computer keyboard and all practiced sport only sporadically.

The time that the participants in the control group spent on motor practice is contrasted with that of the pianists, as these all began their piano studies at around 10 years of age and had had professional careers lasting between 10 and 15 years. Their beginning at an early age and their years of practice was estimated to translate to about 6 h of daily finger practice (more on some days). This amount undoubtedly exceeded that of the control group, not only in terms of motor coordination but also long-term musical orientation, which has been shown to create characteristic brain structures. In the case of the pianist group, the hours spent on other fine motor coordination practices were also collected, but these values were very close to those of the control group and were also irrelevant compared to the hours of musical practice.

3.2. Resources Used and EEG Acquisition

Enobio-8 is the wireless and portable sensor system that we have used for EEG recording. This device consists of a neoprene helmet with 39 holes to cover the main positions of the distribution according to the 10–20 system. The helmet makes it possible to use dry and wet electrodes. The electrodes chosen for this experiment are of the dry type. Eight electrodes are used in this case located at F3, F4, T7, C3, CZ, C4, T8 and Pz according to 10/20 system. Figure 1 shows the location of the electrodes.

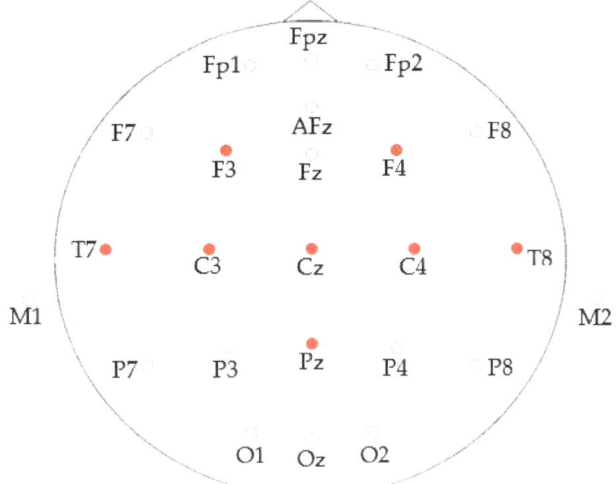

Figure 1. Electrodes location.

We must also take into account two more electrodes which will be used as grounding for the system, which allows the rest of the signals to be correctly referenced. They are attached to an Ag/AgCl EEG measurement patch with a conductive semi-liquid gel with low impedance, placed under the lobe of each ear. The electrical signals from the cortex are collected and subsequently sent to a computer via Bluetooth.

The accuracy of the system is approximately 0.005 μV. Knowing that the voltage we are working with ranges between 10 and 100 μV, the accuracy of the results is at least 0.005%.

The channels were recorded with a sampling frequency of 500 Hz. The recordings had a dynamic range of ±100 μV for all sessions. A notch filter at 50 Hz was enabled. We placed two electrodes at the mastoid bone as an EEG ground.

In addition, different standard software is used in the experiment for signal acquisition and post-processing:

- OpenVibe 2.2.0. This program is used as an interface between the computer and the user since it shows mental tasks and produces feedback. OpenVibe can also be used as a generic real-time EEG acquisition, processing and display system, allowing the online analysis of results, applying a feature extractor and classifier throughout the course of the experiment to perform the feedback task.
- Matlab R2016b. MATLAB, through toolboxes dedicated to Machine Learning, makes it possible to carry out an offline analysis, which will allow us to test the characteristics extractor and the classifier with different time windows.

3.3. Experimental Timeline

In order to find out if the research hypothesis has been fulfilled, the experiment was carried out in which the EEG signal of a group of pianists and a control group was analyzed. To do so, the motor imagery of two dichotomous movements was used, one of the left hand and the other of the right hand, which served to control a BCI system, and the success rate of each individual and each group was analyzed.

The most widely used training method for the BCI (and, therefore, followed in this study) is that based on the Graz Principle [42], developed at the Institute of Neural Engineering of the Technological University of Graz (Austria). This procedure is divided into two stages:

- Stage 1. Training the system to recognize the signal, which allows the computer to record brain information, which will be used to extract some spatial characteristics and divide them into classes. With this training, feature extraction is adjusted, and the classifier algorithm is trained.
- Stage 2. User feedback to teach subjects how to control the BCI. At this stage, the subject receives feedback from the computer and can see how the system perceived their actions, analyzing and modulating their own brain activity. This feedback consists of a blue bar shown in the screen that matches the interpretation of the Machine Learning algorithms trained in the first stage.

Three sessions of three trials each with 40 sequences per trial were carried out with each subject. The first trial of each session corresponds to the first stage described above, followed by the second trial, which develops the second stage. The third trial repeats the second stage in order to increase the number of data collected. Table 3 summarizes the above.

Table 3. Classic training method.

	Trial 1	Trial 2	Trial 3
Session 1	Training	Feedback	Feedback
	40 sequences	40 sequences	40 sequences
Session 2	Training	Feedback	Feedback
	40 sequences	40 sequences	40 sequences
Session 3	Training	Feedback	Feedback
	40 sequences	40 sequences	40 sequences

3.4. EEG Processing

In the first trial, no feedback was produced, since only data collection was carried out, which will be used to train the system through a feature extraction and a classifying algorithm. To do this, a spatial filter obtained by Common Spatial Pattern (CSP) and a Linear Discriminant Analysis (LDA) classifier were used, as will be explained later. However, for trials 2 and 3 of each session, a feedback was performed that consists of showing the user what the BCI system interprets they are thinking, according to the previously trained algorithms (online analysis). However, once the data collection of each subject was completed during the three sessions, an offline analysis was carried out, using as well CSP and LDA.

CSP is a mathematical technique used in signal processing to separate multivariable signals into sub-components with different variances. The CSP method was first proposed under the name Fukunaga-Koontz Transform in [43] as an extension of Principal Component Analysis (PCA) and has been widely used in BCI to maximize the distance between two classes of motions.

A CSP filter maximizes the variance of filtered EEG signals from one motion class while minimizing it for signals from the other class. The development of this technique comes naturally when we try to maximize the difference of variances between the two signals by spatially filtering them.

LDA is a Machine Learning technique used to perform supervised linear classification, based on a range of observations that can be divided into groups or classes [44]. The problem is basically to assign the right class to each observation. Linear classifiers are those that base their decision according to some hyper plane by assigning each class to one side of the subspace evaluated.

LDA is a simple and very stable technique, which will allow us to perform the type of classification we need without the use of any other parameter. This method is based on the assumption that we have two classes that follow a normal distribution. For each class, the parameters of mean and variance are modeled to get the distribution that best describes it and then the Bayes' theorem is used to calculate the probability of belonging to each of the classes.

The combination of CSP and LDA has been widely used in MI-based BCI to maximize the distance between two classes of movements [9,45]. There are some methods which observe the different activity between bilateral sides of hemispheres during the imagery, but some of these methods are too complex or demand too much computing time, so it is hard to apply them in real-time applications. CSP and LDA have also the characteristic of being 2-class methods. A CSP filter maximizes the variance of filtered EEG signals from one motion class while minimizing it for signals from the other class. After using this method, the goal is to train a classifier in such a way that it is able to estimate the class to which an observation belongs from observations of which we know the class, so LDA was used.

3.5. Experiment Deployment

The approximate time necessary for setting up and conducting the experiment was 45 min for each session, which required a certain availability of the subjects. Between each session, it was ensured that there was a rest period for each subject of at least four days. Between trials we provided 10 min in order to let the volunteer rest.

The procedure is similar as that depicted in BCI Competition IV [46]. The subjects were right-handed and had normal vision. All volunteers were sitting in a chair, with a flat screen monitor placed 1 m away at eye level.

Taking the existing literature into account, the subjects were asked to imagine specific movements of the hands and fingers. This movement will consist of a drumming of the fingers of the hand, accompanied by a swinging of the wrist up and down. Before carrying out the experiment, each subject was recommended to practice it in a real way to assimilate the physical sensation and, subsequently, to repeat it in an imagined way. Furthermore, during the experiment, it was recommended that subjects be careful not to make any parasitic movements with their hands, eyes or head, which could have jeopardized the accuracy of the calculation. This fact was explained at the beginning of the experiment and it was discussed in the following sessions.

The structure of the entire experiment can also be described with the diagram shown in Figure 2. Each session would begin with the donning of the helmet for EEG analysis. Subsequently, a first phase would be carried out with each subject that will allow adjustments to both the extraction of characteristics and the training of the classifier (online analysis). Once the parameters are adjusted, the system is ready to be able to present real-time feedback on the screen. Subsequently, two iterations are carried out in which the subject can see in real time the interpretation that the system makes of his movement imagery. This process will be repeated over three days and the data will later be analyzed offline.

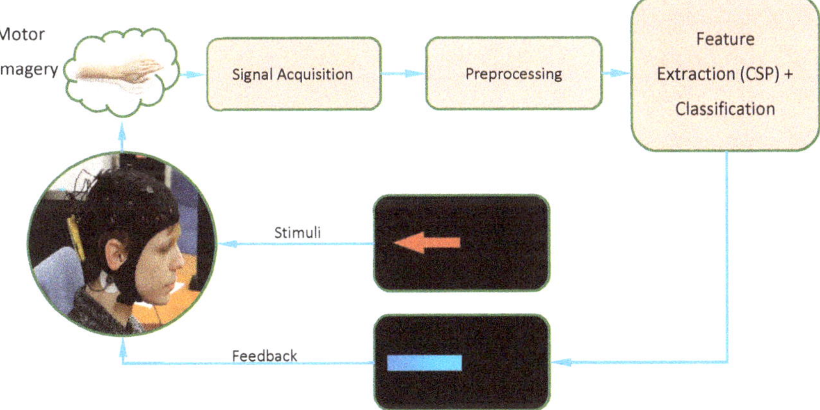

Figure 2. Experiment deployment.

A timing detail of each sequence can be seen in Figure 3. In second 0, the system presents a back screen. In second two presents a green cross and, 2 s later appear and arrow for 1.25 s. The arrow indicates de MI that the user must done. If the arrow point at left, the user must imagine the movement of the right hand and, if the arrow point left the user must imagine the movement with the other hand. Only in Trials 2 and 3 the system shows feedback as a blue line at second 5 and, it changes its longitude according to the output of the classifier. At second 8 the screen turns black and below there is a random time near 2 s in order to avoid synchronization between user and the timing protocol. For the presentation of the stimuli we used the previously named OpenVibe software.

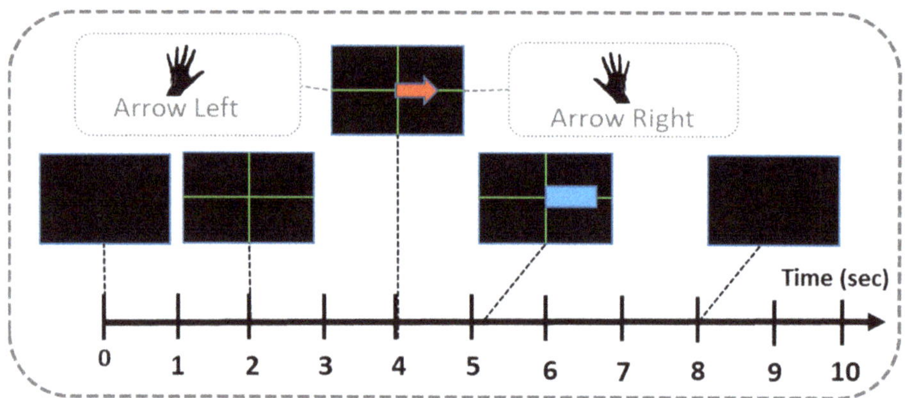

Figure 3. Timing of the brain-computer interface (BCI) System.

3.6. *Offline Analysis*

Once the data is recorded and the experimental phase is ended, we repeated the computation of the EEG offline but this time with a much more elaborate processing, since during the offline analysis the subject is not present, and then a more time-consuming algorithm can be used.

MATLAB is used to preprocess the subjects' EEG data. In the offline analysis, we enabled a notch filter at 50 Hz. This is due to the AC lines of the electrical supplies, which can introduce oscillations at 50 Hz and hence, noise oscillations in the EEG recordings. In addition, the data were low-pass filtered and high-pass filtered with cut-off frequencies of 1 Hz and 100 Hz, respectively, in order to eliminate frequencies which are impossible to be produced by the brain, thereby improving the signal cleanliness. Detection of artifacts is carried out with visually routines. An "artifact" can be described as any component of the EEG signal that is not directly produced by human brain activity, but induced by muscle activity, cardiac activity, respiration, and mainly blinks. The proximity of the eyes to frontal electrodes and the intensity of the blinking can produce a big distortion of the EEG that is sometimes impossible to be cleaned. To the trained eye, it is easy to detect, in an EEG graph, the presence of artifacts and their importance. The most notable artifacts are removed by rejecting the piece of data containing the artifact, forcing to discard in some cases the whole data of some volunteers.

To obtain the performance results of each subject, a combination of the CSP and LDA algorithms executed with leave-one-out cross validation has been used. Leave-one-out cross-validation is a certain cross-validation case where the number of folds is equal to the instances in the data set. Hence, the learning algorithm is applied one time for each instance, using all the other measurements as the training set and using the selected instance as a unique item test-set.

3.7. *Statistical Analysis*

In order to appreciate differences in the performances between groups based on statistical support, some verification must be done. Firstly, a Shapiro-Wilk test has been performed with the goal of checking normal distribution. In this test, the null hypothesis (H0) considers that the data set came from a normally distributed population. The Shapiro-Wilk Test is very appropriate for small sample sizes (<50 samples).

Next, the parametric Welch's T-test has been used to estimate if pianists' performances differ significantly from non-pianists' achievements. This is a two-sample test which is used to test the hypothesis that two populations have equal means. So, the null hypothesis (H0) considers equal means between the two groups under discussion.

In both cases, a significance level $\alpha = 0.05$ is assumed to be appropriate.

4. Results

To carry out the experiment explained above, samples were taken from 8 subjects, 4 of whom were pianists and the other four were not (non-musicians), functioning as a control group.

In a controlled motor imagery BCI experiment, accuracy of between 80% and 90% is expected after 6–9 training sessions of 20 min each [47]. Different investigations have presented different thresholds for the "efficiency" of the BCI, but a reference value for acceptable results is 70% [29]. However, according to the state of the art, certain subjects may have difficulties using BCI systems.

Control of BCI systems requires learning both the system and the user; there must be mutual adaptation. Due to this, it is expected that the yields in the classification will increase with the user's training. Thus, the first two were introductory and learning, achieving the best performance on the third and last day.

These results are obtained from the offline analysis with the MATLAB program, using the 80 sequences together from trials 2 and 3 combined, from each session. Due to this, we will focus on the presentation of the results concerning the last day (session 3). The first trial of each day was used to train the CSP+LDA used in the online analysis in order to provide the closed loop feedback to the user.

Offline processing allows various results to be obtained, since the cross-validation technique can be applied to different subsets of data. This analysis will be done on the second and third trials of the third session, in which the user had feedback. Various authors such as [48] consider the first trial to be a training phase to give participants the opportunity to learn how to carry out the motor imagery, concentrating their analysis on the following phases. Moreover, the performance on the set of the 40 sequences of the second trial will be studied and we will also proceed with the 40 sequences of the third trial. In addition, independently, the 80 sequences of both tests will be taken together. The results corresponding to the group of pianists are summarized in Figure 4, and those of the control group, below (Figure 5).

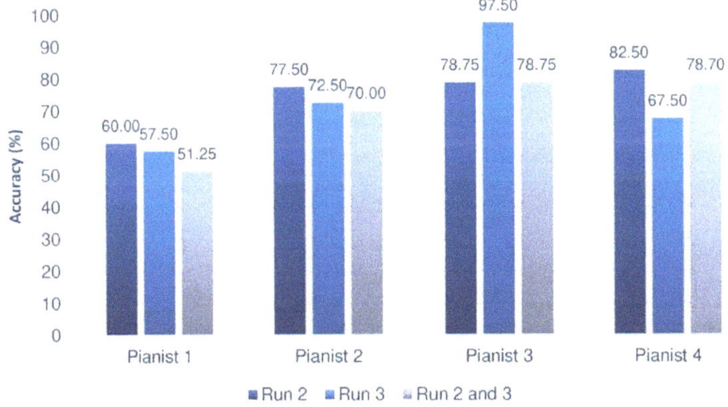

Figure 4. Offline results of the group of pianists in the third session. Measurements in Run 2 and 3 were taken from feedback trials.

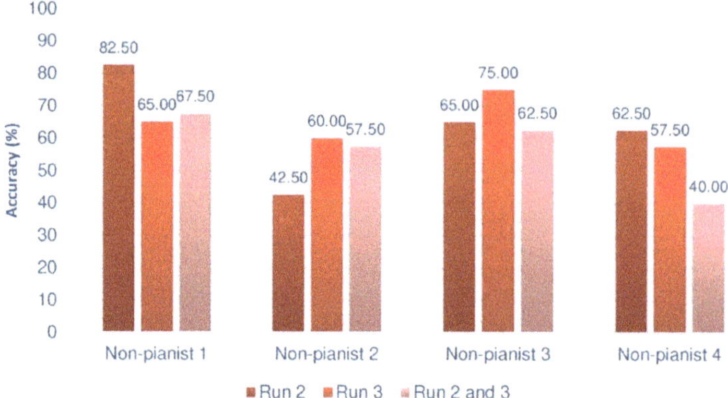

Figure 5. Offline results of the non-pianist group in the third session. Measurements in Run 2 and 3 were taken from feedback trials.

As stated above, a Shapiro–Wilk test can determine whether the data present a normal distribution. The results indicated that the data was normally distributed (p-values > 0.05). To compare the means of both groups, we performed the parametric Welch's T-test. The results indicated in an overall consideration that the performance of the pianists differed significantly from that of the non-pianists (p-value = 0.0344445, $\alpha = 0.05$). Figure 6 indicates the comparison between Runs (2, 3 and 2-and-3), showing that, in all circumstances, the difference between the averages is large enough to be statistically relevant and below the significance level of 0.05.

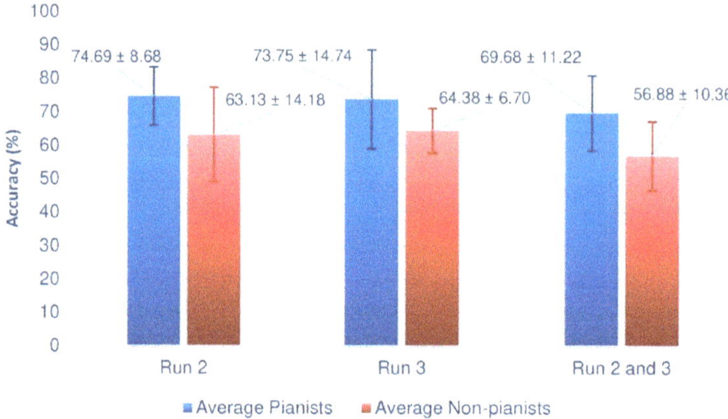

Figure 6. Comparison of offline results in both groups. Measurements in Run 2 and 3 were taken from feedback trials.

The scalp topography illustrates how the physiological sources project to the scalp. Figure 7 shows two examples of projected EEG signal after a CSP filter [49]. Subplots (a) and (b) present topographies of Pianist 3 in right and left MI along Run 2 and 3 at 22 Hz. Subplots (c) and (d) present topographies of Non pianist 2. All of them were computed along Run 2 and 3 at 22 Hz. Dense red or blue areas show where the greatest differences in the projected signals were found. As can be seen, for Pianist 3, channel F4 and Cz were the most actives, being C3 and C4 in the medium scale of colors. However, in Non-Pianist 2, channels C3, C4 and Pz are marked with strong red color. In general, we observe a greater area of activation in the Non-Pianist subject as compared with that in the Pianist participant.

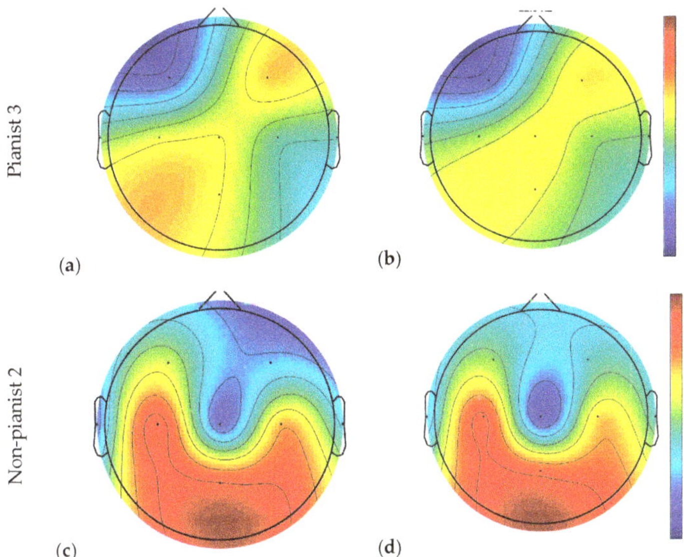

Figure 7. Examples of Projected electroencephalogram (EEG) signal after a Common Spatial Pattern (CSP) filter for Left and Right MI of Pianist 3 and Non-Pianist 2. Subfigures (**a**) and (**b**) belong to Pianist 3 (left/right, respectively), and subfigures (**c**) and (**d**) belong to Non-pianist 2 (left/right, respectively).

5. Discussion

Despite the disparity of results depending on the set of sequences analyzed, some conclusions can be outlined. Paying attention to the group of pianists, we can see that almost all of them, whatever the test taken, achieve yields in the third session greater than 70%. Pianist 1 would be a possible BCI-Inefficient user, although, as we have seen, we could only confirm this after 6 or 9 training sessions. This kind of users would require a different approach in order to improve their performance. In this sense, some studies have been developed with the aim of deploy a specific design of the experiment [50]. This would explain why the subject has not evolved or shown better performance. Among all of them, Pianist 3 stands out, reaching an almost perfect efficiency of 97.5% in the third trial. This result is very striking, considering that these rates are usually reached in more advanced sessions.

A common aspect in both groups is the drop in performance in the third trial, which occurs in 5 of the 8 cases. Mood, motivation, frustration, etc., are factors that determine the performance of the BCI. In total, the trial lasts an hour, and it is quite difficult to maintain the optimal concentration state during that time, carrying out the movement imagery tasks. Tejedor [51] calls it the "maturation effect", so that the optimum point of the results of a group of participants who are undergoing an experiment would be between the beginning (where the procedure is not yet mastered) and the end (where tiredness, fatigue, lack of interest, etc. come into play). In addition to this concept, Tejedor indicates that we have also to bear in mind the "experimental mortality", this is, the dropout of the volunteers along the experiment due to a lack of interest or change in the familiar/labor circumstances. Frustration can also trim the learning curve when we are considering inexperienced subjects. For such reason, motivation is an important issue to take into account in order to get acceptable results [52].

Comparing both figures, it is quite evident to conclude that the performance of pianists is, in general, higher than that of the control group, which seems to suggest that the proposed research hypothesis is fulfilled. Indeed, when calculating the mean of the results in each of the trials of the third session (the second, the third, and the set of second and third), we observe that the average is always higher in the pianists, as shown by the Figure 6, where the standard deviations of each mean value are also indicated.

Not only is it significant that the means are superior, regardless of the way in which the data is processed, but also that the performance is always close to 70%. Furthermore, recall that these results are impoverished by the presence of a supposed subject with BCI illiteracy, which would also explain why the mean of the set of trials 2 and 3 remains a few hundredths away from the optimal value of BCI performance. In the case of the group of non-pianists, except for some partial results, this figure is not reached. It is striking that the fourth subject of the group of non-pianists has an extremely low result when joining the two sessions.

The present study was intended to examine the analysis of the performance of a BCI device by means of the motor imagery carried out by pianists. In this sense, this work has synthesized some relevant results regarding the neurobiological differences that musical practice entails, showing that said functional specialization entails significant anatomical differences. Moreover, the foundations of the technology of the brain-computer interfaces have been exposed. These artifacts are able to detect the electrical activity of the brain with electrodes that perform an EEG and, through processing with machine learning algorithms, this biological signal is transferred into an order that can have different applications, such as the control of smart homes or robots.

The musician's brain is a paradigmatic case of neuroscience. However, in the existing literature, no previous study has been found that analyzes the performance of musicians with BCI systems, using motor imagery. The literature does seem to show that having musical prowess suggests better performance, but in the absence of a specific study in this regard, it was questionable whether, in fact, musical training and structural differences in the brain of musicians entailed better performance. To answer this hypothesis, an experimental procedure based on BCI has been designed for this paper.

Music is a global activity that involves the development of different capacities and intelligences: it improves memory, concentration, reading, self-esteem, emotional intelligence, psychomotricity, etc. In this sense, given the need to control BCI devices in the future, we can affirm that musical practice could improve motor coordination as well as neural plasticity, as revealed in the literature review, hence favoring their optimal use. To conclude that musical training is one of the factors that favors performance with motor imagery does nothing more than claim the importance of musical education in our society.

6. Conclusions

In this work, the performance of a set of pianists as they interacted with an MI-based BCI system was assessed and compared with a control group. The Common Spatial Patterns (CSP) and Linear Discriminant Analysis (LDA) machine learning algorithms were applied to the EEG signals for feature extraction and classification, respectively. The results revealed that the pianists achieved a higher mean level of BCI control—by means of MI—during the final trial (74.69%) in comparison to the control group (63.13%).

Regarding the above, it can be concluded that there seems to be indications that musical training is indeed a factor that improves the performance of a BCI device through movement imagery. As mentioned before, the performances achieved by the pianists in the last trial are on the order of 10 points higher than the non-pianists, regardless of the data set analyzed, which suggests that the previous hypothesis is true.

In future research, these results could be completed, on the one hand, by expanding the sample and, on the other, by supporting training for several sessions in order to reach more definitive conclusions. With this, we could explore the improvement of the performance of the two cases we found that were BCI inefficient. As we cannot venture a hypothesis based on only three sessions, with a greater number of measurements, we would be able to explore the evolution of these participants and deepen our understanding of so-called BCI illiteracy.

Author Contributions: Conceptualization, J.-V.R.-R., I.R.-R., G.R.-B. and J.-V.R.; methodology, J.-V.R.-R., I.R.-R. and G.R.-B.; software, G.R.-B.; validation, J.-V.R.-R., I.R.-R. and G.R.-B.; formal analysis, J.-V.R.-R., I.R.-R. and G.R.-B.; investigation, J.-V.R.-R., I.R.-R. and G.R.-B.; resources, J.-V.R.-R., I.R.-R. and G.R.-B.; data curation,

J.-V.R.-R., I.R.-R. and G.R.-B.; writing—original draft preparation, J.-V.R.-R. and I.R.-R.; writing—review and editing, J.-V.R.-R., I.R.-R., G.R.-B., J.-V.R. and J.-M.M.-G.-P.; visualization, J.-V.R.-R., I.R.-R., G.R.-B., J.-V.R. and J.-M.M.-G.-P.; supervision, I.R.-R., G.R.-B., J.-V.R. and J.-M.M.-G.-P.; project administration, G.R.-B., J.-V.R. and J.-M.M.-G.-P.; funding acquisition, J.-V.R. and J.-M.M.-G.-P. All authors have read and agreed to the published version of the manuscript.

Funding: This work has been funded by the Ministerio de Ciencia e Innovación, Spain (TEC2016-78028-C3-2-P and PID2019-107885GB-C33), and by European Fonds Européen de Développement Économique et Régional (FEDER) funds. This paper has been partially supported by Ministerio de Ciencia, Innovacion y Universidades grant number PGC2018-0971-B-100 and Fundación Séneca de la Región de Murcia grant number 20783/PI/18.

Acknowledgments: Ignacio Rodríguez-Rodríguez would like to thank the support of Programa Operativo FEDER Andalucía 2014–2020 under Project No. UMA18-FEDERJA-023 and Universidad de Málaga, Campus de Excelencia Internacional Andalucía Tech.

Conflicts of Interest: The authors declare no conflict of interest.

References

1. Nicolás-Alonso, L.F.; Gómez-Gil, J. Brain computer interfaces, a review. *Sensors* **2012**, *12*, 1211–1279. [CrossRef] [PubMed]
2. Abdulkader, S.N.; Atia, A.; Mostafa, M.-S.M. Brain computer interfacing: Applications and challenges. *Egypt. Inform. J.* **2015**, *16*, 213–230. [CrossRef]
3. Edelman, B.J.; Meng, J.; Suma, D.; Zurn, C.; Nagarajan, E.; Baxter, B.S.; He, B. Noninvasive neuroimaging enhances continuous neural tracking for robotic device control. *Sci. Robot.* **2019**, *4*, eaaw6844. [CrossRef] [PubMed]
4. Anumanchipalli, G.K.; Chartier, J.; Chang, E.F. Speech synthesis from neural decoding of spoken sentences. *Nature* **2019**, *568*, 493–498. [CrossRef] [PubMed]
5. Biasiucci, A.; Leeb, R.; Iturrate, I.; Perdikis, S.; Al-Khodairy, A.; Corbet, T.; Viceic, D. Brain-actuated functional electrical stimulation elicits lasting arm motor recovery after stroke. *Open Access* **2018**, *9*, 1–13. [CrossRef]
6. Zhang, X.; Yao, L.; Zhang, S.; Kanhere, S.; Sheng, Q.; Liu, Y. Internet of things meets brain-computer interface: A unified deep learning framework for enabling human-thing cognitive interactivity. *J. Latex Cl. Files* **2015**, *14*, 1–8. [CrossRef]
7. Mathe, E.; Spyrou, E. Connecting a consumer brain-computer interface to an internet-of-things ecosystem. In Proceedings of the 9th ACM International Conference on PErvasive Technologies Related to Assistive Environments (PETRA '16), Corfu Island, Greece, 29 June 2016; pp. 1–2. [CrossRef]
8. Münte, T.F.; Altenmüller, E.; Jäncke, L. The musician's brain as a model of neuroplasticity. *Nat. Rev. Neurosci.* **2002**, *3*, 473–478.
9. Rodriguez-Bermudez, G.; Lopez-Belchi, A.; Girault, A. Testing brain–computer interfaces with airplane pilots under new motor imagery tasks. *Int. J. Comput. Intell. Syst.* **2019**, *12*, 937–946. [CrossRef]
10. Soria-Urios, G.; Duque, P.; García-Moreno, J.M. Música y cerebro (II): Evidencias cerebrales del entrenamiento musical. *Neurología* **2011**, *53*, 739–746. [CrossRef]
11. Herholz, S.C.; Zatorre, R.J. Musical training as a framework for brain plasticity: Behavior, function, and structure. *Neuron* **2012**, *76*, 486–502. [CrossRef]
12. Altenmüller, E.; Finger, S.; Boller, F. *Music, Neurology, and Neuroscience: Evolution, the Musical Brain, Medical Conditions, and Therapies*; Elsevier: Amsterdam, The Netherlands, 2015.
13. Panksepp, J. *Affective Neuroscience: The Foundations of Human and Animal Emotions*; Oxford University Press: New York, NY, USA, 2004.
14. Panksepp, J.; Bernatzky, G. Emotional sounds and the brain: The neuro-affective foundations of musical appreciation. *Behav. Process.* **2002**, *60*, 133–155. [CrossRef]
15. Sacks, O. *Musicofilia: Relatos de la música y el cerebro*; Editorial Anagrama: Barcelona, Spain, 2009.
16. Weeks, R.; Horwitz, B.; Aziz-Sultan, A.; Tian, B.; Wessinger, C.M.; Cohen, L.G.; Rauschecker, J.P. A positron emission tomographic study of auditory localization in the congenitally blind. *J. Neurosci.* **2000**, *20*, 2664–2672. [CrossRef]
17. Jäncke, L.; Shah, N.J.; Peters, M. Cortical activations in primary and secondary motor areas for complex bimanual movements in professional pianists. *Cogn. Brain Res.* **2000**, *10*, 177–183. [CrossRef]

18. Krings, T.; Töpper, R.; Foltys, H.; Erberich, S.; Sparing, R.; Willmes, K.; Thron, A. Cortical activation patterns during complex motor tasks in piano players and control subjects. A functional magnetic resonance imaging study. *Neurosci. Lett.* **2000**, *278*, 189–193. [CrossRef]
19. Kami, A.; Meyer, G.; Jezzard, P.; Adams, M.M.; Turner, R.; Ungerleider, L.G. Functional MRI evidence for adult motor cortex plasticity during motor skill learning. *Nature* **1995**, *377*, 155. [CrossRef] [PubMed]
20. Hund-Georgiadis, M.; Von Cramon, D.Y. Motor-learning-related changes in piano players and non-musicians revealed by functional magnetic-resonance signals. *Exp. Brain Res.* **1999**, *125*, 417–425. [CrossRef]
21. Boyke, J.; Driemeyer, J.; Gaser, C.; Buchel, C.; May, A. Training-induced brain structure changes in the elderly. *J. Neurosci.* **2008**, *28*, 7031–7035. [CrossRef]
22. Pascual-Leone, A.; Nguyet, D.; Cohen, L.G.; Brasil-Neto, J.P.; Cammarota, A.; Hallett, M. Modulation of muscle responses evoked by transcranial magnetic stimulation during the acquisition of new fine motor skills. *J. Neurophysiol.* **1995**, *74*, 1037–1045. [CrossRef]
23. Zatorre, R.J.; Halpern, A.R. Mental concerts: Musical imagery and auditory cortex. *Neuron* **2005**, *47*, 9–12. [CrossRef]
24. Bangert, M.; Peschel, T.; Schlaug, G.; Rotte, M.; Drescher, D.; Hinrichs, H.; Altenmüller, E. Shared networks for auditory and motor processing in professional pianists: Evidence from fMRI conjunction. *Neuroimage* **2006**, *30*, 917–926. [CrossRef]
25. Jeannerod, M. The representing brain: Neural correlates of motor intention and imagery. *Behav. Brain Sci.* **1994**, *17*, 187–202. [CrossRef]
26. Zabielska-Mendyk, E.; Francuz, P.; Jaśkiewicz, M.; Augustynowicz, P. The effects of motor expertise on sensorimotor rhythm desynchronization during execution and imagery of sequential movements. *Neuroscience* **2018**, *384*, 101–110. [CrossRef] [PubMed]
27. Lotze, M.; Scheler, G.; Tan, H.R.; Braun, C.; Birbaumer, N. The musician's brain: Functional imaging of amateurs and professionals during performance and imagery. *Neuroimage* **2003**, *20*, 1817–1829. [CrossRef] [PubMed]
28. Dobrea, M.C.; Dobrea, D.M. The selection of proper discriminative cognitive tasks. A necessary prerequisite in high-quality BCI applications. In Proceedings of the 2009 2nd International Symposium on Applied Sciences in Biomedical and Communication Technologies, Bratislava, Slovakia, 24–27 November 2009; pp. 1–6.
29. Jeunet, C.; N'Kaoua, B.; Lotte, F. Advances in user-training for mental-imagery-based BCI control: Psychological and cognitive factors and their neural correlates. *Prog. Brain Res.* **2016**, *228*, 3–35.
30. Ahn, M.; Jun, S.C. Performance variation in motor imagery brain–computer interface: A brief review. *J. Neurosci. Methods* **2015**, *243*, 103–110. [CrossRef]
31. Thompson, M.C. Critiquing the concept of BCI illiteracy. *Sci. Eng. Ethics* **2019**, *25*, 1217–1233. [CrossRef]
32. Makeig, S.; Leslie, G.; Mullen, T.; Sarma, D.; Bigdely-Shamlo, N.; Kothe, C. First demonstration of a musical emotion BCI. In Proceedings of the International Conference on Affective Computing and Intelligent Interaction, Memphis, TN, USA, 9–12 October 2011; Springer: Berlin/Heidelberg, Germany, 2011; pp. 487–496.
33. Randolph, A.B.; Jackson, M.M.; Karmakar, S. Individual characteristics and their effect on predicting mu rhythm modulation. *Int. J. Hum. Comput. Interact.* **2010**, *27*, 24–37. [CrossRef]
34. Bernardi, N.F.; De Buglio, M.; Trimarchi, P.D.; Chielli, A.; Bricolo, E. Mental practice promotes motor anticipation: Evidence from skilled music performance. *Front. Hum. Neurosci.* **2013**, *7*, 451. [CrossRef]
35. Soriano, D.; Silva, E.L.; Slenes, G.F.; Lima, F.O.; Uribe, L.F.; Coelho, G.P.; Anjos, C.A. Music versus motor imagery for BCI systems a study using fMRI and EEG: Preliminary results. In Proceedings of the 2013 ISSNIP Biosignals and Biorobotics Conference: Biosignals and Robotics for Better and Safer Living (BRC), Rio de Janerio, Brazil, 18–20 February 2013; pp. 1–6.
36. McFarland, D.J.; Miner, L.A.; Vaughan, T.M.; Wolpaw, J.R. Mu and beta rhythm topographies during motor imagery and actual movements. *Brain Topogr.* **2000**, *12*, 177–186. [CrossRef]
37. Halder, S.; Agorastos, D.; Veit, R.; Hammer, E.M.; Lee, S.; Varkuti, B.; Kübler, A. Neural mechanisms of brain–computer interface control. *Neuroimage* **2011**, *55*, 1779–1790. [CrossRef]
38. Neuper, C.; Scherer, R.; Wriessnegger, S.; Pfurtscheller, G. Motor imagery and action observation: Modulation of sensorimotor brain rhythms during mental control of a brain–computer interface. *Clin. Neurophysiol.* **2009**, *120*, 239–247. [CrossRef] [PubMed]

39. Berman, B.D.; Horovitz, S.G.; Venkataraman, G.; Hallett, M. Self-modulation of primary motor cortex activity with motor and motor imagery tasks using real-time fMRI-based neurofeedback. *Neuroimage* **2012**, *59*, 917–925. [CrossRef] [PubMed]
40. Birch, G.E.; Bozorgzadeh, Z.; Mason, S.G. Initial on-line evaluations of the LF-ASD brain-computer interface with able-bodied and spinal-cord subjects using imagined voluntary motor potentials. *IEEE Trans. Neural Syst. Rehabil. Eng.* **2002**, *10*, 219–224. [CrossRef] [PubMed]
41. Formaggio, E.; Storti, S.F.; Cerini, R.; Fiaschi, A.; Manganotti, P. Brain oscillatory activity during motor imagery in EEG-fMRI coregistration. *Magn. Reson. Imaging* **2010**, *28*, 1403–1412. [CrossRef] [PubMed]
42. Clerc, M.; Bougrain, L.; Lotte, F. (Eds.) *Brain-computer Interfaces: Foundations and Methods*; ISTE Limited: London, UK, 2016.
43. Fukunaga, K.; Koontz, W.L. Application of the karhunen-loeve expansion to feature selection and ordering. *IEEE Trans. Comput.* **1970**, *100*, 311–318. [CrossRef]
44. Balakrishnama, S.; Ganapathiraju, A. Linear discriminant analysis-A brief tutorial. *Inst. Signal Inf. Process.* **1998**, *18*, 1–8.
45. Wu, S.L.; Wu, C.W.; Pal, N.R.; Chen, C.Y.; Chen, S.A.; Lin, C.T. Common spatial pattern and linear discriminant analysis for motor imagery classification. In Proceedings of the 2013 IEEE Symposium on Computational Intelligence, Cognitive Algorithms, Mind, and Brain (CCMB), Singapore, 16–19 April 2013; pp. 146–151.
46. Leeb, R.; Brunner, C.; Müller-Putz, G.R.; Schlögl, A.; Pfurtscheller, G. *BCI Competition 2008—Graz Data Set B*; Institute for Knowledge Discovery (Laboratory of Brain-Computer Interfaces), Graz University of Technology: Graz, Austria, 2008.
47. Guger, C.; Edlinger, G.; Harkam, W.; Niedermayer, I.; Pfurtscheller, G. How many people are able to operate an EEG-based brain-computer interface (BCI)? *IEEE Trans. Neural Syst. Rehabil. Eng.* **2003**, *11*, 145–147. [CrossRef]
48. Mousavi, M.; Koerner, A.S.; Zhang, Q.; Noh, E.; de Sa, V.R. Improving motor imagery BCI with user response to feedback. *Brain Comput. Interfaces* **2017**, *4*, 74–86. [CrossRef]
49. Ramoser, H.; Muller-Gerking, J.; Pfurtscheller, G. Optimal spatial filtering of single trial EEG during imagined hand movement. *IEEE Trans. Rehabil. Eng.* **2000**, *8*, 441–446. [CrossRef]
50. Vidaurre, C.; Blankertz, B. Towards a cure for BCI illiteracy. *Brain Topogr.* **2010**, *23*, 194–198. [CrossRef] [PubMed]
51. Tejedor, F.T. Validez interna y externa en los diseños experimentales. *Rev. Esp. Pedagog.* **1981**, *39*, 15–39.
52. Lotte, F.; Larrue, F.; Mühl, C. Flaws in current human training protocols for spontaneous brain-computer interfaces: Lessons learned from instructional design. *Front. Hum. Neurosci.* **2013**, *7*, 568. [CrossRef] [PubMed]

Article

Indoor Scene Change Captioning Based on Multimodality Data

Yue Qiu [1,2,*], Yutaka Satoh [1,2], Ryota Suzuki [2], Kenji Iwata [2] and Hirokatsu Kataoka [2]

1 Graduate School of Science and Technology, University of Tsukuba, Tsukuba 305-8577, Japan; yu.satou@aist.go.jp

2 National Institute of Advanced Industrial Science and Technology (AIST), Tsukuba 305-8560, Japan; ryota.suzuki@aist.go.jp (R.S.); kenji.iwata@aist.go.jp (K.I.); hirokatsu.kataoka@aist.go.jp (H.K.)

* Correspondence: s1830151@s.tsukuba.ac.jp

Received: 31 July 2020; Accepted: 20 August 2020; Published: 23 August 2020

Abstract: This study proposes a framework for describing a scene change using natural language text based on indoor scene observations conducted before and after a scene change. The recognition of scene changes plays an essential role in a variety of real-world applications, such as scene anomaly detection. Most scene understanding research has focused on static scenes. Most existing scene change captioning methods detect scene changes from single-view RGB images, neglecting the underlying three-dimensional structures. Previous three-dimensional scene change captioning methods use simulated scenes consisting of geometry primitives, making it unsuitable for real-world applications. To solve these problems, we automatically generated large-scale indoor scene change caption datasets. We propose an end-to-end framework for describing scene changes from various input modalities, namely, RGB images, depth images, and point cloud data, which are available in most robot applications. We conducted experiments with various input modalities and models and evaluated model performance using datasets with various levels of complexity. Experimental results show that the models that combine RGB images and point cloud data as input achieve high performance in sentence generation and caption correctness and are robust for change type understanding for datasets with high complexity. The developed datasets and models contribute to the study of indoor scene change understanding.

Keywords: image captioning; three-dimensional (3D) vision; deep learning; human-robot interaction

1. Introduction

There have been significant improvements in artificial intelligence (AI) technologies for human–robot interaction (HRI) applications. For example, modern intelligent assistants (e.g., Google Assistant [1]) enable the control of household appliances through speech and allow remote home monitoring. HRI experiences can be improved through the use of AI technologies, such as the semantic and geometric understanding of 3D surroundings [2–5], the recognition of human gestures [6,7], actions [8,9], emotions [10,11], speech recognition [12,13], and dialog management [14,15]. A fundamental problem in indoor scene understanding is that scenes often change due to human activities, such as the rearranging of furniture and cleaning. Therefore, understanding indoor scene changes is essential for many HRI applications.

Developments in graphic processing units and convolutional neural network (CNN)-based methods have led to tremendous progress in 3D recognition-related studies. Various 2D approaches have been adapted for 3D data, such as recognition [3–5], detection [16], and segmentation [17]. Researchers have proposed a series of embodied AI tasks that define an indoor scene and an agent that explores the scene and answers vision-related questions (e.g., embodied question answering [18,19]), or navigates based on a given instruction (e.g., vision-language navigation [20,21]). However, most 3D recognition-related studies have focused on static scenes. Scene change understanding is less often discussed despite its importance in real-world applications.

Vision and language tasks, including visual question answering [22–24], image captioning [25–30], and visual dialog [31,32], have received much attention due to their practicality in HRI applications. These tasks correlate visual information with language. The image captioning task aims to describe image information using text, and thus can be used to report scene states to human operators in HRI applications. Several recent image captioning methods describe a scene change based on two images of the scene [33,34]. However, they use single-view image inputs and neglect the geometric information of a scene, which limits their capability in scenes that contain occlusions. Qiu et al. [35] proposed scene change captioning based on multiview observations made before and after a scene change. However, they considered only simulated table scenes with limited visual complexity.

To solve the above problems, we propose models that use multimodality data of indoor scenes, including RGB (red, green, and blue) images, depth images, and point cloud data (PCD), which can be obtained using RGB-D (RGB-Depth) sensors, such as Microsoft Kinect [36], as shown in Figure 1. We automatically generated large-scale indoor scene change caption datasets that contain observations made before and after scene changes in the form of RGB and depth images taken from multiple viewpoints and PCD along with related change captions. These datasets were generated by sampling scenes from a large-scale indoor scene dataset and objects from two object model datasets. We created scene changes by randomly selecting, placing, and rearranging object models in the scenes. Change captions were automatically generated based on the recorded scene information and a set of predefined grammatical structures.

Figure 1. Illustration of indoor scene change captioning from multimodality data. From the input of two observations (consisting of RGB (red, green, and blue) and depth images captured by multiple virtual cameras and point cloud data (PCD)) of a scene observed before and after a change, the proposed approach predicts a text caption describing the change. The multiple RGB and depth images are obtained from multiple viewpoints of the same scene via virtual cameras. Each virtual camera takes an RGB and a depth image from a given viewpoint.

We also propose a unified end-to-end framework that generates scene change captions from observations made before and after a scene change, including RGB images, depth images, and PCD. We conducted extensive experiments on input modalities, encoders, and ensembles of modalities with datasets under various levels of complexity. Experimental results show that the models that combine RGB images and PCD can generate change captions with high scores in terms of conventional image caption evaluation metrics and high correctness in describing detailed change information, including change types and object attributes. The contributions of our work are four-fold:

- We automatically generated the first large-scale indoor scene change caption dataset, which will facilitate further studies on scene change understanding.
- We developed a unified end-to-end framework that can generate change captions from multimodality input, including multiview RGB images, depth images, and PCD, which are available in most HRI applications.

- We conducted extensive experiments on various types of input data and their ensembles. The experimental results show that both RGB images and PCD are critical for obtaining high performance, and that the use of PCD improves change type prediction robustness. These results provide perspectives for enhancing performance for further research.
- We conducted experiments using datasets with various levels of complexity. The experimental results show that our datasets remain challenging and can be used as benchmarks for further exploration.

2. Related Work

2.1. 3D Scene Understanding

CNN-based methods have promising performance in various 3D scene understanding tasks, such as 3D object recognition [3–5], 3D detection [16], 3D semantic segmentation [17], and shape completion [37]. These methods use CNN structures to learn the underlying 3D structures based on data in various formats, such as multiview RGB images, RGB-D images, PCD, and meshes.

Su et al. [38] proposed a network for 3D object recognition based on multiview RGB images. They proposed a multiview CNN (MVCNN) structure for aggregating information via a view pooling operation (max or average pooling) from the CNN features of multiview images. Kanezaki et al. [39] proposed a framework for feature extraction from multiview images that predicts object poses and classes simultaneously to improve performance. Esteves et al. [40] suggested that existing MVCNNs discard useful global information and proposed a group convolutional structure to better extract the global information contained in multiview images. Eslami et al. [2] proposed a generative query network (GQN) that learns 3D-aware scene representations from multiview images via an autoencoder structure. Several studies have focused on 3D understanding based on RGB-D data. Zhang et al. [41] proposed a network for depth completion from a single RGB-D image that predicts pixel-level geometry information. Qi et al. [42] proposed a 3D detection framework that detects objects in RGB images and fuses depth information to compute 3D object regions. Recent CNNs that utilize PCD have also shown promising results. Qi et al. proposed PointNet [3], which is a structure for extracting features from raw PCD via the aggregation of local information by symmetric functions (e.g., global max pooling). They later proposed PointNet++ [4] for obtaining better local information. Zhang et al. [5] proposed a simple yet effective structure that aggregates local information of PCD via a region-aware max pooling operation.

Considering the availability of RGB-D data in HRI applications, we propose models that use multiview RGB and depth images and PCD. We adopt an MVCNN and a GQN for scene understanding based on RGB images, an MVCNN for aggregating multiview depth information, and PointNet for processing PCD.

2.2. Indoor Scene Datasets

Due to the high complexity and diversity of visual information, training a CNN-based indoor scene understanding method usually requires a massive amount of data. SUNCG [43] is a widely used dataset that consists of simulated 3D indoor scenes generated using computer graphics technologies. Several indoor scene datasets with scanned models of real scenes have recently been made publicly available [44–46]. The Gibson [44] dataset consists of 572 scenes and a simulator, which allows training for multiple embodied AI tasks, such as visual navigation. Matterport3D [45] contains 90 indoor scenes that are densely annotated with semantic labels. The Replica [46] dataset consists of 18 high-resolution (nearly photorealistic) scenes.

Several datasets for embodied AI tasks have been built based on the above 3D datasets. The Embodied Question Answering (EQA) v1.0 [18] dataset consists of scenes sampled from the SUNCG dataset with additional question–answer pairs. The authors further extended the EQA task for realistic scene setting by adapting the Matterport3D dataset to their Matterport3D EQA dataset [19]. The Room-to-Room dataset [20] added navigation instruction annotation to the Matterport3D dataset for the vision-language navigation task. In these datasets, the states of the scenes are static. Qiu et al. [35] proposed a simulated dataset for scene change captioning from multiview RGB images. However, they generated scenes with a solid background color, limiting the visual complexity. In contrast, we combine the Matterport3D dataset with two open source object model datasets, namely, NEDO item database [47] and YCB dataset [48], for creating scene change datasets, where scene changes are constructed by rearranging objects in 3D scenes. To the best of our knowledge, our dataset is the first large-scale indoor scene change dataset.

2.3. Change Detection

Change detection is a long-standing task in computer vision due to its practicality in real-world applications, such as scene anomaly detection, and disaster influence analysis. Change detection from street view images or videos has attracted much attention because it allows algorithms to focus on changed regions, decreasing the cost of image or video recognition [49,50]. Alcantarilla et al. [49] proposed a method that first reconstructs 3D geometry from video input and then inputs coarsely registered image pairs into a deconvolutional network for change detection. Zhao et al. [50] proposed a method with an encoder–decoder structure for pixel-level change detection based on street view images.

Change detection is also important in robot applications [51–53]. Ambrus et al. [51] proposed a method that distinguishes static and dynamic objects by reconstructing and comparing PCD of a room scene observed at different time steps. Fehr et al. [52] proposed a 3D reconstruction method that reconstructs static 3D scenes based on RGB-D images of scenes with dynamic objects. Jinno et al. [53] proposed a framework for updating a 3D map by comparing the existing 3D map with newly observed 3D data that may contain new or removed objects.

Existing change detection methods that utilize RGB images lack 3D geometry understanding. Several methods have been proposed for detecting changes from 3D data in various formats, such as RGB-D images and PCD, for robot applications. However, most works are limited to relatively small-scale datasets and do not specify detailed changes, such as the attributes of changed objects. In contrast, we consider change detection based on multimodality input, including RGB and depth images and PCD. Our models describe detailed scene changes, including change types and object attributes.

2.4. Change Captioning

The image captioning task has been widely discussed. Various image captioning methods have been proposed to achieve high-performance sentence construction by using attention mechanisms [25,26] or exploring relationships between vision and language [27,28]. Generating image captions with high diversity has also been widely discussed [29,30]. However, most existing image captioning methods generate descriptions from single-view images.

Several recent works have discussed captioning based on images that include scene changes [33–35]. Difference Description with Latent Alignment (DDLA) [33] generates change descriptions from two video frames observed from different time steps of a given scene. In DDLA, an image indicating the pixel-level difference between input frames is computed and a CNN is used for generating captions from this difference image. The DUal Dynamic Attention model (DUDA) [34] uses a dual attention structure for focusing regions of images before and after a change and a dynamic attention structure, which dynamically selects information from image features before or after a change, or the difference between them. DUDA is more robust to camera transformation compared to DDLA. However, both DDLA and DUDA neglect the 3D geometry information of scenes and thus are less suitable for scenes with occlusions. Qiu et al. [35] proposed a method that generates a compact scene representation from multiview images and then generates captions based on the scene representation. However, they performed experiments using scenes with solid colored backgrounds, and only considered RGB images from a fixed number of cameras.

In contrast, we explore and evaluate several input modalities, namely, RGB and depth images (with random camera position changes), and PCD. We conducted extensive experiments on various ensembles of these modalities. We also conducted experiments on datasets with complex and diverse visual information.

3. Approach

In robot applications, the ability to recognize scene changes is essential. We propose a framework that generates scene change captions from image pairs taken before and after a scene change. Our framework correlates the after-change scene with the before-change scene and provides detailed change descriptions, including change types and object attributes. Due to the availability of RGB-D data in robot applications, we developed models that use RGB and depth images of scenes observed from multiple viewpoints and PCD. Our framework can be trained end-to-end using raw images, PCD, and related change captions. Moreover, our framework enables inputs from one or more modalities. In the following subsections, we give the details of the proposed framework.

3.1. Overall Framework

As shown in Figure 2, our framework generates a change caption from the input of observations taken before and after a scene change. A scene observation consists of RGB and depth images observed from multiple viewpoints and PCD.

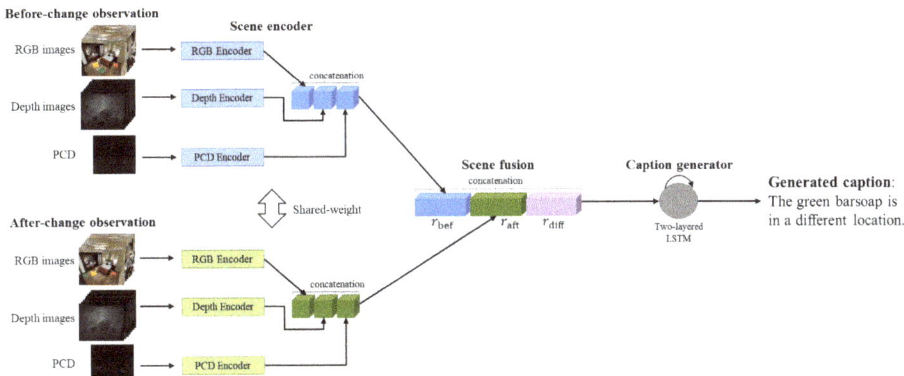

Figure 2. Overall framework. From before- and after-change scene observations of a scene, the proposed framework generates a text caption describing the scene change via three components, namely, a scene encoder that encodes the input composed of various modalities, a scene fusion component that combines the before- and after-change features r_{bef} and r_{aft}, a caption generator that generates a caption from the output of the scene fusion component.

Our framework comprises three components: a scene encoder, which processes input modalities with respective encoders; a scene fusion component, which combines the features of observations taken before and after a scene change; and a caption generator, which generates text captions from fused scene representations. Our framework can be further enhanced by adding more modalities, such as normal maps. Moreover, the scene fusion component and caption generator can also be improved by adopting novel approaches. We give the details of these three components in the remaining subsections.

3.2. Scene Encoder

The scene encoder component transforms an observation of a scene consisting of multiview RGB, depth images, and PCD into a feature vector to express semantic and geometric information for the scene. As shown in Figure 2, we first extract feature vectors from multiview RGB images, multiview depth images, and PCD separately with respective encoders. Then, the information is aggregated via a concatenation operation. We experimented with two encoder structures, namely, MVCNN and GQN, for encoding multiview RGB images and MVCNN for depth images.

RGB/depth encoder (MVCNN): This network is adapted from Su et al. [38]. In our implementation (Figure 3a), we first extract features from each viewpoint (we transformed depth images to RGB images through the applyColorMap function with mapping parameter COLORMAP_JET defined in OpenCV [54]) via ResNet101 [55]. Then, we apply convolution operation (separated weights) to extracted features and compute a weight vector via fully connection and softmax function. Then, we use another convolution layer to ResNet101-extracted features and multiply the output with the weight vector. Finally, a $4 \times 4 \times 128 \times 8$-dimensional feature vector is obtained.

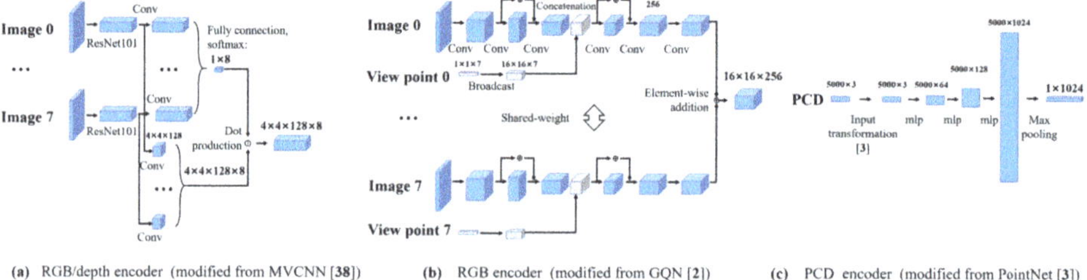

Figure 3. Detailed network structure for the encoders used in this study.

RGB encoder (GQN): Eslami et al. [2] proposed GQN, which recovers an image from a given viewpoint through a scene representation network and a generation network. We adapt GQN (tower-structure [2], Figure 3b) to extract a scene representation from multiview RGB images in two stages. During the pretraining stage, we train the overall GQN using multiview scene images. Then, we discard the generation network of GQN and use the pretrained scene representation network to aggregate information from multiview images.

PCD encoder (PointNet): We use PointNet, proposed by Qi et al. [3], for extracting features from PCD. PointNet transforms raw PCD into a feature vector that can be used for a range of downstream tasks, such as classification and segmentation. The detailed structure used in this work is shown in Figure 3c.

After processing these modalities separately, we resize features to $1 \times 1 \times k$-dimensional vectors (k is different for three modalities) and combine features using a concatenation operation. The concatenation operation makes it possible to change the number of input feature vectors, enabling both single- and multimodality inputs.

3.3. Before- and After-Change Scene Fusion

We process the observation pairs taken before and after a scene change using the process described in the previous subsection. The two feature vectors are combined via the fusion method proposed by Park et al. [34].

Specifically, we denote the feature vector of the before-change scene as r_{bef} and that of the after-change scene as r_{aft}. We first compute the vector difference r_{diff} of r_{bef} and r_{aft} via the following formulation.

$$r_{\text{diff}} = r_{\text{aft}} - r_{\text{bef}} \tag{1}$$

Then, we concatenate r_{bef}, r_{aft}, and r_{diff} to create a compact feature vector as the caption generator's input.

3.4. Caption Generator

The caption generator predicts a change caption from the output of the scene fusion component. As shown in Figure 2, we used a two-layer long short-term memory (LSTM) [56] structure for caption generation. Notably, the caption generator can be replaced by other language models, such as a transformer [57].

The overall network is trained end-to-end with the following loss function to minimize the distance L between generated captions $\mathbf{x}$ and ground truth captions $\mathbf{y}$:

$$L = -\log(\mathrm{P}(\mathbf{y}|\mathbf{x})) \tag{2}$$

4. Indoor Scene Change Captioning Dataset

Due to the high semantic and geometric complexity, training tasks targeting indoor scene understanding require a large amount of training data. Moreover, the construction of indoor scene datasets often requires a lot of manual labor. To the best of our knowledge, there is no existing large-scale indoor scene change captioning dataset.

To solve the above problems, we propose an automatic dataset generation process for indoor scene change captioning based on the existing large-scale indoor scene dataset Matterport3D [45] and two object model datasets, namely, NEDO [47] and YCB [48]. We create the before- and after-change scenes by arranging object models in indoor scenes. We set four atomic change types: add (add an object to a scene), delete (delete an object), move (change the position of an object), and replace (replace an object with a different one). We also set a distractor type in our dataset that indicates only camera position changes compared to original scenes. Implementing changes in the original 3D dataset is an alternative approach for creating datasets; however, this will result in artifacts, such as large holes after an object is deleted or moved from its original position. Thus, we sample object models from existing object model datasets and arrange them in 3D scenes to create before- and after-change scenes.

In the following subsections, we give the details of the automatic dataset generation process and the datasets created for experiments.

4.1. Automatic Dataset Generation

4.1.1. Scene and Object Models

We generated before- and after-change scene observation pairs based on arranging object models in 3D scenes. We used the Matterport3D dataset (consisting of 3D mesh models of 90 buildings with 2056 rooms) as our scene source. We selected 115 cuboid rooms that contain fewer artifacts (e.g., large holes in geometry) from the Matterport3D dataset. The object models used in our dataset generation were sampled from the NEDO and YCB datasets. We list the object class and instances in Table 1.

Table 1. Set-ups of object classes and instances for datasets used in this study.

Setups	Class (Number of Instances)
10-object setup	barsoap (2); cup (2); dishwasher (2); minicar (2); snack (2)
85-object setup	snack (12); dishwasher (11); tv dinner (11); minicar (7); barsoap (5); cup (5); plate (3); soft drink (3); sponge (3); air brush (2); baseball (2); bowl (2); facial tissue (2); magic marker (2); sauce (2); water bottle (2); weight (2); glass (1); glue (1); shampoo (1); soccer ball (1); tape (1); teddy bear (1); timer (1); toilet tissue (1); tooth paste (1)

4.1.2. Virtual Camera Setups

We took RGB and depth images from multiple camera viewpoints along with PCD for each scene observation. To obtain an overall observation for each room scene, as shown in Figure 4, we used cuboid rooms and set eight virtual cameras (four corners and four centers of edges of the ceiling) for observing scenes. Each virtual camera was set to look at the center of the room. In addition, to enhance robustness to camera position transformation, we added random offsets of [−10.0 cm, +10.0 cm] in three dimensions for each camera during the dataset acquisition.

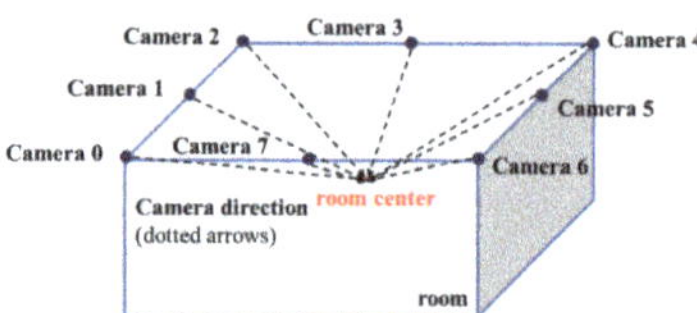

Figure 4. Virtual camera set-ups. Eight virtual cameras (four corners and four centers of edges of the ceiling) are set to look at the center of the room.

The data acquisition process can be implemented by using a single RGBD camera to observe a scene multiple times from various camera viewpoints.

4.1.3. Generation Process

We use AI Habitat [58] as the simulator for data acquisition. AI habitat enables generating RGB and depth images of given viewpoints from a mesh file. We generated each before- and after-change scene observation pair and related change captions in four steps. We first randomly selected a room scene from the scene sets and object models (three to five) from the object sets. The AI Habitat simulator provides a function named "get_random_navigable_point()", which computes a random position where the agent can walk on based on the mesh data and semantic information (semantic label information, such as "floor" and "wall", for each triangle vertex, provided in Matterport3D dataset). In the second step, we utilized the function to obtain random navigable positions and arranged objects on those positions. We took eight RGB and depth images and generated PCD as the original scene observation through the AI Habitat simulator. The Matterport3D dataset provides mesh data of every building and position annotation for each room. We generated PCD by transforming vertices of mesh data (triangular mesh) into points of PCD. We extracted PCD for each room from PCD of building based on room position annotation (3D bounding box annotation of rooms provided by the Matterport3D dataset). Next, we implemented the four change types (add, delete, move, and replace) for the original scene along with a distractor (only camera position transformation) and obtained scene observations. The change information, including change type and object attributes, was recorded. Finally, we generated five change captions for each change type and the distractor based on the recorded change information and predefined sentence structure templates (25 captions in total for each scene). We show an example of our dataset in Figure 5. The above process makes it easy to generate datasets with various levels of complexity by adjusting scene and object numbers, change types, and sentence templates.

We currently used PCD generated from meshes, which contains fewer artifacts, such as holes and less occlusion. To further improve the practicality of our method, we plan to use PCD generated from RGBD images and conduct experiments to discuss the effects of occlusion and artifacts.

4.2. Dataset Statistics

We generated dataset s15_o10 with 9000 scenes for training and 3000 scenes for testing. In s15_o10, we used 15 scenes and 10 object models (10-object set-up in Table 1). We used the s15_o10 dataset to evaluate the performance obtained with various input modalities, encoders, and ensembles.

To evaluate model performance under more complex scene settings, we adjusted the number of scenes and objects and generated dataset s15_o85 with 85 object models (85-object set-up in Table 1) and dataset s100_o10 with 100 scenes. The other settings of s15_o85 and s100_o10 are the same as those for s15_o10. The detailed dataset statistics are shown in Table 2. Experiments with these three datasets are presented below.

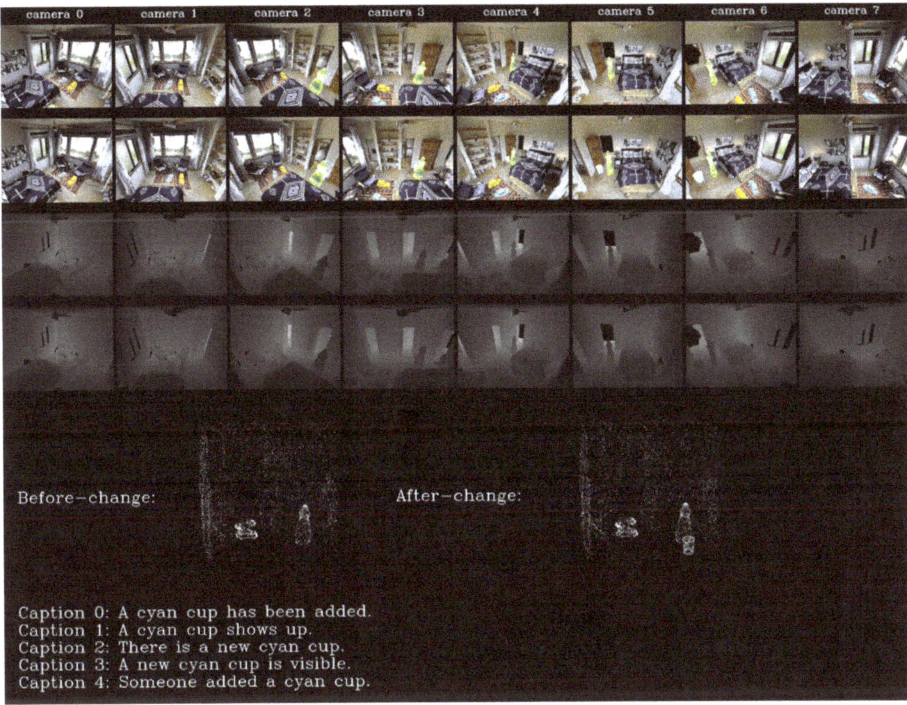

Figure 5. Dataset instance example of adding an object. From the top row: before-change RGB images observed from eight virtual cameras; after-change RGB images; before-change depth images; after-change depth images; before- and after-change PCD; five ground truth change captions.

Table 2. Statistics for datasets used in this study.

Dataset	No. of Scenes (Train/Test)	No. of Captions (Train/Test)	Change Types	Viewpoints	Scene Types	Object Classes	Objects Per Scene
s15_o10	(9000/3000)	(225,000/75,000)	5	8	15	10	2–6
s15_o85	(9000/3000)	(225,000/75,000)	5	8	15	85	2–6
s100_o10	(9000/3000)	(225,000/75,000)	5	8	100	10	2–6

5. Experiments

We used datasets s15_o10, s15_o85, and s100_o10 for training and evaluation. Specifically, we first used s15_o10 for the comparison of different input modalities, encoders (MVCNN and GQN for RGB images), and ensembles of input modalities. We then used s15_o85 and s100_o10 for assessing the models' abilities under more complex scene setups with an increased number of objects and scenes.

We adopted several conventional image captioning evaluation metrics in each experiment. In addition to these metrics, we conducted a caption correctness evaluation to examine the detailed information given by the generated captions (change types and object attributes).

5.1. Evaluation Metrics

We used four conventional evaluation metrics widely adopted in image captioning: BLEU-4 [59], ROUGE [60], SPICE [61], and METEOR [62]. These metrics evaluate the similarities between the generated captions and the ground truth captions. BLEU-4 is used to evaluate the recall of words or phrases (multiple words) of generated captions in the ground truth captions. ROUGE evaluates the recall of ground truth captions in generated captions. SPICE considers the correctness of the sentence structures of generated captions. METEOR introduces the similarity between words to encourage the generation of captions with diverse words.

The correctness of change type and object attributes is important in the change captioning task. Therefore, in addition to the above metrics, we conducted a caption correctness evaluation. We neglect the correctness of the sentence structure and extract change type, class, color, and object (including class and color, such as "red cup") from the generated captions and compute the accuracy when compared to the ground truth captions. This evaluation indicates how well the generated captions reflect the detailed change information.

5.2. Implementation Details

Here, we give the details of all the implementations. We set the input image size of both MVCNN and GQN to 64×64. We set the point number of PCD to 5000 for PointNet by random selecting points from PCD of rooms. For the pretraining process of GQN, we set the learning rate to 10^{-4} and trained the overall GQN network for 10 epochs in all experiments. For the overall framework training (including all single modalities and ensembles), we set the learning rate to 10^{-3} for PointNet and 10^{-4} for MVCNN and the decoder. All ablations were trained for 40 epochs. We used the Adam optimizer in all experiments.

5.3. Experiments on Input Modality and Model Ablations

We first used dataset s15_o10 to evaluate the performance of various input modalities, encoders, and ensembles. Here, we implemented four single-modality ablations, namely, depth images with the MVCNN encoder, RGB images with MVCNN and GQN encoders, and PCD with the PointNet encoder. We implemented five two-modality ablations containing two different input modalities, where the RGB images were processed using MVCNN and GQN encoders. We also implemented two ensembles with three modalities, where different RGB encoders were adopted.

Evaluation results on the test split of s15_o10 in terms of conventional evaluation metrics are shown in the top 11 rows of Table 3. The four single-modality ablation results show that PCD with the PointNet encoder obtained the best performance and that depth with the MVCNN encoder obtained the lowest scores for all metrics. RGB images with the GQN encoder outperformed the ablation with the MVCNN encoder. PCD contain geometric and object edge information, which is an advantage in a task that requires recognizing object change and detailed object attributes. GQN was trained to obtain a compact scene representation from multiview images, which likely made it better at correlating multiview information compared to MVCNN. Depth images do not contain color information and it is difficult to obtain object shapes from them, making it challenging to understand a scene change from only depth images. We found that although depth images alone performed poorly, ensembles containing RGB (MVCNN or GQN encoder) and depth images outperformed RGB images alone. We think that this resulted from the geometric information in depth images, which is difficult to extract from RGB images. The two ensembles with three input modalities outperformed all single-modalities and ensembles with two input modalities composed of their subsets. Scene change captioning performance can thus be enhanced by using both geometric and RGB information.

The caption correctness evaluation results are shown in the top 11 rows in Table 4. We found that ablation with RGB input (GQN encoder) outperformed PCD input in terms of object correctness, whereas PCD obtained higher accuracy in change type prediction (single modality). The abundant geometric and edge information contained in PCD is beneficial for change type prediction. We also found that models with both RGB and PCD (or depth) input obtained higher object correctness than that of single modalities. This result indicates that combining geometric and RGB information leads to a better understanding of detailed object information, which is critical for obtaining high performance in this task.

Table 3. Results of evaluation using s15_o10 (top 11 rows), s15_o85 (middle 6 rows), and s100_o10 (bottom 6 rows). The highest scores are shown in bold.

Modality (Encoder)	ROUGE [60]	SPICE [61]	METEOR [62]	BLEU-4 [59] Overall	Add	Delete	Move	Replace	Distractor
Depth (MVCNN)	58.36	18.46	26.50	36.93	21.30	39.15	42.22	32.90	45.11
RGB (MVCNN)	65.53	26.52	34.80	49.43	25.73	49.83	54.18	51.83	60.60
RGB (GQN)	81.98	36.02	48.49	71.01	59.59	77.31	59.52	68.52	85.80
PCD (PointNet)	89.75	35.32	49.34	76.17	62.89	79.19	77.98	71.73	96.19
Depth,RGB (MVCNN,MVCNN)	80.74	34.44	46.30	67.58	55.04	71.82	69.25	63.41	75.80
Depth,RGB (MVCNN,GQN)	84.52	36.51	50.71	74.46	63.99	77.12	65.49	71.48	90.22
Depth,PCD (MVCNN,PointNet)	84.41	27.74	43.03	65.91	54.84	68.73	59.37	58.00	98.85
RGB,PCD (MVCNN,PointNet)	89.31	36.40	50.50	75.65	65.97	80.02	75.76	70.93	91.00
RGB,PCD (GQN,PointNet)	92.36	41.36	57.00	84.74	81.65	**89.87**	70.67	80.77	**99.41**
Depth,RGB,PCD (MVCNN,MVCNN,PointNet)	89.87	35.99	52.11	79.02	64.18	82.38	77.06	73.49	96.51
Depth,RGB,PCD (MVCNN,GQN,PointNet)	**93.38**	**44.29**	**58.46**	**86.33**	**83.32**	89.69	**80.21**	**82.84**	98.55
Depth (MVCNN)	57.56	15.27	24.19	36.46	12.44	33.70	30.21	19.83	74.73
RGB (MVCNN)	57.31	15.17	24.80	35.67	18.08	31.78	32.46	29.78	59.91
RGB (GQN)	63.90	18.94	29.05	43.75	23.83	42.51	34.60	34.96	77.00
PCD (PointNet)	80.44	21.04	34.98	52.02	33.26	52.79	**55.59**	53.63	52.66
Depth,RGB,PCD (MVCNN,MVCNN,PointNet)	**81.26**	23.76	**38.72**	**59.79**	**43.52**	**60.90**	43.64	**55.96**	95.98
Depth,RGB,PCD (MVCNN,GQN,PointNet)	78.83	**24.49**	36.86	57.11	39.74	59.96	41.75	43.27	**97.69**
Depth (MVCNN)	56.06	16.51	24.56	33.46	16.45	31.00	35.64	29.72	48.53
RGB (MVCNN)	65.72	25.02	33.80	47.01	31.74	43.49	51.31	51.69	50.32
RGB (GQN)	65.23	23.91	32.24	47.10	35.31	46.79	47.36	39.92	58.85
PCD (PointNet)	82.93	26.29	41.48	63.30	47.22	66.04	49.33	62.50	**99.60**
Depth,RGB,PCD (MVCNN,MVCNN,PointNet)	**87.99**	35.72	**49.68**	**75.31**	**60.49**	**77.45**	**73.79**	**68.44**	96.72
Depth,RGB,PCD (MVCNN,GQN,PointNet)	86.07	**35.97**	47.23	73.10	60.46	74.30	63.08	67.83	99.02

Table 4. Results of change caption correctness evaluation using s15_o10 (top 11 rows), s15_o85 (middle 6 rows), and s100_o10 (bottom 6 rows). The highest scores are shown in bold.

Modality (Encoder)	Accuracy (%) Change Type	Object	Color	Class
Depth (MVCNN)	44.54	26.82	35.11	36.78
RGB (MVCNN)	49.26	49.52	62.16	53.18
RGB (GQN)	74.48	69.41	79.54	71.88
PCD (PointNet)	**97.24**	53.65	59.20	65.44
Depth,RGB (MVCNN,MVCNN)	73.59	62.53	73.74	66.38
Depth,RGB (MVCNN,GQN)	78.44	72.46	81.84	74.91
Depth,PCD (MVCNN,PointNet)	93.03	39.64	48.75	52.06
RGB,PCD (MVCNN,PointNet)	93.37	58.52	70.68	65.29
RGB,PCD (GQN,PointNet)	93.94	74.39	83.85	77.39
Depth,RGB,PCD (MVCNN,MVCNN,PointNet)	93.09	63.02	73.32	69.32
Depth,RGB,PCD (MVCNN,GQN,PointNet)	94.96	**75.35**	**84.05**	**78.30**
Depth (MVCNN)	52.42	9.13	22.10	19.73
RGB (MVCNN)	46.38	17.23	38.99	23.34
RGB (GQN)	59.68	17.33	34.45	26.77
PCD (PointNet)	**96.17**	14.62	27.53	27.28
Depth,RGB,PCD (MVCNN,MVCNN,PointNet)	92.55	**23.51**	**44.92**	**32.18**
Depth,RGB,PCD (MVCNN,GQN,PointNet)	90.75	20.29	37.65	**32.18**
Depth (MVCNN)	42.88	19.85	29.70	31.29
RGB (MVCNN)	50.66	43.58	58.18	47.73
RGB (GQN)	52.00	40.28	52.13	45.72
PCD (PointNet)	90.49	33.73	44.60	47.13
Depth,RGB,PCD (MVCNN,MVCNN,PointNet)	**91.54**	**55.28**	**69.60**	**61.64**
Depth,RGB,PCD (MVCNN,GQN,PointNet)	90.07	48.20	59.93	55.77

We show two example results in Figure 6. For the first example (object moved), all ensembles predict change captions correctly. Ablations with depth images and RGB images (MVCNN encoder) correctly determined the related object attributes (cyan minicar) but failed to predict the correct change type (move). In contrast, the single-modality ablation with PCD predicted the correct change type but gave the wrong color. For the second example (distractor with no object change), all models with

PCD and the ensemble with RGB (GQN encoder) and depth images gave correct change captions. However, the other models predicted wrong change captions. These results indicate that combining modalities enhances performance, and that PCD can provide geometry information, which is beneficial for predicting change type.

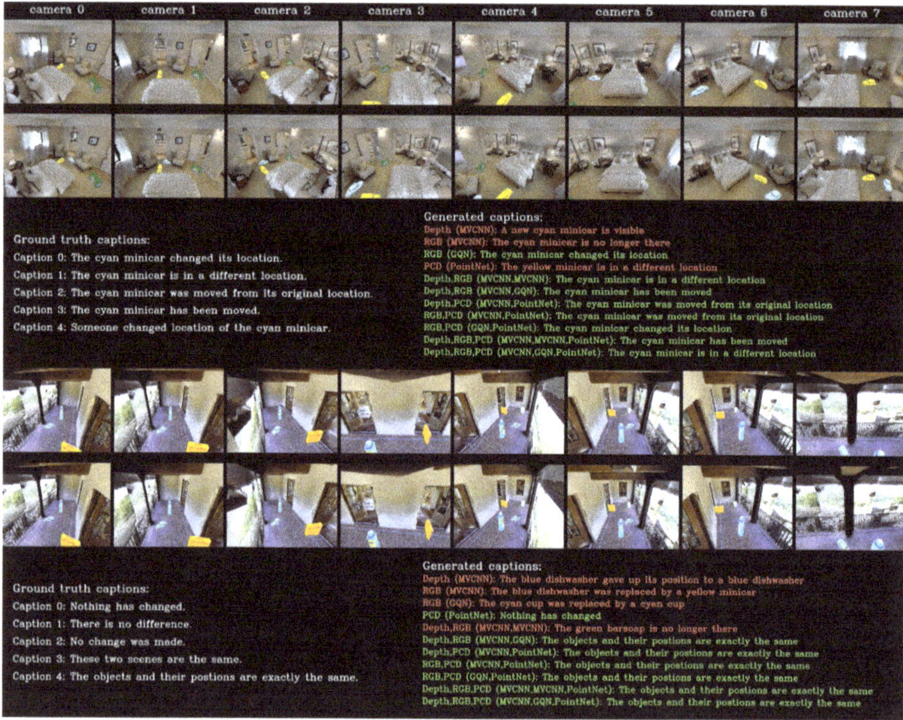

Figure 6. Example results for s15_o10. From the top row: before-change RGB images observed from eight virtual cameras for first example; after-change RGB images; ground truth and generated captions for various models; before-change RGB images for second example; after-change RGB images (distractor with only camera position transformation); ground truth and generated captions. Correct captions are shown in green and false captions are shown in red.

5.4. Experiments on Dataset Complexity (Object Class Number)

In this experiment, we evaluated model performance under a more complex set-up s15_o85, where the object instance number was 85 (10 objects in s15_o10). We conducted experiments on four single-modality and two three-modality ablations.

The results of conventional evaluation metrics are shown in the middle six rows of Table 3. The performance for all ablations is lower than that for s15_o10. Notably, the performance of models with RGB images and the GQN encoder degraded significantly. This result indicates that the GQN encoder is less robust to scene set-ups with high complexity. Similar to s15_o10, for single-modality input, PCD with PointNet performed the best. Ensembles outperformed single modalities.

The caption correctness evaluation is shown in the middle six rows of Table 4. The performance of all modalities degraded compared to that for s15_o10 in terms of object correctness. However, models with PCD, including both single modalities and ensembles, tended to be more robust for change type prediction. This result indicates that the geometry and edge information make change type prediction more consistent.

We show one example result (object deleted) in Figure 7 (top) . All single modalities failed to give correct captions, whereas the two ensembles predicted the correct caption. Single-modality models with depth or RGB images gave the wrong change type, whereas that with PCD correctly predicted the change type (delete). s15_o85 is more challenging than s15_o10 because it included more objects (85 vs. 10). Combining different modalities is effective for handling datasets with relatively high complexity.

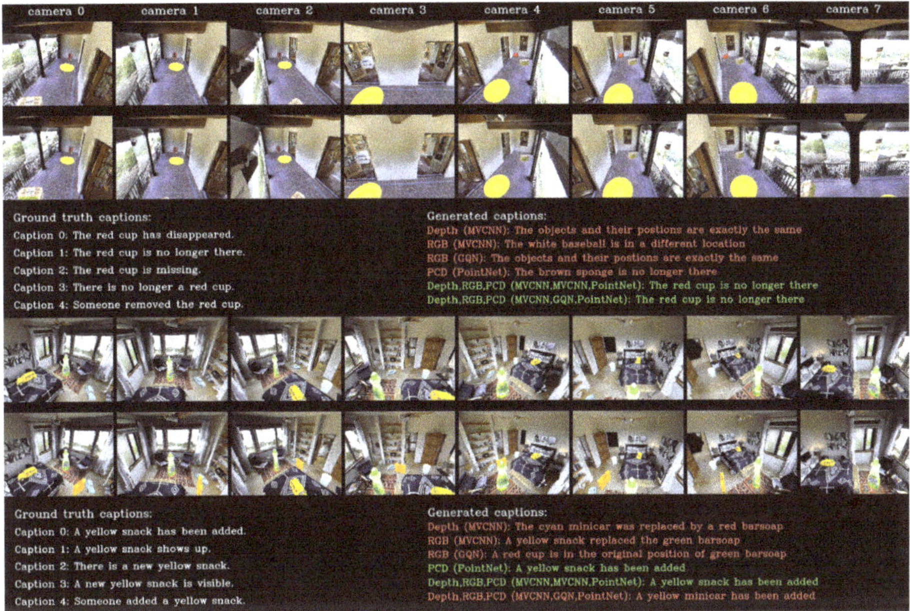

Figure 7. Example results for s15_o85 and s100_o10. From the top row: before-change RGB images observed from eight virtual cameras for example from s15_o85; after-change RGB images; ground truth and generated captions for various models; before-change RGB images for example from s100_o10; after-change RGB images; ground truth and generated captions. Correct captions are shown in green and false captions are shown in red.

5.5. Experiments on Dataset Complexity (Scene Number)

Here, we evaluate four single-modality and two three-modality ablations with dataset s100_o10, which included 100 scenes (15 scenes in s15_o10 and s15_o85).

The experimental results (bottom six rows of Table 3) for conventional evaluation metrics show that the performance of all ablations degraded compared to that for s15_o10. This is especially true for RGB images with the GQN encoder, which indicates that GQN is less suitable for large-scale scene datasets. Similar to s15_o10, PCD with PointNet obtained the highest scores and depth images showed poor performance. Ensembles tended to outperform single modalities. The caption correctness results (bottom six rows of Table 4) also show that performance degraded (especially for object correctness) compared to that for s15_o10. PCD with the PointNet encoder and ensembles tended to be more robust for caption type prediction.

One example result (object addition) for s100_o10 is shown in Figure 7 (bottom). Here, PCD with the PointNet encoder and the ensemble model with PCD, depth, and RGB images (MVCNN encoder) correctly predicted the change caption. For s100_o10, which contains 100 scenes, the performance of GQN encoder dramatically degraded, which may have influenced the performance of the ensemble model.

5.6. Discussion

The experiments using s15_o10, s15_o85, and s100_o10 indicate that for single modalities, PCD with the PointNet encoder consistently obtained the highest scores for most conventional caption evaluation metrics and the depth images with MVCNN encoder obtained the low scores in most experiments. A further evaluation of caption correctness indicated that models with RGB images (both MVCNN and GQN encoders) performed well in recognizing object attributes and those with PCD performed well in predicting change types. Model performance can be enhanced by adopting ensembles. In addition, both the RGB images and PCD are crucial for obtaining high performance with ensembles. Additional depth images improve the performance of models with RGB images but degrade that of models with PCD alone.

We found that for all modalities, performance degraded in experiments using datasets with more objects (s15_o85) and more scenes (s100_o10). However, models with PCD and the PointNet encoder tended to be relatively robust for change type prediction. Regarding the two types of RGB encoder, GQN outperformed MVCNN for s15_o10. For s15_o85 and s100_o10, the performance of GQN encoder significantly degraded, becoming worse than that of MVCNN. This result indicates that compared to MVCNN, the GQN network is less suitable for in large-scale scene-setting.

The experimental results reported here will facilitate future research on scene change understanding and captioning. To understand and describe scene changes, both geometry and color information are critical. Because we used the concatenation operation to aggregate the information of various modalities, introducing an attention mechanism to dynamically determine the needed features could help enhance model performance. We evaluated model performance under complex scene settings through experiments using s15_o85 and s100_o10. It is important to further study the adaptiveness of models to scene complexity by conducting more experiments using diverse dataset setups. PCD used in this study consist of scenes and object models with integral shapes that are beneficial in change captioning task. However, in real-world applications, obtaining PCD with integral object shapes is challenging. One way to enhance the practicality of our work is to conduct experiments on partially observed PCD further.

To the best of our knowledge, our work is the first attempt for indoor scene change captioning. It is an essential future direction to adapt existing indoor scene change detection methods, such as those in [51–53], to the change detection task and conduct comparison experiments between our work and existing change detection methods.

6. Conclusions

This study proposes an end-to-end framework for describing scene change based on before- and after-change scenes observed by multiple modalities, including multiview RGB and depth images and PCD. Because indoor scenes are constantly changing due to human activities, the ability to automatically understand scene changes is crucial for HRI applications. Previous scene change detection methods do not specify detailed scene changes, such as change types or attributes of changed objects. Existing scene change captioning methods use RGB images and conduct experiments using small-scale datasets with limited visual complexity. We automatically generated large-scale indoor scene change captioning datasets with high visual complexity and proposed a unified framework that handles multiple input modalities. For all experiments, models with PCD input obtained the best performance among single-modality models, which indicates that the geometry information contained in PCD is beneficial for change understanding. The experimental results show that both geometry and color information are critical for better understanding and describing scene changes. Models with the RGB images and PCD have promising performance in scene change captioning and exhibit high robustness for change type prediction. Because we used a concatenation operation for aggregating information from various modalities, model performance could be enhanced by introducing an attention mechanism to determine the required features. Experiments on datasets with high levels of complexity show that there is still room for improvement, especially for object attribute understanding. We plan to conduct more experiments on the adaptiveness to scene complexity.

Author Contributions: Y.Q. proposed and implemented the approaches and conducted the experiments. She also wrote the paper, together with Y.S., R.S., K.I., and H.K. All authors have read and agreed to the published version of the manuscript.

Funding: Computational resource of AI Bridging Cloud Infrastructure (ABCI) provided by National Institute of Advanced Industrial Science and Technology (AIST) was used.

Acknowledgments: The authors would like to acknowledge the assistance and comments of Hikaru Ishitsuka, Tomomi Satoh, Akira Noguchi, and Kodai Nakashima.

Conflicts of Interest: The authors declare no conflict of interest.

Abbreviations

The following abbreviations are used in this manuscript.

3D	Three-Dimensional
AI	Artificial Intelligence
HRI	Human–Robot Interaction
CNN	Convolutional Neural Network
RGB	Red Green Blue
PCD	Point Cloud Data
RGB-D	RGB-Depth
MVCNN	Multiview CNN
GQN	Generative Query Network
EQA	Embodied Question Answering
DDLA	Difference Description with Latent Alignment
DUDA	DUal Dynamic Attention
LSTM	Long Short-Term Memory

References

1. Google Assistant Site. Available online: https://assistant.google.com/ (accessed on 31 July 2020).
2. Eslami, S.A.; Rezende, D.J.; Besse, F.; Viola, F.; Morcos, A.S.; Garnelo, M.; Reichert, D.P. Neural scene representation and rendering. *Science* **2018**, *360*, 1204–1210. [CrossRef] [PubMed]
3. Qi, C.R.; Su, H.; Mo, K.; Guibas, L.J. Pointnet: Deep learning on point sets for 3d classification and segmentation. In Proceedings of the IEEE Conference on Computer Vision and Pattern Recognition, Honolulu, HI, USA, 21–26 July 2017; pp. 652–660.
4. Qi, C.R.; Yi, L.; Su, H.; Guibas, L.J. Pointnet++: Deep hierarchical feature learning on point sets in a metric space. In Proceedings of the Advances in Neural Information Processing Systems, Long Beach, CA, USA, 4–9 December 2017; pp. 5099–5108.
5. Zhang, Z.; Hua, B.S.; Yeung, S.K. Shellnet: Efficient point cloud convolutional neural networks using concentric shells statistics. In Proceedings of the IEEE International Conference on Computer Vision, Seoul, Korea, 27 October–3 November 2019; pp. 1607–1616.
6. Nickel, K.; Stiefelhagen R. Visual recognition of pointing gestures for human–robot interaction. *Image Vis. Comput.* **2007**, *25*, 1875–1884. [CrossRef]
7. Pavlovic, V.I.; Sharma, R.; Huang, T.S. Visual interpretation of hand gestures for human-computer interaction: A review. *IEEE Trans. Pattern Anal. Mach. Intell.* **1997**, *19*, 677–695. [CrossRef]
8. Ji, S.; Xu, W.; Yang, M.; Yu, K. 3D convolutional neural networks for human action recognition. *IEEE Trans. Pattern Anal. Mach. Intell.* **2012**, *35*, 221–231. [CrossRef] [PubMed]
9. Xia, L.; Chen, C.C.; Aggarwal, J.K. View invariant human action recognition using histograms of 3d joints. In Proceedings of the 2012 IEEE Computer Society Conference on Computer Vision and Pattern Recognition Workshops, Providence, RI, USA, 16–21 June 2012; pp. 20–27.
10. Hassan, M M.; Alam, M.G.R.; Uddin, M.Z.; Huda, S.; Almogren, A.; Fortino, G. Human emotion recognition using deep belief network architecture. *Inf. Fusion* **2019**, *51*, 10–18. [CrossRef]

11. Tzirakis, P.; Trigeorgis, G.; Nicolaou, M.A.; Schuller, B.W.; Zafeiriou, S. End-to-end multimodal emotion recognition using deep neural networks. *IEEE J. Sel. Top. Signal Process.* **2017**, *11*, 1301–1309. [CrossRef]
12. Afouras, T.; Chung, J.S.; Senior, A.; Vinyals, O.; Zisserman, A. Deep audio-visual speech recognition. *IEEE Trans. Pattern Anal. Mach. Intell.* **2018**. [CrossRef] [PubMed]
13. Dong, L.; Xu, S.; Xu, B. Speech-transformer: A no-recurrence sequence-to-sequence model for speech recognition. In Proceedings of the 2018 IEEE International Conference on Acoustics, Speech and Signal Processing (ICASSP), Calgary, AB, Canada, 15–20 April 2018; pp. 5884–5888.
14. Zhao, T.; Eskenazi, M. Towards end-to-end learning for dialog state tracking and management using deep reinforcement learning. *arXiv* **2016**, arXiv:1606.02560.
15. Asadi, K.; Williams, J.D. Sample-efficient deep reinforcement learning for dialog control. *arXiv* **2016**, arXiv:1612.06000.
16. Zhou, Y.; Tuzel, O. Voxelnet: End-to-end learning for point cloud based 3d object detection. In Proceedings of the IEEE Conference on Computer Vision and Pattern Recognition, Salt Lake City, UT, USA, 18–23 June 2018; pp. 4490–4499.
17. Graham, B.; Engelcke, M.; Van Der Maaten, L. 3d semantic segmentation with submanifold sparse convolutional networks. In Proceedings of the IEEE Conference on Computer Vision and Pattern Recognition, Salt Lake City, UT, USA, 18–22 June 2018; pp. 9224–9232.
18. Das, A.; Datta, S.; Gkioxari, G.; Lee, S.; Parikh, D.; Batra, D. Embodied question answering. In Proceedings of the IEEE Conference on Computer Vision and Pattern Recognition, Salt Lake City, UT, USA, 18–22 June 2018; pp. 2054–2063.
19. Wijmans, E.; Datta, S.; Maksymets, O.; Das, A.; Gkioxari, G.; Lee, S.; Essa, I.; Parikh, D.; Batra, D. Embodied question answering in photorealistic environments with point cloud perception. In Proceedings of the IEEE Conference on Computer Vision and Pattern Recognition, Long Beach, CA, USA, 15–21 June 2019; pp. 6659–6668.
20. Anderson, P.; Wu, Q.; Teney, D.; Bruce, J.; Johnson, M.; Sünderhauf, N.; van den Hengel, A. Vision-and-language navigation: Interpreting visually-grounded navigation instructions in real environments. In Proceedings of the IEEE Conference on Computer Vision and Pattern Recognition, Long Beach, CA, USA, 15–21 June 2019; pp. 3674–3683.
21. Fried, D.; Hu, R.; Cirik, V.; Rohrbach, A.; Andreas, J.; Morency, L.P.; Darrell, T. Speaker-follower models for vision-and-language navigation. In Proceedings of the Advances in Neural Information Processing Systems, Montreal, QC, Canada, 3–8 December 2018; pp. 3314–3325.
22. Antol, S.; Agrawal, A.; Lu, J.; Mitchell, M.; Batra, D.; Lawrence Z.C.; Parikh, D. Vqa: Visual question answering. In Proceedings of the IEEE International Conference on Computer Vision, Santiago, Chile, 7–13 December 2015; pp. 2425–2433.
23. Goyal, Y.; Khot, T.; Summers-Stay, D.; Batra, D.; Parikh, D. Making the V in VQA matter: Elevating the role of image understanding in Visual Question Answering. In Proceedings of the IEEE Conference on Computer Vision and Pattern Recognition, Honolulu, HI, USA, 21–26 July 2017; pp. 6904–6913.
24. Johnson, J.; Hariharan, B.; van der Maaten, L.; Fei-Fei, L.; Lawrence Zitnick, C.; Girshick, R. Clevr: A diagnostic dataset for compositional language and elementary visual reasoning. In Proceedings of the IEEE Conference on Computer Vision and Pattern Recognition, Honolulu, HI, USA, 21–26 July 2017; pp. 2901–2910.
25. Anderson, P.; He, X.; Buehler, C.; Teney, D.; Johnson, M.; Gould, S.; Zhang, L. Bottom-up and top-down attention for image captioning and visual question answering. In Proceedings of the IEEE Conference on Computer Vision and Pattern Recognition, Salt Lake City, UT, USA, 18–22 June 2018; pp. 6077–6086.
26. Lu, J.; Xiong, C.; Parikh, D.; Socher, R. Knowing when to look: Adaptive attention via a visual sentinel for image captioning. In Proceedings of the IEEE Conference on Computer Vision and Pattern Recognition, Honolulu, HI, USA, 21–26 July 2017; pp. 375–383.
27. Yao, T.; Pan, Y.; Li, Y.; Mei, T. Exploring visual relationship for image captioning. In Proceedings of the European Conference on Computer Vision, Munich, Germany, 8–14 September 2018; pp. 684–699.
28. Lee, K.H.; Palangi, H.; Chen, X.; Hu, H.; Gao, J. Learning visual relation priors for image-text matching and image captioning with neural scene graph generators. *arXiv* **2019**, arXiv:1909.09953.
29. Deshpande, A.; Aneja, J.; Wang, L.; Schwing, A.G.; Forsyth, D. Fast, diverse and accurate image captioning guided by part-of-speech. In Proceedings of the IEEE Conference on Computer Vision and Pattern Recognition, San Francisco, CA, USA, 13–18 June 2019; pp. 10695–10704.

30. Chen, F.; Ji, R.; Ji, J.; Sun, X.; Zhang, B.; Ge, X.; Wang, Y. Variational Structured Semantic Inference for Diverse Image Captioning. In Proceedings of the Advances in Neural Information Processing Systems, Montreal, QC, Canada, 3–8 December 2019; pp. 1931–1941.
31. Das, A.; Kottur, S.; Gupta, K.; Singh, A.; Yadav, D.; Moura, J.M.; Batra, D. Visual dialog. In Proceedings of the IEEE Conference on Computer Vision and Pattern Recognition, Honolulu, HI, USA, 21–26 July 2017; pp. 326–335.
32. Das, A.; Kottur, S.; Moura, J.M.; Lee, S.; Batra, D. Learning cooperative visual dialog agents with deep reinforcement learning. In Proceedings of the IEEE International Conference on Computer Vision, Venice, Italy, 22–29 October 2017; pp. 2951–2960.
33. Jhamtani, H.; Berg-Kirkpatrick, T. Learning to describe differences between pairs of similar images. *arXiv* **2018**, arXiv:1808.10584.
34. Park, D.H.; Darrell, T.; Rohrbach, A. Robust change captioning. In Proceedings of the IEEE International Conference on Computer Vision, Seoul, Korea, 27 October–3 November 2019; pp. 4624–4633.
35. Qiu, Y.; Satoh, Y.; Suzuki, R.; Iwata, K.; Kataoka, H. 3D-Aware Scene Change Captioning From Multiview Images. *IEEE Robot. Autom. Lett.* **2020**, *5*, 4743–4750. [CrossRef]
36. Kinect Site. Available online: https://www.xbox.com/en-US/kinect/ (accessed on 31 July 2020).
37. Dai, A.; Ruizhongtai Q.C.; Nießner, M. Shape completion using 3d-encoder-predictor cnns and shape synthesis. In Proceedings of the IEEE Conference on Computer Vision and Pattern Recognition, Honolulu, HI, USA, 21–26 July 2017; pp. 5868–5877.
38. Su, H.; Maji, S.; Kalogerakis, E.; Learned-Miller, E. Multi-view convolutional neural networks for 3d shape recognition. In Proceedings of the IEEE International Conference on Computer Vision, Santiago, Chile, 7–13 December 2015; pp. 945–953.
39. Kanezaki, A.; Matsushita, Y.; Nishida, Y. Rotationnet: Joint object categorization and pose estimation using multiviews from unsupervised viewpoints. In Proceedings of the IEEE Conference on Computer Vision and Pattern Recognition, Salt Lake City, UT, USA, 18–22 June 2018; pp. 5010–5019.
40. Esteves, C.; Xu, Y.; Allen-Blanchette, C.; Daniilidis, K. Equivariant multi-view networks. In Proceedings of the IEEE International Conference on Computer Vision, Long Beach, CA, USA, 16–20 June 2019; pp. 1568–1577.
41. Zhang, Y.; Funkhouser, T. Deep depth completion of a single rgb-d image. In Proceedings of the IEEE Conference on Computer Vision and Pattern Recognition, Salt Lake City, UT, USA, 18–22 June 2018; pp. 175–185.
42. Qi, C.R.; Liu, W.; Wu, C.; Su, H.; Guibas, L.J. Frustum pointnets for 3d object detection from rgb-d data. In Proceedings of the IEEE Conference on Computer Vision and Pattern Recognition, Salt Lake City, UT, USA, 18–22 June 2018; pp. 918–927.
43. Song, S.; Yu, F.; Zeng, A.; Chang, A.X.; Savva, M.; Funkhouser, T. Semantic scene completion from a single depth image. In Proceedings of the IEEE Conference on Computer Vision and Pattern Recognition, Honolulu, HI, USA, 21–26 July 2017; pp. 1746–1754.
44. Xia, F.; Zamir, A.R.; He, Z.; Sax, A.; Malik, J.; Savarese, S. Gibson env: Real-world perception for embodied agents. In Proceedings of the IEEE Conference on Computer Vision and Pattern Recognition, Salt Lake City, UT, USA, 18–22 June 2018; pp. 9068–9079.
45. Chang, A.; Dai, A.; Funkhouser, T.; Halber, M.; Niebner, M.; Savva, M.; Zhang, Y. Matterport3D: Learning from RGB-D Data in Indoor Environments. In Proceedings of the 2017 International Conference on 3D Vision (3DV), Qingdao, China, 10–12 October 2017; pp. 667–676.
46. Straub, J.; Whelan, T.; Ma, L.; Chen, Y.; Wijmans, E.; Green, S.; Clarkson, A. The Replica dataset: A digital replica of indoor spaces. *arXiv* **2019**, arXiv:1906.05797.
47. NEDO Item Database. Available online: http://mprg.cs.chubu.ac.jp/NEDO_DB/ (accessed on 31 July 2020).
48. Calli, B.; Singh, A.; Walsman, A.; Srinivasa, S.; Abbeel, P.; Dollar, A.M. The ycb object and model set: Towards common benchmarks for manipulation research. In Proceedings of the 2015 International Conference on Advanced Robotics (ICAR), Istanbul, Turkey, 27–31 July 2015; pp. 510–517.
49. Alcantarilla, P.F.; Stent, S.; Ros, G.; Arroyo, R.; Gherardi, R. Street-view change detection with deconvolutional networks. *Auton. Robot.* **2018**, *42*, 1301–1322. [CrossRef]
50. Zhao, X.; Li, H.; Wang, R.; Zheng, C.; Shi, S. Street-view Change Detection via Siamese Encoder-decoder Structured Convolutional Neural Networks. In Proceedings of the VISIGRAPP, Prague, Czech Republic, 25–27 February 2019; pp. 525–532.

51. Ambruş, R.; Bore, N.; Folkesson, J.; Jensfelt, P. Meta-rooms: Building and maintaining long term spatial models in a dynamic world. In Proceedings of the 2014 IEEE/RSJ International Conference on Intelligent Robots and Systems, Chicago, IL, USA, 14–18 September 2014; pp. 1854–1861.
52. Fehr, M.; Furrer, F.; Dryanovski, I.; Sturm, J.; Gilitschenski, I.; Siegwart, R.; Cadena, C. TSDF-based change detection for consistent long-term dense reconstruction and dynamic object discovery. In Proceedings of the 2017 IEEE International Conference on Robotics and Automation, Singapore, 29 May–3 June 2017; pp. 5237–5244.
53. Jinno, I.; Sasaki, Y.; Mizoguchi, H. 3D Map Update in Human Environment Using Change Detection from LIDAR Equipped Mobile Robot. In Proceedings of the 2019 IEEE/SICE International Symposium on System Integration (SII), Paris, France, 14–16 January 2019; pp. 330–335.
54. Bradski, G. The OpenCV Library. *Dr. Dobb's J. Softw. Tools* **2000**, *12*, 120–125.
55. He, K.; Zhang, X.; Ren, S.; Sun, J. Deep residual learning for image recognition. In Proceedings of the IEEE Conference on Computer Vision and Pattern Recognition, Las Vegas, NV, USA, 27–30 June 2016; pp. 770–778.
56. Gers, F.A.; Schmidhuber, J.; Cummins, F. Learning to forget: Continual prediction with LSTM. *Neural Comput.* **2000**, *12*, 2451–2471. [CrossRef] [PubMed]
57. Vaswani, A.; Shazeer, N.; Parmar, N.; Uszkoreit, J.; Jones, L.; Gomez, A.N.; Polosukhin, I. Attention is all you need. In Proceedings of the Advances in Neural Information Processing Systems, Long Beach, CA, USA, 4–9 December 2017; pp. 5998–6008.
58. Savva, M.; Kadian, A.; Maksymets, O.; Zhao, Y.; Wijmans, E.; Jain, B.; Parikh, D. Habitat: A platform for embodied ai research. In Proceedings of the IEEE International Conference on Computer Vision, Seoul, Korea, 27 October–2 November 2019; pp. 9339–9347.
59. Papineni, K.; Roukos, S.; Ward, T.; Zhu, W.J. BLEU: A method for automatic evaluation of machine translation. In Proceedings of the 40th Annual Meeting of the Association for Computational Linguistics, Philadelphia, PA, USA, 7–12 July 2002; pp. 311–318.
60. Lin, C.Y. Rouge: A package for automatic evaluation of summaries. In Proceedings of the ACL-02 Workshop on Automatic Summarization, Phildadelphia, PA, USA, 11–12 July 2002; pp. 45–51.
61. Anderson, P.; Fernando, B.; Johnson, M.; Gould, S. Spice: Semantic propositional image caption evaluation. In Proceedings of the European Conference on Computer Vision, Munich, Germany, 8–14 September 2016; pp. 382–398.
62. Denkowski, M.; Lavie, A. Meteor universal: Language specific translation evaluation for any target language. In Proceedings of the Ninth Workshop on Statistical Machine Translation, Baltimore, MA, USA, 26–27 June 2014; pp. 376–380.

Article

Nonlocal Total Variation Using the First and Second Order Derivatives and Its Application to CT image Reconstruction

Yongchae Kim and Hiroyuki Kudo *

Department of Computer Science, Graduate School of Systems and Information Engineering, University of Tsukuba, Tennoudai 1-1-1, Tsukuba 305-8573, Japan; acevip12@hotmail.com

* Correspondence: kudo@cs.tsukuba.ac.jp

Received: 16 May 2020; Accepted: 18 June 2020; Published: 20 June 2020

Abstract: We propose a new class of nonlocal Total Variation (TV), in which the first derivative and the second derivative are mixed. Since most existing TV considers only the first-order derivative, it suffers from problems such as staircase artifacts and loss in smooth intensity changes for textures and low-contrast objects, which is a major limitation in improving image quality. The proposed nonlocal TV combines the first and second order derivatives to preserve smooth intensity changes well. Furthermore, to accelerate the iterative algorithm to minimize the cost function using the proposed nonlocal TV, we propose a proximal splitting based on Passty's framework. We demonstrate that the proposed nonlocal TV method achieves adequate image quality both in sparse-view CT and low-dose CT, through simulation studies using a brain CT image with a very narrow contrast range for which it is rather difficult to preserve smooth intensity changes.

Keywords: image reconstruction; computed tomography; compressed sensing; nonlocal total variation; sparse-view CT; low-dose CT; proximal splitting; row-action; brain CT image

1. Introduction

Nonlocal Total Variation (TV) [1–6] was proposed as an improved version of ordinary TV. Nonlocal TV can use a weighting function (e.g., the weight of nonlocal means filter) by taking the intensity difference between the pixel pair into account, and can obtain higher image quality than the ordinary TV that uses only pairs of adjacent pixels.

Since G. Gilboa and S. Osher (2009) [5] proposed nonlocal operator, nonlocal TV has been widely applied to image reconstruction problems in sparse-view CT and low-dose CT [1–4].

H. Kim et al. (2016) [2] applied nonlocal TV to sparse-view CT and showed that nonlocal TV improves image quality over ordinary TV and incorporating the reweighted L1 norm into nonlocal TV further improves tissue contrast and structural details. Following this, K. Kim et al. (2017) [3] applied nonlocal TV to low-dose CT and showed nonlocal TV is effective for low-dose noise (Poisson noise).

D. Lv et al. (2019) [4] proposed a hybrid prior distribution (called NLTG prior) that combines nonlocal TV with Gaussian distribution. Additionally, they showed the possibility of applying NLTG prior to a large class of image reconstruction problems, especially when reference images were available.

However, the most existing nonlocal TV studies [2–4] are based on the first-order derivative, and still contain the staircase artifact problem as a potential drawback. Since the first-order derivative is too sensitive to the pixel values, even linear intensity changes are detected as false edges, which leads to staircase artifacts in the same way as local TV does [7–9].

On the other hand, higher-order derivatives (the second-order or more) possesses a potential risk that, as the order of differentiation is larger, its ability enhance image edges is smaller leading to an image blurring problem.

In previous studies, to overcome the disadvantage of using only the first-order or only the second-order, Bredies et al. (2010) [10] proposed Total Generalized Variation (TGV) that involves and balances higher-order derivatives and showed impressive results in the denoising problem. After that work, Ranftl et al. (2014) [11] proposed a nonlocal version of TGV.

Our proposed method is also a combined method of the first and second order derivatives. We define the regularization term as a weighted sum of two terms, where the first term is the ordinary nonlocal TV based on the first-order and the second term is based on the second-order. Additionally, in this paper, we introduce a newly discovered idea called nonlocal Total K-Split Variation, which is based on the second-order derivative using nonlocal regularization. This idea allows us to preserve smooth intensity changes well.

Furthermore, to accelerate the iterative algorithm associated with the proposed nonlocal TV, we propose a specially designed proximal splitting based on Passty's framework. In this proximal splitting, as the number of dividing the cost function into small subfunctions increases, the faster convergence is achieved [12–14]. The structure of final iterative algorithm is row-action type with respect to both the data-fidelity term and the regularization term [15–17], which converges to a minimizer very quickly [18–22].

Finally, in order to demonstrate the performances of combining the first and second order derivatives in reconstructed images, we use a brain CT image, in which a very narrow contrast range is used to display the image, where it is very difficult to preserve smooth intensity changes. Also, simulation studies are performed for both the sparse-view CT and the low-dose CT. We demonstrate that the proposed nonlocal TV method achieves adequate image quality within a small number of iterations.

2. Methodology

2.1. Problem Definition

We define the following unconstrained cost function:

$$\underset{\vec{x}\geq 0}{argmin} J(\vec{x}) = f(\vec{x}) + u(\vec{x}) = \left\|A\vec{x} - \vec{b}\right\|_2^2 + \beta\omega\left\|W\vec{x}\right\|_1^1, \quad (1)$$

where $f(\vec{x}) = \|A\vec{x} - \vec{b}\|_2^2$ is the data-fidelity term, and $u(\vec{x}) = \beta\omega\|W\vec{x}\|_1^1$ is the regularization term, and $A = \{a_{ij}\}$ is the $I \times J$ system matrix, β is the hyper-parameter to control regularization strength, and ω is the weight of regularization term, and W is the sparsifying transform to make $W\vec{x}$ sparse. Image reconstruction is an inverse problem to recover the image of attenuation coefficients $\vec{x} = (x_1, x_2, \ldots, x_J)^T$ from the measured projection data $\vec{b} = (b_1, b_2, \ldots, b_I)^T$.

In the sparse-view CT [23,24], by using the projection data corresponding to less than 100 directions (the conventional CT uses 1000–2000 projection data), the equation $A\vec{x} = \vec{b}$ becomes severely underdetermined, i.e., the dimension J of unknowns $\vec{x}$ is larger than the dimension I of measurements $\vec{b}$ $(I < J)$. In this case, the regularization term acts to avoid the ill-posed problem by introducing the prior knowledge that most components of the vector $W\vec{x}$ are close to 0.

On the other hand, in the low-dose CT, the equation $A\vec{x} = \vec{b}$ becomes inconsistent due to Poisson noise $\vec{n}$ $(A\vec{x} - \vec{b} = \vec{n})$. In this case, the regularization term helps to reduce the effect of Poisson noise $\vec{n}$ by a smoothing.

First, we begin by the modified anisotropic nonlocal TV based on the first-order derivative expressed as

$$Nonlocal\ TV = u^{TV}(\vec{x}) = \beta t \sum_{j=1}^{J} \sum_{j' \in \Omega} \omega_{jj'} |x_j - x_{j'}|, \tag{2}$$

where x_j is the intensity in the pixel j, and $x_{j'}$ is the intensity in a distant pixel j', and $\omega_{jj'}$ is the weight of smoothing assigned for each pixel pair $(x_j, x_{j'})$.

Next, we describe a newly discovered regularization term based on the second-order derivative called nonlocal Total K-split Variation (TKV). The main idea is to consider more derivatives around x_j and $x_{j'}$ as

$$Nonlocal\ TKV = u^{TKV}(\vec{x}) = \frac{\beta(1-t)}{8} \sum_{j=1}^{J} \sum_{j' \in \Omega} \sum_{k=1}^{8} \omega_{jj'} \left| (x_j - x_{j_k}) - (x_{j'} - x_{j'_k}) \right|, \tag{3}$$

where x_{j_k} is the adjacent pixel value of x_j, $x_{j'_k}$ is the adjacent pixel value of $x_{j'}$. We remark that the TKV term is divided into a sum of eight terms dependent of the direction to take the pixel difference as shown in Figure 1.

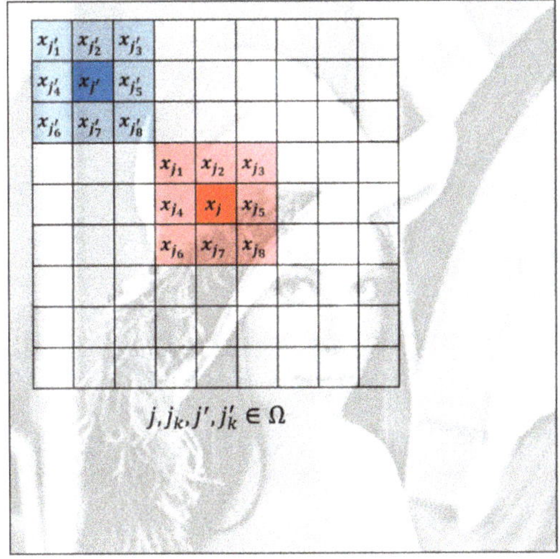

Figure 1. Definition of the pixel location in the proposed regularization term corresponding to $k = 1, 2, 3, \cdots, 7, 8$.

In the proposed method, we assume that the reconstructed image $\vec{x}$ is close to piecewise-polynomial of first order. Under this assumption, our proposed regularization term is designed to include both the first and second order derivatives as

$$u(\vec{x}) = \beta \sum_{j=1}^{J} \sum_{j' \in \Omega} \omega_{jj'} \left[t|x_j - x_{j'}| + \frac{(1-t)}{8} \sum_{k=1}^{8} \left| (x_j - x_{j_k}) - (x_{j'} - x_{j'_k}) \right| \right], (0 < t < 1), \tag{4}$$

where the proposed regularization term is a combination of two terms $(u(\vec{x}) = u^{TV}(\vec{x}) + u^{TKV}(\vec{x}))$. Additionally t is the trade-off parameter between nonlocal TV and TKV. If t is large, the reconstructed image becomes closer to that of nonlocal TV. If t is small, the reconstructed image becomes closer to that of nonlocal TKV. To match the strength of the first term into that of the second term, we divide the second-order derivative based nonlocal TKV into 8 directions. Figure 1 shows the location of pixels appearing in $u^{TV}(\vec{x})$ and $u^{TKV}(\vec{x})$ for $k = 1, 2, 3, \cdots, 7, 8$.

2.2. Accelerated Algorithm Using the Proximal Splitting with Passty's Framework

To accelerate the iterative algorithm to minimize the cost function, we propose a specially designed proximal splitting with Passty's framework. The ordinary proximal splitting is able to minimize a cost function consisting of a sum of only two component terms, where the proximity operator corresponding to each subfunction is applied alternately [12,13]. On the other hand, by using Passty's framework which is not well-known in the image reconstruction community, a cost function can be divided into a number of much simpler subfunctions [14]. This results in considerable benefits for optimization problems appearing in CT reconstruction. By applying the proximity operator corresponding to each subfunction alternately, the proposed algorithm possesses a form of row-action type [15–17], which converges to a minimizer very quickly [18–22].

We begin by a brief review the proximal splitting with Passty's framework. Let us consider a convex minimization problem formulated as

$$\underset{\vec{x}}{\operatorname{argmin}} J(\vec{x}), \tag{5}$$

where $J(\vec{x})$ is a lower semi-continuous (lsc) convex function.

The proximity operator corresponding to the function $J(\vec{x})$ is defined as

$$\vec{x} = prox_{\alpha J}(\vec{z}) \equiv \underset{\vec{x}}{\operatorname{argmin}}(J(\vec{x}) + \frac{1}{2\alpha}\|\vec{x} - \vec{z}\|_2^2), \tag{6}$$

where α is the parameter called step-size. We note that $J(\vec{x})$ can be a non-differentiable function such as component terms of TV or nonlocal TV. The proximity operator is a non-expansive mapping such that its fixed-points $\vec{x}$ satisfying $\vec{x} = prox_{\alpha J}(\vec{x})$ coincides with a minimizer of $J(\vec{x})$ for any $\alpha > 0$. Therefore, the minimization problem of $J(\vec{x})$ can be solved by the iterative formula expressed as $\vec{x}^{(n+1)} = prox_{\alpha J}(\vec{x}^{(n)})$ (i.e., proximal algorithm). Next, we are going to explain the proximal splitting with Passty's framework.

[Passty's framework] Let us consider the case where $J(\vec{x})$ can be divided into a sum of subfunctions as:

$$J(\vec{x}) = \sum_{i=1}^{I} J_i(\vec{x}). \tag{7}$$

The iterative algorithm can be constructed by applying the proximity operator corresponding to each subfunction $J_i(\vec{x})$ $(i = 1, 2, \ldots, I)$ as below

$$\vec{x}^{(n+1)} = prox_{\alpha^{(n)} J_I} \cdot \cdots \cdot prox_{\alpha^{(n)} J_2} \cdot prox_{\alpha^{(n)} J_1}(\vec{x}^{(n)}). \tag{8}$$

Furthermore, let us consider the case where $J_i(\vec{x})$ is a sum of two subfunctions, like our cost function including data-fidelity term and regularization term as

$$J_i(\vec{x}) = f_i(\vec{x}) + u(\vec{x}). \tag{9}$$

The update can be constructed by applying two operators corresponding to each subfunction alternately as below

$$\vec{x}^{(n+1)} = prox_{\alpha^{(n)}u} \cdot prox_{\alpha^{(n)}f_i}\left(\vec{x}^{(n)}\right). \tag{10}$$

Finally, we show in Algorithm 1 the optimization model applied to this paper:

Algorithm 1: Proximal splitting with Passty's framework

Give an initial vector $\vec{x}^{(0,1)}$. Execute the following.

For $n = 0, 1, 2, \ldots,$ (n is the main iteration)

For $i = 1, 2, \ldots, I$

$\vec{x}^{(n,i+1)} = prox_{\alpha^{(n)}J_i}\left(\vec{x}^{(n,i)}\right)$

$\vec{x}^{(n+1,1)} = \vec{x}^{(n,I+1)}$

In Passty's framework, by increasing the number to divide the cost function into smaller subfunctions, the better convergence can be expected. In this paper, this division is performed as follows. First, the data-fidelity term is divided as shown in Equation (7) in such a way that each subfunction $f_i\left(\vec{x}\right)$ contains only a single term corresponding to projection data b_i. Therefore, the final algorithm can be designed in the form of a row-action type algorithm such as ART method.

We mention that $u\left(\vec{x}\right)$ can also be divided into a sum of many subfunctions. In this paper, we divide $u\left(\vec{x}\right)$ as finely as possible similarly to the case of the data-fidelity term. This idea leads to a significant benefit to simplify the processing of nonlocal TV+TKV term as well as improving the convergence speed. With respect to the regularization term $u\left(\vec{x}\right)$, we perform a division described in Section 2.3.2.

2.3. Optimization

In this section, we focus on how to divide the cost function and how to derive the resulting iterative algorithm.

2.3.1. Update the Data-Fidelity Term

The data-fidelity term can be divided into I subfunctions as below

$$f(\vec{x}) = \left\|A\vec{x} - \vec{b}\right\|_2^2 = \sum_{i=1}^{I} f_i(\vec{x}) = \sum_{i=1}^{I} (\vec{a}_i^T\vec{x} - b_i)^2, \tag{11}$$

where I is the number of projection data, and $\vec{a}_i$ is i-th row vector of the system matrix A, and b_i is i-th component of projection data. Furthermore, we note that $f_i\left(\vec{x}\right)$ is a subfunction corresponding to the data-fidelity term.

The minimization problem for the data-fidelity term can be defined as

$$\vec{x}^{(n,i+1)} = prox_{\alpha^{(n)}f_i}\left(\vec{x}^{(n,i)}\right) = \underset{\vec{x}}{\operatorname{argmin}}\left\{\left(\vec{a}_i^T\vec{x} - b_i\right)^2 + \frac{1}{2\alpha^{(n)}}\|\vec{x} - \vec{x}^{(n,i)}\|_2^2\right\}. \tag{12}$$

By introducing a slack variable $z = \vec{a}_i^T\vec{x}$, the above minimization problem for each subfunction $f_i\left(\vec{x}\right)$ can be converted into the constrained minimization below

$$\min_{x,z} \ (z - b_i)^2 + \frac{1}{2\alpha^{(n)}}\left\|\vec{x} - \vec{x}^{(n,i)}\right\|_2^2 \quad s.t. \ z = \vec{a}_i^T\vec{x}. \tag{13}$$

The Lagrange function can be defined by

$$L_{f_i}\left(\vec{x}, z, \lambda\right) = (z - b_i)^2 + \frac{1}{2\alpha^{(n)}}\|\vec{x} - \vec{x}^{(n,i)}\|_2^2 + \lambda\left(z - \vec{a}_i^T \vec{x}\right), \tag{14}$$

where λ is the Lagrange multiplier called the dual variable.

Finally, by optimizing the Lagrange function, we obtain the following expression of row-action type iteration.

$$\vec{x}^{(n,i+1)} = \vec{x}^{(n,i)} + \alpha^{(n)} \frac{b_i - \vec{a}_i^T \vec{x}^{(n,i)}}{1/2 + \alpha^{(n)}\|\vec{a}_i\|_2^2} \vec{a}_i,\ \alpha^{(n)} = \frac{\alpha_0}{1+\varepsilon n}, \left(i \in R_{[I]}\right),$$
$$\left(\alpha_0 : \text{ initial value of step size, } \varepsilon : \text{ deceleration rate of step-size, } R_{[I]} : \text{ ordered subsets}\right) \tag{15}$$

where $\alpha^{(n)}$ is the step-size parameter to control the convergence, and n is the number of main iterations. After updating all elements of projection data, n is increased by 1. The step-size $\alpha^{(n)}$ is diminished gradually to zero as the iteration proceeds (i.e. diminishing step-size rule). In Passty's framework, it is known that this diminishing contributes to ensuring an exact convergence to a minimizer, thereby, avoiding the so-called limit cycle problem. Furthermore, we introduce a random-access order of projection data $R_{[I]}$ to enable a fast convergence within 20–30 iterations [19,20]. The mathematical detail to derive the update formula in Equation (15) has been already described in our previous studies [19–22].

2.3.2. Update the Regularization Term

We describe how to divide the regularization term. The modified anisotropic nonlocal TV possesses the following structure and is called L1 based group LASSO here. The L1 based group LASSO possesses a very simple structure, where each absolute value term can be considered a group element.

$$u^{TV}\left(\vec{x}\right) = \beta t \sum_{j=1}^{J} \sum_{j' \in \Omega} \omega_{jj'} \left|x_j - x_{j'}\right| = \beta t \sum_{g=1}^{G} \left[\omega_{jj'} \|\vec{x}_j - \vec{x}_{j'}\|_1^1\right]_g = \beta t \sum_{g=1}^{G} u_g^{TV}\left(\vec{x}\right), \quad (j = g). \tag{16}$$

where g is the group index, and the group index g corresponds to the pixel index j, and the TV term can be divided into G groups (G subfunctions). Furthermore, the group itself becomes a subfunction $u_g^{TV}\left(\vec{x}\right)$.

In the case of the TKV term, it can be divided into $(G \times 8)$ subfunctions as below

$$u^{TKV}\left(\vec{x}\right) = \frac{\beta(1-t)}{8} \sum_{k=1}^{8} \sum_{g=1}^{G} \left[\omega_{jj'} \left\|\left(\vec{x}_j - \vec{x}_{j_k}\right) - \left(\vec{x}_{j'} - \vec{x}_{j'_k}\right)\right\|_1^1\right]_g = \frac{\beta(1-t)}{8} \sum_{k=1}^{8} \sum_{g=1}^{G} u_g^{TKV}\left(\vec{x}\right). \tag{17}$$

The detailed structure of the TV term is shown in Figure 2. Among the elements of the group, the pixel j is common and distant pixel j' has different values for each other.

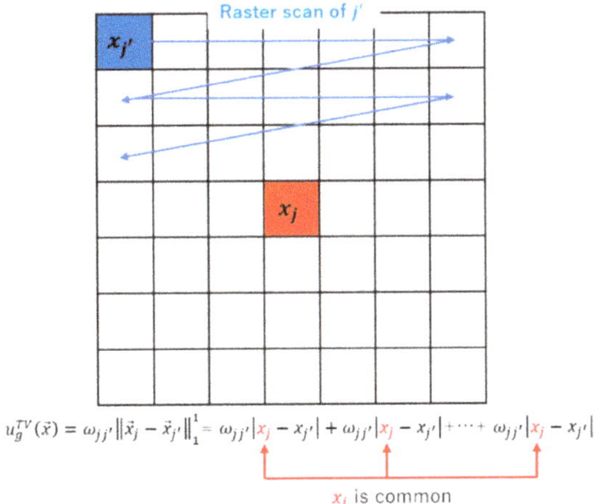

Figure 2. Raster scanning during the update ($j' = 1, 2, 3, \ldots$).

The sequential update is related to raster scanning.

In the TV term, when assuming that x_j and $x_{j'}$ are updated simultaneously, x_j is updated sequentially J' times, and $x_{j'}$ is updated once respectively.

In the case of TKV term, when assuming that x_j, x_{j_k}, $x_{j'}$, $x_{j'_k}$ are updated simultaneously, x_j is updated sequentially $(J' \times 8)$ times, and $x_{j'}$ is updated sequentially eight times. Following this, x_{j_k} and $x_{j'_k}$ (the adjacent pixels) are updated once respectively.

[Update the TV term] First, we consider updating the pixel j.

The proximity operator for a subfunction can be defined as follows

$$\underset{\vec{x}_j}{\operatorname{argmin}}(\omega_{jj'}\|\vec{x}_j - \vec{x}_{j'}^{(n)}\|_1^1 + \frac{1}{2\alpha^{(n)}}\|\vec{x}_j - \vec{x}_j^{(n,j')}\|_2^2), \tag{18}$$

where $x_{j'}^{(n)}$ is the current updated solution as a constant approximation, which is the image updated from the data-fidelity term. The above L1 norm minimization problems can be solved with the following soft-thresholding.

$$\begin{gathered} S_\nabla\left(x_j^{(n,j')}\right) = x_j^{(n,j')} - \begin{cases} \nabla, & \tau > \nabla \\ -\nabla, & \tau < -\nabla \\ \tau_{TV} & (otherwise) \end{cases}, \\ \left(\tau_{TV} = (x_j^{(n,j')} - x_{j'}^{(n)}), \nabla = \alpha^{(n)}\beta t\omega_{jj'}\right), \end{gathered} \tag{19}$$

where $S_\nabla\left(x_j^{(n,j')}\right)$ is the soft-thresholding function. The solution ($S_\nabla\left(x_j^{(n,j')}\right)$) corresponding to otherwise is simply $x_{j'}^{(n)}$.

For better convergence, although it is possible to update only pixel j, we update the pixel j and j' simultaneously.

For updating the pixel j and j', we further divide a subfunction $u_g^{TV}(\vec{x})$ into two subfunctions as below

$$\begin{aligned} u^{TV}(\vec{x}) &= \beta t \sum_{g=1}^{G} u_g^{TV}(\vec{x}) = \beta t \sum_{g=1}^{G} \left[\omega_{jj'} \|\vec{x}_j - \vec{x}_{j'}\|_1^1\right]_g \\ &= \beta t \sum_{g=1}^{G} \left[\left[\omega_{jj'} \left\| \frac{\vec{x}_j - \vec{x}_{j'}}{2} \right\|_1^1\right]_{x_j} + \left[\omega_{jj'} \left\| \frac{\vec{x}_j - \vec{x}_{j'}}{2} \right\|_1^1\right]_{x_{j'}}\right]_g \\ &= \beta t \sum_{g=1}^{G} \left[p(\vec{x}_j) + p(\vec{x}_{j'})\right]_g. \end{aligned} \tag{20}$$

where $p(\vec{x}_j) = [\,]_{x_j}, p(\vec{x}_{j'}) = [\,]_{x_{j'}}$.

The proximity operator for each subfunction ($prox_{\alpha^{(n)}p}\left(\vec{x}_j^{(n,j')}\right)$, $prox_{\alpha^{(n)}p}\left(\vec{x}_{j'}^{(n)}\right)$) can be defined as follows

$$\begin{aligned} &\underset{\vec{x}_j}{\text{argmin}}\left(\omega_{jj'} \left\| \vec{x}_j - \frac{\vec{x}_j^{(n,j')} + \vec{x}_{j'}^{(n)}}{2} \right\|_1^1 + \frac{1}{2\alpha^{(n)}} \left\| \vec{x}_j - \vec{x}_j^{(n,j')} \right\|_2^2\right), \\ &\underset{\vec{x}_{j'}}{\text{argmin}}\left(\omega_{jj'} \left\| -\vec{x}_{j'} + \frac{\vec{x}_j^{(n,j')} + \vec{x}_{j'}^{(n)}}{2} \right\|_1^1 + \frac{1}{2\alpha^{(n)}} \left\| \vec{x}_{j'} - \vec{x}_{j'}^{(n)} \right\|_2^2\right), \end{aligned} \tag{21}$$

where $x_j^{(n,j')}$ and $x_{j'}^{(n)}$ are the current updated solution as a constant approximation, which is the image updated from the data-fidelity term.

Finally, the above L1 norm minimization problems can be solved with the following soft-thresholding.

$$\begin{aligned} &S_\nabla\left(x_j^{(n,j')}\right) = x_j^{(n,j')} - \begin{cases} \nabla, & \tau > \nabla \\ -\nabla, & \tau < -\nabla \\ \tau_{TV} & (otherwise) \end{cases}, \\ &S_\nabla\left(x_{j'}^{(n)}\right) = x_{j'}^{(n)} + \begin{cases} \nabla, & \tau > \nabla \\ -\nabla, & \tau < -\nabla \\ \tau_{TV} & (otherwise) \end{cases}, \\ &\left(\tau_{TV} = \left(x_j^{(n,j')} - x_{j'}^{(n)}\right)/2, \quad \nabla = \alpha^{(n)} \beta t \omega_{jj'}\right), \end{aligned} \tag{22}$$

where $S_\nabla\left(x_j^{(n,j')}\right)$ and $S_\nabla\left(x_{j'}^{(n)}\right)$ are soft-thresholding functions. By dividing a subfunction and updating for each variable as shown in Equations (20)–(22), the solution ($S_\nabla\left(x_j^{(n,j')}\right)$, $S_\nabla\left(x_{j'}^{(n)}\right)$) corresponding to otherwise becomes the average of $x_j^{(n,j')}$ and $x_{j'}^{(n)}$ ($(x_j^{(n,j')} + x_{j'}^{(n)})/2$). Compared to Equation (19), this weak average that occurs otherwise can reduce the error in convergence and improve the stability of convergence.

[Update the TKV term] We update the pixel j, j_k, j', j'_k simultaneously. For updating the pixel j, j_k, j', j'_k, we further divide a subfunction $u_g^{TKV}(\vec{x})$ into four subfunctions as below

$$\begin{aligned} u^{TKV}(\vec{x}) &= \tfrac{\beta(1-t)}{8}\sum_{k=1}^{8}\sum_{g=1}^{G} u_g^{TKV}(\vec{x}) = \tfrac{\beta(1-t)}{8}\sum_{k=1}^{8}\sum_{g=1}^{G}\left[\omega_{jj'}\left\|\left(\vec{x}_j-\vec{x}_{j_k}\right)-\left(\vec{x}_{j'}-\vec{x}_{j'_k}\right)\right\|_1^1\right]_g \\ &= \tfrac{\beta(1-t)}{8}\sum_{k=1}^{8}\sum_{g=1}^{G}\left[\left[\omega_{jj'}\left\|\tfrac{(\vec{x}_j-\vec{x}_{j_k})-(\vec{x}_{j'}-\vec{x}_{j'_k})}{4}\right\|_1^1\right]_{x_j}+\left[\omega_{jj'}\left\|\tfrac{(\vec{x}_j-\vec{x}_{j_k})-(\vec{x}_{j'}-\vec{x}_{j'_k})}{4}\right\|_1^1\right]_{x_{j_k}}\right. \\ &\quad \left.+\left[\omega_{jj'}\left\|\tfrac{(\vec{x}_j-\vec{x}_{j_k})-(\vec{x}_{j'}-\vec{x}_{j'_k})}{4}\right\|_1^1\right]_{x_{j'}}+\left[\omega_{jj'}\left\|\tfrac{(\vec{x}_j-\vec{x}_{j_k})-(\vec{x}_{j'}-\vec{x}_{j'_k})}{4}\right\|_1^1\right]_{x_{j'_k}}\right]_g \\ &= \tfrac{\beta(1-t)}{8}\sum_{k=1}^{8}\sum_{g=1}^{G}\left[q(\vec{x}_j)+q(\vec{x}_{j_k})+q(\vec{x}_{j'})+q(\vec{x}_{j'_k})\right]_g, \end{aligned} \tag{23}$$

where $q(\vec{x}_j)=[]_{x_j}$, $q(\vec{x}_{j_k})=[]_{x_{j_k}}$, $q(\vec{x}_{j'})=[]_{x_{j'}}$, $q(\vec{x}_{j'_k})=[]_{x_{j'_k}}$.

The proximity operator for each subfunction ($prox_{\alpha^{(n)}q}\left(\vec{x}_j^{(n,j',k)}\right)$, $prox_{\alpha^{(n)}q}\left(\vec{x}_{j_k}^{(n)}\right)$, $prox_{\alpha^{(n)}q}\left(\vec{x}_{j'}^{(n,k)}\right)$, $prox_{\alpha^{(n)}q}\left(\vec{x}_{j'_k}^{(n)}\right)$) can be defined as follows

$$\begin{aligned} &\underset{\vec{x}_j}{\operatorname{argmin}}\left(\omega_{jj'}\left\|\vec{x}_j-\tfrac{3\vec{x}_j^{(n,j',k)}+\vec{x}_{j_k}^{(n)}+\vec{x}_{j'}^{(n,k)}-\vec{x}_{j'_k}^{(n)}}{4}\right\|_1^1+\tfrac{1}{2\alpha^{(n)}}\|\vec{x}_j-\vec{x}_j^{(n,j',k)}\|_2^2\right), \\ &\underset{\vec{x}_{j_k}}{\operatorname{argmin}}\left(\omega_{jj'}\left\|-\vec{x}_{j_k}+\tfrac{\vec{x}_j^{(n,j',k)}+3\vec{x}_{j_k}^{(n)}-\vec{x}_{j'}^{(n,k)}+\vec{x}_{j'_k}^{(n)}}{4}\right\|_1^1+\tfrac{1}{2\alpha^{(n)}}\|\vec{x}_{j_k}-\vec{x}_{j_k}^{(n)}\|_2^2\right), \\ &\underset{\vec{x}_{j'}}{\operatorname{argmin}}\left(\omega_{jj'}\left\|-\vec{x}_{j'}+\tfrac{\vec{x}_j^{(n,j',k)}-\vec{x}_{j_k}^{(n)}+3\vec{x}_{j'}^{(n,k)}+\vec{x}_{j'_k}^{(n)}}{4}\right\|_1^1+\tfrac{1}{2\alpha^{(n)}}\|\vec{x}_{j'}-\vec{x}_{j'}^{(n,k)}\|_2^2\right), \\ &\underset{\vec{x}_{j'_k}}{\operatorname{argmin}}\left(\omega_{jj'}\left\|\vec{x}_{j'_k}-\tfrac{-\vec{x}_j^{(n,j',k)}+\vec{x}_{j_k}^{(n)}+\vec{x}_{j'}^{(n,k)}+3\vec{x}_{j'_k}^{(n)}}{4}\right\|_1^1+\tfrac{1}{2\alpha^{(n)}}\|\vec{x}_{j'_k}-\vec{x}_{j'_k}^{(n)}\|_2^2\right), \end{aligned} \tag{24}$$

where $\vec{x}_j^{(n,j',k)}$, $\vec{x}_{j_k}^{(n)}$, $\vec{x}_{j'}^{(n,k)}$, $\vec{x}_{j'_k}^{(n)}$ are the current updated solution as a constant approximation, which is the image updated from the TV term.

Finally, the above L1 norm minimization problems can be solved with the following soft-thresholding.

$$
\begin{aligned}
S_\nabla\left(x_j^{(n,j',k)}\right) &= x_j^{(n,j',k)} - \begin{cases} \nabla, & \tau > \nabla \\ -\nabla, & \tau < -\nabla \\ \tau_{TKV} & (otherwise) \end{cases}, \\
S_\nabla\left(x_{j_k}^{(n)}\right) &= x_{j_k}^{(n)} + \begin{cases} \nabla, & \tau > \nabla \\ -\nabla, & \tau < -\nabla \\ \tau_{TKV} & (otherwise) \end{cases}, \\
S_\nabla\left(x_{j'}^{(n,k)}\right) &= x_{j'}^{(n,k)} + \begin{cases} \nabla, & \tau > \nabla \\ -\nabla, & \tau < -\nabla \\ \tau_{TKV} & (otherwise) \end{cases}, \\
S_\nabla\left(x_{j'_k}^{(n)}\right) &= x_{j'_k}^{(n)} - \begin{cases} \nabla, & \tau > \nabla \\ -\nabla, & \tau < -\nabla \\ \tau_{TKV} & (otherwise) \end{cases}, \\
\Big(\tau_{TKV} = \Big(x_j^{(n,j',k)} &- x_{j_k}^{(n)} - x_{j'}^{(n,k)} + x_{j'_k}^{(n)}\Big)/4, \quad \nabla = \alpha^{(n)}\beta(1-t)\omega_{jj'}/8\Big),
\end{aligned}
\tag{25}
$$

where $S_\nabla\left(x_j^{(n,j',k)}\right)$, $S_\nabla\left(x_{j_k}^{(n)}\right)$, $S_\nabla\left(x_{j'}^{(n,k)}\right)$, $S_\nabla\left(x_{j'_k}^{(n)}\right)$ are soft-thresholding function.

2.3.3. The Weight

In this paper, we used the weight of nonlocal means filter [25] as

$$
\omega_{jj'} = \frac{exp(-\max(\|B(x_j) - B(x_{j'})\|_2^2 - 2\sigma^2, 0)/h^2)}{\sum_{j' \in \Omega} exp(-max(\|B(x_j) - B(x_{j'})\|_2^2 - 2\sigma^2, 0)/h^2)}, \tag{26}
$$

where $\|B(x_j) - B(x_{j'})\|_2^2$ means the average Euclidean distance between patches $(B(x_j), B(x_{j'}))$ centered in an interest pixel x_j and a distant pixel $x_{j'}$.

Theoretically, the weight must be a fixed value as a hyper-parameter of the regularization term. However, there have been previous studies showing that reweighting at each iteration contributes to better image quality [2,9]. Additionally, the larger the size of weight (Ω), the better the performance of removing artifacts or noise. As long as the computer is capable of processing, we recommend increasing the size of the weight. However, if the size of the weight is too large, the calculation cost will increase enormously as compared with the image quality improvement. Therefore, it is important to decide the size of the weight while paying attention to cost performance considering the level of artifacts or noise. Figure 3 shows the cost performance of changing the size of weight.

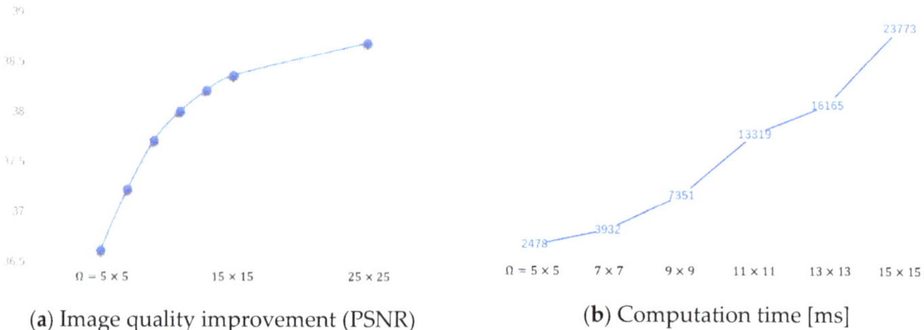

(**a**) Image quality improvement (PSNR) (**b**) Computation time [ms]

Figure 3. Cost performance of changing the size of weight. (**a**) Image quality improvement (PSNR), (**b**) Computation time [ms].

Next, we describe the proper estimation of the weight. If the ground truth image is known, the weight can be calculated from the ground truth image. However, in the image reconstruction, only the projection data is given and the reconstructed image and the weight must be estimated simultaneously from the projection data. Therefore, we construct the optimization by alternating the estimation of the reconstructed image and the weight.

We show the reweighting process of optimization including the data-fidelity term and regularization term:

$$\begin{cases} For\ i = 1,2,3,\ldots,I\ (I \text{ is the number of projection data}) \\ \quad \text{1) Update the data term by Equation (15): } \vec{x}^{(n,i+1)} = prox_{\alpha^{(n)} f_i}\left(\vec{x}^{(n,i)}\right); \\ \quad\quad if\left((i\ mod\ S) == 0\right) \\ \quad\quad\quad \left| \begin{array}{l} \text{2) Calculate the weight } \omega_{jj'} \text{ from } \vec{x}^{(n,i+1)}; \\ \text{3) Update the TV term by Equation (22);} \\ \text{4) Update the TKV term by Equation (25);} \end{array} \right. \\ \vec{x}^{(n+1,1)} \leftarrow \vec{x}^{(n,I+1)} \end{cases} \tag{27}$$

where the weight is calculated once as common to nonlocal TV and TKV and the weight is calculated from the image updated from the data-fidelity term. The span parameter S determines how often regularization is performed in the main iteration n. If S is small, many regularization updates are performed in one iteration. Theoretically, the smaller the S, the more accurate the convergence. However, since nonlocal TV has a large amount of computational complexity, it is desirable to determine an appropriate value of S.

Figure 4 shows how the size of weight (Ω) influences image quality and computation time.

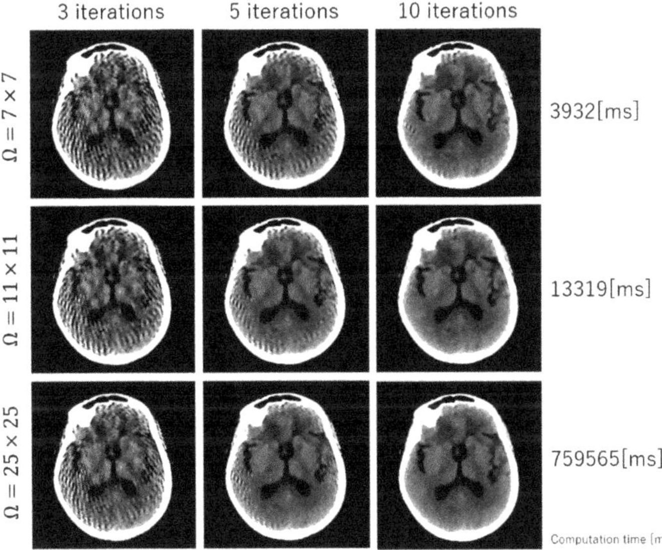

Figure 4. Demonstration of how the size of weight (Ω) influences image quality and computation time.

3. Experimental Results

We performed simulation studies using a brain CT image. The reason behind using the brain image is as follows. In the brain CT imaging, the reconstructed images are shown with a compressed gray scale range much larger compared to the other CT imaging like chest imaging or abdominal CT imaging. Therefore, preserving smooth intensity changes and avoiding the staircase artifacts are much important in the brain case. Additionally, simulation studies were performed for both the sparse-view CT (the number of projection data was 64) and the low-dose CT (the number of photons was 3×10^6). The reconstructed image consisted of 512×512 pixels, where the pixel size was 0.0585 (cm^2). We compressed the range showing the reconstructed images to [7.82, 62.30] HU, where this contrast range was determined based on the contrast range used in clinical brain CT imaging. To evaluate image quality, standard RMSE, PSNR, SSIM values were used as metrics. The number of iterations in image reconstructions was 20 for nonlocal TV, TKV, and TV+TKV, which was determined by the fact that changes in image were small enough with this iteration number. We also showed the reconstructed images by the standard Filtered Back-Projection (FBP), and differences in image quality by changing values of the hyper-parameter t (i.e., the trade-off parameter between nonlocal TV (first derivative) term and the TKV (second derivative) term). The ground truth image and the FBP reconstructions are shown in Figure 5. The reconstructed images in the case of sparse-view CT are shown in Figure 6. In Figure 7, we show the used brain CT image with three display gray-scale ranges, from which we observe that the staircase artifacts are severe when the range of display gray-scale range is small. The reconstructed images in the case of low-dose CT are shown in Figure 8. In Figures 9–11, we show convergence properties of our iterative algorithm based on Passty's proximal splitting framework. In Figures 12 and 13, to show the effect of acceleration by Passty's proximal splitting, we incorporated the TV+TKV term into SIRT (simultaneous iterative reconstruction technique) which is a non-row-action method (a type of the standard iterative algorithm) and compared this with row-action based on our proposed nonlocal TV+TKV. From these figures, it can be observed that our algorithm converged very quickly. It is well-known that the standard iterative algorithms such as Chambolle–Pock [26] and proximal gradient algorithms require several hundreds of iteration up to the convergence. The benefit of our iterative algorithm mainly originates from the fact that our algorithm is of row-action type.

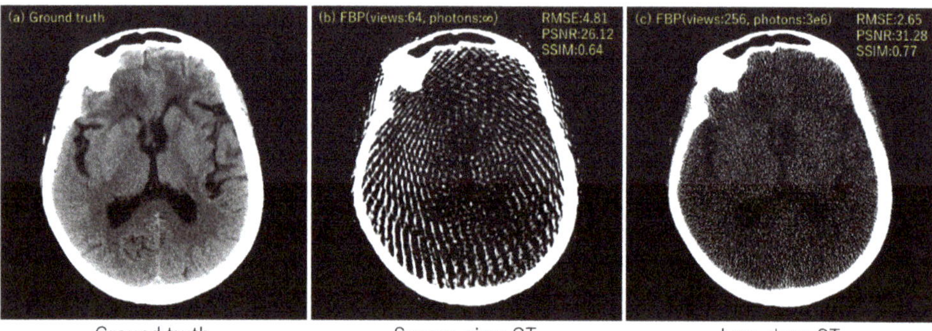

Figure 5. (**a**) Ground truth, (**b**) FBP with 64 projection data with no noise, (**c**) FBP with 256 projection data with the number of photon counts 3×10^6. All images are displayed with the same window of [7.82, 62.30] HU.

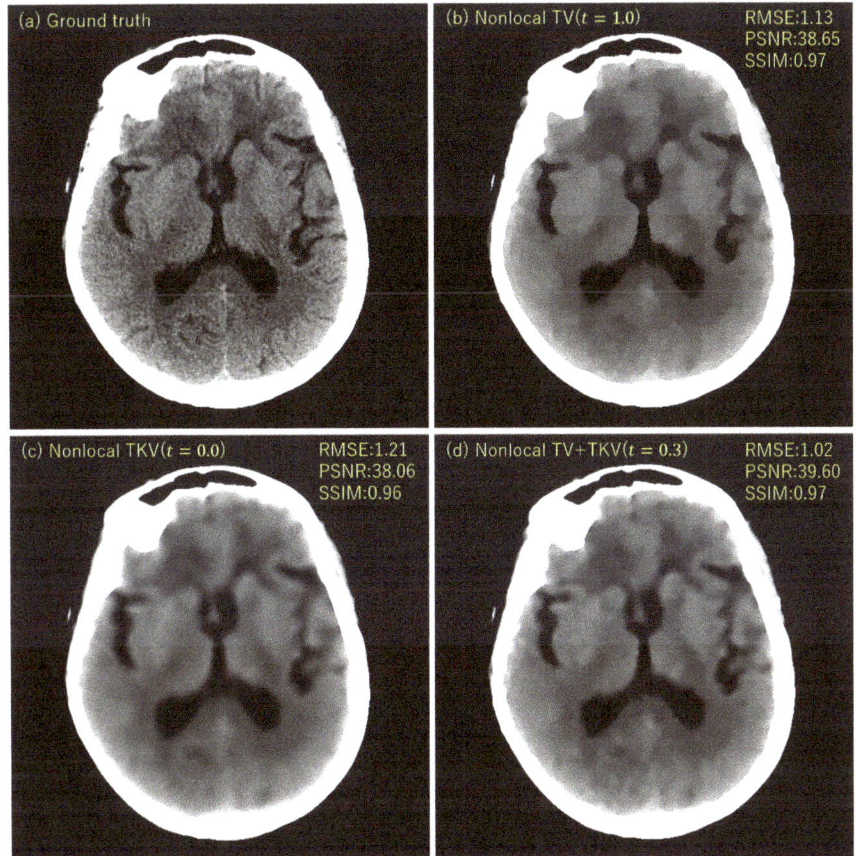

Figure 6. The reconstructed images of sparse-view CT (64 projection data with no noise). (**a**) Ground truth, (**b**) Nonlocal TV ($t = 1.0$), (**c**) Nonlocal TKV ($t = 0.0$), (**d**) Nonlocal TV+TKV ($t = 0.3$) were compared. All images are displayed with the same window of [7.82, 62.30] HU.

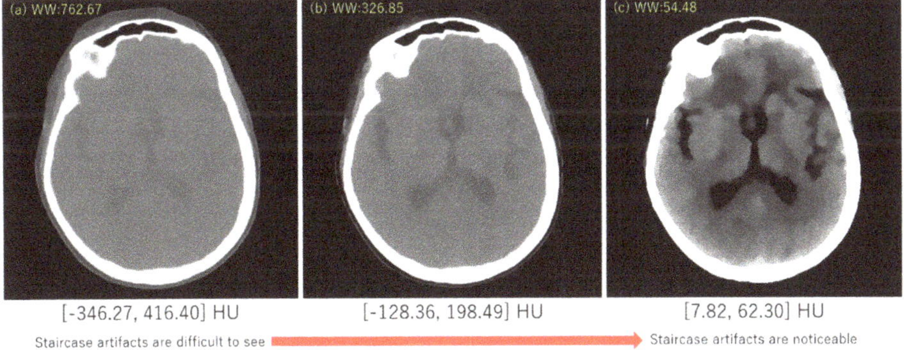

Figure 7. Demonstration of appearance of the staircase artifacts with various gray-scale ranges in displaying the brain CT image. (**a**) Window Width [−346.27, 416.40] HU, (**b**) Window Width [−128.36, 198.49] HU, (**c**) Window Width [7.82, 62.30] HU.

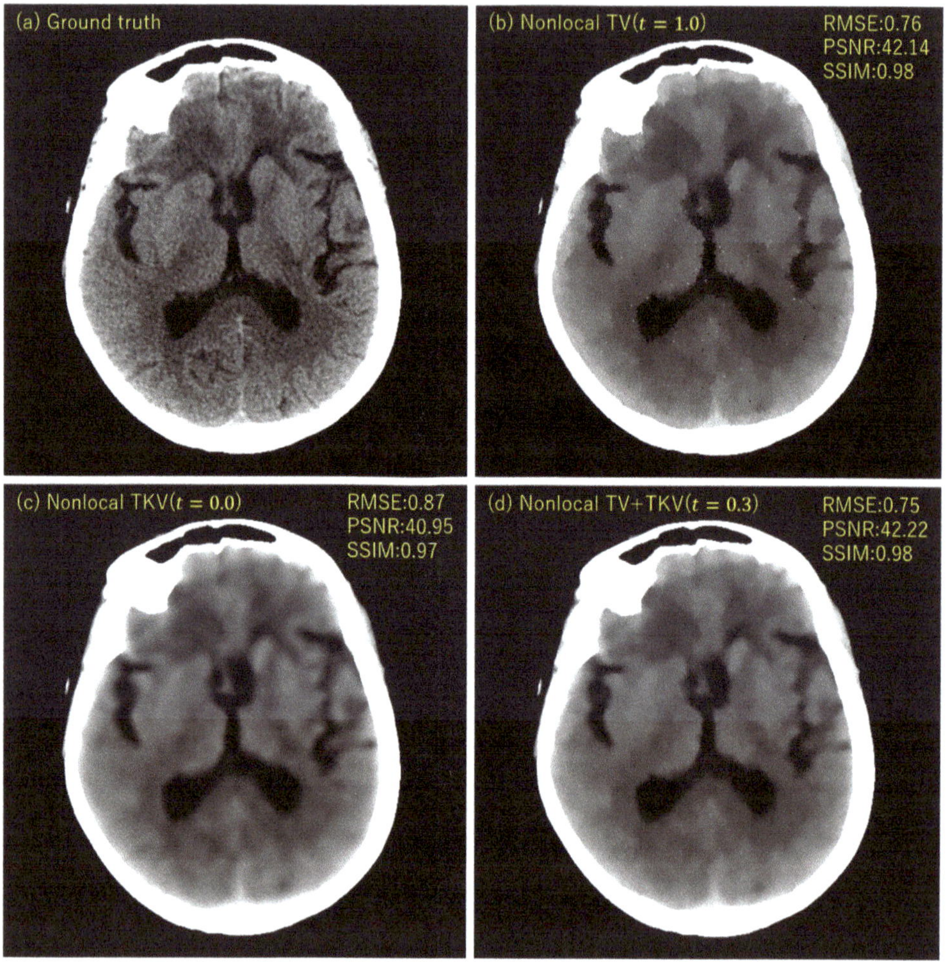

Figure 8. The reconstructed images of low-dose CT (256 projection data and the number of photon counts 3×10^6). (**a**) Ground truth, (**b**) Nonlocal TV ($t = 1.0$), (**c**) Nonlocal TKV ($t = 0.0$), (**d**) Nonlocal TV+TKV ($t = 0.3$) were compared. All images are displayed with the same window of [7.82, 62.30] HU.

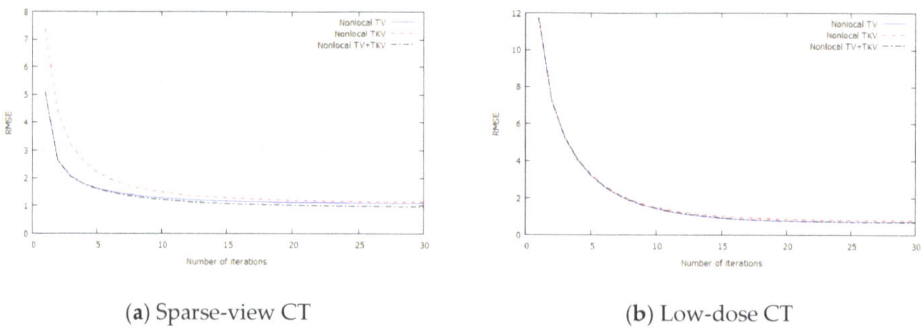

(**a**) Sparse-view CT (**b**) Low-dose CT

Figure 9. RMSE for each iteration. (**a**) Sparse-view CT, (**b**) Low-dose CT.

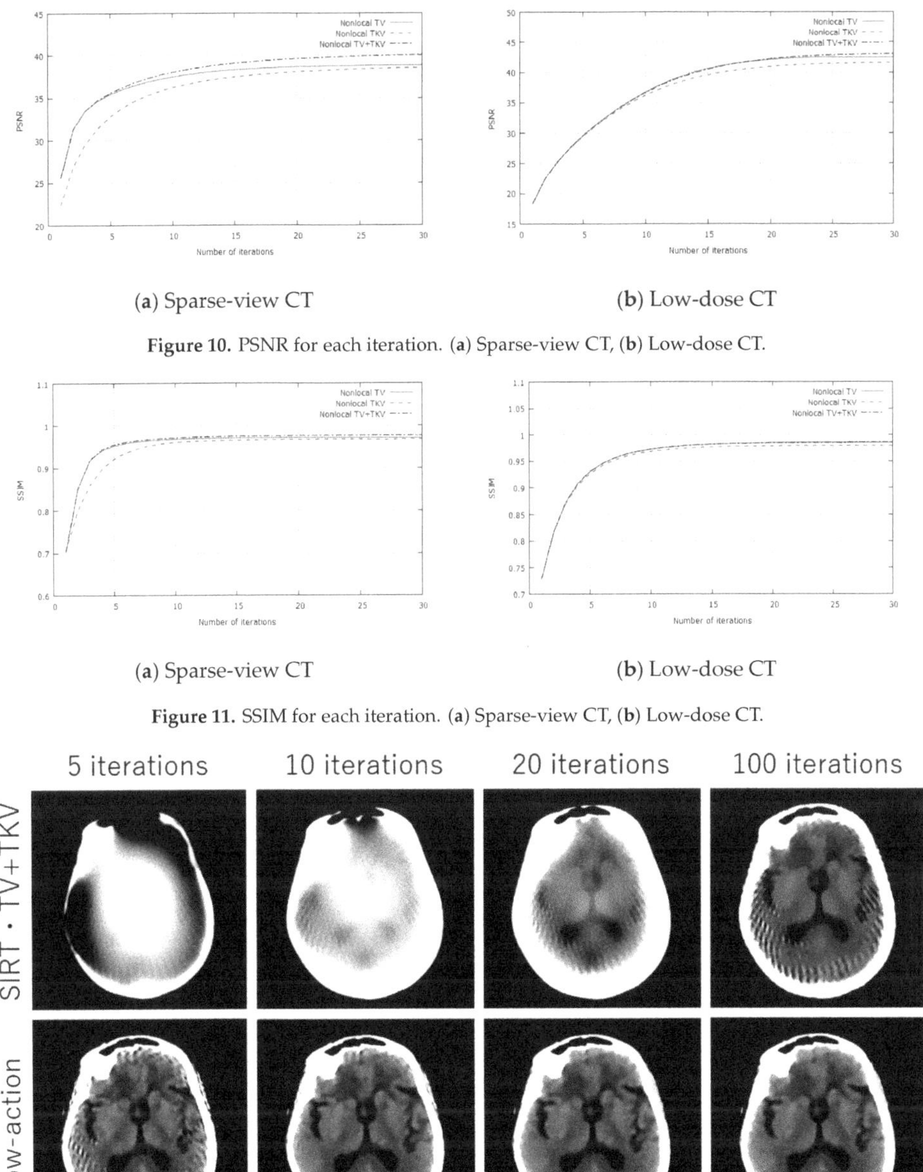

(a) Sparse-view CT (b) Low-dose CT

Figure 10. PSNR for each iteration. (**a**) Sparse-view CT, (**b**) Low-dose CT.

(a) Sparse-view CT (b) Low-dose CT

Figure 11. SSIM for each iteration. (**a**) Sparse-view CT, (**b**) Low-dose CT.

Figure 12. The reconstructed images of sparse-view CT (64 projection data with no noise). SIRT nonlocal TV+TKV and row-action accelerated nonlocal TV+TKV (our proposed method) were compared. All images are displayed with the same window of [7.82, 62.30] HU.

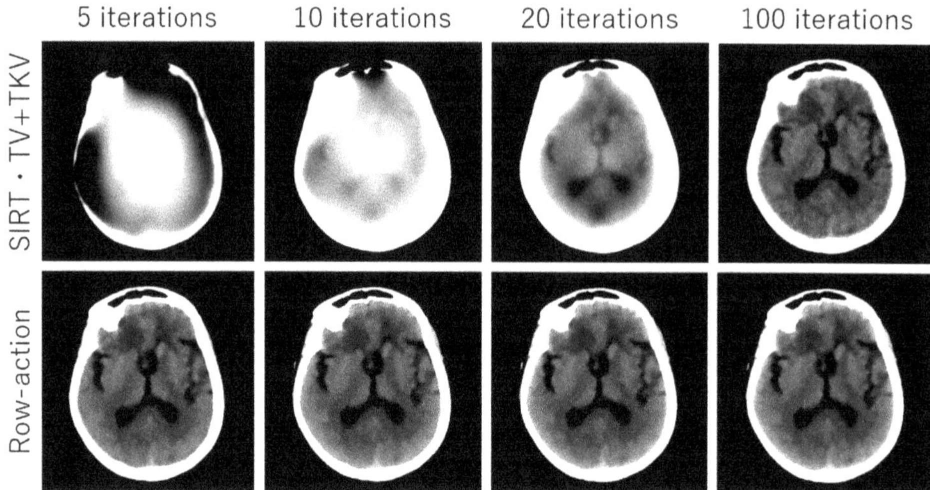

Figure 13. The reconstructed images of low-dose CT (256 projection data and the number of photon counts 3×10^6). SIRT nonlocal TV+TKV and row-action accelerated nonlocal TV+TKV (our proposed method) were compared. All images are displayed with the same window of [7.82, 62.30] HU.

4. Discussion

The experimental results Table 1 showed that the reconstructed image by nonlocal TV+TKV was closest to the ground truth image, with good RMSE, PSNR and SSIM values. Furthermore, no isolated points caused by outliers of the soft-thresholding, which often appear in the TV-class reconstruction methods, were visible. In the sparse-view CT, the result of nonlocal TV ($t = 1.0$) using only the first-order derivative was of very high-contrast, but the staircase artifacts appeared in the smooth intensity changes. In other words, the region with small intensity changes was over-smoother by the regularization as if the oil-painting. In the low-dose CT, the result of nonlocal TV showed isolated points in the region closer to the center of the image. In both the sparse-view CT and the low-dose CT, nonlocal TKV ($t = 0.0$) using only the second-order derivative was able to preserve fine soft tissues well including textures and low-contrast objects, however, it suffered from the blurring in the edge parts. This is because, by using a large weight in the second-order derivative, the threshold value (τ_{TKV}) of the soft-thresholding operation becomes very small (i.e., $\tau_{TV} \gg \tau_{TKV}$).

Table 1. Summary of each method.

	Nonlocal TV	Nonlocal TKV	Nonlocal TV + TKV
Convergence	**Good**	Not bad	**Good**
High contrast	**Yes**	No	**Yes**
Smooth intensity change	No	**Yes**	**Yes**

5. Conclusions

In this paper, we proposed a new concept in nonlocal TV, in which the first and second order derivatives are combined in the regularization term. By combining the two terms, we were able to compensate for each other's weaknesses, i.e., staircase artifact and loss in smooth intensity changes in the first derivative and image blurring in the second derivative. Furthermore, we proposed a specially designed proximal splitting algorithm that is based on Passty's framework. The key idea is to split the original cost function to minimize as finely as possible to accelerate convergence and simplify necessary computations. This allows as to make the final iterative algorithm into a form of row-action type, which is known to converge very quickly compared to other standards such as Chambolle–Pock algorithm and proximal gradient. In our experiments, we experimentally confirmed that the proposed algorithm converges within 20 iterations even for the case of brain CT imaging in which the requirement of image contrast is very severe. The simulation results with the brain CT image were performed for both the sparse-view CT and the low-dose CT. We showed that our proposed algorithm works well in practice.

As future work, our proposed nonlocal TV can be compared with the latest technology e.g., deep leaning [27,28] or other applied methods such as low-rank minimization [29,30].

Recently, image reconstruction methods using deep learning have been actively investigated. Our proposed method can be compared with existing deep leaning [27,28] as advanced compressed sensing. Additionally, our proposed method can be applied to low-rank TV, which can improve image quality by combing low-rank minimization and Total Variation [29,30].

Author Contributions: Conceptualization, Y.K. and H.K.; methodology, Y.K.; software, Y.K.; validation, Y.K. and H.K.; formal analysis, Y.K. and H.K.; investigation, Y.K. and H.K.; resources, Y.K. and H.K.; data curation, Y.K. and H.K.; writing—original draft preparation, Y.K.; writing—review and editing, Y.K. and H.K.; visualization, Y.K.; supervision, H.K.; project administration, H.K.; funding acquisition, H.K. All authors have read and agreed to the published version of the manuscript.

Funding: This work was supported by JST CREST Grant Number JPMJCR1765, Japan.

Conflicts of Interest: The authors declare no conflicts of interest.

References

1. Kim, Y.; Kudo, H.; Chigita, K.; Lian, S. Image reconstruction in sparse-view CT using improved nonlocal total variation regularization. In Proceedings of the SPIE Optical Engineering + Applications, San Diego, CA, USA, 10 September 2019.
2. Kim, H.; Chen, J.; Wang, A.; Chuang, C.; Held, M.; Pouliot, J. Non-local total-variation (NLTV) minimization combined with reweighted L1-norm for compressed sensing CT reconstruction. *Phys. Med. Biol.* **2016**, *61*, 6878–6891. [CrossRef]
3. Kim, K.; El Fakhri, G.; Li, Q. Low-dose CT reconstruction using spatially encoded nonlocal penalty. *Med. Phys.* **2017**, *44*, e376–e390. [CrossRef] [PubMed]
4. Lv, D.; Zhou, Q.; Choi, J.K.; Li, J.; Zhang, X. NLTG Priors in Medical Image: Nonlocal TV-Gaussian (NLTG) prior for Bayesian inverse problems with applications to Limited CT Reconstruction. *Inverse Prob. Imaging* **2019**, *14*, 117–132. [CrossRef]
5. Gilboa, G.; Osher, S. Nonlocal Operators with Applications to Image Processing. *Multiscale Model. Simul.* **2009**, *7*, 1005–1028. [CrossRef]
6. Bresson, X. A Short Note for Nonlocal TV Minimization. Technical Report. Available online: http://citeseerx.ist.psu.edu/viewdoc/download?doi=10.1.1.210.471&rep=rep1&type=pdf (accessed on 15 June 2020).
7. Chambolle, A. An algorithm for Total variation Regularization and Denoising. *J. Math. Imaging.* **2004**, *20*, 89–97.
8. Rudin, L.; Osher, S.; Fatemi, E. Nonlinear total variation based noise removal algorithms. *Physica D* **1992**, *60*, 259–268. [CrossRef]
9. Zhang, X.; Xing, L. Sequentially reweighted TV minimization for CT metal artifact reduction. *Med. Phys.* **2013**, *40*, 1–12. [CrossRef]
10. Bredies, K.; Kunisch, K.; Pock, T. Total generalized variation. *SIAM J. Imag. Sci.* **2010**, 3, 492–526. [CrossRef]
11. Ranftl, R.; Bredies, K.; Pock, T. Non-local total generalized variation for optical flow estimation. *Lect. Notes Comput. Sci.* **2014**, *8689*, 439–454.
12. Parikh, N.; Boyd, S. Proximal Algorithms. Available online: https://web.stanford.edu/~{}boyd/papers/pdf/prox_algs.pdf (accessed on 15 June 2020).
13. Combettes, P.L.; Pesquet, J.C. Proximal splitting methods in signal processing. In *Fixed-Point Algorithms for Inverse Problems in Science and Engineering*; Springer: New York, NY, USA, 2011.
14. Passty, G.B. Ergodic convergence to a zero of the sum of monotone operators in Hilbert space. *J. Math. Anal. Appl.* **1979**, *72*, 383–390. [CrossRef]

15. Herman, G.T.; Meyer, L.B. Algebraic reconstruction techniques can be made computationally efficient (positron emission tomography application). *IEEE Trans. Med. Imaging* **1993**, *12*, 600–609. [CrossRef]
16. Tanaka, E.; Kudo, H. Subset-dependent relaxation in block-iterative algorithms for image reconstruction in emission tomography. *Phys. Med. Biol.* **2003**, *48*, 1405–1422. [CrossRef] [PubMed]
17. Wang, G.; Jiang, M. Ordered subset simultaneous algebraic reconstruction techniques (OS SART). *J. of X Ray Sci. Technol.* **2004**, *12*, 169–177.
18. Dong, J.; Kudo, H. Proposal of Compressed Sensing Using Nonlinear Sparsifying Transform for CT Image Reconstruction. *Med. Imaging Technol.* **2016**, *34*, 235–244.
19. Dong, J.; Kudo, H. Accelerated Algorithm for Compressed Sensing Using Nonlinear Sparsifying Transform in CT Image Reconstruction. *Med. Imaging Technol.* **2017**, *35*, 63–73.
20. Dong, J.; Kudo, H.; Kim, Y. Accelerated Algorithm for the Classical SIRT Method in CT Image Reconstruction. In Proceedings of the 5th International Conference on Multimedia and Image Processing, Nanjing, China, 10–12 January 2020; pp. 49–55.
21. Kudo, H.; Takaki, K.; Yamazaki, F.; Nemoto, T. Proposal of fault-tolerant tomographic image reconstruction. In Proceedings of the SPIE Optical Engineering + Applications, San Diego, CA, USA, 3 October 2016.
22. Kudo, H.; Dong, J.; Chigita, K.; Kim, Y. Metal artifact reduction in CT using fault-tolerant image reconstruction. In Proceedings of the SPIE Optical Engineering + Applications, San Diego, CA, USA, 10 September 2019.
23. Li, M.; Yang, H.; Kudo, H. An accurate iterative reconstruction algorithm for sparse objects: Application to 3 D blood vessel reconstruction from a limited number of projections. *Phys. Med. Biol.* **2002**, *47*, 2599–2609. [CrossRef] [PubMed]
24. Kudo, H.; Suzuki, T.; Rashed, E.A. Image reconstruction for sparse-view CT and interior CT-introduction to compressed sensing and differentiated backprojection. *Quant. Imaging Med. Surg.* **2013**, *3*. [CrossRef]
25. Buades, A.; Coll, B.; Morel, J.M. On image denoising methods. Available online: http://citeseerx.ist.psu.edu/viewdoc/download?doi=10.1.1.100.81&rep=rep1&type=pdf (accessed on 15 June 2020).
26. Sidky, E.Y.; Jørgensen, J.H.; Pan, X. Convex optimization problem prototyping for image reconstruction in computed tomography with the Chambolle–Pock algorithm. *Phys. Med. Biol.* **2012**, *57*, 3065–3091. [CrossRef]
27. Han, Y.; Ye, J.C. Framing U-Net via Deep Convolutional Framelets: Application to Sparse-View CT. *IEEE Trans. Med. Imaging* **2018**, *37*, 1418–1429. [CrossRef]
28. Jin, K.H.; McCann, M.T.; Froustey, E.; Unser, M. Deep Convolutional Neural Network for Inverse Problems in Imaging. *IEEE Trans. Image Process.* **2017**, *26*, 4509–4522. [CrossRef]
29. Shi, F.; Cheng, J.; Wang, L.; Yap, P.-T.; Shen, D. Low-Rank Total Variation for Image Super-Resolution. *Med. Image Comput. Comput. Assist. Interv.* **2013**, *16*, 155–162. [PubMed]
30. Niu, S.; Yu, G.; Ma, J.; Wang, J. Nonlocal low-rank and sparse matrix decomposition for spectral CT reconstruction. *Inverse Probl.* **2018**, *34*. [CrossRef] [PubMed]

Article

Non-Rigid Multi-Modal 3D Medical Image Registration Based on Foveated Modality Independent Neighborhood Descriptor

Feng Yang [1,2], Mingyue Ding [1] and Xuming Zhang [1,*]

1 Department of Biomedical Engineering, School of Life Science and Technology, Ministry of Education Key Laboratory of Molecular Biophysics, Huazhong University of Science and Technology, Wuhan 430074, China; fyang@foxmail.com (F.Y.); myding@hust.edu.cn (M.D.)

2 School of Computer and Electronics and Information, Guangxi University, Nanning 530004, China

* Correspondence: zxmboshi@hust.edu.cn; Tel.: +86-27-877-923-66

Received: 15 August 2019; Accepted: 23 October 2019; Published: 28 October 2019

Abstract: The non-rigid multi-modal three-dimensional (3D) medical image registration is highly challenging due to the difficulty in the construction of similarity measure and the solution of non-rigid transformation parameters. A novel structural representation based registration method is proposed to address these problems. Firstly, an improved modality independent neighborhood descriptor (MIND) that is based on the foveated nonlocal self-similarity is designed for the effective structural representations of 3D medical images to transform multi-modal image registration into mono-modal one. The sum of absolute differences between structural representations is computed as the similarity measure. Subsequently, the foveated MIND based spatial constraint is introduced into the Markov random field (MRF) optimization to reduce the number of transformation parameters and restrict the calculation of the energy function in the image region involving non-rigid deformation. Finally, the accurate and efficient 3D medical image registration is realized by minimizing the similarity measure based MRF energy function. Extensive experiments on 3D positron emission tomography (PET), computed tomography (CT), T1, T2, and proton density (PD) weighted magnetic resonance (MR) images with synthetic deformation demonstrate that the proposed method has higher computational efficiency and registration accuracy in terms of target registration error (TRE) than the registration methods that are based on the hybrid L-BFGS-B and cat swarm optimization (HLCSO), the sum of squared differences on entropy images, the MIND, and the self-similarity context (SSC) descriptor, except that it provides slightly bigger TRE than the HLCSO for CT-PET image registration. Experiments on real MR and ultrasound images with unknown deformation have also be done to demonstrate the practicality and superiority of the proposed method.

Keywords: medical image registration; similarity measure; non-rigid transformation; computational efficiency; registration accuracy

---

1. Introduction

In recent years, the non-rigid three-dimensional (3D) multi-modal medical image registration has attached significant attention [1–4]. This mainly stems from two aspects. Firstly, the different 3D imaging modalities are often fused to produce the precise diagnosis, since they can provide complementary information for interpreting the anatomy, tissue, and organ. As the necessary prerequisite for image fusion, multi-modal medical image registration is significant in relating clinically significant information from the different images. However, the relationship of intensity values in multi-modal 3D medical images might be highly complicated due to differences between the imaging principles, which leads to the difficulty in the construction of the appropriate similarity measure.

Secondly, non-rigid deformation generally cannot be ignored for the soft organs that are easy to deform. Accordingly, the non-rigid transformation must be used as a deformation model in the non-rigid multi-modal medical image registration. However, the non-rigid transformation often involves numerous parameters, which will render accurate image registration difficult [5–8]. Therefore, the non-rigid multi-modal 3D medical image registration has become a challenging task [9–12].

The grey information and spatial information of 3D images are generally considered at the same time in order to construct a suitable similarity measure for non-rigid multi-modal 3D medical image registration. The typical similarity measure construction method is to combine the mutual information (MI) and spatial information [13]. Rueckert et al. [14] proposed using the second-order MI to encode the local information by considering both intensity information and structural information of images. However, this method needs to use a four-dimensional (4D) histogram to calculate the similarity measure. The number of grey levels cannot be too large in order to avoid the curse of dimensionality of high dimensional histograms. Pluim et al. [15] put forward a method that combines the normalized MI with the gradient amplitude and direction for rigid multi-modal image registration. Loeckx et al. [16] presented the image registration method that is based on the conditional MI. This method adopted the 3D joint histogram including the grey levels and the spatial information distribution of the reference and float images. However, the MI for 3D images in itself is computationally complicated and its combination with the spatial information will further lead to high computational complexity for the above registration methods.

The structural representation methods have been presented to more effectively measure the similarity between the different images [17–22]. Wachinger et al. [17] presented the entropy images based structural representation method. In this method, the entropy images were produced by calculating the histogram of image blocks, and then the sum of squared differences (SSD) on the entropy images was used as the similarity measure for image registration. As this method tends to produce the blurred entropy images, it cannot ensure the satisfactory registration results. Heinrich et al. [18] proposed a modality independent neighborhood descriptor (MIND) for the non-rigid multi-modal image registration. Based on the concept of image self-similarity that was introduced in non-local means image denoising, the MIND first extracted the distinctive and multi- dimensional features based on the intensity differences within a search region around each voxel in each modality. Subsequently, the SSD between MIND representations of two images was used as the similarity metric within a standard non-rigid registration framework. Although the MIND is robust to non-functional intensity relations and image noise, it cannot provide the effective structural representation for the complicated medical images with the weak, discontinuous, and complex details, because it only utilizes the similarity of image intensities. The self-similarity context (SSC) descriptor, an improved version of MIND, was proposed in [19]. The SSC descriptor was designed to find the context around the voxel of interest. The point-wise distance between SSC descriptors was used as the metric for the deformable registration on a minimum spanning tree while using dense displacement sampling (deeds) [20]. Zhu et al. [21] explored the self-similarity inspired local descriptor for structural representation based on the low-order Zernike moments with good robustness to image noise. This method cannot work well for ultrasound (US) images and positron emission tomography (PET) images with blurred features due to the ignorance of high-order Zernike moments with better feature representation ability than the low-order ones.

The solution of deformation parameters involved in the transformation model, as a high dimensional optimization problem, is a very challenging task apart from the difficulty in the construction of similarity measure for the non-rigid multi-modal 3D medical image registration. One approach to solve this problem is to use the local optimization methods (e.g., the L-BFGS-B method [23]), the global optimization methods (e.g., the evolutionary strategies [24] and the particle swarm optimization (PSO) [25]), as well as the combined methods (e.g., the hybrid L-BFGS-B and cat swarm optimization (HLCSO) method [26]). However, these methods cannot produce the satisfactory registration results in the case of the high-dimensional optimization problem. Another popular method is to reduce the dimension of transformation parameters while using the geometric transform models that are based on knowledge [27–30]. In these methods, it is required to have enough understanding of material properties of organs or tissues to establish a suitable geometric transform. However, some organs and tissues are so complicated that the existing methods cannot accurately characterize their material properties. Meanwhile, when determining the geometry and the boundary conditions, it is necessary to accurately segment the anatomy of medical images, which indeed is a very challenging task. Some alternative methods can be adopted to address this challenging problem. For example, by means of the mask image, the areas in the images that involve no non-rigid deformation can be covered up to reduce the number of the deformation field variables that are involved in the optimization process. However, the shape of such areas might often be irregular, thereby accurately leading to the difficulty in determining the mask image.

We have proposed a novel registration method using an improved modality independent neighborhood descriptor (MIND) based on the foveated nonlocal self-similarity to address these problems in the construction of similarity measure and the solution of non-rigid transformation parameters. The contributions of our work lie in the two aspects. For one thing, we have designed the foveated MIND (FMIND) for the effective structural representations of 3D medical images, thereby ensuring accurate image registration. One the other hand, the spatial constraint method based on the FMIND is proposed and introduced into the Markov random field (MRF) optimization to reduce the number of non-rigid transformation parameters and restrict the calculation of the energy function in the image regions involving local non-rigid deformation, thereby ensuring efficient image registration. Extensive experiments on multi-modal medical images demonstrate that the proposed method is provided with higher registration accuracy, except for computed tomography-positron emission tomography (CT-PET) images and higher computational efficiency than other evaluated registration methods.

2. Methods

2.1. The Framework of the FMIND Based Image Registration Method

Figure 1 shows the flowchart of the proposed image registration based on the FMIND. Firstly, the FMIND is constructed based on the foveated nonlocal self-similarity and it is applied to the reference image I_R and the float image I_F to produce the corresponding structural representations FMIND (I_R) and FMIND (I_F), respectively. Afterwards, the objective function, i.e., the energy function, is established based on the free-from deformation (FFD) model and the similarity measure defined as the sum of absolute differences (SAD) between FMIND(I_R) and FMIND(I_F). Finally, the FMIND based spatial constraint is introduced to produce the mask image for the MRF discrete optimization. During the iterative optimization, the deformation vector, which is a vector of parameters defining the deformation field, is produced at each iteration. The final optimal deformation vector T' will be obtained once the optimization procedure is terminated, and it is utilized to produce the registration result.

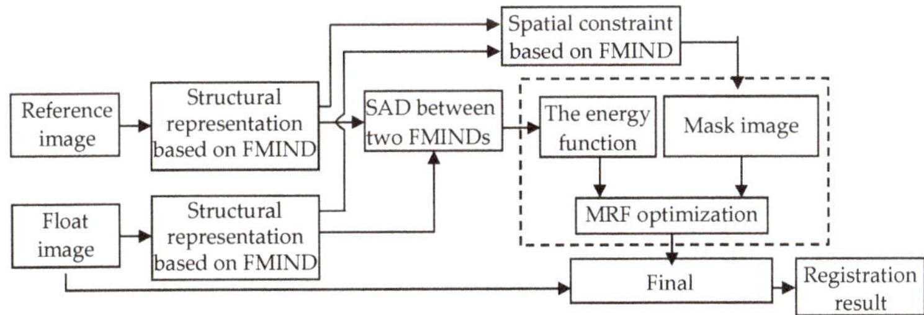

Figure 1. Flowchart of the foveated modality independent neighborhood descriptor (FMIND) based image registration.

2.2. The Foveated Modality Independent Neighborhood Descriptor

The FMIND is presented based on the characteristics of human visual system (HVS). In the HVS, the distribution of cone cells is uneven. The foveation has the highest density. If the foveation is taken as the center, the cell density will fall fast when it is extended around. The optic nerve cells have similar characteristics. Therefore, when we watch a point in an image, this point will have the highest sensitivity and the sensitivity will drop with the increasing distance to the point. Inspired by the characteristics of the HVS, Alessandro Foi et al. [31] have proposed calculating the patch similarity based on the Euclidean distance d^{FOV} between the the foveated patches, defined as:

$$d^{\text{FOV}}(I,\, x_1,\, x_2) = \|I_{x_1}^{\text{FOV}} - I_{x_2}^{\text{FOV}}\|_2^2 \tag{1}$$

where $I_{x_1}^{\text{FOV}}$ and $I_{x_2}^{\text{FOV}}$ denote the foveated patches that were obtained by foveating the image I at the two fixation points x_1 and x_2. By applying the foveation operator F to the image I, the foveated patch I_x^{FOV} is produced as:

$$I_x^{\text{FOV}}(u) = F[I,\, x](u),\ u \in S \tag{2}$$

where u denotes the location of any pixel in the foveated image patch S. In [31], the designed foveation operators mainly include the isotropic and anisotropic foveation operators. As the latter has more advantages than the former in describing the image edges and textures, it will be used as the foveation operator. This operator is defined as:

$$F_{\rho,\theta}[I,\, x](u) = \sum_{\xi \in Z^2} I(\xi + x) v_u^{\rho,\theta}(\xi - u),\ \forall u \in S \tag{3}$$

where $v_u^{\rho,\theta}$ denotes the blur kernel and it is mainly structured by the elliptical Gaussian probability density function (PDF), ρ determines the elongation of the Gaussian PDF, and θ denotes the angular offset, respectively. The blur kernel $v_u^{\rho,\theta}$ is defined as [31]:

$$v_u^{\rho,\theta} = \begin{cases} \sqrt{K(0)}\, g^{\rho,\angle u+\theta}_{\frac{1}{2\sqrt{\pi}}\sqrt{\frac{K(0)}{K(u)}}} & u \neq 0 \\ \sqrt{K(0)}\, g_{\frac{1}{2\sqrt{\pi}}} & u = 0 \end{cases} \tag{4}$$

where $\sqrt{K(0)} = \|v_u\|_1$, $\sqrt{K(u)} = \|v_u\|_2$, $g_{\frac{1}{2\sqrt{\pi}}}$ denote the elliptical Gaussian PDF with the standard deviation of $\frac{1}{2\sqrt{\pi}}$ and $\angle u+\theta$ determines the orientation of the axes of the elliptical Gaussian PDF.

Figure 2 gives an example of two anisotropic foveation operators, where S is a 7×7 foveated patch, $\theta = 0$, and the different kernel elongation parameters $\rho = 2$ and $\rho = 6$ are used, respectively. Clearly, this radial design of these anisotropic foveation operators is consistent with HVS features, which thereby leads to the effective structural representation of images for the FMIND.

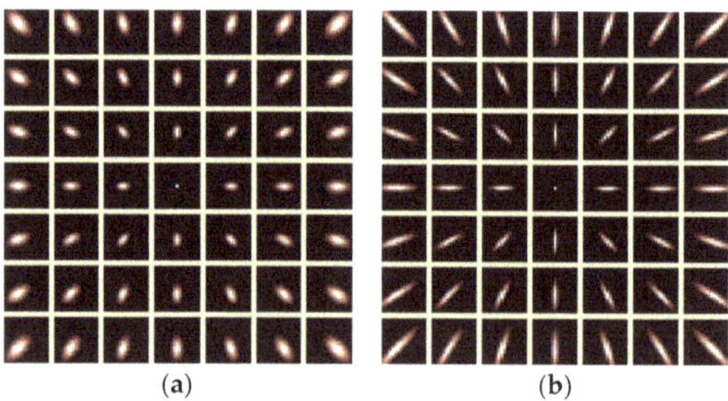

Figure 2. Anisotropic foveation operators with a 7×7 foveated patch and $\theta = 0$. (**a**) $\rho = 2$; and, (**b**) $\rho = 6$.

We will propose the FMIND based on the foveated nonlocal self-similarity between different image patches in the same image borrowing the idea of self-similarity in the non-local means denoising. The FMIND is expressed as:

$$\mathrm{FMIND}(I,\ x,\ r) = \frac{1}{n}\exp\left(-\frac{d^{\ \mathrm{FOV}}(I,\ x,\ x+r)}{V^{\ \mathrm{FOV}}(I,\ x)}\right)\ r \in R \tag{5}$$

where R denotes a search window centered at x, $d^{\ \mathrm{FOV}}(I,\ x,\ x+r)$ denotes the distance between the foveated image patches I_x^{FOV} and I_{x+r}^{FOV}; n is a normalization constant to ensure that the maximum of $\mathrm{FMIND}(I,\ x,\ r)$ is 1; $V^{\ \mathrm{FOV}}(I,\ x)$ denotes the variance of the foveated image patch I_x^{FOV} centered at x in the image I, and it controls the attenuation degree of this function in Equation (5). The variance $V^{\ \mathrm{FOV}}(I,\ x)$ is estimated as the mean of foveated distances for all the pixels in the foveated patch S.

$$V^{\ \mathrm{FOV}}(I,\ x) = \frac{1}{|S|}\sum_{m\in S} d^{\ \mathrm{FOV}}(I,\ x,\ x+m) \tag{6}$$

where $|S|$ denotes the number of pixels in S.

The structural information around the pixel x in the image I will be described by one-dimensional vector of size $|R|$, where $|R|$ denotes the number of pixels in the search window R by means of the FMIND. After obtaining the FMIND for the reference and float images, the similarity metric $\mathrm{SADF}(I(x),\ J(x))$ between two pixels at the same position x in the images I and J can be expressed as the mean SAD between $\mathrm{FMIND}(I,\ x,\ r)$ and $\mathrm{FMIND}(J,\ x,\ r)$ of pixels in R.

$$\mathrm{SADF}(I(x),\ J(x)) = \frac{1}{|R|}\sum_{r\in R}\left|\mathrm{FMIND}(I,\ x,\ r) - \mathrm{FMIND}(J,\ x,\ r)\right| \tag{7}$$

where R takes a six-neighborhood in this paper.

2.3. MRF Optimization Based on the Spatial Constraint

2.3.1. Discrete Optimization Based on the MRF

After obtaining the similarity measure for the two different modal images, we will use the FFD as the transformation model and use Markov random field (MRF) optimization [32] to obtain the transformation parameters in the FFD. The reason for choosing this discrete optimization method is that it does not need to calculate the gradient of the energy function in the process of optimization, which thereby facilitates producing the good registration result by avoiding falling into the local minimum. In this method, the image registration problem will be converted into the MRF based discrete optimization problem.

$$E_{\mathrm{MRF}}(\mathbf{l}) = \frac{1}{|G|}\sum_{p\in G}\left(V_p(l_p) + \lambda\sum_{q\in N(p)} V_{pq}(l_p,\ l_q)\right) \tag{8}$$

$$V_p(l_p) = \int_{\Omega} SADF\left(I(x),\ J \circ T_{l_p}(x)\right)dx \tag{9}$$

$$V_{pq}(l_p,\ l_q) = \|T_{l_p} - T_{l_q}\|_1 \tag{10}$$

where E denotes the general form of a first-order MRF, i.e., the energy function and λ is a constant; G is the set of vertices and $|G|$ denotes the number of vertices in G, where G can be regarded as the vertex set in the FFD, because this method uses the FFD as the deformation model; $N(p)$ and $N(q)$ refer to the neighborhood of vertices p and q, respectively; $\mathbf{l}$ is the discrete labelling while l_p and l_q are the labels that are assigned to the vertices p and q, respectively; $V_p(l_p)$ denotes the data item of the energy function $E_{\text{MRF}}(\mathbf{l})$, while $V_{pq}(l_p, l_q)$ represents its smooth regularization and it takes the L_1-norm to encourage the neighboring nodes p and q to keep the displacement.

Accordingly, the MRF optimization, actually, is to seek to assign a label that is associated with the deformation to each vertex, so that the energy function in Equation (8) is minimized. In this paper, the fast primal-dual (Fast-PD) [33] algorithm will be used for the MRF optimization to produce the registration result. More details about the Fast-PD algorithm can be found in [33].

2.3.2. Spatial Constraint Based on the FMIND

When the above registration method is applied to three-dimensional (3D) medical images, the number of deformation field variables will be large. If all pixels' displacement along the x, y, and z directions is considered, the number of dense deformation field variables will be $3\cdot|I_x|\cdot|I_y|\cdot|I_z|$, where $|I_x|$, $|I_y|$ and $|I_z|$ denote the number of pixels in the x, y, and z dimensions of the image I, respectively. For example, there will be 50,331,648 deformation field variables when $|I_x| = |I_y| = |I_z| = 256$. It is indeed very time-consuming to address such a high dimensional optimization problem.

In the reference and float images, sometimes only some areas involve non-rigid deformation. In addition, the non-rigid registration is unnecessary for some smooth areas. For these regions, the mask can be used to indicate that they will be excluded from the registration process. In this way, we can not only reduce the number of variables for describing the deformation field, but also focus the calculation of the energy function in the image areas that indeed involve the local non-rigid deformation. However, the shape of areas without non-rigid deformation is often irregular. Generally, manual intervention or image segmentation is needed for obtaining the appropriate mask image. However, these technologies cannot ensure that the satisfactory mask image can be produced for US and PET images due to their low image contrast, blurriness, and edge discontinuousness.

We will put forward the spatial constraint method based on the FMIND to address the above problem. From Equation (7), it can be seen that $\text{SADF}(I(x), J(x))$ contains the corresponding relationship of the local spatial information at the pixel x in the images I and J. This information can be used to reduce the number of variables for describing the deformation field and limit the control nodes in the deformation field to move in the areas with the local non-rigid deformation. In the FMIND based spatial constraint method, the vertex set G will be divided into the set G_s of static vertices and the set G_d of dynamic vertices based on the local spatial information included in the FMIND. The vertices in G_s are similar to those in the smooth areas and the areas that involve no deformation. Meanwhile, the vertices in G_d are similar to those in the non-smooth areas involving the deformation. The calculation of the energy function can be restricted in the areas involving the non-rigid deformation through the movement of these dynamic vertices with the local non-rigid deformation. In this way, the number of deformation field variables will decrease from $3\cdot|G_x|\cdot|G_y|\cdot|G_z|$ to $3\cdot|G_{d,x}|\cdot|G_{d,y}|\cdot|G_{d,z}|$, where $|G_{d,x}|$, $|G_{d,y}|$, and $|G_{d,z}|$ denote the number of vertices in x, y, and z dimensions of G_d. By utilizing the FMIND based spatial constraint, we can obtain the division of vertices, thereby generating the mask image without manual intervention and image segmentation.

There will be two requirements that no vertices of the whole MRF model will be omitted and no repeated division of vertices will be done to ensure the effective division of G in the FMIND based spatial constraint method. Correspondingly, the logical relationship among G, G_s, and G_d can be expressed as $G = G_s \cup G_d$ and $\varnothing = G_s \cap G_d$, where $\varnothing$ denotes the empty set. We have designed the following vertex partition algorithm according to the above requirements. For any vertex $p_{i,\,j,\,k}$ of the set G in the MRF model, i.e., $G = \left\{p_{i,\,j,\,k}|1 \le i \le |G_x|,\ 1 \le j \le |G_y|,\ 1 \le k \le |G_z|\right\}$, we will check the similarity metric SADF for each pixel x in the local image patch $LP(p_{i,\,j,\,k})$ with radius R_{LP}, which takes $p_{i,\,j,\,k}$ in G as the center, as shown in Figure 3. Let *con* denote the number of pixels in $LP(p_{i,\,j,\,k})$ whose similarity 1-$\text{SADF}(I(x), J(x))$ is greater than a certain threshold δ. If the ratio of *con* to the patch size $\left|LP(p_{i,\,j,\,k})\right|$ is greater than the static factor ε, $p_{i,\,j,\,k}$ will be regarded as a static vertex. In a similar way, we can determine other static vertices to generate the final set G_s. Accordingly, G_d will be easily computed as $G_d = G - G_s$.

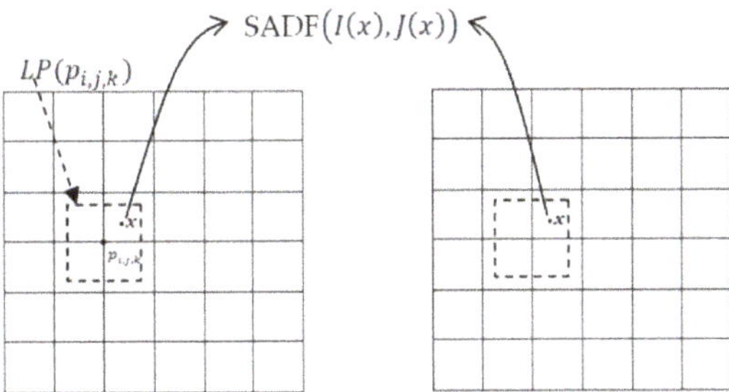

Figure 3. The illustration of the local image block $LP(p_{i,\ j,\ k})$.

Obviously, the performance of the vertex partition algorithm depends on three key parameters R_{LP}, δ, and ε. Here, δ will influence the decision of whether two pixels are similar and ε is used to adjust the probability that $p_{i,\ j,\ k}$ is divided into G_s. Section 3.1 discusses the choice of these parameters. Algorithm 1 shows the detailed implementation of the proposed vertex partition algorithm.

Algorithm 1. Partition of the vertex set G

Input: FMIND(I, x, r), FMIND(J, x, r), G, δ, ε, R_{LP}
Output: G_s, G_d
(1) $G_s = \varnothing$;
(2) ***for*** $(i = 1; i \leq |G_x|; i = i{+}{+})$
(3) ***for*** $(j = 1; j \leq |G_y|; j = j{+}{+})$
(4) ***for*** $(k = 1; k \leq |G_z|; k = k{+}{+})$
(5) $con = 0$;
(6) ***while*** $x \in LP\left(p_{i,\ j,\ k}\right)$
(7) ***if*** (1-SADF($I(x)$, $J(x)$)> δ)
(8) $con{+}{+}$;
(9) ***end if***
(10) ***end while***
(11) ***if*** $\left(\frac{con}{\left|LP(p_{i,\ j,\ k})\right|} > \varepsilon\right)$
(12) $G_s = p_{i,\ j,\ k} \cup G_s$;
(13) ***end if***
(14) ***end for***
(15) ***end for***
(16) ***end for***
(17) $G_d = G - G_s$;
(18) ***return*** G_s, G_d;

3. Results

In this section, we will first discuss the selection of several key parameters in the FMIND method, and then use the method based on the anatomical landmarks selected by doctors to compare registration accuracy and efficiency of the proposed FMIND method with those of the entropy images based SSD (ESSD) [17], MIND [18], SSC [19], and HLCSO [26] methods. For the appreciation of registration performance, we have used four datasets with synthetic deformation, including simulated 3D MR images in BrainWeb database [34], 3D CT and MR images in NA-MIC database [35], 3D CT and PET images in NA-MIC database [36], and real 3D MR images from Retrospective Image Registration Evaluation project [37]. Besides, we have used the real MR and US images with unknown real deformation in the Brain Images of Tumors for Evaluation (BITE) database [38] available at [39] to appreciate the practicality of the proposed method. Here, the implementation efficiency of the evaluated registration methods is appreciated by their running time. For all evaluated methods, they are implemented on the personal computer with 2.40 GHz CPU and 4 GB RAM while using the mixed programming of Matlab and C++.

In the case of synthetic deformation, the registration accuracy is appreciated by the target registration error (TRE) [40], defined as:

$$\text{TRE} = \frac{1}{N}\sum_{i=1}^{N} \sqrt{(T_{L_x} - T_{D_x})^2 + \left(T_{L_y} - T_{D_y}\right)^2 + (T_{L_z} - T_{D_z})^2} \tag{11}$$

where T_L is the synthetic deformation (i.e., the ground truth generated by using a linear combination of radial basis functions), T_D is the deformation that is estimated by the registration methods, and N denotes the number of landmarks selected manually based on doctors' advice from the reference images. For each pair of reference and float images, different synthetic deformations will be applied to the float image for 25 times and we will manually select 90 ($N = 90$) landmarks from each 3D reference image to compute the TRE. The mean of TREs values for registering 25 deformed images will be used to appreciate the registration accuracy. Figure 4 gives an example of chosen landmarks in one slice of simulated 3D PD weighted image and real 3D T1 weighted image for MR image registration, 3D CT image for CT-MR image registration and 3D CT image for CT-PET image registration.

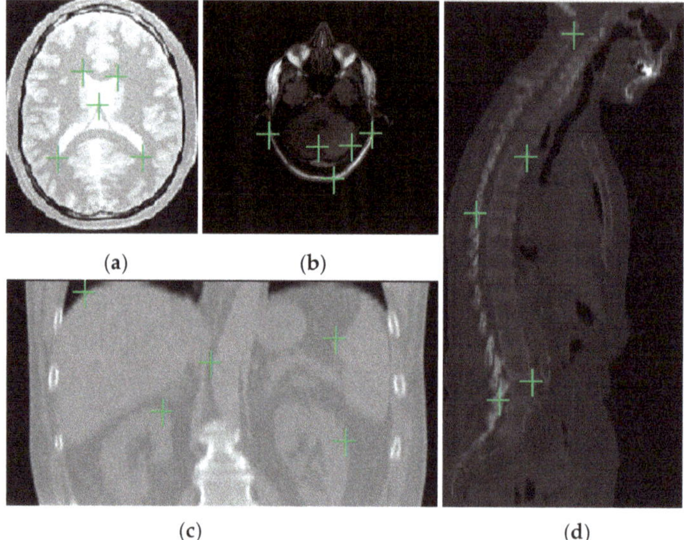

Figure 4. Landmarks in one slice of three-dimensional (3D) medical images. (**a**) simulated proton density (PD) weighted image; (**b**) real T1 weighted image; (**c**) abdomen computed tomography (CT) image; and, (**d**) whole-body CT image.

3.1. Parameter Setting

In the FMIND method, we will fix $\theta = 0$, S to be a 5×5 foveated patch in Equation (3) and $\lambda = 0.01$ in Equation (8). The remaining parameters include the kernel elongation parameter ρ, the image patch radius R_{LP}, the similarity threshold δ, and the static factor ε. We will conduct experiments on three pairs of simulated MR images (T2-T1, PD-T2, and PD-T1) from BrainWeb database in order to effectively determine these parameters, where the former and the latter will be used as the reference and float images, respectively. These simulated MR images are realistic MRI data volumes that are produced by an MRI simulator while using three sequences (T1, T2, and PD weighted) and a variety of slice thicknesses, noise levels, and levels of intensity non-uniformity.

3.1.1. The Kernel Elongation Parameter ρ

Figure 5 shows the TRE values of the FMIND method using different ρ values. For the purpose of evaluating the influence of ρ on registration accuracy, we have set the significant level $\alpha = 0.05$ in one-way Analysis of Variance (ANOVA) [41]. The obtained significance value P is 0.001, which means that ρ has a significant impact on registration accuracy of the FMIND method. From Figure 5, we can see that the TRE achieves the minimum value when $\rho = 2$. Thus, we have fixed $\rho = 2$ in the proposed method.

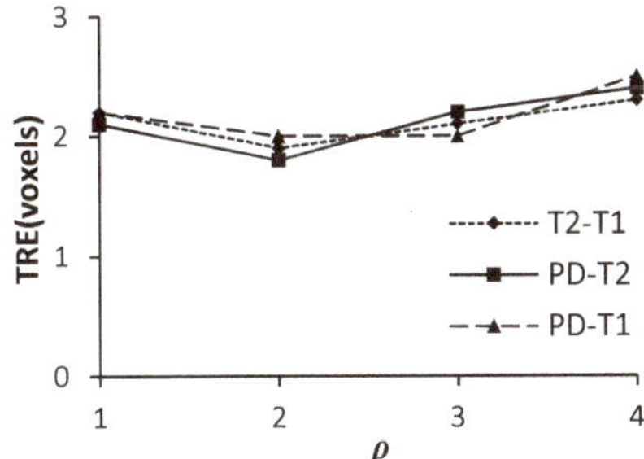

Figure 5. The target registration error (TRE) with different ρ values.

3.1.2. The Image Patch Radius R_{LP}

Figure 6 shows the TRE values and computational time of the FMIND method while using the different R_{LP} values. Likewise, one-way ANOVA with $\alpha = 0.05$ is used to evaluate the influence of R_{LP} on the registration results. The obtained significance value P is 0.001 for registration accuracy and P is 0.007 for registration efficiency. Therefore, R_{LP} has the significant impact on both registration accuracy and efficiency. It can be seen from Figure 6a that the TRE significantly declines when R_{LP} varies from 2 to 7, and it tends to be stable when R_{LP} varies from 7 to 10. Besides, Figure 6b indicates that the registration time gradually increases for the increasing R_{LP}. The reason is that, for a larger R_{LP}, the more pixels need to be processed in the vertex partition algorithm. Therefore, we have set R_{LP} as 7 to achieve the trade-off between registration accuracy and efficiency.

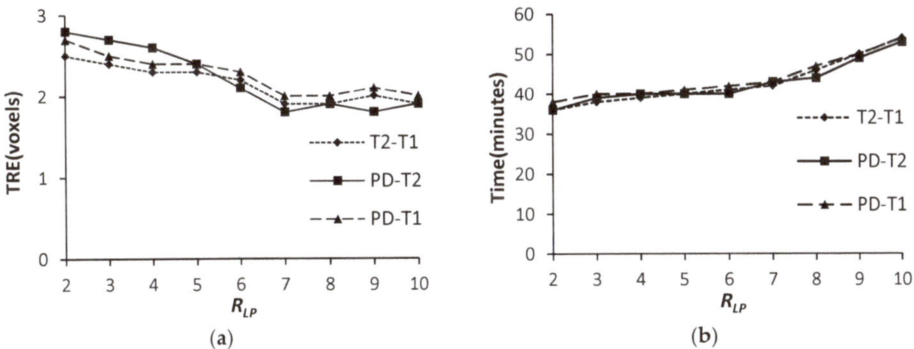

Figure 6. The TRE and computation time with different R_{LP}. (**a**) TRE (voxels); (**b**) Time (minutes).

3.1.3. The Similarity Threshold δ

Figure 7 shows the effect of δ. For one-way ANOVA with $\alpha = 0.05$, the obtained P values are 0.001 for both registration accuracy and efficiency, which means the significant impact of δ on the registration performance of the FMIND method. From Figure 7a, we can see that the TRE significantly declines when δ varies from 0.3 to 0.8. The reason is that for a larger δ in this range, fewer control vertices are divided into the static ones and the number of dynamic control vertices increases, thereby resulting in a smaller TRE. Meanwhile, the TRE tends to be stable when δ varies from 0.8 to 1.0. The reason is that, for the threshold δ in this range, the number of dynamic control vertices will increase to a certain value, so that the variation of δ will have little effect on the TRE. Besides, Figure 7b indicates that the registration time significantly increases with the increasing δ. It is easy to understand that, for a larger δ, the increasing dynamic control vertices will lead to more processing time. Therefore, we set δ as 0.8 to balance the registration accuracy and efficiency.

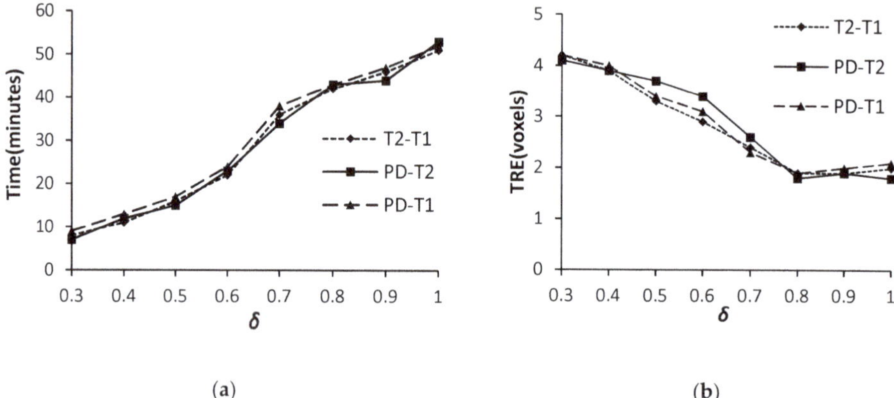

Figure 7. The TRE and computation time with different δ. (**a**) TRE (voxels); and, (**b**) Time (minutes).

3.1.4. The Static Factor ε

Figure 8 shows the effect of the static factor ε on registration accuracy and efficiency. As ε is also used to divide the control vertices, it has a similar effect on registration performance to the similarity threshold δ. According to Figure 8a,b, ε has the opposite effect on TRE and computational time. We have chosen $\varepsilon = 0.9$ for the FMIND method based on the comprehensive consideration of registration performance.

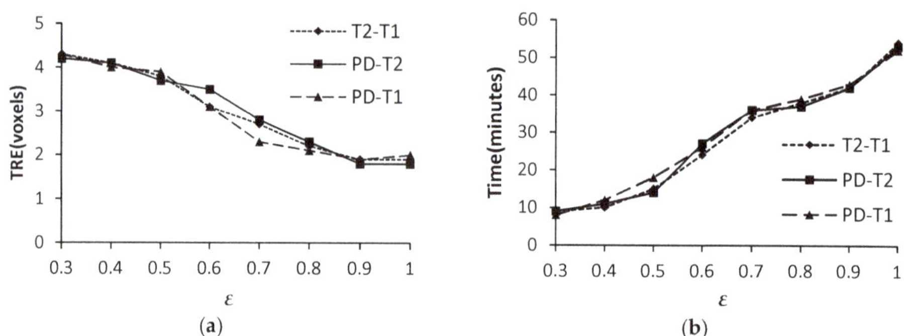

Figure 8. The TRE and computation time with different ε. (**a**) TRE (voxels); and, (**b**) Time (minutes).

3.2. Comparison of Registration Performance

3.2.1. Registration Results of Simulated T1, T2 and PD Images

In order to quantitatively and qualitatively compare the registration performance of the FMIND method and other methods on 3D T1, T2 and PD weighted MR images, we will test them on three pairs of simulated T2-T1, PD-T2, and PD-T1 images of size 256 × 256 × 32. For all evaluated methods, the mean and the standard deviation (std) of TRE values as well as the P values for the t-test with the significance level $\alpha = 0.05$ are computed and are shown in Table 1. In Table 1, "/" means that no registration is implemented. It is shown that all of the P values are less than 0.002, which indicates that there exists significant difference between the FMIND method and any other compared method in terms of TRE. Specifically, as regards the registration of T2-T1 images, the mean and the standard deviation of TRE values for the MIND method are 2.2 voxel and 0.5 voxel, respectively. By comparison, the FMIND method has the lower mean (1.8 voxel) and standard deviation (0.2 voxel) of TRE values than the MIND method. This is mainly due to the advantage of the proposed FMIND in describing the structural information of multi-modal MR images over the MIND method.

Table 1. The TRE for all evaluated methods and the P values for the t-test between the FMIND method and other compared methods operating on the T2-T1, PD-T2, and PD-T1 image pairs.

Methods	TRE (Voxels)								
	T2-T1			PD-T2			PD-T1		
	Mean	Std	P	Mean	Std	P	Mean	Std	P
/	4.8	2.7		4.8	2.7		4.9	2.9	
ESSD	2.7	0.8	2.8×10^{-4}	2.8	0.8	4.4×10^{-4}	2.9	0.9	8.2×10^{-4}
MIND	2.2	0.5	6.4×10^{-4}	2.3	0.6	5.2×10^{-4}	2.3	0.5	3.2×10^{-4}
HLCSO	2.0	0.2	1.2×10^{-3}	2.1	0.3	1.7×10^{-3}	2.2	0.4	2.6×10^{-4}
FMIND	1.8	0.2		1.9	0.3		2.0	0.3	

Figure 9 visually shows the registration results of 3D PD-T1 images for all the evaluated methods. Here, it should be noted that the background regions in these images are removed and the same operation will be implemented for other experiments in the rest of this paper. The comparison among Figures 9f and 9c–e shows that the registration result of the FMIND method is more similar to the reference image that is shown in Figure 9a than those of the ESSD, MIND, and HLCSO methods. Especially for the tissue indicated by the red boxes in Figure 9, the FMIND can recover its deformation better than other evaluated methods.

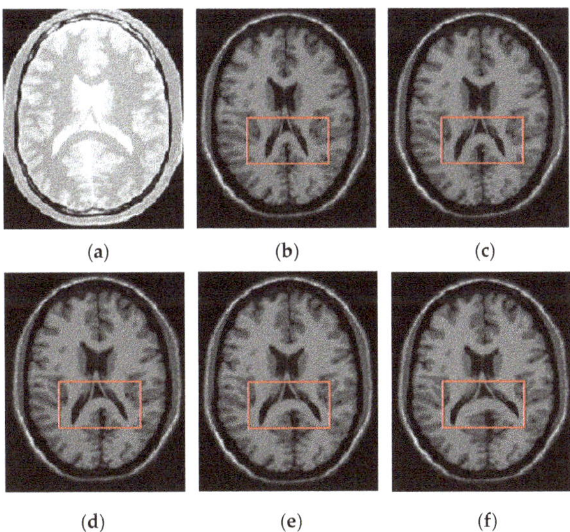

Figure 9. The registration results of all evaluated methods operating on 3D PD-T1 images. (**a**) PD image (reference image); (**b**) T1 image (float image); (**c**) ESSD; (**d**) MIND; (**e**) hybrid L-BFGS-B and cat swarm optimization (HLCSO); and, (**f**) FMIND.

Table 2 lists the implementation time of all evaluated methods on 3D T1, T2, and PD weighted MR images. It can be observed that the FMIND method, on average, takes approximately 44 min to produce the registration results, and it has the highest computational efficiency among all of the methods. The reason lies in that the FMIND method can generally reduce the number of deformation field variables by utilizing the FMIND based spatial constraint for MRF optimization.

Table 2. Computation time for all evaluated methods operating on the T2-T1, PD-T2, and PD-T1 image pairs.

Methods	Time (Minutes)					
	T2-T1		PD-T2		PD-T1	
	Mean	Std	Mean	Std	Mean	Std
ESSD	53.8	7.4	52.2	9.6	55.3	8.8
MIND	62.2	11.4	61.7	11.1	63.5	13.9
HLCSO	98.4	18.8	97.2	17.4	102.6	21.5
FMIND	44.2	7.2	45.4	6.8	46.3	8.0

3.2.2. Registration Results of CT and MR Images

To appreciate registration performance of all evaluated methods operating on CT and MR images from NA-MIC database, these methods are implemented to correct the synthetic deformation that is applied to the float image, where the CT and MR images will be used as the reference and float images, respectively. Here, the liver CT and MR images of size 256 × 256 × 32 are intra-operatively and pre-operatively acquired, respectively. Due to strong differences in image contrast between CT and MR images, their registration is difficult.

Table 3 lists the TRE and P values of t-test for the FMIND method and other methods. As you can see, among all the compared methods, the FMIND method has the highest registration accuracy by providing the lower TRE than other methods. Meanwhile, all the P values are less than 0.003, which indicates the significant difference between the FMIND method and any other method in terms of TRE.

Table 3. The TRE for all evaluated methods and the P values for the t-test between the FMIND method and other compared methods operating on the three-dimensional CT-magnetic resonance (3D CT–MR) image pairs.

Methods	TRE (Voxels)		
	Mean	Std	P
/	6.7	2.9	
ESSD	3.3	1.0	4.7×10^{-4}
MIND	2.7	0.8	2.9×10^{-3}
HLCSO	2.5	0.7	1.6×10^{-3}
FMIND	2.3	0.7	

Figure 10 shows the registration results of 3D CT-MR images for all the evaluated methods. As shown in Figure 10c,d, the ESSD and MIND method cannot effectively correct the deformation that is involved in the MR image. The FMIND method can produce a more similar registration result to the reference image that is shown in Figure 10a than the ESSD and MIND methods. When compared with the most competitive HLCSO method, the proposed method performs better in that it can correct the deformation of some tissues more effectively, as indicated by the three red boxes that are shown in Figure 10e,f.

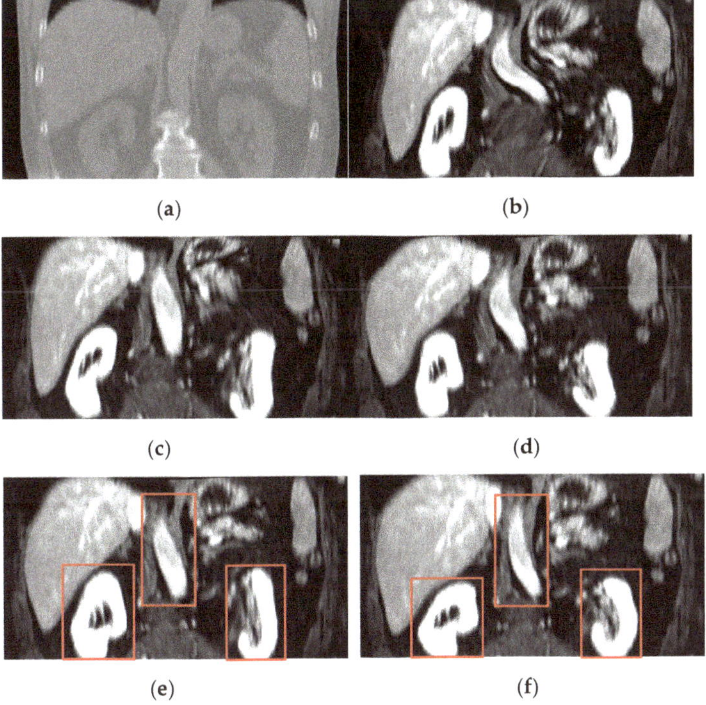

Figure 10. The registration results of all evaluated methods operating on 3D CT–MR images. (**a**) CT image (reference image); (**b**) MR image (float image); (**c**) ESSD; (**d**) MIND; (**e**) HLCSO; and, (**f**) FMIND.

Table 4 lists the implementation time of all the evaluated methods. The comparison indicates the advantage of the FMIND method in computational efficiency. Here, it should be noted that the implementation time for all evaluated registration methods in Table 4 is very similar to that in Table 2, because the used CT and MR images have the same size (256 × 256 × 32) to T1, T2, and PD images.

Table 4. Computation time for all evaluated methods operating on the 3D CT–MR image pairs.

Methods	Time (Minutes)	
	Mean	Std
ESSD	54.6	7.6
MIND	64.4	11.8
HLCSO	101.2	19.2
FMIND	45.0	7.3

3.2.3. Registration Results of CT and PET Images

The 3D whole body CT-PET images from NA-MIC database are also used to demonstrate the advantage of the FMIND method. Here, the CT and PET images of size 168 × 168 × 149 are the reference image and the float image, respectively. It is difficult to realize accurate registration of CT images and blurry PET images of low resolution.

Figure 11 shows the registration results of the 3D CT-PET images for the ESSD, MIND and HLCSO, and FMIND methods. It can be observed that the ESSD and MIND methods cannot correct the deformation in the regions that are marked with the red boxes in Figure 11c,d well. By comparison, the registration results of the HLCSO and FMIND methods are more similar to the reference image shown in Figure 11a than the ESSD and MIND methods.

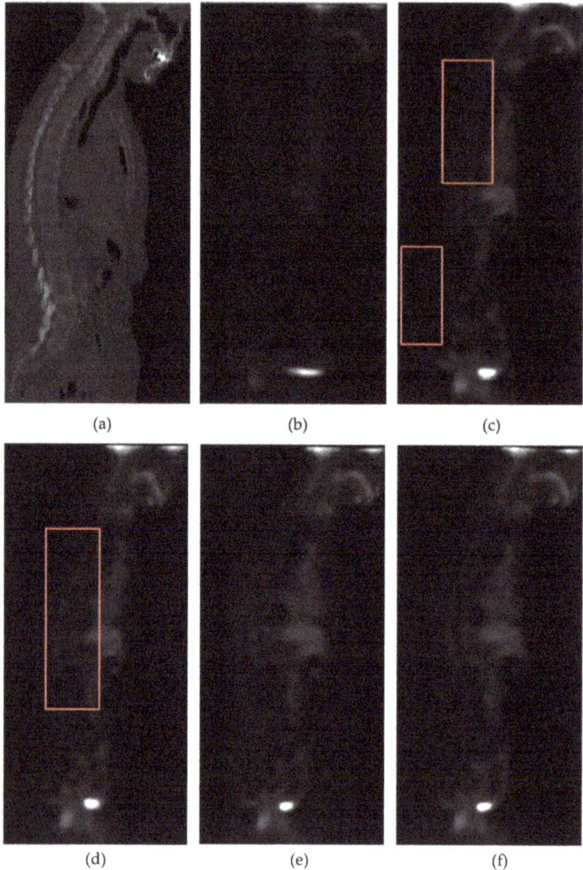

Figure 11. Registration results of all evaluated methods operating on the 3D CT–PET images. (**a**) CT (reference image); (**b**) PET (float image); (**c**) ESSD; (**d**) MIND; (**e**) HLCSO; and, (**f**) FMIND.

Table 5 lists the mean and standard deviation of TRE for all evaluated methods operating on 3D CT and PET images. The comparison of TRE values shows that the HLCSO method provides the minimum mean (2.6 voxel) and standard deviation (0.7 voxel) of TRE among all the compared methods. However, the mean (2.8 voxel) and standard deviation (0.9 voxel) of TRE for the FMIND method are lower than those for the ESSD and MIND methods. The reason can be explained in this way. For the PET image, its contrast and resolution are poor and the edge features are not obvious. Therefore, the FMIND method is slightly inferior to the HLCSO method in the registration of 3D CT-PET images. However, the proposed FMIND method can still provide better structural representation results than the ESSD and MIND methods, thereby leading to its improved registration accuracy than the latter.

Table 5. TRE for all evaluated methods operating on the 3D CT–PET image pairs.

Methods	TRE (Voxels)	
	Mean	Std
/	5.7	2.8
ESSD	3.6	1.6
MIND	3.1	1.2
HLCSO	2.6	0.7
FMIND	2.8	0.9

Table 6 lists the calculation time of the various methods operating on CT and PET images. It can be seen from Table 6 that, as compared with other methods, the calculation time of the FMIND method is significantly reduced because the spatial constraint based on the FMIND, to a certain extent, helps to reduce the number of variables that are required by the deformation model. Especially, when compared with the HLCSO method, although the FMIND method has slightly lower registration accuracy, its computational efficiency is more than two times higher. Besides, as compared with the calculation time listed in Table 2, more calculation time will be involved in the registration of CT-PET images because their size (168 × 168 × 149) is bigger than that of T1, T2, and PD images.

Table 6. Computation time for all evaluated methods operating on the 3D CT–PET image pairs.

Methods	Time (Minutes)	
	Mean	Std
ESSD	106.2	19.4
MIND	124.8	22.2
HLCSO	198.6	30.2
FMIND	88.2	13.2

3.2.4. Registration Results of Real MR Images

The MR images from RIRE database are chosen to verify the superiority of the FMIND method in registering the real MR images, where the T1 and PD weighted MR images are used as the reference and float images, respectively. These MR images were acquired while using a Siemens SP Tesla scanner, among which the T1 and PD image volumes were obtained with an echo time of 15 ms and 20 ms, respectively [42].

Here, we will only compare the proposed method with the MIND and SSC methods, which are most similar to our method. Table 7 lists the TRE values of the three methods. Clearly, the SSC method generally provides slightly smaller TRE values than the MIND methods. The two methods are outperformed by the FMIND method in terms of registration accuracy. The comparison of TRE values indeed demonstrates the effectiveness and advantage of the FMIND method in correcting the deformation of real MR images.

Table 7. The TRE for the MIND, self-similarity context (SSC), and FMIND methods operating on the real T1-PD image pairs.

Methods	TRE (Voxels)	
	Mean	Std
/	4.0	1.8
MIND	2.4	0.4
SSC	2.3	0.4
FMIND	2.1	0.3

3.2.5. Registration Results of Real MR and US Images

We will use the pre-operative T1 weighted MR and intra-operative post-resection US images of 13 patients [43] from BITE database for registration performance appreciation to further demonstrate the practicality of the FMIND method. In [43], the MR images were obtained a few days before the surgery while the post-resection 2D US images were acquired while using Philips HDI 5000 ultrasound machine with a P7-4 MHz phased array transducer and they were reconstructed into ultrasound volume with a voxel size of 1 mm. The used MR data contain the tumor, which is replaced by the resection cavity, and thus will not exist in the post-resection US images. Therefore, to register 3D MR to 3D US images is highly challenging. For each patient, 15 landmarks in average selected in [43] are used for TRE evaluation.

Table 8 lists the TRE values of the MIND, SSC, and FMIND methods. Clearly, the SSC method provides smaller TRE values than the MIND method. The FMIND method also performs better than the MIND method, in that the introduction of foveated nonlocal self-similarity ensures more effective structural representations of MR and US images. Please note that the proposed method cannot significantly outperform the SSC method for registration of US-MR images due to the disadvantageous influence of speckle noise that is inherent in US images.

Table 8. The TRE for the MIND, SSC, and FMIND methods operating on the real US-MR images of 13 patients.

Methods	TRE (mm)	
	Mean	Std
/	5.9	3.2
MIND	3.6	1.0
SSC	3.3	0.9
FMIND	3.2	0.9

4. Conclusions

In this paper, we have proposed a novel non-rigid multi-modal 3D medical image registration method that is based on the foveated independent neighborhood descriptor. The advantages of the proposed method lie in two aspects. Firstly, the proposed FMIND can effectively capture the structural feature information of 3D medical images, thereby providing better structural representations than the existing approaches. Secondly, the FMIND based spatial constraint method can help to reduce the number of non-rigid transformation parameters because the FMIND contains the corresponding relationship of the local spatial information at the same pixel in the reference and float images, thereby providing an effective means for solving the high-dimensional optimization problem that is involved in the medical image registration. Experiments on 3D ultrasound, CT, PET, T1, T2, and PD weighted MR images demonstrate that our method can provide higher computational efficiency and higher registration accuracy as compared with the HLCSO, ESSD, MIND and SSC methods, except that its TRE is slightly bigger than that of the HLCSO for CT-PET image registration. Future work will be focused on the acceleration of the method without compromising registration accuracy by using sparse data sampling and parallel data processing strategies to facilitate its clinical applications.

Author Contributions: F.Y. performed the experiments, analyzed the data and drafted the manuscript. X.Z. and M.D. supervised the research and contributed to the article's revision.

Funding: This work was partly supported by the National Natural Science Foundation of China (NSFC) (Grant No.: 61871440, 61861004), and the National Key Research and Development Program of China (Grant No.: 2017YFB1303100).

Conflicts of Interest: The authors declare no conflict of interest.

References

1. Maintz, J.A.; Viergever, M.A. A survey of medical image registration. *Med. Image Anal.* **1998**, *2*, 1–36. [CrossRef]
2. Zitova, B.; Flusser, J. Image registration methods: A survey. *Image Vis. Comput.* **2003**, *21*, 977–1000. [CrossRef]
3. Sotiras, A.; Davatzikos, C.; Paragios, N. Deformable medical image registration: A survey. *IEEE Trans. Med. Imaging* **2013**, *32*, 1153–1190. [CrossRef]
4. Viergever, M.A.; Maintz, J.B.A.; Klein, S.; Murphy, K.; Staring, M.; Pluim, J.P.W. A survey of medical image registration—under review. *Med. Image Anal.* **2016**, *33*, 140–144. [CrossRef] [PubMed]
5. Yang, F.; Ding, M.; Zhang, X.; Wu, Y.; Hu, J. Two phase non-rigid multi-modal image registration using weber local descriptor-based similarity metrics and normalized mutual information. *Sensors* **2013**, *13*, 7599–7617. [CrossRef] [PubMed]
6. Zhang, Z.; Han, D.; Dezert, J.; Yang, Y. A new image registration algorithm based on evidential reasoning. *Sensors* **2019**, *19*, 1091. [CrossRef] [PubMed]
7. Ferreira, D.P.L.; Ribeiro, E.; Barcelos, C.A.Z. A variational approach to non-rigid image registration with Bregman divergences and multiple features. *Pattern Recognit.* **2018**, *77*, 237–247. [CrossRef]
8. Darkner, S.; Pai, A.; Liptrot, M.G.; Sporring, J. Collocation for diffeomorphic deformations in medical image registration. *IEEE Trans. Pattern Anal. Mach. Intell.* **2018**, *40*, 1570–1583. [CrossRef]
9. Studholme, C.; Hill, D.; Hawkes, D. An overlap invariant entropy measure of 3D medical image alignment. *Pattern Recognit.* **1999**, *32*, 71–86. [CrossRef]
10. Zhu, X.; Ding, M.; Huang, T.; Jin, X.; Zhang, X. PCANet-based structural representation for nonrigid multimodal medical image registration. *Sensors* **2018**, *18*, 1477. [CrossRef]
11. Öfverstedt, J.; Lindblad, J.; Sladoje, N. Fast and robust symmetric image registration based on distances combining intensity and spatial information. *IEEE Trans. Image Process.* **2019**, *28*, 3584–3597. [CrossRef] [PubMed]
12. Nie, Z.; Yang, X. Deformable image registration using functions of bounded deformation. *IEEE Trans. Med. Imaging* **2019**, *38*, 1488–1500. [CrossRef] [PubMed]
13. Rohlfing, T. Image similarity and tissue overlaps as surrogates for image registration accuracy: Widely used but unreliable. *IEEE Trans. Med. Imaging* **2012**, *31*, 153–163. [CrossRef] [PubMed]
14. Rueckert, D.; Clarkson, M.J.; Hill, D.L.G.; Hawkes, D.J. Non-rigid registration using higher-order mutual information. In Proceedings of the SPIE Medical Imaging, San Diego, CA, USA, 16–21 February 2000; pp. 438–447.
15. Pluim, J.P.; Maintz, J.A.; Viergever, M.A. Image registration by maximization of combined mutual information and gradient information. *IEEE Trans. Med. Imaging* **2000**, *19*, 409–814. [CrossRef]
16. Loeckx, D.; Slagmolen, P.; Maes, F.; Vandermeulen, D.; Suetens, P. Nonrigid image registration using conditional mutual information. *IEEE Trans. Med. Imaging* **2010**, *29*, 19–29. [CrossRef]
17. Wachinger, C.; Navab, N. Entropy and Laplacian images: Structural representations for multi-modal registration. *Med. Image Anal.* **2012**, *16*, 1–17. [CrossRef]
18. Heinrich, M.P.; Jenkinson, M.; Bhushan, M.; Matin, T.; Gleeson, F.V.; Brady, M.; Schnabel, J.A. MIND: Modality independent neighbourhood descriptor for multi-modal deformable registration. *Med. Image Anal.* **2012**, *16*, 1423–1435. [CrossRef]
19. Heinrich, M.P.; Jenkinson, M.; Papiez, B.W.; Brady, S.M.; Schnabel, J.A. Towards realtime multimodal fusion for image-guided interventions using self-similarities. In Proceedings of the 16th International Conference on Medical Image Computing and Computer-Assisted Intervention, Nagoya, Japan, 22–26 September 2013; pp. 187–194.
20. Heinrich, M.P.; Jenkinson, M.; Brady, S.M.; Schnabel, J.A. Globally optimal deformable registration on a minimum spanning tree using dense displacement sampling. In Proceedings of the 15th International Conference on Medical Image Computing and Computer-Assisted Intervention, Nice, France, 1–5 October 2012; pp. 115–122.
21. Zhu, F.; Ding, M.; Zhang, X. Self-similarity inspired local descriptor for non-rigid multi- modal image registration. *Inf. Sci.* **2016**, *372*, 16–31. [CrossRef]
22. Piella, G. Diffusion maps for multimodal registration. *Sensors* **2014**, *14*, 10562–10577. [CrossRef]

23. Morales, J.L.; Nocedal, J. Remark on "algorithm 778: L-BFGS-B: Fortran subroutines for large-scale bound constrained optimization". *ACM Trans. Math. Softw.* **2011**, *38*, 71–74. [CrossRef]
24. Klein, S.; Staring, M.; Pluim, J.P.W. Evaluation of optimization methods for nonrigid medical image registration using mutual information and B-splines. *IEEE Trans. Image Process.* **2007**, *16*, 2879–2890. [CrossRef] [PubMed]
25. Wachowiak, M.P.; Smolikova, R.; Zheng, Y.; Zurada, J.M.; Elmaghraby, A.S. An approach to multimodal biomedical image registration utilizing particle swarm optimization. *IEEE Trans. Evol. Comput.* **2004**, *8*, 289–301. [CrossRef]
26. Yang, F.; Ding, M.; Zhang, X.; Hou, W.; Zhong, C. Non-rigid multi-modal medical image registration by combining L-BFGS-B with cat swarm optimization. *Inf. Sci.* **2015**, *316*, 440–456. [CrossRef]
27. Camara, O.; Delso, G.; Colliot, O.; Moreno-Ingelmo, A.; Bloch, I. Explicit incorporation of prior anatomical information into a nonrigid registration of thoracic and abdominal CT and 18-FDG whole-body emission PET images. *IEEE Trans. Med. Imaging* **2007**, *26*, 164–178. [CrossRef] [PubMed]
28. Tang, S.; Fan, Y.; Wu, G.; Kim, M.; Shen, D. RABBIT: Rapid alignment of brains by building intermediate templates. *NeuroImage* **2009**, *47*, 1277–1287. [CrossRef]
29. Zacharaki, E.I.; Hogea, C.S.; Shen, D.; Biros, G.; Davatzikos, C. Non-diffeomorphic registration of brain tumor images by simulating tissue loss and tumor growth. *Neuroimage* **2009**, *46*, 762–774. [CrossRef]
30. Brun, C.C.; Leporé, N.; Pennec, X.; Chou, Y.Y.; Lee, A.D.; De Zubicaray, G.; Thompson, P.M. A nonconservative lagrangian framework for statistical fluid registration—Safira. *IEEE Trans. Med. Imaging* **2011**, *30*, 184–202. [CrossRef]
31. Foi, A.; Boracchi, G. Foveated nonlocal self-similarity. *Int. J. Comput. Vis.* **2016**, *120*, 78–110. [CrossRef]
32. Glocker, B.; Komodakis, N.; Tziritas, G.; Navab, N.; Paragios, N. Dense image registration through MRFs and efficient linear programming. *Med. Image Anal.* **2008**, *12*, 731–741. [CrossRef]
33. Komodakis, N.; Tziritas, G.; Paragios, N. Performance vs computational efficiency for optimizing single and dynamic MRFs: Setting the state of the art with primal-dual strategies. *Comput. Vis. Image Underst.* **2008**, *112*, 14–29. [CrossRef]
34. Brainweb. Available online: http://www.bic.mni.mcgill.ca/brainweb/ (accessed on 10 August 2018).
35. NA-MIC Data. Available online: http://na-mic.org/Wiki/index.php/Projects:RegistrationLibrary:RegLib_C47. (accessed on 10 August 2018).
36. NA-MIC Data. Available online: http://na-mic.org/Wiki/index.php/Projects:RegistrationLibrary:.RegLib_C20 (accessed on 10 August 2018).
37. Retrospective Image Registration Evaluation Project. Available online: https://www.insight-journal.org/rire/ (accessed on 10 August 2018).
38. Mercier, L.; Del Maestro, R.; Petrecca, K.; Araujo, D.; Haegelen, C.; Collins, D. Online database of clinical MR and ultrasound images of brain tumors. *Med. Phys.* **2012**, *39*, 3253. [CrossRef] [PubMed]
39. BITE Database. Available online: http://nist.mni.mcgill.ca/?page_id=248 (accessed on 10 September 2019).
40. Maurer, C.R., Jr.; Fitzpatrick, J.M.; Wang, M.Y.; Galloway, R.L.; Maciunas, R.J.; Allen, G.S. Registration of head volume images using implantable fiducial markers. *IEEE Trans. Med. Imaging* **1997**, *16*, 447–462. [CrossRef] [PubMed]
41. Wang, C.W.; Chen, H.C. Improved image alignment method in application to X-ray images and biological images. *Bioinformatics* **2013**, *29*, 1879–1887. [CrossRef] [PubMed]
42. West, J.; Fitzpatrick, J.M.; Wang, M.Y.; Dawant, B.M.; Maurer, C.R., Jr.; Kessler, R.M.; Maciunas, R.J.; Barillot, C.; Lemoine, D.; Collignon, A.; et al. Comparison and evaluation of retrospective intermodality brain image registration techniques. *J. Comput. Assist. Tomogr.* **1997**, *21*, 554–566. [CrossRef]
43. Rivaz, H.; Chen, S.; Collins, D.L. Automatic deformable MR-ultrasound registration for image- guided neurosurgery. *IEEE Trans. Med. Imaging* **2015**, *34*, 366–380. [CrossRef]

Article

Semantically Guided Large Deformation Estimation with Deep Networks

In Young Ha [1,*], Matthias Wilms [2] and Mattias Heinrich [1]

1 Institute of medical informatics, University of Luebeck, 23558 Luebeck, Germany; heinrich@imi.uni-luebeck.de
2 Department of Radiology, University of Calgary, Calgary, AB T2N 4N1, Canada; matthias.wilms@ucalgary.ca
* Correspondence: ha@imi.uni-luebeck.de

Received: 9 January 2020; Accepted: 1 March 2020; Published: 4 March 2020

Abstract: Deformable image registration is still a challenge when the considered images have strong variations in appearance and large initial misalignment. A huge performance gap currently remains for fast-moving regions in videos or strong deformations of natural objects. We present a new semantically guided and two-step deep deformation network that is particularly well suited for the estimation of large deformations. We combine a U-Net architecture that is weakly supervised with segmentation information to extract semantically meaningful features with multiple stages of nonrigid spatial transformer networks parameterized with low-dimensional B-spline deformations. Combining alignment loss and semantic loss functions together with a regularization penalty to obtain smooth and plausible deformations, we achieve superior results in terms of alignment quality compared to previous approaches that have only considered a label-driven alignment loss. Our network model advances the state of the art for inter-subject face part alignment and motion tracking in medical cardiac magnetic resonance imaging (MRI) sequences in comparison to the FlowNet and Label-Reg, two recent deep-learning registration frameworks. The models are compact, very fast in inference, and demonstrate clear potential for a variety of challenging tracking and/or alignment tasks in computer vision and medical image analysis.

Keywords: image registration; large deformation; weakly supervised

---

1. Introduction

Estimation of motion and deformation between images continues to play an important role for multiple vision tasks. While the computation of optical flow for 3D scene motion of smaller magnitude can be performed very effectively and accurately using a range of variational (coarse-to-fine) as well as deep-learning models (e.g., FlowNet [1]), larger deformations still pose a significant challenge. We hypothesize that the large performance gap is due to two reasons. First, current approaches have a limited capture range, in particular for low-textured regions and remain prone to local optima despite extensive multi-scale processing. Second, even supervised approaches that are trained on huge datasets with ground truth flow fields fail to learn sufficient scene understanding and tend to rely on low-level visual clues that become ambiguous for large deformations. Furthermore, learning primarily from synthetically generated ground truth flow fields limits the practical application of deep-learning optical flow models, because adaptation to a new unseen domain is usually not possible and supervised dense correspondences hard to define for real world images.

Weakly supervised approaches [2,3], in contrast, only require semantic segmentation labels to define a loss directly based on the warped segmentation masks for the training of a convolutional neural network (CNN)-based registration model. Yet, these previous works have not addressed large deformation problems and the learned features may again be prone to ambiguous matches (even in combination with smoothness regularization). We therefore propose to use semantic guidance information directly into the feature learning of a two-step deep spatial transformer network for face alignment. Our model comprises a U-Net part that extracts semantic object information for both considered images. This information is fed into flow predictors that rely on a B-spline parameterization for smooth deformable transformations. By employing more than one spatial transformer module within a multi-iteration warping framework [1], we can also avoid the use of memory-intensive cost volume approaches [4] that are hard to extend to high-resolution images or 3D problems.

1.1. Contributions

Our work is most closely related to the work of Qin et al. [2] that also proposed to learn segmentation and registration jointly, but was limited to a single (recurrent) transformation network and uses shared weights for both segmentation and registration networks. In contrast to the FlowNet [1,5] or the SVF-Net [6] no (pseudo) ground truth deformation fields are required for training our network, but a loss based on the alignment of labels is used instead. Hu et al. [3] also considered this weakly supervised learning based on segmentation information, but did not include this guidance in the feature extraction U-Net, which is crucial for large deformations as our experiments show. (Example code available at: https://github.com/multimodallearning/semantically-guided).

1.2. Related Works

Dataset with ground truth deformation fields are difficult to find. Most supervised methods use automatically generated ground truth deformation fields or estimate parameters for deformable transformations [6–8] to train the network. They optimize the mean residual distance between ground truth and estimated deformation fields. In the method of Krebs et al. [7], U-Net-like architecture is used to estimate a dense deformation field from the concatenated fixed and moving images, while Sokooti et al. [8] gave image patches as inputs to a Siamese network to estimate a patchwise 3D deformation field. The obvious drawbacks of these methods are that the estimated deformation fields can only be as good as the automatically generated deformation fields and that the trustworthiness of the generated ground truth field cannot be guaranteed.

Some methods deal with this problem by training the network indirectly using available annotation data, such as segmentation or in an unsupervised manner (only based on a predefined similarity metric) using a spatial transformer network to warp the segmentation of the input image [3,9–12]. De Vos et al. [9] compare the difference between warped moving image and fixed image directly as loss, where a spatial transformer was employed for image warping. In addition to the image intensity difference, the Label-Reg approach of [3] warps the segmentation using spatial transformer and minimizes the difference between the warped moving and fixed segmentation. Various segmentations or labels are given as input to the network and the soft probabilistic Dice is computed as loss at different image scales. In VoxelMorph [12], in addition to unsupervised losses segmentation can also be used as an auxiliary information to improve the accuracy. The use of segmentation for the estimation of the deformation field can be beneficial especially to counter ambiguous correspondences of object parts with locally similar appearance.

Within computer vision, there are several methods which jointly estimate semantic segmentation and optical flow for videos [13–16]. Recent works integrate learning into optical flow frameworks to improve the flow estimation using semantic segmentation. Sevilla et al. [13] use localized layers to deal with the movement of different objects in videos. Cheng et al. [15] combine two existing network architectures with a small modification into a network with two branches. Feature maps generated by each branch (one for segmentation, the other for optical flow) are propagated to the other branch to fuse the information for a final output. Hur et al. [14] employ superpixels to connect optical flow and semantic segmentation. Tsai et al. [16] use an iterative scheme to jointly optimize optical flow and segmentation to obtain video segmentation with improved object boundaries. In the medical image domain, Qin et al. [2] have adopted the joint estimation of segmentation and motion for cardiac MR image sequences. For both segmentation and motion estimation, the same network, i.e., with shared weights, is used for feature extraction.

2. Materials and Methods

A general overview of our proposed method is shown in Figure 1. Our deep network architecture consists of three parts. The first part aims to extract semantically meaningful information using a U-Net structured network [17] and is supervised using a soft constraint on pixel-wise class probabilities using multi-label segmentation. The estimation of the deformation field is carried out in two consecutive convolutional networks with nonlinear spatial transformer modules. Similar to the method of Hu et al. [3], the transformation is only weakly supervised based on the alignment of the warped moving segmentation in comparison with the fixed segmentation. To reduce the number of predicted parameters and ensure a spatially smooth transformation, we use a coarse cubic B-spline transformation model [18] with an additional regularization term.

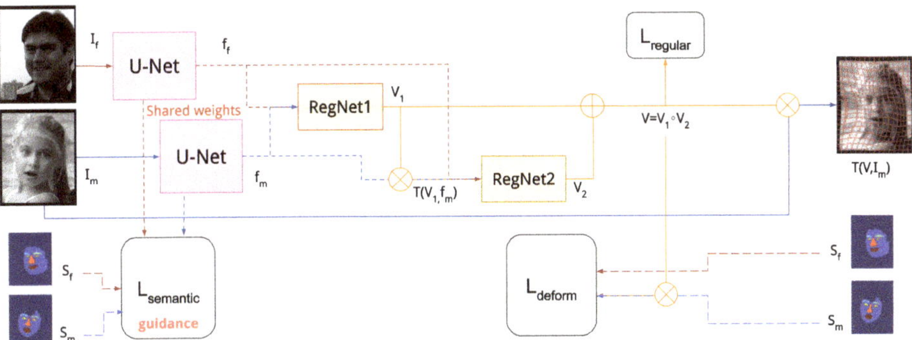

Figure 1. Overview of our framework. From input images I_f, I_m the semantic features f_f, f_m are extracted by U-Net and passed on to the registration networks (RegNet) after the concatenation. Deformation field V_1, V_2 are then estimated by the first and the second RegNet, respectively. As input of the second RegNet, the semantic features of the moving image are transformed using the output of the first RegNet. Finally, two deformation fields are combined to obtain the final deformation field V.

2.1. Semantic Feature Extraction

To cope with large deformation, we strongly believe that semantic scene information is of great importance. While previous methods [3,6] have implicitly used segmentation information to derive a loss with respect to the estimated transform, we make additional explicit use of semantic clues within the feature extraction for each image individually. Figure 2 shows the U-Net architecture of the feature extraction part of our method, which contains 11 convolution filters with 3×3 kernels, two skip connections and roughly 200k trainable parameters. Given a greyscale image $\mathsf{I} \in \mathbb{R}^{H \times W}$, the network produces a SoftMax prediction $\mathbf{f} \in [0,1]^N$ for each pixel and each label $l \in 1, \ldots L$. The semantic loss is computed using weighted cross-entropy:

$$\mathcal{L}_{\text{semantic}} = -\sum_{j}^{N} \mathbf{w}\hat{\mathbf{f}}(x_j) \log \mathbf{f}(x_j), \tag{1}$$

where $\hat{\mathbf{f}}$ is the ground truth segmentation with N sample points and x_j is the j-th sample point. The label weights vector $\mathbf{w} \in \mathbb{R}^L$ is determined by computing the square root of the inverse class frequency for each label. For our task of registering two input images I_f and I_m, the network weights are shared.

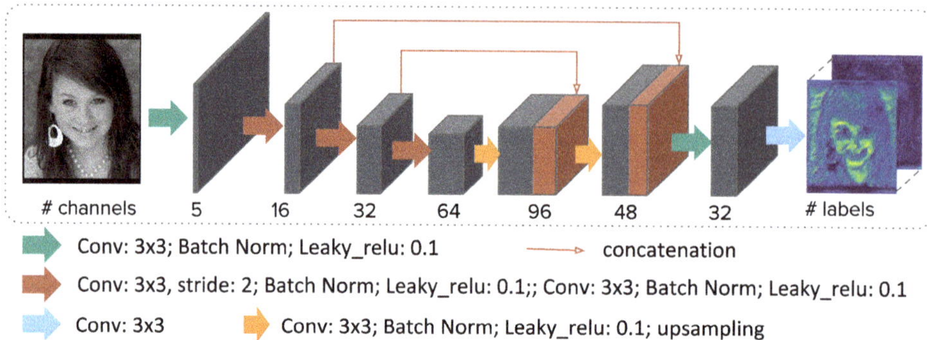

Figure 2. U-Net architecture to extract semantic image features for the subsequent spatial transformer registration network.

2.2. Registration Network

Estimation of the deformation field is performed in two steps, given the network outputs $\mathbf{f}_f$ and $\mathbf{f}_m$ of the feature extraction network. The extracted features contain semantic information of the input images, which is the class probability of each pixel. These are passed on to the registration network after concatenation, which is trained to learn a low-parametric deformation field representation $\tilde{\mathsf{V}}$ in a smaller dimension. Deformation field $\mathsf{V} \in \mathbb{R}^{2\times H\times W}$ in the original image size can be retrieved by upsampling using a cubic B-spline function. The evaluation of a dense displacement field using B-splines (also called free form deformations [18]) can be efficiently calculated using three consecutive average pooling layers without stride, following the theory of recursive cardinal splines.

In our two-step setting, we obtain two different deformation fields V_1 and V_2 from the first and second registration network, respectively. Both networks have the same architecture; however, the first network is trained with the immediate output of the feature extraction network $\mathbf{f}_f$ and $\mathbf{f}_m$, while the second network is trained with the warped moving features $\mathsf{V}_1 \circ \mathbf{f}_m$ instead of $\mathbf{f}_m$. The final deformation field V is generated by combining the two fields. In this paper, we combine the fields in two different ways; addition and transformation.

Finally, with the resulting dense displacement field V, the ground truth segmentation of the moving image S_m is warped and the deformation loss is computed as follows:

$$\mathcal{L}_{deform} = \frac{1}{L}\sum_{l=1}^{L} w_l \mid \mathsf{S}_f(l) - \mathcal{T}(\mathsf{V}, \mathsf{S}_m(l)) \mid, \tag{2}$$

where w_l denote the label weights and $\mathcal{T}$ is the spatial transformer for the B-spline transformation. An additional loss term $\mathcal{L}_{\text{regular}}$ for regularization loss is computed for the estimated deformation field, which penalizes deviations between the final estimation of deformation field V and a locally smoothed version of the same displacements field V_{smooth}:

$$\mathcal{L}_{\text{regular}} = ||\mathsf{V} - \mathsf{V}_{smooth}||_2 \,. \tag{3}$$

Finally, overall loss for semantically guided deformation estimation is computed as follows:

$$\mathcal{L} = \lambda_s\,\mathcal{L}_{semantic} + \mathcal{L}_{deform} + \lambda_r\,\mathcal{L}_{regular}\,, \tag{4}$$

where λ_s and λ_r are the weight parameters.

2.3. Experiments

We have performed ablation studies for our framework and potential variants using the Helen dataset [19]. Our method is also tested on a publicly available medical cardiac dataset (ACDC [20]). The Helen dataset consists of 2330 face RGB images in different image sizes as well as ten segmentation labels for face, eyes, eyebrows, nose, mouth and hair provided by Smith et al. [21]. The ACDC dataset consists of 100 images of 4D MRI for training, for which three segmentation labels of right and left ventricle and myocardium are given.

2.3.1. Data Preprocessing

For the face dataset, we cropped the images to have the same image size of 320 $\times$ 260 using an enlarged face bounding box and converted them into the greyscale images. For training, 2000 images were used as in the work of Le et al. [19] and for the test, the remaining 330 images are used. For our experiment only 7 labels are used, which excludes the hair label, due to their obscure appearance and the mouth structures (upper lip, lower lip and inner mouth) are combined as one single mouth label, since the inner mouth is not given in many samples.

The medical cardiac MRI dataset consists of 100 training image pairs (end-diastolic and end-systolic) with the ground truth segmentation, which we divide into two subsets with 70 and 30 image pairs for training and testing respectively. The original images have a pixel spacing of 1.56 mm for in-plain slice and 5–10 mm of slice thickness. We resampled and cropped all images to have the same voxel dimension 128 $\times$ 128 $\times$ 64 with the slice thickness of $\approx$1.56 mm.

2.3.2. Implementation Details

The U-Net architecture used to extract semantic features is outlined in Section 2.1 and visualized in Figure 2. To account for small structures in the images, a weighted cross-entropy loss was used. Eleven 3 $\times$ 3 convolutional layers and two skip connections with a relatively small total number of 200'000 network parameters were used to avoid overfitting. Downsampling of the input is done using a stride size of 2 in three convolutional layers and two upsampling layers are used to obtain output. The output has half the input size. The network receives an image as input and returns SoftMax probabilities for each structure as a channel.

For the registration part, a convolutional network was implemented as shown in Figure 3. As described in Section 2.2, the same architecture is used twice to build a two-step framework with substantially more channels to accommodate the challenging matching problem and optimize both spatial transformations simultaneously during training. The output of registration network has a smaller parametrization than the image. Therefore, we upsample the control-point displacement and apply three cardinal B-spline smoothing steps before warping using average pooling layers. The kernel size of the average pooling layers and thus the scale of the B-spline transformation for the output is chosen to be 19 for the first network part (to capture large and coarse deformations) and 11 for output of the second network (for a refinement of small structure alignment). For the medical dataset, the scales of 5 and 3 are used, because the motion was smaller on average for this dataset. The scale was chosen to be approximately 5% and 3% respectively (for first and second registration network output) of the largest image dimension for the B-spline transformation, based on empirical tests. The smoothing kernel size was also set relative to image size and object size (in particular the following values were chosen: 5 and 3 for the face and the medical dataset respectively).

Throughout our experiment, we use Adam optimizer with the learning rate of 0.001 and the momentum of 0.97. The weight of the regularization loss λ_r was chosen to be 0.001 and the weight of semantic loss λ_s 1.0, which were determined empirically. For the regularization, the estimated dense field is smoothed using two average pooling layers with the kernel size of 3 for the medical dataset and 5 for the face dataset, also chosen empirically. The training batch size for the face dataset was 20 and for the medical dataset 5. The training was performed for 300 epochs for both datasets.

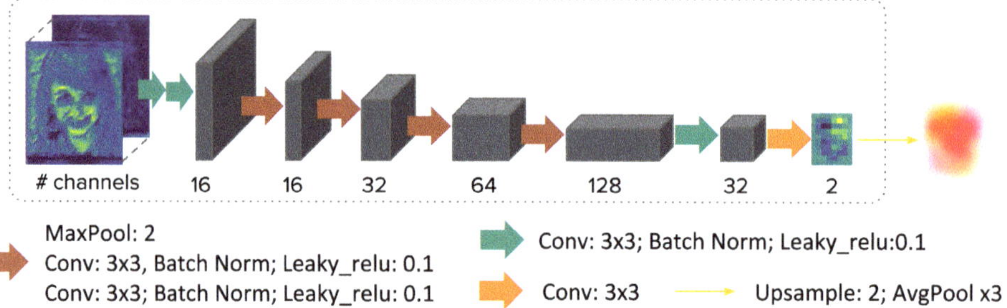

Figure 3. Architecture of our registration network, which estimates displacements of a nonlinear B-spline grid for spatial transformations.

2.3.3. Evaluation Metric

Given a deformation field for a pair of test images, we use the mean of the Dice coefficient (equivalent to F1 score) across individual face parts as well as the contour distance of each structure to evaluate the alignment (or registration) quality. Dice score for a label is computed as:

$$Dice = \frac{2\sum_i |a_i b_i|}{\sum_i |a_i| + \sum_i |b_i|} \tag{5}$$

where a_i, b_i are the labels in pixel i of reference image and target image, respectively. Contour distance of each structure is computed as:

$$D_{contour} = \frac{\sum_{s_a \in S(A)} d(s_a, S(B)) + \sum_{s_b \in S(B)} d(s_b, S(A))}{|S(A)| + |S(B)|} \tag{6}$$

where $S(A)$ and $S(B)$ are the total contour pixels of the reference and target images, respectively. The mean Dice values are computed between the ground truth segmentation of fixed images and warped ground truth segmentation of moving images. Please note that the segmentation labels for guidance are only available for training images and new test pairs are registered without any manual labels. The labels are only used to evaluate the registration accuracy. In addition to the alignment accuracy, we have evaluated the quality of the estimated transformations, which is crucial for subsequent vision applications such as style transfer. We calculate the Jacobian determinant (also *Jacobian*) to provide a quantitative evaluation on the topology of the deformation field.

For each pixel (x, y), $D(x, y)$ is the estimated displacement vector.

$$Jacobian = det\left[I + \begin{pmatrix} \frac{\delta D_x(x,y)}{\delta x} & \frac{\delta D_x(x,y)}{\delta y} \\ \frac{\delta D_y(x,y)}{\delta x} & \frac{\delta D_y(x,y)}{\delta y} \end{pmatrix}\right] \tag{7}$$

where I is the identity matrix, $D_x(x, y)$ and $D_y(x, y)$ are the x and y component of $D(x, y)$ respectively. A Jacobian in each pixel provides the characteristic of the deformation, i.e., a Jacobian of 1 for no change, >1 indicate expansion, 0–1 shrinkage and a Jacobian ≤ 0 indicates singularity. We report standard deviation of the Jacobian, where smaller values indicate smoother transforms. The number of negative Jacobian determinants indicates the number of singularity points, i.e., with smaller mean number of pixels with a negative Jacobian, the deformation is more plausible.

In the following, different settings are compared to analyze the effect and importance of each setting.

3. Results and Discussion

3.1. Ablation Study

In this section, different parts of our framework are evaluated and compared to determine the importance of each part. Our experiments compare minor modifications to the training scheme of our proposed method: with or without guidance, with or without shared weights in the feature extraction, and diffeomorphic transformation models.

3.1.1. Single vs. Two-Step Registration

In many cases, the estimation of deformation field is performed in a single step. We believe that dividing the registration process into smaller steps by using two networks with the same total number of parameters is more beneficial than estimating deformation in a single step. In the two-step registration, one of the input features is transformed with the output deformation field from the first registration network before being forwarded to the second one. We compare the performance of the two-step network with the performance of the single network, where both networks have a total of approximately 2.3 million network parameters. As shown in Figure 4, the two-step network (pre-trained unet, two-step) outperforms the single network (pre-trained unet, single).

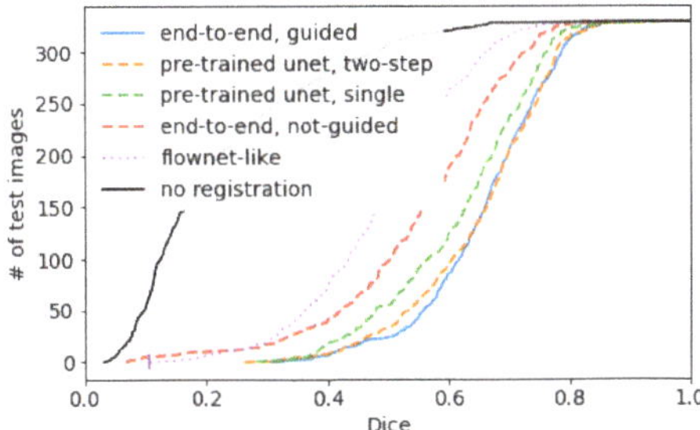

Figure 4. Sorted Dice scores (averaged across face structures) for all test images for various network configurations (lower line is better). Details in Section 3.1.

3.1.2. Regularization

As described in Equation (3), our regularization term penalizes the difference between non-smoothed and smoothed output fields and is scaled by λ_r. Without the regularization loss the networks could not learn to estimate any plausible deformation field. We use average pooling layers for computation of smoothed deformation fields, where the kernel sizes of the pooling layers are chosen to reduce the standard deviation of the Jacobian of deformation fields. Using a too large kernel size might deteriorate the accuracy of the deformation fields with respect to smaller structures. Dice values for each structure are shown in Figure 5 for the empirically chosen kernel size of 5. Although we found this size to be usually most appropriate, for small structures such as eyebrows and eyes approximately 10% are not registered at all. However, this might also have been influenced by the occlusion due to hair, especially when the forehead is covered.

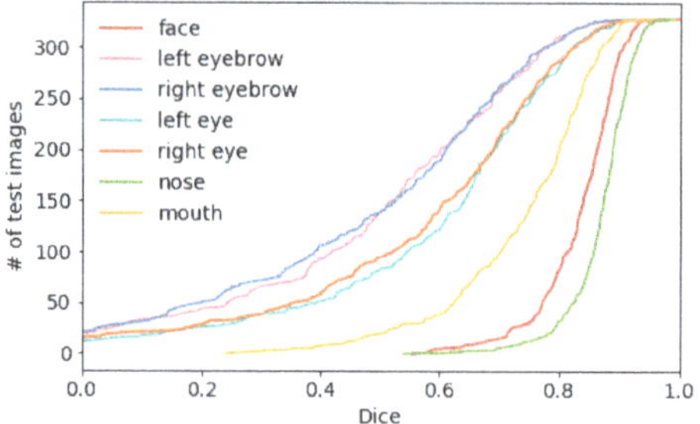

Figure 5. Sorted Dice scores for different face structures. The alignment of eyes and eyebrows is challenging due to occlusions caused by hair in many images.

3.1.3. Semantically Guided Deformation Estimation

Instead of using a pre-trained segmentation network to start the deformation estimation, we jointly train the semantic feature extraction U-Net with the two-step networks using semantic loss (see Figure 4: end-to-end, guided and Table 1: ours) and compare with the same configuration without semantic loss using only the deformation and regularization loss (see Figure 4: end-to-end, not-guided and Table 1: ours (without guidance)). This has the advantage that only a soft constraint is used for the semantic guidance loss and the deformation network can use clues based on more generic image features (e.g., edges). However, this only led to a small quantitative improvement of the alignment in terms of Dice score (0.5%), the warped images appear to be more realistically transformed. Example results of our method is shown in Figure 6 and the training curves are shown in Figure 7.

Figure 6. Example results of our approach. (Top) fixed image with the ground truth fixed labels, (middle) target image with warped ground truth fixed labels, (bottom) warped fixed images.

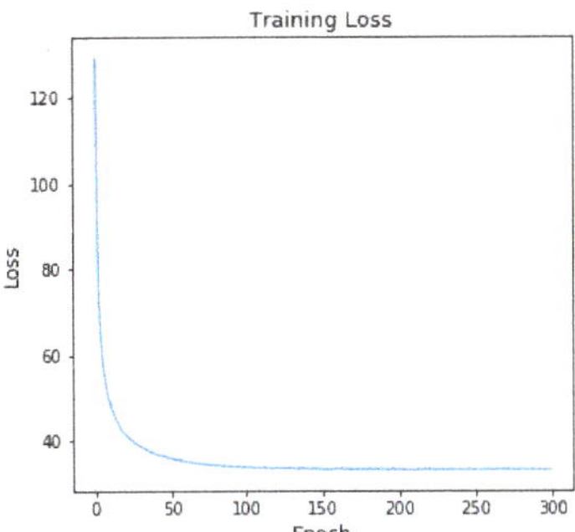

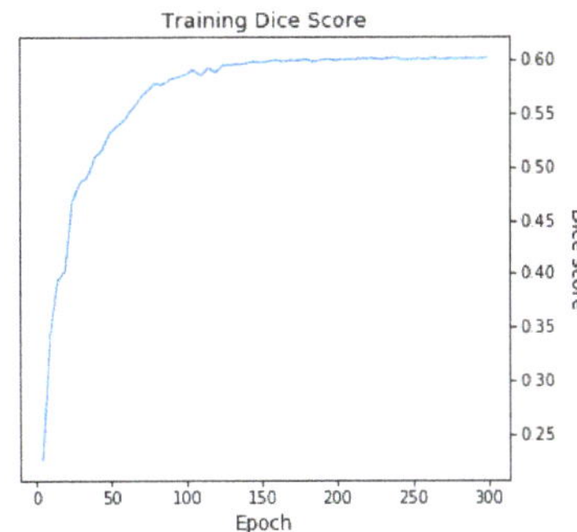

Figure 7. Training loss and accuracy curves of the guided network (two-step, end-to-end). The training loss and accuracy graph of other variations of our experiment show similar curves and therefore left out.

3.1.4. Diffeomorphic Transformation

The final deformation fields from our method are smooth and the percentage of pixels with negative Jacobian is relatively small (≈0.1%), albeit the fields are not diffeomorphic. We have compared two cases where diffeomorphic deformation fields are computed. First, we estimate stationary velocity fields instead of deformation fields and compute the final field using scaling and squaring to obtain diffeomorphic transformation (see Table 1: ours (diffeomorphic)). Second, we apply a poly-affine transformation as post-processing step to reduce the singularities in the resulting field (see Table 1: ours (poly-affine)). The result shows that with the decreased singularity in the resulting field, the accuracy of the transformation is also reduced. The singularity present in the final fields could be necessary, e.g., to compensate for occlusion in the images.

Table 1. Dice scores, contour distances in pixel and the standard deviation of Jacobian determinant of estimated deformation fields using different methods (for the Helen dataset). The standard deviation of Jacobian determinants of deformation fields show the smoothness of the deformation fields, where small values indicate more plausible, regular results. The mean of Jacobian negatives is the number of singular points in the estimated field divided by the total number of pixels. For B-Spline FlowNet, ours (without guidance) and ours (shared RegNet weights), we have implemented our own version of the method from the each reference (details in Section 3.2). Explicit shape regression (ESR) requires in addition corresponding manual landmarks for training. The experiment using pre-trained FlowNet was performed without fine-tuning and using downsampled images, which we first affine transformed based on the face landmarks. Approximated inference time is also given in seconds per image.

Method	Dice (%)	Contour Distance (px)	Jacobian Std.	Jacobian Negatives	Inference Time (s/img)
no registration	23.0	15.55	-	-	-
ESR ([22]) + CPD ([23])	65.6	**1.96**	0.154	-	-
B-Spline FlowNet	49.4	5.96	0.579	0.03124	≈0.007
FlowNet w/ smaller images ([1])	30.6	12.83	-	-	-
ours (without guidance, ∼[3])	55.5	5.41	0.257	0.00062	≈0.007
ours (shared RegNet weights, ∼[2])	60.4	5.02	0.269	0.00061	≈0.007
ours (diffeomorphic)	52.0	5.66	0.240	0.00062	≈0.007
ours (poly-affine)	65.8	3.95	0.281	0.00093	≈0.024
ours	**66.0**	4.01	0.285	0.00106	≈0.007

3.2. Comparison with Other Methods

The registration results with different quantitative evaluation measures are shown in Table 1. As long as it was possible, we have evaluated the methods using Dice score, contour distance, standard deviation of Jacobian and the mean number of pixels with negative Jacobian.

We compare our method with various methods; however for this experiment we have implemented our own version of the networks, where we use same loss function as in the original methods except for FlowNet. We use the original FlowNet version and used the pre-trained model. Solely for this case, we have downsampled the input images and performed an affine transformation beforehand (which constitutes an easier registration task). For the reproduction of Qin et al.'s method [2], we used shared weights for registration networks, which are however, not shared with the segmentation network. The reason for this was to compare the effect of using a two-step registration network against recursive training of registration networks.

A new variant that combines the correlation layer of the FlowNet with our proposed B-spline parameterization was implemented. This B-Spline FlowNet (Table 1: B-Spline FlowNet) yields significant improvements compared to the original FlowNet (Table 1: FlowNet w/ smaller images) but less accurate alignment (more than 10% points gap) than our proposed method. This clearly demonstrates that large deformations are very hard to estimate without meaningful semantic guidance during the training process.

We also compare our networks that are learned with segmentation information with strongly supervised landmark models. As an example approach of this category, we used explicit face regression of Cao et al. [22] to predict landmarks for all test images. Subsequently, we calculated a dense warp field using the coherent point drift algorithm [23] (Table 1: ESR + CPD). Please note that corresponding landmarks are much harder to annotate than segmentation labels and in many applications, in particular 3D medical images, no meaningful points can be selected at all. This means that landmark-model-based registration is somewhat out of competition and could be seen as an upper bound to employing supervision with segmentations only. Nevertheless, our approach achieves comparably high Dice overlap scores to the method of Cao et al. [22] with only slightly more complex transformations (reflected by higher Jacobi determinants).

3.3. Medical Cardiac Image Registration

We evaluate our semantically guided registration network for the preprocessed MRI images from the ACDC challenge [20]. The difficulty here lies in the large motion and strong imaging artefacts. In addition, the substantial differences in appearance and contrast across subjects make learning a model that generalizes well for a population difficult. Therefore, many previous approaches have employed unsupervised learning strategies [24,25] or classical optimization-based methods [26], which can rely on comparable intensity levels within one patient. We compare the results of these three methods (Dice scores of compared methods are reported by Krebs et al. [25]) with our semantically guided framework. As shown in Table 2, our method outperforms all these state-of-the-art methods for the alignment of all structure. The resulting deformations are also plausible with a standard deviation of Jacobian determinant of 0.3. Example results of intra-patient image registration are shown in Figure 8.

Table 2. Mean Dice scores (%) of different methods for medical cardiac dataset averaged across 30 test subjects by labelled structures (R.V.: right ventricle, L.V.: left ventricle).

	Unregistered	LCC-Demons [26]	Voxel-Morph [24]	Krebs et al. [25]	Ours
R.V.	65.1	70.6	68.1	68.4	**77.4**
L.V.	66.0	77.6	74.3	75.6	**82.5**
Myocard	52.5	73.0	69.7	70.4	**73.4**
Mean	61.2	73.7	70.7	71.5	**77.8**

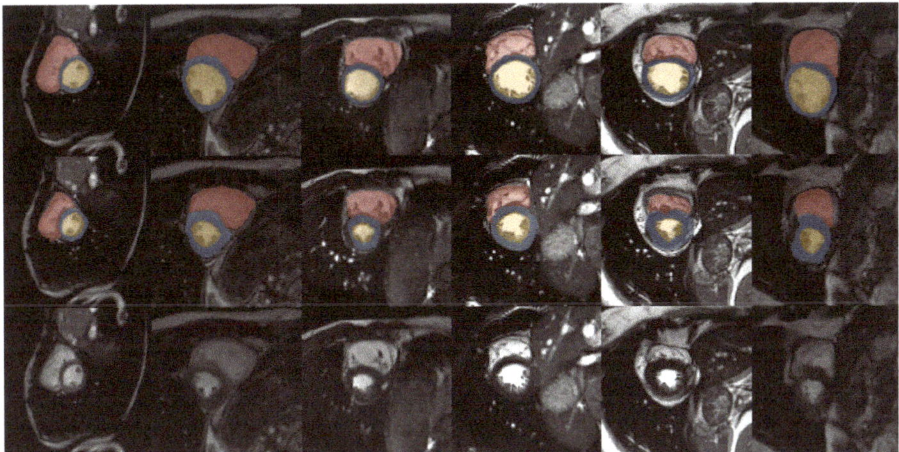

Figure 8. Example results (example slices) of our approach for intra-patient registration. (**Top**) end-systolic slices with the corresponding ground truth labels, (**middle**) end-diastolic slices with warped ground truth labels of end-systolic slices, (**bottom**) warped end-systolic slices.

4. Conclusions

We have presented a new semantically guided and two-stage deep deformation network that can be trained end-to-end and excels at registering image pairs with large initial misalignments. Our extensive experimental validation shows that employing semantic labels available only during training for both an alignment loss and a soft constraint on correct segmentation prediction yields superior results compared to previous approaches that have only considered the former one. Moreover, the use of two-stage networks improves the accuracy compared to a single network or two networks with shared weights. This guidance can also be used beneficially in a series of multiple spatial transformers to improve the alignment of particularly challenging image pairs. Our results on both the Helen face dataset and the medical cardiac ACDC data improve upon the state of the art including FlowNet [1] and Label-Reg [3]—two very recent deep-learning registration frameworks—as well as several unsupervised approaches. Our resulting models are compact and very fast in inference ($\approx$0.009 s per image pair) and can be employed for a variety of challenging tracking and/or alignment tasks in computer vision and medical image analysis.

Author Contributions: Conceptualization, M.H., M.W and I.Y.H.; methodology, M.H., M.W and I.Y.H.; software, I.Y.H. and M.H.; validation, I.Y.H; formal analysis, I.Y.H.; investigation, I.Y.H.; resources, M.H.; data curation, I.Y.H. and M.H.; writing-original draft preparation, I.Y.H.; writing-review and editing, M.H., M.W. and I.Y.H.; visualization, I.Y.H.; supervision, M.H; project administration, M.H.; funding acquisition, M.H. All authors have read and agreed to the published version of the manuscript.

Funding: This research was funded by German research funding organization (DFG) grant number HE7364/1-2.

Conflicts of Interest: The authors declare no conflict of interest.

References

1. Ilg, E.; Mayer, N.; Saikia, T.; Keuper, M.; Dosovitskiy, A.; Brox, T. Flownet 2.0: Evolution of optical flow estimation with deep networks. In Proceedings of the IEEE Conference on Computer Vision and Pattern Recognition (CVPR), Honolulu, HI, USA, 21–26 July 2017; Volume 2, p. 6.
2. Qin, C.; Bai, W.; Schlemper, J.; Petersen, S.E.; Piechnik, S.K.; Neubauer, S.; Rueckert, D. Joint Learning of Motion Estimation and Segmentation for Cardiac MR Image Sequences. *arXiv* **2018**, arXiv:1806.04066.

3. Hu, Y.; Modat, M.; Gibson, E.; Ghavami, N.; Bonmati, E.; Moore, C.M.; Emberton, M.; Noble, J.A.; Barratt, D.C.; Vercauteren, T. Label-driven weakly-supervised learning for multimodal deformarle image registration. In Proceedings of the 2018 IEEE 15th International Symposium on Biomedical Imaging (ISBI 2018), Washington, DC, USA, 4–7 April 2018; pp. 1070–1074.
4. Sun, D.; Yang, X.; Liu, M.Y.; Kautz, J. PWC-Net: CNNs for optical flow using pyramid, warping, and cost volume. In Proceedings of the IEEE Conference on Computer Vision and Pattern Recognition, Salt Lake City, UT, USA, 19–21 June 2018; pp. 8934–8943.
5. Dosovitskiy, A.; Fischer, P.; Ilg, E.; Hausser, P.; Hazirbas, C.; Golkov, V.; Van Der Smagt, P.; Cremers, D.; Brox, T. Flownet: Learning optical flow with convolutional networks. In Proceedings of the IEEE International Conference on Computer Vision, Santiago, Chile, 13–16 December 2015; pp. 2758–2766.
6. Rohé, M.M.; Datar, M.; Heimann, T.; Sermesant, M.; Pennec, X. SVF-Net: Learning deformable image registration using shape matching. In Proceedings of the International Conference on Medical Image Computing and Computer-Assisted Intervention, Quebec City, QC, Canada, 11–13 September 2017; pp. 266–274.
7. Krebs, J.; Mansi, T.; Delingette, H.; Zhang, L.; Ghesu, F.C.; Miao, S.; Maier, A.K.; Ayache, N.; Liao, R.; Kamen, A. Robust non-rigid registration through agent-based action learning. In Proceedings of the International Conference on Medical Image Computing and Computer-Assisted Intervention, Quebec City, QC, Canada, 11–13 September 2017; pp. 344–352.
8. Sokooti, H.; de Vos, B.; Berendsen, F.; Lelieveldt, B.P.; Išgum, I.; Staring, M. Nonrigid image registration using multi-scale 3D convolutional neural networks. In Proceedings of the International Conference on Medical Image Computing and Computer-Assisted Intervention, Quebec City, QC, Canada, 11–13 September 2017; pp. 232–239.
9. de Vos, B.D.; Berendsen, F.F.; Viergever, M.A.; Staring, M.; Išgum, I. End-to-end unsupervised deformable image registration with a convolutional neural network. In *Deep Learning in Medical Image Analysis and Multimodal Learning for Clinical Decision Support*; Springer: Quebec City, QC, Canada, 2017; pp. 204–212.
10. Li, H.; Fan, Y. Non-rigid image registration using fully convolutional networks with deep self-supervision. *arXiv* **2017**, arXiv:1709.00799.
11. Jaderberg, M.; Simonyan, K.; Zisserman, A.; Kavukcuoglu, K. Spatial transformer networks. In *Advances in Neural Information Processing Systems*; Neural Information Processing Systems Foundation, Inc. (NIPS): Montreal, QC, Canada, 2015; pp. 2017–2025.
12. Balakrishnan, G.; Zhao, A.; Sabuncu, M.R.; Guttag, J.; Dalca, A.V. VoxelMorph: A learning framework for deformable medical image registration. *IEEE Trans. Med. Imaging* **2019**, *38*, 1788–1800. [CrossRef] [PubMed]
13. Sevilla-Lara, L.; Sun, D.; Jampani, V.; Black, M.J. Optical flow with semantic segmentation and localized layers. In Proceedings of the IEEE Conference on Computer Vision and Pattern Recognition, Las Vegas, NV, USA, 26 June–1 July 2016; pp. 3889–3898.
14. Hur, J.; Roth, S. Joint optical flow and temporally consistent semantic segmentation. In Proceedings of the European Conference on Computer Vision, Amsterdam, The Netherlands, 8–16 October 2016; pp. 163–177.
15. Cheng, J.; Tsai, Y.H.; Wang, S.; Yang, M.H. SegFlow: Joint learning for video object segmentation and optical flow. In Proceedings of the 2017 IEEE International Conference on Computer Vision (ICCV), Venice, Italy, 22–29 October 2017; pp. 686–695.
16. Tsai, Y.H.; Yang, M.H.; Black, M.J. Video segmentation via object flow. In Proceedings of the IEEE Conference on Computer Vision and Pattern Recognition, Las Vegas, NV, USA, 26 June–1 July 2016; pp. 3899–3908.
17. Ronneberger, O.; Fischer, P.; Brox, T. U-net: Convolutional networks for biomedical image segmentation. In Proceedings of the International Conference on Medical Image Computing and Computer-Assisted Intervention, Munich, Germany, 5–9 October 2015; pp. 234–241.
18. Rueckert, D.; Sonoda, L.I.; Hayes, C.; Hill, D.L.; Leach, M.O.; Hawkes, D.J. Nonrigid registration using free-form deformations: Application to breast MR images. *IEEE Trans. Med. Imaging* **1999**, 18, 712–721. [CrossRef] [PubMed]
19. Le, V.; Brandt, J.; Lin, Z.; Bourdev, L.; Huang, T.S. Interactive facial feature localization. In Proceedings of the European Conference on Computer Vision, Florence, Italy, 7–13 October 2012; pp. 679–692.
20. Bernard, O.; Lalande, A.; Zotti, C.; Cervenansky, F.; Yang, X.; Heng, P.A.; Cetin, I.; Lekadir, K.; Camara, O.; Ballester, M.A.G.; et al. Deep Learning Techniques for Automatic MRI Cardiac Multi-structures Segmentation and Diagnosis: Is the Problem Solved? *IEEE Trans. Med. Imaging* **2018**, *37*, 2514–2525. [CrossRef] [PubMed]

21. Smith, B.M.; Zhang, L.; Brandt, J.; Lin, Z.; Yang, J. Exemplar-based face parsing. In Proceedings of the IEEE Conference on Computer Vision and Pattern Recognition, Portland, OR, USA, 25–27 June 2013; pp. 3484–3491.
22. Cao, X.; Wei, Y.; Wen, F.; Sun, J. Face alignment by explicit shape regression. *Int. J. Comput. Vis.* **2014**, 107, 177–190. [CrossRef]
23. Myronenko, A.; Song, X. Point set registration: Coherent point drift. *IEEE Trans. Pattern Anal. Mach. Intell.* **2010**, *32*, 2262–2275. [CrossRef] [PubMed]
24. Balakrishnan, G.; Zhao, A.; Sabuncu, M.R.; Guttag, J.; Dalca, A.V. An Unsupervised Learning Model for Deformable Medical Image Registration. In Proceedings of the IEEE Conference on Computer Vision and Pattern Recognition, Granada, Spain, 16–20 September 2018; pp. 9252–9260.
25. Krebs, J.; Mansi, T.; Mailhé, B.; Ayache, N.; Delingette, H. Unsupervised Probabilistic Deformation Modeling for Robust Diffeomorphic Registration. In *Deep Learning in Medical Image Analysis and Multimodal Learning for Clinical Decision Support*; Springer: Granada, Spain, 2018; pp. 101–109.
26. Lorenzi, M.; Ayache, N.; Frisoni, G.B.; Pennec, X.; Alzheimer's Disease Neuroimaging Initiative (ADNI). LCC-Demons: A robust and accurate symmetric diffeomorphic registration algorithm. *NeuroImage* **2013**, 81, 470–483. [CrossRef] [PubMed]

Article

Compressive Sensing Spectroscopy Using a Residual Convolutional Neural Network

Cheolsun Kim, Dongju Park and Heung-No Lee *

School of Electrical Engineering and Computer Science, Gwangju Institute of Science and Technology, Gwangju 61005, Korea; csk0315@gist.ac.kr (C.K.); toriving@gist.ac.kr (D.P.)
* Correspondence: heungno@gist.ac.kr; Tel.: +82-62-715-2237

Received: 20 December 2019; Accepted: 20 January 2020; Published: 21 January 2020

Abstract: Compressive sensing (CS) spectroscopy is well known for developing a compact spectrometer which consists of two parts: compressively measuring an input spectrum and recovering the spectrum using reconstruction techniques. Our goal here is to propose a novel residual convolutional neural network (ResCNN) for reconstructing the spectrum from the compressed measurements. The proposed ResCNN comprises learnable layers and a residual connection between the input and the output of these learnable layers. The ResCNN is trained using both synthetic and measured spectral datasets. The results demonstrate that ResCNN shows better spectral recovery performance in terms of average root mean squared errors (RMSEs) and peak signal to noise ratios (PSNRs) than existing approaches such as the sparse recovery methods and the spectral recovery using CNN. Unlike sparse recovery methods, ResCNN does not require *a priori* knowledge of a sparsifying basis nor prior information on the spectral features of the dataset. Moreover, ResCNN produces stable reconstructions under noisy conditions. Finally, ResCNN is converged faster than CNN.

Keywords: spectroscopy; compressed sensing; deep learning; inverse problems; sparse recovery; dictionary learning

---

1. Introduction

There has been considerable interest in producing compact spectrometers having a high spectral resolution, wide working range, and short measuring time. Such a spectrometer can be used in a broad range of fields such as remote sensing [1], forensics [2], and medical applications [3]. Spectrometers that exploit advanced signal-processing methods are promising candidates. The compressive sensing (CS) [4,5] framework makes it possible for a spectrometer to improve its spectral resolution while retaining its compact size. CS spectroscopy comprises two parts: Capturing a spectrum with a small number of compressed measurements and reconstructing the spectrum from the compressed measurements using reconstruction techniques.

To date, for effective signal recovery in CS spectroscopy, three requirements should be satisfied. First, the spectrum should be a sparse signal or capable of sparse representation on a certain basis. Second, the sensing patterns of optical structures should be designed to have a small mutual coherence [6]. Third, appropriate reconstruction algorithms are required. Note that several sparsifying bases have been used in CS spectroscopy such as a family of orthogonal Daubechies wavelets [7], a Gaussian line shape matrix [8,9], and a learned dictionary [10]. Furthermore, numerous optical structures have been proposed to attain the necessary small mutual coherence for sensing patterns such as thin-film filters [11,12], a liquid crystal phase retarder [13], Fabry–Perot filters [7,14], and photonic crystal slabs [15,16]. As algorithms for reconstructing the original signal, two types of basic reconstruction techniques have been developed: greedy iterative algorithms [17,18] and convex relaxation [19,20]. In CS spectroscopy, the reconstruction algorithms have been used with a sparsity constraint. Additionally, a non-negativity constraint is used in Reference [16,21]. Combining these three considerations, CS spectrometers have shown stable performance for light-emitting diodes (LEDs) and monochromatic lights.

Since not all signals can be represented as sparse on a fixed basis, prior information on structural features of the spectral dataset is therefore required to generate a best-fit sparsifying basis. Furthermore, a high computational cost is required for reconstruction techniques. Recently, deep learning [22] has been emerging as a promising alternative framework for reconstructing the original signal from the compressed measurements.

Mousavi et al. [23] was the first study on image recovery from structured measurements using deep learning. Moreover, a deep-learning framework for inverse problems has been applied in biomedical imaging for imaging through scattering media [24], magnetic resonance imaging [25,26], and X-ray computed tomography [27]. Kim et al. [28] reported the first attempt to use deep learning in CS spectroscopy. They trained a convolutional neural network (CNN) to output the reconstructed signal from the network. From here on the network reported by Kim et al. will be referred to as CNN.

Unlike CNN [28] in which learnable layers were simply stacked and trained to directly reconstruct the original spectrum, we make a residual connection [29] between the input and output of CNN and train the network to reconstruct the original spectrum by referring the input of the network. As a result, the network learns residuals between the input of the network and the original spectrum. It has been reported that it is easier to train a network when using residual connections than to train a plain network that was simply stacked with learnable layers [25,29]. Lee et al. [25] analyzed the topological structure of magnetic resonance (MR) images and the residuals of MR images. They showed that the residuals possessed a simpler topological structure, thus making learning residuals easier than learning the original MR images. In addition, He et al. [29] demonstrated with empirical results that the residual networks are easy to optimize and they achieved improvements in image-recognition tasks. From these works, we gain insights such that adding residual connections to CNN would improve the spectral reconstruction performance in CS spectroscopy.

In this paper, we aim to propose a novel residual convolutional neural network (ResCNN) for recovering an input spectrum from the compressed sensing measurements in CS spectroscopy. The novelty lies in the proposed ResCNN structure, with a moderate depth of learnable layers and a single residual connection, which provides the desired spectral reconstruction performance. The desired performance here means that the proposed ResCNN offers a performance which is better than that of CNNs as well as that of CS reconstruction with its sparsifying base known. In CS reconstruction, the prior knowledge of a fixed sparsifying basis is useful and offers good sparse representation results. However, in general it is a difficult problem to identify a sparsifying basis for various kinds of spectra and apply the identified basis to have the recovery performance improved. In this regard, it is an important advance to find a simple ResCNN which offers good enough performance. It is also worth to note that the proposed ResCNN is tested with the array type CS spectroscopy, discussed in Section 2, which we have designed with an array of multilayer thin-film filters.

The previous works on CS spectroscopy [7,11,13,14,16] have shown decent reconstruction performance but on limited simple sources such as LEDs and monochromatic lights. Using ResCNNs, we are now able to reconstruct more complex spectra, such as spectra with multiplicity of peaks mixed with a gradual rise-and-fall.

The remainder of this paper is organized as follows. In Section 2, we model the optical structure which is used for CS spectroscopy. In Section 3, we describe the system of CS spectroscopy and the proposed ResCNN. In Section 4, simulated experiments are described. Section 5 presents the results of experiments. In Section 6, we discuss the results. Finally, we conclude this paper in Section 7.

2. Optical Structure

Numerous optical structures have been proposed for CS spectroscopy. It has been reported that CS spectrometers, which have various spectral features in the transmission spectrum, show high spectral-resolving performance [16]. In this work, we used thin-film filters to model CS spectrometers. Thin-film filters demonstrate a variety of spectral features depending on the materials used, the number of layers, and the thicknesses of the layers. Once the structure of thin-film is determined, a transmission value at a given wavelength λ is defined as follows [30]:

$$T(\lambda) = 1 - \frac{1}{2}\left(\left|\rho_{TE}(\lambda)\right|^2 + \left|\rho_{TM}(\lambda)\right|^2\right), \tag{1}$$

where $\rho_{TE}(\lambda)$ and $\rho_{TM}(\lambda)$ are amplitude reflection coefficients. The coefficients represent the fraction of the power reflected by a multilayer thin-film in the transverse electric (TE) and transverse magnetic (TM) modes of an incident light, respectively. We summarized recursive processes for calculating amplitude reflection coefficients in Algorithm 1 [11,12,31].

Algorithm 1: Recursive processes for amplitude reflection coefficients.

Input:	λ
	Structure parameters: θ_1,$\mathbf{n} = \{n_1, n_2, \cdots, n_l\}$,$\mathbf{d} = \{d_2, d_3, \cdots, d_l\}$.
Step 1:	Calculate θ_k,β_k, and N_k using structure parameters.
	$\theta_k = \sin^{-}\left(\frac{n_{k-1}}{n_k}\sin\theta_{k-1}\right)$, for k = 2, 3, ..., l.
	$\beta_k = 2\pi\cos(\theta_k)n_k d_k/\lambda$, for k = 2, 3, ..., l.
	$N_k = \begin{cases} n_k/\cos\theta_k & for \quad TE \\ n_k\cos\theta_k & for \quad TM \end{cases}$, for k = 2, 3, ..., l.
Step 2:	Obtain η_2 by setting $\eta_l = N_l$.
	For k = l-1 to 2
	$\eta_k = N_k \frac{\eta_{k+1}\cos\beta_k + jN_k\sin\beta_k}{N_k\cos\beta_k + j\eta_{k+1}\sin\beta_k}$.
Step 3:	Compute $\rho = (N_1 - \eta_2)/(N_1 + \eta_2)$.
Output:	ρ

Here, θ_k is the angle of an incident light passing from k^{th} to k+1th layer. The refractive index of kth layer is denoted as n_k. d_k denotes the thickness of the kth layer. Given a wavelength vector $\boldsymbol{\lambda} = (\lambda_1\,\lambda_2\,\ldots\,\lambda_N) \in \mathbb{R}^{1\times N}$ in the range of interest, i.e., $\lambda_{\max} - \lambda_{\min}$. Let $\Delta\lambda = \frac{\lambda_{\max}-\lambda_{\max}}{N}$. Then, evaluating the function at the integer multiple of $\Delta\lambda$, i.e., $T(\lambda = \lambda_{\min} + n\Delta\lambda)$ for $n = 0, 1, \cdots, N-1$, we obtain the vector of transmission spectrum $\mathbf{T}_m \in \mathbb{R}^{1\times N}$ for the wavelength range. Then, the sensing pattern matrix of optical structures $\mathbf{T} \in \mathbb{R}^{M\times N}$ is obtained by repeating the calculation of $\mathbf{T}_m$ for $m = 1, 2, \cdots, M$.

We have used SiNx and SiO_2 for high- and low-refractive index materials, respectively. We numerically generated thin-film filters by alternately stacking high- and low- refractive index materials, changing the number of layers, and varying the thickness of each layer. The number of layers in each filter is in the interval of (19, 24), and the thickness (nm) of each layer is in the interval of (50, 300). Initially, we randomly generated reference filters and compute the mutual coherence among the filters. Then, new filters were generated by changing thicknesses of the layers and the mutual coherence of the filters is compared to the mutual coherence of reference filters. Filters with a smaller mutual coherence then became the new reference filters. This process is repeated until reasonable reference filters with the required small mutual coherence are obtained.

Figure 1 shows the heatmap for the transmission spectra of the reference filters and two selected transmission spectra. In Figure 1a, each of the transmission spectra shows a unique sensing pattern because of the iterative modeling process of the reference filters based on mutual coherence. Figure 1b shows two transmission spectra that correspond to the 15th and 30th rows in the heatmap of reference filters. The transmission spectrum reveals a deep spectral modulation depth and various features such as broadband backgrounds, multiple peaks with a small full width at half maximums (FWHMs), and irregular fluctuations.

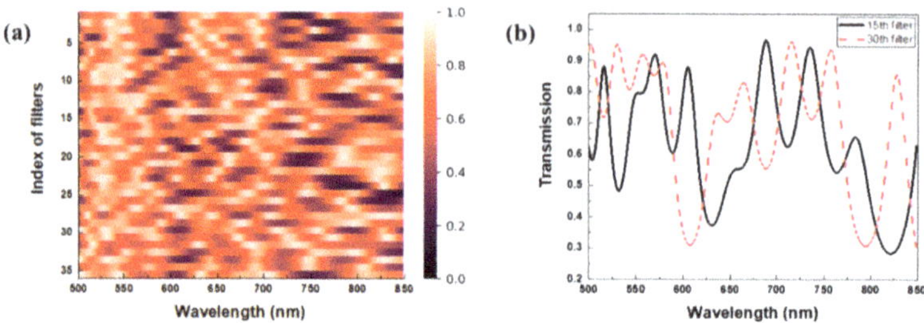

Figure 1. (**a**) Heatmap of the sensing matrix: each row represents the transmission spectrum of the designed thin-film filter. (**b**) Two transmission spectra corresponding to the 15th and 30th rows in the sensing matrix.

3. Compressive Sensing (CS) Spectrometers Using the Proposed Residual Convolutional Neural Network (ResCNN)

3.1. CS Spectrometers

In CS spectroscopy, the measurement column vector $\mathbf{y} \in \mathbb{R}^{M\times1}$ is represented using the following relation:

$$\mathbf{y} = \mathbf{T}\mathbf{x}, \tag{2}$$

where $\mathbf{x} \in \mathbb{R}^{N\times1}$ is the spectrum column vector of incident light and $\mathbf{T} \in \mathbb{R}^{M\times N}$ is the sensing matrix of the optical structure. Each row of $\mathbf{T}$ represents a transmission spectrum. Because the length of the measurement vector is smaller than the length of the spectrum vector ($M < N$), the system is underdetermined. Conventionally, if $\mathbf{x}$ is a sparse signal or can be sparsely represented in a certain basis, i.e., $\mathbf{x} = \mathbf{\Phi}\mathbf{s}$, reconstruction algorithms can determine a unique sparse solution $\hat{\mathbf{S}}$ from the following optimization problem:

$$\min_{\mathbf{s}} \left\|\mathbf{T}\mathbf{\Phi}\mathbf{s} - \mathbf{y}\right\|_2^2 + \tau\|\mathbf{s}\|_1, \tag{3}$$

where $\mathbf{\Phi} \in \mathbb{R}^{N\times N}$ is a sparsifying basis and τ is a regularization parameter. Here, $\mathbf{s}$ is a sparse signal whose components are zero except for a small number of non-zero components. Then, the recovered spectrum $\hat{\mathbf{x}}$ is $\mathbf{\Phi}\hat{\mathbf{s}}$. In this paper, we refer to the methods of solving the optimization problem using Equation (3) as sparse recovery.

Typically, except for narrow-band spectra, a spectrum is not a sparse signal, and a fixed sparsifying basis cannot transform all spectra into sparse signals. Clearly, the use of a fixed basis may lead the sparse recovery to struggle, as no fixed basis will transform every signal into a sparse signal. In addition, the sparse recovery is time-consuming and takes a high computational cost.

Our goal is to overcome the limitations of the sparse recovery in CS spectroscopy and recover various kinds of spectra using ResCNN. Figure 2 shows the schematic of the CS spectroscopy system using ResCNN. This system consists of two parts: compressive sampling and dimension extension, and the reconstruction using ResCNN. In the compressive sampling and dimension extension, the measurement vector $\mathbf{y}$ is obtained from Equation (1), which then transforms into $\widetilde{\mathbf{x}} \in \mathbb{R}^{N\times1}$ using a linear transformation. A transform matrix $\mathbf{A} \in \mathbb{R}^{N\times M}$ extends the M dimension of $\mathbf{y}$ to N dimension of $\widetilde{\mathbf{x}}$, where $\widetilde{\mathbf{x}}$ is a representative spectrum corresponding to $\mathbf{x}$. We used $\widetilde{\mathbf{x}}$ as the input for the reconstruction. ResCNN learnt a non-linear mapping between $\widetilde{\mathbf{x}}$ and $\mathbf{x}$, and afforded a reconstructed spectrum $\hat{\mathbf{x}} \in \mathbb{R}^{N\times1}$. The dimension extension by the transform matrix was used to make it easier for ResCNN to extract features and reconstruct spectra from the non-linear mapping.

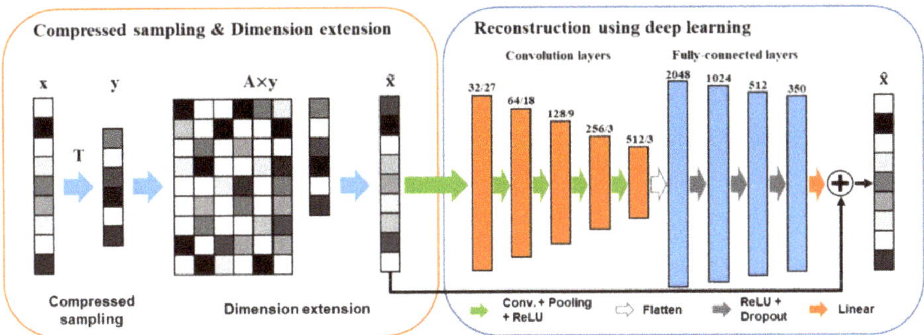

Figure 2. Overview of compressive sensing (CS) spectrosocopy system including the proposed residual convolutional neural network (ResCNN): An input spectrum is compressively sampled by the sensing matrix, and the dimension of measurements is extended by the transform matrix. ResCNN is trained to recover the original spectrum from the extented measurements.

3.2. The Proposed ResCNN

As depicted in Figure 2, ResCNN comprises nine learnable layers, five of which are convolution layers, four are fully-connected layers, and one is a residual connection. Convolution layers are used for the feature extraction in the non-linear mapping between $\widetilde{\mathbf{x}}$ and $\mathbf{x}$. Fully-connected layers are used for the spectra reconstruction. Each of the convolution layers has a set of one-dimensional learnable kernels with specific window sizes. The number of kernels and the window sizes are indicated in Figure 2. After every convolutional layer, the rectified linear unit (ReLU) is used as an activation function, and the subsampling is then applied. We use non-overlapping max-pooling to down-sample the output of the activation function. We stack the convolutional layer, the ReLU, and the subsampling five times. The output of the last subsampling is flattened and then fed into the subsequent four fully-connected layers. The first three layers are followed by the ReLU and dropout in sequence. The dropout is introduced to reduce the overfitting of ResCNN. The output of the last fully-connected layer is fed into a linear activation function. The number of units in each of the fully-connected layers is noted in Figure 2. Unlike CNN [28] in which learnable layers are simply stacked, we make the residual connection that the representative spectrum $\widetilde{\mathbf{x}}$ and the output of the linear activation function are added up to the reconstructed spectrum $\hat{\mathbf{x}}$. Consequently, $\hat{\mathbf{x}}$ is trained to become $\mathbf{x}$. Given training data $\left\{\mathbf{x}_t{}^i\right\}_{i=1}^{k}$, we train ResCNN to minimize a loss function L. We use the mean squared error between the original $\mathbf{x}_t$ and recovered $\hat{\mathbf{x}}_t$ as the loss function:

$$L = \frac{1}{k}\sum_{i=1}^{k}\left\|\mathbf{x}_t{}^i - \hat{\mathbf{x}}_t{}^i\right\|_2^2. \tag{4}$$

The non-linear mapping that $\widetilde{\mathbf{x}}$ becomes $\mathbf{x}$ can be defined as $H(\widetilde{\mathbf{x}}) = \mathbf{x}$. Because of the residual connection in ResCNN, $H(\widetilde{\mathbf{x}})$ can be rewritten as $H(\widetilde{\mathbf{x}}) = F(\widetilde{\mathbf{x}}) + \widetilde{\mathbf{x}}$, where $F(\widetilde{\mathbf{x}})$ is the mapping of the learnable layers. The representative spectrum $\widetilde{\mathbf{x}}$ is referenced by the residual connection, and then,$F(\widetilde{\mathbf{x}}) = H(\widetilde{\mathbf{x}}) - \widetilde{\mathbf{x}}$. In particular, the mapping of $F(\widetilde{\mathbf{x}})$ is called a residual mapping; therefore, the learnable layers learn the residual of $\mathbf{x}$ and $\widetilde{\mathbf{x}}$.

The previous researches [25,29] have used numerous residual connections in very deep neural networks in order to make networks converge faster by avoiding vanishing gradient problems. We use one residual connection between input and output of the moderate depth network. Figure 3 depicts the manner in which a spectrum is recovered in CNN and ResCNN. The learnable layers of CNN directly reconstruct the spectrum from the representative spectrum $\widetilde{\mathbf{x}}$. Alternatively, ResCNN reconstructs the spectrum by passing the representative spectrum $\widetilde{\mathbf{x}}$ through the residual connection shown in Figure 3b. Consequently, the learnable layers of ResCNN learn to reconstruct residuals.

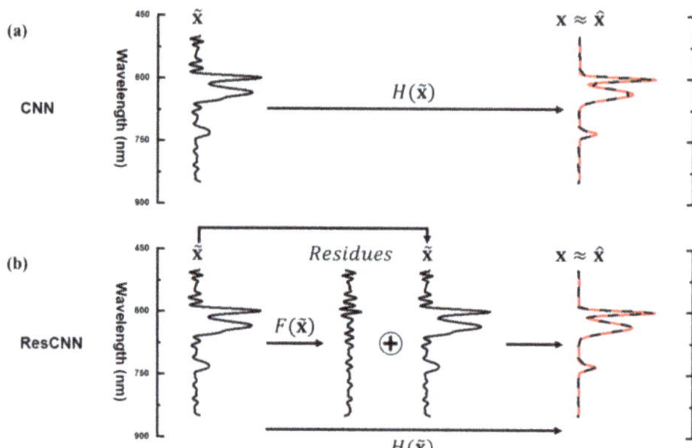

Figure 3. Descriptions of the spectrum recovery process: (**a**) convolutional neural network (CNN), (**b**) ResCNN.

4. Simulated Experiments

We reconstructed 350 spectral bands (N = 350) using 36 thin-film filters (M = 36) whose sensing patterns have a spacing of 1 nm for wavelengths from 500 to 850 nm. We determined the sensing matrix **T**, assuming that the incident light falls onto the filters with normal incidence. As the transform matrix **A**, we used the Moore–Penrose inverse of the sensing matrix **T**, i.e., $\mathbf{A} = \mathbf{T}^T\left(\mathbf{T}\mathbf{T}^T\right)^{-1}$.

4.1. Spectral Datasets

To evaluate the performance of ResCNN, we used two synthetic spectral datasets and two measured spectral datasets. The first synthetic dataset is composed of Gaussian distribution functions while the other is composed of Lorentzian distribution functions. These two synthetic datasets were selected as generally these types of functions are used to represent spectral line shapes. As shown in Figure 4, component functions are added to produce the spectra. We generated 12,000 spectra for each dataset. For each spectrum, the number of component functions was generated using a geometric distribution with the probability parameter p set to 0.3. We added one to the number of component functions to prevent the number of component functions from becoming zero. Then, we randomly set a location, a height, and an FWHM of each peak. To set a peak location (nm), an integer number was randomly selected from a uniform distribution with the interval (500, 849). A random number from a uniform distribution in the interval (0, 1) was used for the height. An integer number for an FWHM (nm) was randomly drawn from a uniform distribution with the interval (2, 50). Finally, all of the component functions were summed to generate the spectrum. The height of each generated spectrum was normalized such that it was mapped from zero to one.

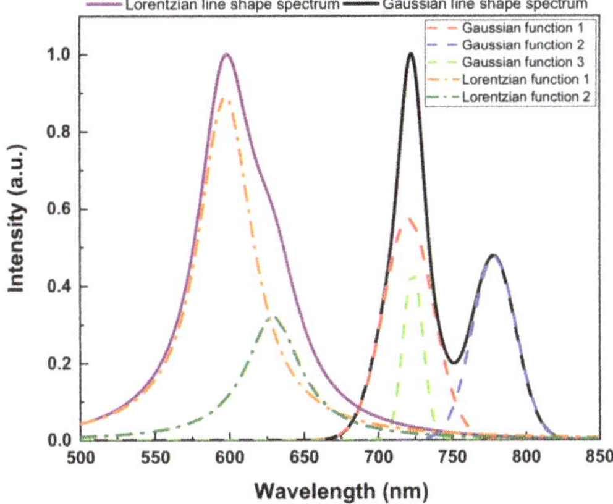

Figure 4. Examples of two synthetic spectra: the solid purple line is composed of two Lorentzian distribution functions (dash-dotted orange and olive lines), and the solid black line is composed of three Gaussian distribution functions (dashed red, blue, and green lines).

As measured datasets, we used the US Geological Survey (USGS) spectral library version 7 [32], and the glossy Munsell colors spectral dataset [33]. The USGS spectral library provides reflectance spectra for artificial materials, coatings, liquids, minerals, organic compounds, soil mixtures, and vegetation. We discarded any spectrum that has missing spectral bands. Then, we extracted the spectrum in the wavelength range of interest (500 to 849 nm) from the wavelength range of the original spectrum (350 to 2500 nm). The measured wavelength range for the glossy Munsell colors spectral dataset, which contains the reflectance spectra of the glossy Munsell color chips, was 380 to 780 nm. The wavelength range of the original spectrum was different from the wavelength range of interest. We decided to use the wavelength range from 400 to 749 nm to ensure each spectrum was set to 350 spectral bands. This selection of wavelengths is reasonable because the wavelengths were located in the center of the wavelength range of the original spectrum, and showed different spectral features with respect to each spectrum. In addition, our aim was to show the reconstruction performance with respect to various kinds of spectra. Finally, each spectrum was normalized such that the height varies from 0 to 1. Overall, 1473 spectra from USGS spectral dataset and 1600 spectra from Munsell color spectral dataset were used for our simulated experiments. Table 1 lists the details of each of the spectral datasets.

Table 1. Description of the spectral datasets.

Dataset	Training/Validation/Test	Avg. Number of Nonzero Values	Description
Gaussian dataset	8000/2000/2000	336.8/350	FWHM (nm) on the interval [2, 50], Height on the interval [0, 1]
Lorentzian dataset	8000/2000/2000	349/350	FWHM (nm) on the interval [2, 50], Height on the interval [0, 1]
US Geological Survey [32]	982/246/245	348.9/350	350–2500 nm, 2151 spectral bands (we use 350 spectral bands in 500–849 nm)
Munsell colors [33]	1066/267/267	349/350	380–780 nm, 401 spectral bands (we use 350 spectral bands in 400–749 nm)

4.2. Data Preprocessing and Training

Given the sensing matrix, the spectral data are compressively sampled as the measurement vector $\mathbf{y}$ shown in Equation (1), and then transformed into the representative spectrum $\widetilde{\mathbf{x}}$ by multiplying the transform matrix $\mathbf{A}$ and $\mathbf{y}$.

In each spectral dataset, the number of training, validation, and test spectra are randomly assigned using a ratio of 4:1:1 for the synthetic and measured data sets, respectively. The validation spectra are used for estimating the number of epochs and tuning the hyper-parameters. To train ResCNN, we used the Adam optimizer [34] implemented in Tensorflow with the batch size of 16 and 250 epochs. The experiments were conducted on an NVIDIA GeForce RTX 2060 graphics processing unit (GPU). Training the architecture can be done in half an hour for each dataset.

4.3. Sparsifying Bases for Spare Recovery

Using sparse recovery, we evaluated the performance of conventional CS reconstructions to benchmark the performance of ResCNN. As shown in Table 1, the spectra for both the synthetic and measured datasets are dense spectra. Therefore, we must transform the spectra into sparse signals to solve Equation (3). In this section, we considered methods to make a sparsifying basis $\mathbf{\Phi}$.

First, we considered a Gaussian line shape matrix as a sparsifying basis. Each column of the matrix comprises a Gaussian distribution function whose length is N. A collection of N Gaussian functions works as a sparsifying basis $\mathbf{\Phi} \in \mathbb{R}^{N \times N}$. We generate two Gaussian line shape matrices. Figure 5 a shows the heatmap images for two Gaussian line shape matrices. Seven different FWHMs are used to generate the Gaussian distributions. Given an FWHM, Gaussian distributions are generated by shifting the peak location using uniform spacing. To create a small dissimilarity between the two Gaussian line shape matrices, two of the seven FWHMs in Gaussian 1 were replaced with other FWHMs, thus producing Gaussian 2, as shown in Figure 5a.

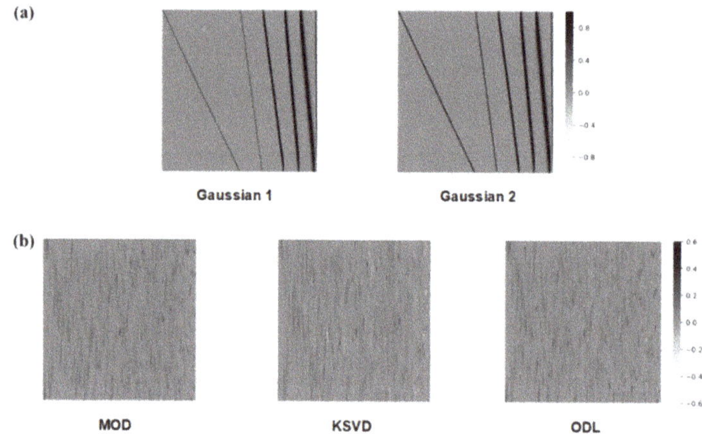

Figure 5. Heatmap images of sparsifying bases that were used in simulated experiments: (**a**) Gaussian line shape matrices, (**b**) the learned dictionaries which are from the Gaussian training dataset.

Second, a learned dictionary [35–38] is used as a sparsifying basis. Given a training dataset $\{\mathbf{x}_t{}^i\}_{i=1}^{k}$, we can derive a learned dictionary $\mathbf{\Phi}$ that sparsely represents the training data $\mathbf{x}_t$ by solving the following optimization problem, known as the dictionary learning problem:

$$\min_{\mathbf{\Phi}, \mathbf{s}_t{}^1, \ldots, \mathbf{s}_t{}^k} \sum_{i=1}^{k} \left\|\mathbf{x}_t{}^i - \mathbf{\Phi}\mathbf{s}_t{}^i\right\|_2^2 + \tau\left\|\mathbf{s}_t{}^i\right\|_1, \tag{5}$$

where τ is a regularization parameter and $\mathbf{s}_t{}^i$ is ith sparse signal over the training dataset. By fixing an initial guess for the dictionary $\mathbf{\Phi}$ in Equation (5), we obtain a solution for the sparse signals $\left\{\mathbf{s}_t{}^i\right\}_{i=1}^{k}$. The dictionary is then updated by solving Equation (5) using the sparse signals obtained. This process is iteratively repeated until convergence is reached and we derive the learned dictionary. We used three dictionary learning methods: method of optimal directions (MOD) [36], K-SVD [37], and online dictionary learning (ODL) [38]. The learned dictionaries are generated for each of the training datasets, and the reconstruction performances are evaluated for each test dataset. Figure 5b shows learned dictionaries identified using the Gaussian training dataset. The learned dictionaries clearly depend on the dictionary-learning methods used. Nevertheless, each column of the dictionaries shows a learned spectral feature from the training dataset.

5. Results

To demonstrate the ability of ResCNN to reconstruct spectra, we evaluated its performance using three different datasets: Synthetic datasets, noisy synthetic datasets, and measured datasets. We used the same hyper-parameters of ResCNN for each of these datasets. Moreover, we adopted *l1_ls* [39] as the fixed reconstruction algorithm in the sparse recovery. We compared the recovered signal with the original signal by calculating the root mean squared error (RMSE) and the peak signal to noise ratio (PSNR). In addition, the performance of five conventional sparse recovery methods, described in Section 4.3 and CNN was calculated.

5.1. Synthetic Datasets

The two synthetic data sets described in Table 1 were used to perform the signal recovery using sparse recovery and deep learning. Table 2 shows the average RMSE and PSNR for each of the seven methods evaluated. ResCNN shows the smallest average RMSE for both the Gaussian and Lorentzian datasets of 0.0094 and 0.0073, respectively. Moreover, ResCNN shows the largest average PSNR of 49.0 dB for the Lorentzian dataset. For the Gaussian dataset, the sparse recovery method with Gaussian 2 shows the largest average PSNR, 49.7 dB, which is slightly higher than the 47.2 dB for ResCNN. Note that the minor difference between the two Gaussian line shape matrices results in considerable performance difference. However, reconstruction using the learned dictionaries show similar performance across all of the synthetic datasets.

Table 2. Average root mean squared errors (RMSEs) and peak signal to noise ratios (PSNRs) over synthetic datasets.

	Sparse Recovery					Deep Learning	
Dataset	Gaussian 1	Gaussian 2	K-SVD	MOD	ODL	CNN	ResCNN
Gaussian dataset	0.0226 (43.1 dB)	0.0112 (**49.7 dB**)	0.0172 (40.3 dB)	0.0174 (40.3 dB)	0.0161 (41.1 dB)	0.0132 (40.5 dB)	**0.0094** (47.2 dB)
Lorentzian dataset	0.0146 (44.9 dB)	0.0094 (47.5 dB)	0.0136 (42.3 dB)	0.0137 (42.3 dB)	0.0127 (42.9 dB)	0.0101 (42.8 dB)	**0.0073** **(49.0 dB)**

Figure 6 shows the reconstructed test spectra from each of the synthetic datasets. The solid red line (i) is the input spectra from each dataset. ResCNN is shown in dashed black line (ii), while CNN is shown in solid orange lines (iii). The reconstructed spectra using sparse recovery with Gaussian 1 (iv), Gaussian 2 (v), and ODL (vi) are shown in solid green, blue, and purple lines in respectively. Because of the similar performance from each of the learned dictionaries, only the ODL method is shown. The RMSE and PSNR of ResCNN are 0.0138 (37.2 dB) for the spectrum from the Gaussian dataset and 0.0096 (40.4 dB) for the spectrum from the Lorentzian dataset. For the selected spectra, ResCNN achieves superior reconstruction performance compared with the other four reconstructions.

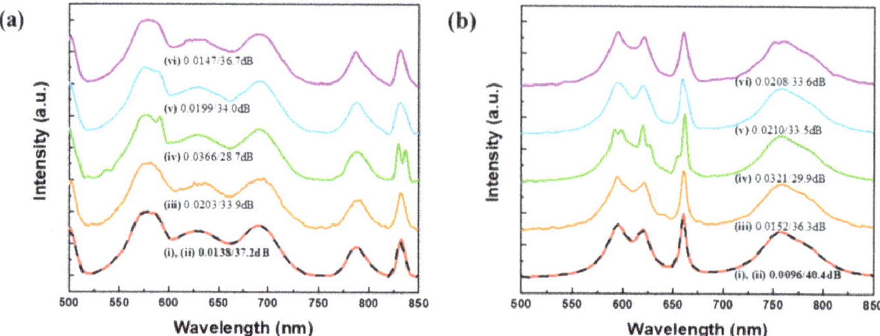

Figure 6. Spectral reconstructions of test spectra in synthetic datasets, (**a**) Gaussian dataset, (**b**) Lorentzian dataset. An input spectrum (solid red (i)) is compared with ResCNN (dashed black (ii)), CNN (orange (iii)), sparse recovery: Gaussian 1 (green (iv)), Gaussian 2 (blue (v)), and online dictionary learning (ODL) (purple (vi)). The baselines are shifted for clarity.

Only sparse recovery with Gaussian 1 fails to recover the fine details of the input spectrum. One example of the poor ability of sparse recovery with Gaussian 1 to resolve the signal is the recovery of the peak at ~830 and 590 nm being recovered as two neighboring peaks in Figure 6a,b, respectively. CNN was unable to capture the smoothness of the spectral features compared to the other methods.

5.2. Noisy Synthetic Datasets

To verify the stability of ResCNN, we evaluated the accuracy of the reconstruction at various noise levels. Gaussian white noise was added to the measurement vector $\mathbf{n} \in \mathbb{R}^{M\times 1}$ to Equation (2), i.e., $\mathbf{y} = \mathbf{Tx}+\mathbf{n}$. We considered six different noise levels whose signal-to-noise ratios (SNRs) are 15, 20, 25, 30, 35, and 40 dB. The SNR (dB) is defined as $10 \cdot \log_{10}\left(\|\mathbf{x}\|_2^2/N\sigma^2\right)$, where σ is the standard deviation of the noise. Using Gaussian and Lorentzian datasets, we compared the reconstruction performance of ResCNN with the sparse recovery using Gaussian 2, which shows the best reconstruction performances among sparse recovery methods in synthetic datasets. ResCNN was evaluated with the same hyper-parameters that were used for the noise-free datasets. The average RMSE and PSNR for each of the six noise levels are shown in Table 3. While ResCNN was trained using noise-free data, it outperformed the sparse recovery with Gaussian 2 at every noise level, which indicates that ResCNN remains stable even with noisy datasets.

Table 3. Average RMSE and PSNR under various signal-to-noise ratios (SNRs, dB) with synthetic datasets.

		SNR (dB)					
Dataset	Method	15 dB	20 dB	25 dB	30 dB	35 dB	40 dB
Gaussian Dataset	Sparse recovery + Gaussian 2	0.0796 (22.7 dB)	0.0482 (27.1 dB)	0.0308 (31.2 dB)	0.0215 (34.8 dB)	0.0166 (37.9 dB)	0.0138 (40.7 dB)
	ResCNN	**0.0671 (24.2 dB)**	**0.0401 (28.7 dB)**	**0.0251 (32.9 dB)**	**0.0171 (36.6 dB)**	**0.0130 (39.8 dB)**	**0.0110 (42.4 dB)**
Lorentzian Dataset	Sparse recovery + Gaussian 2	0.0817 (22.6 dB)	0.0483 (27.1 dB)	0.0300 (31.2 dB)	0.0201 (35.0 dB)	0.0147 (38.5 dB)	0.0119 (41.4 dB)
	ResCNN	**0.0689 (24.1 dB)**	**0.0404 (28.7 dB)**	**0.0243 (33.1 dB)**	**0.0157 (37.1 dB)**	**0.0113 (40.6 dB)**	**0.0091 (43.4 dB)**

5.3. Measured Datasets

ResCNN was trained using the two measured datasets listed in Table 1, USGS and Munsell colors, and its reconstruction performance was evaluated. In addition, the signal reconstruction was performed using CNN and sparse recovery with five different sparsifying bases. Table 4 reports the average RMSE and PSNR for each of the seven methods. ResCNN achieves the smallest average RMSE and the largest average PSNR for both datasets. In the USGS dataset, the average RMSE and PSNR of ResCNN are 0.0048 and 52.4 dB, respectively. In addition, ResCNN achieves 0.0040 for the average RMSE and 50.0 dB for the average PSNR in the Munsell colors dataset. Similar to synthetic datasets, all of the learned dictionaries provided similar reconstruction performances. In addition, the small differences between Gaussian 1 and 2 show large differences in the RMSE and PSNR. The average RMSE and PSNR of the learned dictionary methods approach the values of ResCNN for Munsell colors dataset because the Munsell colors dataset has simpler spectral features than the other datasets.

Table 4. Average RMSEs and PSNRs for the measured datasets.

	Sparse Recovery					Deep Learning	
Dataset	Gaussian 1	Gaussian 2	K-SVD	MOD	ODL	CNN	ResCNN
USGS [32]	0.0081 (45.3 dB)	0.0061 (48.4 dB)	0.0070 (48.5 dB)	0.0081 (47.4 dB)	0.0074 (47.6 dB)	0.0116 (40.8 dB)	**0.0048 (52.4 dB)**
Munsell colors [33]	0.0068 (44.6 dB)	0.0050 (47.5 dB)	**0.0040** (49.8 dB)	**0.0040** (49.9 dB)	0.0042 (49.5 dB)	0.0076 (43.0 dB)	**0.0040 (50.0 dB)**

Figure 7 shows the reconstruction results of one test spectra from each of the measured datasets. The spectrum for the organic compound dibenzothiophene in the USGS dataset is reconstructed in Figure 7a. The spectrum of Munsell color 5 PB 2/2 is shown in Figure 7b. The solid red lines are the input spectra (i). ResCNN are shown in dashed black lines (ii), and CNN are shown in solid black lines (iii). The spectra of (iv) to (vi) are reconstructed spectra using the sparse recovery with Gaussian 1, Gaussian 2, and K-SVD. Because of the best performance of the K-SVD among the learned dictionaries only the K-SVD method is shown.

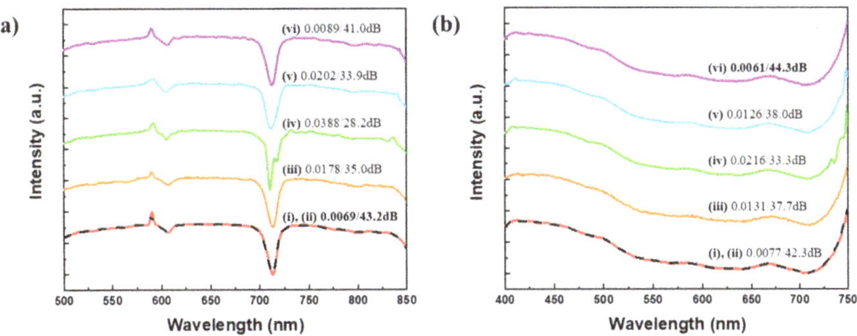

Figure 7. Spectral reconstructions of test spectra in measured datasets: (**a**) spectrum of organic compound dibenzothiophene in USGS dataset, (**b**) spectrum of Munsell color 5PB 2/2. The input spectrum (solid red line (i)) is compared with ResCNN (dashed black (ii)), CNN (orange (iii)), sparse recovery: Gaussian 1 (green (iv)), Gaussian 2 (blue (v)), and K-SVD (purple (vi)). The baselines are shifted for clarity.

The RMSE and PSNR for ResCNN are 0.0069 (43.2 dB) for the spectrum from the USGS dataset and 0.0077 (42.3 dB) for the spectrum from the Munsell colors dataset. ResCNN outperforms other approaches for the spectrum from USGS dataset. However, for the spectrum from Munsell colors dataset, the sparse recovery with K-SVD outperforms ResCNN. ResCNN achieves slightly larger RMSE and smaller PSNR.

The performances of sparse recovery with Gaussian 2 is degraded for measured datasets compared with the performance for synthetic datasets. The measured datasets have rough spectral features unlike the smooth spectral features observed in the synthetic datasets. As a result, the sparse recovery with Gaussian 2 performs worse, because of its inability to represent rough spectral features using Gaussian distribution functions. The performance of sparse recovery with dictionary learning methods are improved for measured datasets compared with the performance of synthetic datasets. Because the number of spectra in measured datasets are smaller than the number of spectra in synthetic datasets. Therefore, finding the best-fit sparsifying basis for measured datasets is easier than finding the best-fit sparsifying basis for synthetic datasets using dictionary-learning methods. Meanwhile, ResCNN shows superior reconstruction performances regardless of spectral features of datasets and the size of datasets.

6. Discussion

As shown in the results, we demonstrate empirically that ResCNN outperforms the sparse recovery methods and the CNN over all datasets. The sparse recovery shows unstable performance because it is highly dependent on the sparsifying basis and spectral features of dataset. This is a direct result of being unable to identify a fixed sparsifying basis that can transform any spectra into a sparse signal, which means the *a priori* structural information such as line shapes and FWHMs is required to select a consistent sparsifying basis. Learned dictionaries are used to cope with the problem of identifying a consistent sparsifying basis. The columns of learned dictionaries are composed of learned spectral features from the training dataset. While this shows an improvement in measured datasets, a learned dictionary is still limited to representing all the spectral features in the large dataset (i.e., synthetic datasets) using linear combinations of columns of the learned dictionary.

Compression approaches for summarizing information with a small number of sensors were proposed in [40]. These approaches can be exploited to generate a sparsifying basis by reducing the loss of spectral information in large datasets.

To improve the reconstruction performance in sparse recovery, pre-defined structure information and side information of unknown target signals were used in [41,42]. The reconstruction of three-dimensional electrical impedance tomography was improved by updating three-dimensional structural correlations using pre-defined structured signals [41]. To recover multi-modal data, a reconstruction framework is proposed in [42] that uses side information in unrolled optimization. Unrolled optimization approaches using deep learning were proposed in [43,44]. Deep-learning architectures were used to train hyper-parameters, such as a gradient regularizer and a step size. Using learned hyper-parameters, it was shown optimized solutions can be obtained within a fixed number of iterations. These proposed approaches for image reconstruction have assumed random sensing matrix and structured or sparse signals. In this work, however, we consider dense spectra and the sensing matrix from thin-film filters for the real implementation. Moreover, the reconstruction performance may change to a sparsifying basis as shown in results because a reconstructed spectrum $\hat{\mathbf{x}}$ should be represented as a linear combination of columns of a fixed sparse basis $\mathbf{\Phi}$ as $\mathbf{\Phi}\hat{\mathbf{s}}$.

For recovering spectra, ResCNN does not require the *a priori* knowledge of a sparsifying basis or prior information of spectral features. During training, ResCNN learns the spectral features using learnable layers, which enable it to recover the fine details for various kinds of spectra without identifying a sparsifying basis.

ResCNN is directly compared with CNN for the synthetic Gaussian dataset in Figure 8a where the mean squared error (Equation (4)) is plotted with respect to the epoch. The mean squared error for CNN and ResCNN are shown in solid black line and solid red line with square symbols, respectively. ResCNN shows a lower mean squared error than that of CNN. Moreover, ResCNN converges faster than CNN, indicating that ResCNN optimizes the learnable layers quicker, as expected based on previous research using residuals [25,29]. In contrast to the previous research that numerous residual connections were used in very deep neural networks to converge networks faster by avoiding vanishing gradient problem, we achieve spectral reconstruction improvements even with one residual connection in a moderate depth CNN.

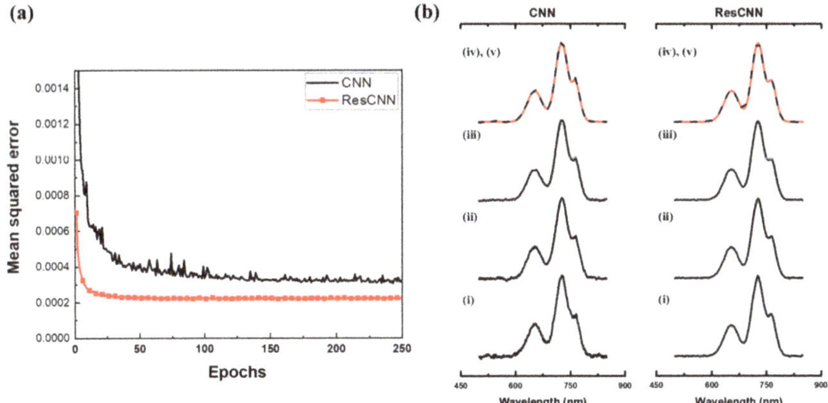

Figure 8. (**a**) Mean squared error of Gaussian dataset with respect to epochs. Solid black line denotes validation error of CNN, and solid red line with square symbols denotes validation error of ResCNN. (**b**) Reconstructions of a spectrum with respect to epochs where (i) to (iv) are epochs 1, 50, 150, and 250, respectively. Red line (v) denotes the original spectrum.

The reconstruction of an example spectrum with respect to the number of epochs is shown in Figure 8b. Black lines ((i) to (iv)) are the reconstructed spectra at 1, 50, 150, and 250 epochs, respectively. The solid red line (v) is the original spectrum, and the series of reconstructed spectrum for ResCNN show that the reconstruction converged earlier than CNN. The increased rate of convergence is because of the residual connection in ResCNN. Overall, the reconstruction performance of ResCNN is an improvement over CNN.

Note that both ResCNN and dictionary learning for sparse recovery require a training dataset and an optimization process to learn the spectral features. While this is a time-consuming process, remember that when using a learned dictionary to recover spectra, an iterative reconstruction algorithm is required, which needs additional time and incurs a high computational cost. The benefit of ResCNN is that it gives a reconstructed spectrum immediately once the training is completed.

7. Conclusions

In this paper, we propose a novel ResCNN for recovering the input spectrum from the compressed measurements in CS spectroscopy. As the optical structure for CS spectroscopy, we numerically generated multilayer thin-film filters which have a small mutual coherence. Therefore, we could compressively measure input spectra with unique sensing patterns. To reconstruct the input spectra from the compressively sampled measurements, we modeled ResCNN, which has a moderate-depth of learnable layers and a residual connection. We stacked nine learnable layers: five convolutional layers and four fully-connected layers with a single residual connection between the input and output of the learnable layers. The measurements were extended by a linear transformation and then fed into ResCNN. Finally, ResCNN reconstructed the input spectra. We demonstrated the empirical reconstruction results for ResCNN using synthetic and measured datasets. We compared the reconstruction performance of ResCNN with sparse recovery using five different sparsifying bases and CNN. Compared with sparse recovery methods, ResCNN shows better reconstruction performance without the *a priori* knowledge of either a sparsifying basis or any spectral features of the spectral datasets. On the other hand, the sparse recovery methods show deviation of reconstruction performances to sparsifying bases and spectral datasets, meaning that a fixed sparsifying basis cannot represent all spectral features of input spectra. Furthermore, ResCNN shows stable reconstruction performances under noisy environments. Compared with CNN, ResCNN shows significant improvement in reconstruction performance and converges faster than CNN. In future work, we will explore compression approaches [40] and unrolled optimization approaches [43,44] for generating a sparsifying basis $\mathbf{\Phi}$ from the training dataset to fully represent spectra without loss of spectral features.

Author Contributions: Conceptualization, C.K. and H.-N.L.; methodology, C.K.; software, C.K. and D.P.; formal analysis, C.K. and D.P.; investigation, C.K.; data curation, C.K.; writing—original draft preparation, C.K.; writing—review and editing, C.K., D.P. and H.-N.L.; project administration, H.-N.L.; funding acquisition, H.-N.L. All authors have read and agreed to the published version of the manuscript.

Funding: This work was supported by the National Research Foundation of Korea (NRF) grant funded by the Korean government (MSIP) (NRF-2018R1A2A1A19018665).

Conflicts of Interest: The authors declare no conflict of interest.

References

1. Clark, R.N.; Roush, T.L. Reflectance spectroscopy: Quantitative analysis techniques for remote sensing applications. *J. Geophys. Res. Solid Earth* **1984**, *89*, 6329–6340. [CrossRef]
2. Izake, E.L. Forensic and homeland security applications of modern portable Raman spectroscopy. *Forensic Sci. Int.* **2010**, *202*, 1–8. [CrossRef] [PubMed]
3. Kim, S.; Cho, D.; Kim, J.; Kim, M.; Youn, S.; Jang, J.E.; Je, M.; Lee, D.H.; Lee, B.; Farkas, D.L.; et al. Smartphone-based multispectral imaging: System development and potential for mobile skin diagnosis. *Biomed. Opt. Express* **2016**, *7*, 5294–5307. [CrossRef] [PubMed]
4. Donoho, D.L. Compressed sensing. *IEEE Trans. Inf. Theory* **2006**, *52*, 1289–1306. [CrossRef]
5. Eldar, Y.C.; Kutyniok, G. *Compressed Sensing: Theory and Applications*; Cambridge University Press: Cambridge, UK, 2012.
6. Candes, E.J.; Eldar, Y.C.; Needell, D.; Randall, P. Compressed sensing with coherent and redundant dictionaries. *Appl. Comput. Harm. Anal.* **2011**, *31*, 59–73. [CrossRef]
7. Oiknine, Y.; August, I.; Blumberg, D.G.; Stern, A. Compressive sensing resonator spectroscopy. *Opt. Lett.* **2017**, *42*, 25–28. [CrossRef]
8. Kurokawa, U.; Choi, B.I.; Chang, C.-C. Filter-based miniature spectrometers: Spectrum reconstruction using adaptive regularization. *IEEE Sens. J.* **2011**, *11*, 1556–1563. [CrossRef]
9. Cerjan, B.; Halas, N.J. Toward a Nanophotonic Nose: A Compressive Sensing-Enhanced, Optoelectronic Mid-Infrared Spectrometer. *ACS Photonics* **2018**, *6*, 79–86. [CrossRef]
10. Oiknine, Y.; August, I.; Stern, A. Multi-aperture snapshot compressive hyperspectral camera. *Opt. Lett.* **2018**, *43*, 5042–5045. [CrossRef]
11. Kim, C.; Lee, W.-B.; Lee, S.K.; Lee, Y.T.; Lee, H.-N. Fabrication of 2D thin-film filter-array for compressive sensing spectroscopy. *Opt. Lasers Eng.* **2019**, *115*, 53–58. [CrossRef]
12. Oliver, J.; Lee, W.-B.; Lee, H.-N. Filters with random transmittance for improving resolution in filter-array-based spectrometers. *Opt. Express* **2013**, *21*, 3969–3989. [CrossRef] [PubMed]
13. August, Y.; Stern, A. Compressive sensing spectrometry based on liquid crystal devices. *Opt. Lett.* **2013**, *38*, 4996–4999. [CrossRef] [PubMed]
14. Huang, E.; Ma, Q.; Liu, Z. Etalon Array Reconstructive Spectrometry. *Sci. Rep.* **2017**, *7*, 40693. [CrossRef] [PubMed]
15. Wang, Z.; Yu, Z. Spectral analysis based on compressive sensing in nanophotonic structures. *Opt. Express* **2014**, *22*, 25608–25614. [CrossRef] [PubMed]
16. Wang, Z.; Yi, S.; Chen, A.; Zhou, M.; Luk, T.S.; James, A.; Nogan, J.; Ross, W.; Joe, G.; Shahsafi, A. Single-shot on-chip spectral sensors based on photonic crystal slabs. *Nat. Commun.* **2019**, *10*, 1020. [CrossRef]
17. Pati, Y.C.; Rezaiifar, R.; Krishnaprasad, P.S. Orthogonal matching pursuit: Recursive function approximation with applications to wavelet decomposition. In Proceedings of the 27th Asilomar Conference on Signals, Systems and Computers, Pacific Grove, CA, USA, 1–3 November 1993; pp. 40–44.

18. Dai, W.; Milenkovic, O. Subspace pursuit for compressive sensing signal reconstruction. *IEEE Trans. Inf. Theory* **2009**, *55*, 2230–2249. [CrossRef]
19. Chen, S.S.; Donoho, D.L.; Saunders, M.A. Atomic decomposition by basis pursuit. *SIAM Rev.* **2001**, *43*, 129–159. [CrossRef]
20. Candes, E.; Tao, T. Decoding by linear programming. *arXiv* **2005**, arXiv:math/0502327. [CrossRef]
21. Oliver, J.; Lee, W.; Park, S.; Lee, H.-N. Improving resolution of miniature spectrometers by exploiting sparse nature of signals. *Opt. Express* **2012**, *20*, 2613–2625. [CrossRef]
22. LeCun, Y.; Bengio, Y.; Hinton, G. Deep learning. *Nature* **2015**, *521*, 436. [CrossRef]
23. Mousavi, A.; Baraniuk, R.G. Learning to invert: Signal recovery via deep convolutional networks. In Proceedings of the 2017 IEEE International Conference on Acoustics, Speech and Signal Processing (ICASSP), New Orleans, LA, USA, 5–9 March 2017; pp. 2272–2276.
24. Li, Y.; Xue, Y.; Tian, L. Deep speckle correlation: A deep learning approach toward scalable imaging through scattering media. *Optica* **2018**, *5*, 1181–1190. [CrossRef]
25. Lee, D.; Yoo, J.; Ye, J.C. Deep residual learning for compressed sensing MRI. In Proceedings of the 2017 IEEE 14th International Symposium on Biomedical Imaging (ISBI 2017), Melbourne, VIC, Australia, 18–21 April 2017; pp. 15–18.
26. Mardani, M.; Gong, E.; Cheng, J.Y.; Vasanawala, S.S.; Zaharchuk, G.; Xing, L.; Pauly, J.M. Deep Generative Adversarial Neural Networks for Compressive Sensing MRI. *IEEE Trans. Med. Imaging* **2019**, *38*, 167–179. [CrossRef] [PubMed]
27. Jin, K.H.; McCann, M.T.; Froustey, E.; Unser, M. Deep convolutional neural network for inverse problems in imaging. *IEEE Trans. Image Process.* **2017**, *26*, 4509–4522. [CrossRef] [PubMed]
28. Kim, C.; Park, D.; Lee, H.-N. Convolutional neural networks for the reconstruction of spectra in compressive sensing spectrometers. In *Optical Data Science II*; International Society for Optics and Photonics: Bellingham, WA, USA, 2019; Volume 10937, p. 109370L.
29. He, K.; Zhang, X.; Ren, S.; Sun, J. Deep residual learning for image recognition. In Proceedings of the IEEE Conference on Computer Vision and Pattern Recognition, Las Vegas, NV, USA, 27–30 June 2016; pp. 770–778.
30. Macleod, H.A. *Thin-Film Optical Filters*; CRC Press: Boca Raton, FL, USA, 2010.
31. Barry, J.R.; Kahn, J.M. Link design for nondirected wireless infrared communications. *Appl. Opt.* **1995**, *34*, 3764–3776. [CrossRef] [PubMed]
32. Kokaly, R.F.; Clark, R.N.; Swayze, G.A.; Livo, K.E.; Hoefen, T.M.; Pearson, N.C.; Wise, R.A.; Benzel, W.M.; Lowers, H.A.; Driscoll, R.L. *USGS Spectral Library Version 7 Data: US Geological Survey Data Release*; United States Geological Survey (USGS): Reston, VA, USA, 2017.
33. University of Eastern Finland. Spectral Color Research Group. Available online: http://www.uef.fi/web/spectral/-spectral-database (accessed on 2 August 2019).
34. Kingma, D.P.; Ba, J. Adam: A method for stochastic optimization. *arXiv* **2014**, arXiv:1412.6980.
35. Chen, G.; Needell, D. Compressed sensing and dictionary learning. *Finite Fram. Theory* **2016**, *73*, 201.
36. Engan, K.; Aase, S.O.; Husoy, J.H. Method of optimal directions for frame design. In Proceedings of the 1999 IEEE International Conference on Acoustics, Speech, and Signal Processing. Proceedings. ICASSP99 (Cat. No. 99CH36258), Phoenix, AZ, USA, 15–19 March 1999; Volume 5, pp. 2443–2446.
37. Aharon, M.; Elad, M.; Bruckstein, A. K-SVD: An algorithm for designing overcomplete dictionaries for sparse representation. *IEEE Trans. Signal. Process.* **2006**, *54*, 4311–4322. [CrossRef]
38. Mairal, J.; Bach, F.; Ponce, J.; Sapiro, G. Online dictionary learning for sparse coding. In Proceedings of the 26th Annual International Conference on Machine Learning, Montreal, QC, Canada, 14–17 June 2009; pp. 689–696.
39. Koh, K.; Kim, S.-J.; Boyd, S. An interior-point method for large-scale l1-regularized logistic regression. *J. Mach. Learn. Res.* **2007**, *8*, 1519–1555.
40. Martino, L.; Elvira, V. Compressed Monte Carlo for distributed Bayesian inference. *arXiv* **2018**, arXiv:1811.0505.
41. Liu, S.; Wu, H.; Huang, Y.; Yang, Y.; Jia, J. Accelerated Structure-Aware Sparse Bayesian Learning for 3D Electrical Impedance Tomography. *IEEE Trans. Ind. Inform.* **2019**. [CrossRef]
42. Tsiligianni, E.; Deligiannis, N. Deep coupled-representation learning for sparse linear inverse problems with side information. *IEEE Signal. Process. Lett.* **2019**, *26*, 1768–1772. [CrossRef]

43. Diamond, S.; Sitzmann, V.; Heide, F.; Wetzstein, G. Unrolled optimization with deep priors. *arXiv* **2017**, arXiv:1705.08041.
44. Gilton, D.; Ongie, G.; Willett, R. Neumann Networks for Linear Inverse Problems in Imaging. *IEEE Trans. Comput. Imaging* **2019**. [CrossRef]

Article

A Deep Learning Framework for Signal Detection and Modulation Classification

Xiong Zha *, Hua Peng, Xin Qin, Guang Li and Sihan Yang

PLA Strategic Support Force Information Engineering University, Zhengzhou 450001, Henan, China; Peng_hua@outlook.com (H.P.); qinxin_0920@163.com (X.Q.); GL_for_study@outlook.com (G.L.); courage32@163.com (S.Y.)

* Correspondence: mici0928@163.com

Received: 20 July 2019; Accepted: 16 September 2019; Published: 19 September 2019

Abstract: Deep learning (DL) is a powerful technique which has achieved great success in many applications. However, its usage in communication systems has not been well explored. This paper investigates algorithms for multi-signals detection and modulation classification, which are significant in many communication systems. In this work, a DL framework for multi-signals detection and modulation recognition is proposed. Compared to some existing methods, the signal modulation format, center frequency, and start-stop time can be obtained from the proposed scheme. Furthermore, two types of networks are built: (1) Single shot multibox detector (SSD) networks for signal detection and (2) multi-inputs convolutional neural networks (CNNs) for modulation recognition. Additionally, the importance of signal representation to different tasks is investigated. Experimental results demonstrate that the DL framework is capable of detecting and recognizing signals. And compared to the traditional methods and other deep network techniques, the current built DL framework can achieve better performance.

Keywords: deep learning; signal detection; modulation classification; the single shot multibox detector networks; the multi-inputs convolutional neural networks

---

1. Introduction

Cognitive radio (CR) [1–3] has been used to refer to radio devices that are capable of learning and adapting to their environment. Due to the increasing requirements for wireless bandwidth of radio spectrum, automatic signal detection and modulation recognition techniques are indispensable. It can help users to identify the modulation format and estimate signal parameters within operating bands, which will benefit communication reconfiguration and electromagnetic environment analysis. Besides, it is widely used in both military and civilian applications, which have attracted much attention in the past decades [4–7].

Multi-signals detection is a task to detect the existing signals in a specific wideband, which is one of the essential components of CR. The most significant difference between signal and non-signal is energy. Hence, many wideband multi-signals detection algorithms are based on energy detector (ED). Some threshold-based wideband signal detection methods, such as [8–13], reduce the probability of false alarm or missed alarm. However, these methods are sensitive to noise changes and challenging to ensure the detection accuracy of all detection scenarios. Therefore, many non-threshold-based detection algorithms have been proposed [14–17]. However, these algorithms have high computational complexity, which results in poor online detection performance.

For automatic modulation recognition, algorithms based on signal phase, frequency, and amplitude have been widely used [18]. However, these algorithms are significantly affected by noise, and the performance can be substantially degraded in low SNR condition. High-order statistical-based algorithms [19–22], such as signal high-order cumulants and cyclic spectrum, have excellent anti-noise performance. The computational complexity of these methods is relatively low, but the selection of features relies too much on expert experience. It is difficult to obtain features that can adapt to non-ideal conditions. In particular, it is challenging to set the decision threshold when there are plenty of modulation formats to be classified.

Deep Learning (DL) techniques [23,24] have made outstanding achievements in Computer Vision [25,26] (CV) and Natural Language Processing [27,28] (NLP) for their strong self-learning ability. Recently, more and more researchers use DL techniques to solve signal processing problems. For signal detection, many DL-based methods, such as [29–31], detect signals in narrowband environment. These methods only detect the existing of signal, but can not estimate the relevant parameters. Therefore, developing a technique leverages deep learning to detect signal efficiently and effectively is still a challenging problem. For DL-based modulation classification, there has been some reported work, including [32–36]. For example, some researchers used the signal IQ waveform as data representation and learned the sample using CNNs [32–34]. Other researchers focused on developing methods to represent modulated signals in data formats for CNNs. Among these methods, constellation-based algorithms [35,36] have been widely utilized, where signal prior knowledge is fully considered.

In this study, DL techniques are fully utilized in multi-signals detection and modulation recognition. For multi-signals detection, we use the deep learning target detection network to detect the location of each signal. In our initial research, the used model is SSD networks, which is a relatively advanced target detection network. Furthermore, we use the time-frequency spectrum as the signal characteristic expression. Due to the time-frequency characteristic of the M-ary Frequency Shift Keying (MFSK) format signals, we can identify the modulation format while the signal is detected. Meanwhile, for M-ary Phase Shift Keying (MPSK), M-ary Amplitude Phase Shift Keying (MAPSK), and M-ary Quadrature Amplitude Modulation (MQAM) signal, the difference in the time-frequency spectrum is not sufficient to identify the signal modulation. Therefore, during the signal detection procession, we identify them in the same format, and only detect the signal presence or absence. Through the signal detection network, we can roughly get the signal carrier frequency and start-stop time. After that, we use a series of traditional methods to convert these signals from the wideband into the baseband. To recognize MPSK, MAPSK, and MQAM signals, a multi-inputs CNNs is designed. Moreover, we adopt the signal vector diagram and eye diagram as the network inputs, which are more robust than in-phase and quadrature (IQ) waveform data and constellation diagram.

This paper addresses the topic of DL based multi-signals detection and modulation classification. The main contributions of this paper are summarized as follows: (1) We propose a relatively complete DL framework for signal detection and modulation recognition, which is more intelligent than traditional algorithms. (2) We establish different signal representation schemes for several tasks, which facilitate the use of the built DL framework for detection and classification. (3) We propose a multi-inputs CNNs model to extract and map the features from different dimensions.

The rest of this paper is presented as follow. In Section 2, we offer a detail introduction to the signal model and the dataset generation. Section 3 shows the DL framework for signal detection and modulation recognition. Section 4 confirms our initial experiment result from different aspects. Finally, our conclusions and directions for further research are given in Section 5.

2. Communication Signal Description and Dataset Generation

In realistic communication processing, the signal may be distorted by the effect of non-linear amplifier and channel. In actual situation, the received signal in the communication system can be expressed as:

$$r(t) = \int_{\tau=0}^{\tau_0} s(n_{clk}(t-\tau))h(\tau)\mathrm{d}\tau + n_{add}(t) \quad (1)$$

where $s(t)$ is the transmission signal, $n_{Clk}(t)$ is timing deviation, $h(t)$ represents the transmitted wireless channel, $n_{add}(t)$ is additive white Gaussian noise.

In this section, we will describe different modulated signals and their sample representation for our DL framework. We will also explain the reason why we use it and the method we enhance it.

2.1. Modulation Signal Description

For any digital modulation signal, the transmission signal can be presented as

$$s(t) = \sum_{n} a_n e^{j(w_n t+\phi)} g(t - nT_b) \tag{2}$$

where w_n is the signal angular frequency, ϕ is the carrier initial phase, T_b is the symbol period, a_n is the symbol sequence, $g(t)$ is the shaping filter.

For MFSK signal, it can be presented as

$$a_n = 1, w_n = w_0 + \frac{2\pi}{M} i, i = 0, 1, \ldots, M-1 \tag{3}$$

For MPSK signal, it can be presented as

$$a_n = e^{j2\pi i/M}, i = 0, 1, \ldots, M-1, w_n = w_0 \tag{4}$$

For MQAM signal, it can be presented as

$$\left.\begin{array}{c} a_n = I_n + jQ_n \\ I_n, Q_n = 2i - \frac{M}{4} + 1, i = 0, 1, \ldots, \frac{M}{4} - 1, w_n = w_0 \end{array}\right\} \tag{5}$$

MAPSK constellations are robust against nonlinear channels due to their lower peak-to-average power ratio (PAPR), compared with QAM constellations. Therefore, APSK was mainly employed and optimized over nonlinear satellite channels during the last two decades. As recommended in DVB-S2 [37], it can be presented as:

$$a_n = r_k \exp\left[j\left(\frac{2\pi}{n_k} i_k + \theta_k\right)\right] \tag{6}$$

where r_k is the radius of the kth circle, n_k is the number of constellations in kth circle, i_k is the ordinal number of constellation points in the kth circle, θ_k is the initial phase of the kth circle.

2.2. Signal Time-Frequency Description

For multi-signals detection task, we use the wideband signal time-frequency spectrum as the neural network input. To prove the feasibility of this method, we theoretically prove the time-frequency visual characteristic of each modulation. Here, we use the short-time Fourier transform [38] (STFT) to analyze the signal time-frequency characteristic.

2.2.1. MFSK Signal Time-frequency Description

The STFT of MFSK signal can be expressed as

$$STFT_{s_{FSK}}(t, w) = \int_{-\infty}^{+\infty} [s_{FSK}(\tau)\gamma^*(\tau - t)] e^{-j\omega\tau} d\tau = \int_{-\infty}^{+\infty} \left[\sum_{k=-\infty}^{+\infty} Ag(\tau - kT_b) e^{j(\omega_k \tau + \phi_k)} \gamma^*(\tau - t)\right] e^{-j\omega\tau} d\tau \tag{7}$$

where $\gamma(t)$ is the window function, whose duration is T. When $\gamma(t)$ is in a symbol duration, Equation (7) can be simplified as

$$\begin{array}{rl} STFT_{s_{FSK}}(t, w) = & \int_{-T/2}^{T/2} Ae^{j(w_k(\tau+t)+\phi_k)} e^{-jw(\tau+t)} d\tau = ATe^{-jwt} e^{j(w_k t+\phi_k)} Sa(\frac{w-w_k}{2} T), \\ & kT_b + T/2 < t < (k+1)T_b - T/2, k = 0, 1, 2, \ldots \end{array} \tag{8}$$

where $Sa(w) = \sin(w)/w$. When $\gamma(t)$ spans two symbols, Equation (7) can be simplified as

$$\begin{aligned} STFT_{s_{FSK}}(t,w) &= \int_{-T/2}^{d} Ae^{j(w_k(\tau+t)+\phi_k)}e^{-jw(\tau+t)}d\tau + \int_{d}^{T/2} Ae^{j(w_{k+1}(\tau+t)+\phi_{k+1})}e^{-jw(\tau+t)}d\tau = \\ &= Ae^{j((w_k-w)t+\phi_k)}\int_{-T/2}^{d} e^{-j(w-w_k)\tau}d\tau + Ae^{j((w_{k+1}-w)t+\phi_{k+1})}\int_{d}^{T/2} e^{-j(w-w_{k+1})\tau}d\tau = \\ &= A\tfrac{T+2d}{2}e^{j((w_k-w)t+\phi_k)}e^{\frac{j(w-w_k)(T-2d)}{4}}Sa(\tfrac{(w-w_k)(T+2d)}{4})+ \\ &+A\tfrac{T-2d}{2}e^{j((w_{k+1}-w)t+\phi_{k+1})}e^{\frac{j(w-w_{k+1})(T+2d)}{4}}Sa(\tfrac{(w-w_{k+1})(T-2d)}{4}), \\ (k+1)T_b &-T/2 < t < (k+1)T_b + T/2, d = (k+1)T_b - t, k = 0,1,2,\ldots \end{aligned} \tag{9}$$

where w_{k+1} is the carrier angular frequency of the k+1-th symbol. If $w_{k+1} = w_k$, it indicates that the carrier angular frequency does not jump, so Equation (8) is same as Equation (9). We take the modulus square of Equation (8). The result can be expressed as

$$\begin{aligned} SPEC_{s_{FSK}}(t,w) = \left|STFT_{s_{FSK}}(t,w)\right|^2 = A^2T^2Sa^2(\tfrac{w-w_k}{2}T), \\ kT_b + T/2 < t < (k+1)T_b - T/2, k = 0,1,2,\ldots \end{aligned} \tag{10}$$

And for Equation (9), it can be expressed as:

$$\begin{aligned} SPEC_{s_{FSK}}(t,w_k) \approx \tfrac{A^2(T+2d)^2}{4} \le A^2T^2, -T/2 < d < T/2, \\ (k+1)T_b - T/2 < t < (k+1)T_b + T/2, k = 0,1,2,\ldots \end{aligned} \tag{11}$$

Obviously, the value of $SPEC_{s_{FSK}}(t,w_k)$ will increase as the increase of jumping time d. The energy decreases gradually as $\gamma(t)$ slips away from the symbol. So when $d = T/2$, the window is completely within one symbol, and the maximum value is obtained.

$$\begin{aligned} SPEC_{s_{FSK}}(t,w_k)_{\max} = A^2T^2, \\ (k+1)T_b - T/2 < t < (k+1)T_b + T/2, k = 0,1,2,\ldots \end{aligned} \tag{12}$$

When $d = -T/2$, the window completely spans to next symbol, and the minimum value is obtained

$$\begin{aligned} SPEC_{s_{FSK}}(t,w_k)_{\min} = 0, \\ (k+1)T_b - T/2 < t < (k+1)T_b + T/2, k = 0,1,2,\ldots \end{aligned} \tag{13}$$

From our analysis, we can easily get the characteristics of FSK modulation: (1) There will be sharp brightness changes in the time-frequency image at the frequency change moment. (2) The signal modulation number M and frequency spacing are important parameters for the MFSK time-frequency characteristics, which determine the value of w_k.

2.2.2. Amplitude–Phase Modulation Signal Time-frequency Description

For MPSK, MAPSK, and MQAM signal, since they all belong to amplitude-phase modulation, the derivation processing of the signal time-frequency characteristics is the same as MPSK. Hence, we specify the time-frequency characteristics of the MPSK signal, and the STFT can be expressed as:

$$\begin{aligned} STFT_{s_{PSK}}(t,w) &= \int_{-\infty}^{+\infty} [s_{PSK}(\tau)\gamma^*(\tau-t)]e^{-jw\tau}d\tau = \\ &= \int_{-\infty}^{+\infty}\left[\sum_{k=-\infty}^{+\infty} Ag(\tau-kT_b)e^{j(w_c\tau+\phi_c+\phi_k)}\gamma^*(\tau-t)\right]e^{-jw\tau}d\tau \end{aligned} \tag{14}$$

As the derivation of MFSK signal time-frequency characteristics, when $\gamma(t)$ is in a symbol duration, the Equation (14) can be simplified as:

$$\begin{aligned} STFT_{s_{PSK}}(t,w) = \int_{-T/2}^{T/2} Ae^{j(w_c(\tau+t)+\phi_c+\phi_k)}e^{-jw(\tau+t)}d\tau = ATe^{-jwt}e^{j(w_ct+\phi_c+\phi_k)}Sa(\tfrac{w-w_c}{2}T), \\ kT_b + T/2 < t < (k+1)T_b - T/2, k = 0,1,2,\ldots \end{aligned} \tag{15}$$

When $\gamma(t)$ spans two symbols, the Equation (14) can be simplified as:

$$\begin{aligned} STFT_{s_{PSK}}(t,w) &= \int_{-T/2}^{d} Ae^{j(w_c(\tau+t)+\phi_c+\phi_k)}e^{-jw(\tau+t)}d\tau + \int_{d}^{T/2} Ae^{j(w_c(\tau+t)+\phi_c+\phi_{k+1})}e^{-jw(\tau+t)}d\tau = \\ &= Ae^{j((w_c-w)t+\phi_c+\phi_k)}\int_{-T/2}^{d} e^{-j(w-w_c)\tau}d\tau + Ae^{j((w_c-w)t+\phi_c+\phi_{k+1})}\int_{d}^{T/2} e^{-j(w-w_c)\tau}d\tau = \\ &= A\tfrac{T+2d}{2}e^{j((w_c-w)t+\phi_c+\phi_k)}e^{\frac{j(w-w_c)(T-2d)}{4}}Sa(\tfrac{(w-w_c)(T+2d)}{4}) + \\ &+ A\tfrac{T-2d}{2}e^{j((w_c-w)t+\phi_c+\phi_{k+1})}e^{\frac{j(w-w_c)(T+2d)}{4}}Sa(\tfrac{(w-w_c)(T-2d)}{4}), \\ &(k+1)T_b - T/2 < t < (k+1)T_b + T/2, d = (k+1)T_b - t, k = 0,1,2,\ldots \end{aligned} \tag{16}$$

where ϕ_{k+1} is the phase of the k+1-th symbol. And if $\phi_{k+1} = \phi_k$, Equation (15) is equal to Equation (16). We take the modulus square of (15), and the result can be expressed as:

$$\begin{aligned} SPEC_{s_{PSK}}(t,w) = \left|STFT_{s_{PSK}}(t,w)\right|^2 = A^2T^2Sa^2(\tfrac{w-w_c}{2}T), \\ kT_b + T/2 < t < (k+1)T_b - T/2, k = 0,1,2,\ldots \end{aligned} \tag{17}$$

And for (16), it can be expressed as:

$$\begin{aligned} SPEC_{s_{PSK}}(t,w_c) = \tfrac{A^2T^2}{2}(1+\cos(\phi_k-\phi_{k+1})) + 2A^2d^2(1-\cos(\phi_k-\phi_{k+1})) \le A^2T^2, \\ (k+1)T_b - T/2 < t < (k+1)T_b + T/2, d = (k+1)T_b - t, k = 0,1,2,\ldots \end{aligned} \tag{18}$$

We take the partial derivative for Equation (18):

$$\begin{aligned} \frac{\partial SPEC_{s_{PSK}}(t,w_c)}{\partial d} = 4A^2d(1-\cos(\phi_k-\phi_{k+1})), -T/2 < d < T/2, \\ (k+1)T_b - T/2 < t < (k+1)T_b + T/2, k = 0,1,2,\ldots \end{aligned} \tag{19}$$

From Equation (19), we can easily learn that $SPEC_{s_{PSK}}(t,w_c)$ get the minimum value when $\phi_{k+1} = \phi_k$ or d = 0. But the minimum value is much greater than 0, which is greatly different for the MFSK signal.

$$\begin{aligned} SPEC_{s_{PSK}}(t,w_c)_{\min} = \tfrac{A^2T^2}{2}(1+\cos(\phi_k-\phi_{k+1})) \gg 0, \\ (k+1)T_b - T/2 < t < (k+1)T_b + T/2, k = 0,1,2,\ldots \end{aligned} \tag{20}$$

Hence, for MPSK signal, there is only one wide frequency band in the time-frequency diagram, and the brightness fluctuation appears in a small range, which is different from MFSK. And from derivation processing, we can know that the MPSK time-frequency characteristics are less affected by M, so it is hard to distinguish PSK signals with different M. Figure 1 presents different modulation signals in the wideband.

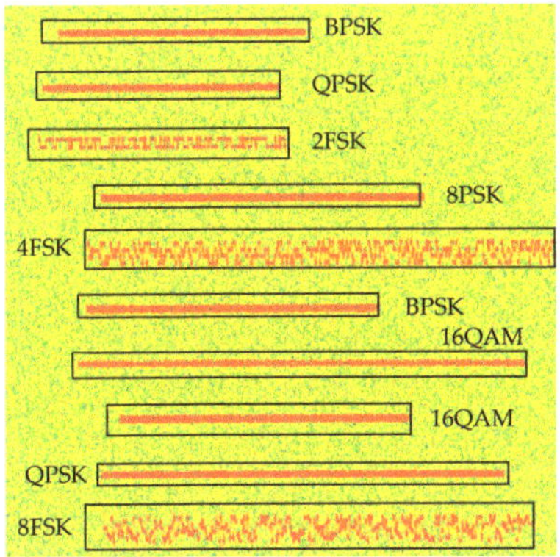

Figure 1. Different modulation signals in the wideband.

2.3. Signal Eye Diagram and Vector Diagram Description

The function of the eye diagram is to observe the baseband signal waveform by an oscilloscope. Through the eye-diagram, we can adjust the receiver filter to improve system performance. Besides, due to the characteristics of the modulated signal itself, different modulation modes have apparent visual differences in the eye diagram. As shown in Figure 2, because of the different modulation scales, there are different eye numbers in each eye diagram. For OQPSK, since the two orthogonal signals stagger for half a symbol period, the eye-opening position is always staggered, while other modulated signals always appear at the same time.

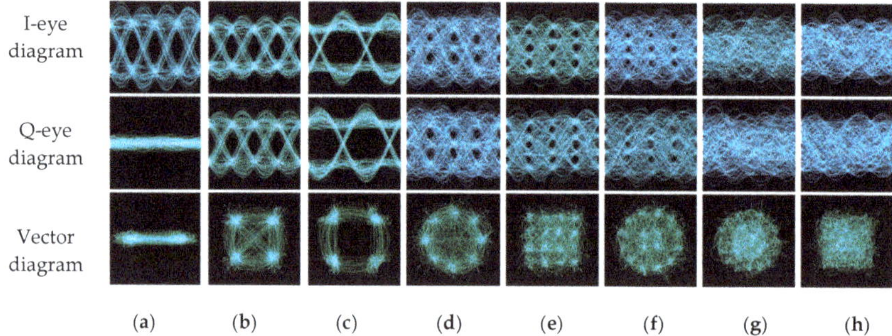

Figure 2. The eye diagram and vector diagram of different modulation signals in 15dB (**a**) BPSK; (**b**) QPSK; (**c**) OQPSK; (**d**) 8PSK; (**e**) 16QAM; (**f**) 16APSK; (**g**) 32APSK; (**h**) 64QAM.

By reconstructing the signal IQ waveforms in the corresponding time, the signal vector diagram shows the symbol trajectory. From its formation mechanism, it is similar to the signal constellation diagram. However, unlike the constellation diagram, the vector diagram can reflect the signal phase information. For example, it can easily distinguish QPSK from OQPSK with the same initial phase, because there is no 180° phase shift in OQPSK, while it exists in QPSK. Meanwhile, compared with the constellation diagram, the vector diagram is more convenient to obtain and requires less prior information.

2.4. The Generation Processing of the Dataset

Figure 3 presents the processing of our dataset construction and annotation. To make samples more diverse, we set sampling phase offset, frequency offset, phase offset, and amplitude attenuation in sample generation processing.

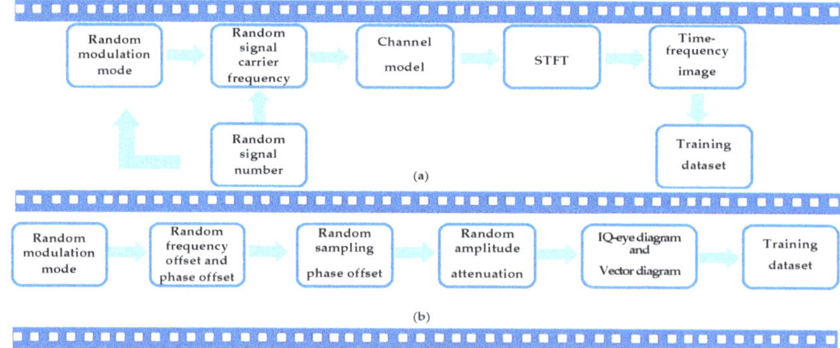

Figure 3. The generation processing of the dataset. (**a**) dataset for signal detection; (**b**) dataset for modulation recognition.

For signal detection, we need to determine the reconnaissance frequency range and set the signal number in the wideband at first. We set different frequency offset for each signal, and ensure that the signals do not overlap in the frequency domain. Then we perform the STFT on the wideband. Not only we record the modulation format of each signal, but also record the start-stop time, carrier frequency, and bandwidth. Then we convert them into the coordinates on time-frequency image, which are the label information for the network.

For modulation recognition, traditional eye diagram and vector diagram are binary images, which do not consider the signal aggregation degree at a particular location. Hence, we consider the signal aggregation degree and enhance the traditional eye diagram and vector diagram. Figure 4 presents the enhancement processing of the dataset. In our initial research, since CNNs are insensitive to edge information, the signal amplitude is quantified between [−1.05, 105] by 128 after normalizing the amplitude. Furthermore, the parameter settings are obtained by experiments. For example, we choose 800 symbols and 4 symbols as a waveform group to generate the eye diagram and the vector diagram, and related experiments will also be described in detail in subsequent chapters. Moreover, we perform the following operations on the images to make the image details more prominent, where $\mathbf{Im}_0$ is the original image, $\mathbf{Im}_1, \mathbf{Im}_2, \mathbf{Im}_3$ are the channels of the enhanced image and α, β are scaling factors.

$$\begin{aligned} \mathbf{Im}_1 &= \text{unit8}\left(\frac{\mathbf{Im}_0 - \min(\mathbf{Im}_0)}{\max(\mathbf{Im}_0) - \min(\mathbf{Im}_0)} \times 255\right), \\ \mathbf{Im}_2 &= \text{unit8}(\alpha \times \log 2(\mathbf{Im}_1 + 1)), \\ \mathbf{Im}_3 &= \text{unit8}(\exp(\mathbf{Im}_1 / \beta)) \end{aligned} \tag{21}$$

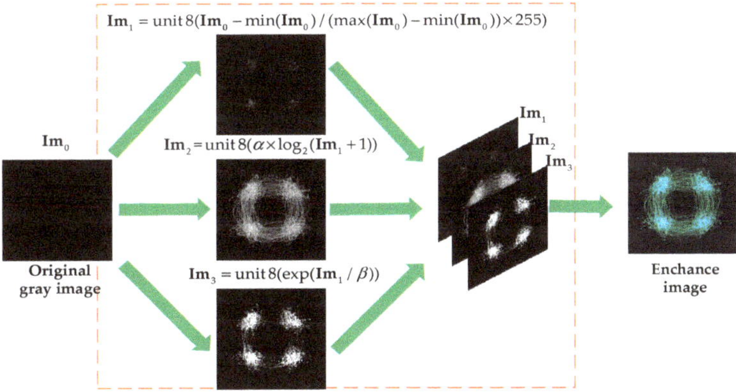

Figure 4. The enhancement processing of the dataset.

3. Deep Learning Framework for Signal Detection and Modulation Recognition

DL networks aim at learning different hierarchies of features from data. As one of the branches of DL techniques, the CNNs perform well in the field of image recognition. It performs feature learning via non-linear transformations implemented as a series of nested layers. Each layer consists of several kernels that perform a convolution over the input. Generally, the kernels are usually multidimensional arrays which can be updated by some algorithms [39]. Our DL framework achieves multi-signals detection and modulation recognition. We use different deep neural networks for different tasks. For signal detection, we use SSD networks. For modulation recognition, we design a multi-inputs CNNs.

3.1. SSD Networks for Signal Detection

We use SSD networks to achieve multi-signals detection. For DL target detection techniques, the existing algorithms are mainly divided into two kinds: algorithms based on region recommendation (two-stage methods) and algorithms based on regression (one-stage methods). Regression-based algorithms include YOLO series algorithms [40–43] and SSD series algorithms [43,44], while region recommendation-based algorithms include RCNN [45], Fast RCNN [46], and Faster RCNN [47]. In our research, since the regression-based algorithms are faster than region recommendation-based algorithms, we use SSD networks as our signal detection model. SSD networks can generate a series of fixed-size borders and the possibility of the containing target in each border. Finally, the final detection and recognition results are calculated by the non-maximum suppression algorithm [48]. The structure of SSD networks is shown in Figure 5, and it can be divided into four parts.

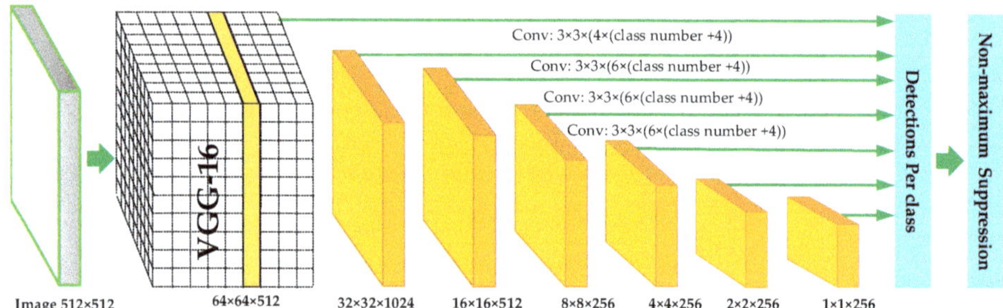

Figure 5. The network model for signal modulation recognition.

Part 1: The networks for feature extraction. The initial part of SSD networks is the first layers of VGG16 network, which is used as the primary network to extract the deep features of the whole input image. Behind the primary network, the model structure is the pyramid networks, which contains a series of simple convolution layers to make feature maps smaller and smaller. With the pyramid network structure, we can get several feature maps with different scales.

Part 2: The design of the default box. In this part, we will design several feature default box for different scales of feature maps. Each feature map at the top of the VGG16 networks is associated with a set of feature default box. As shown in Figure 6, there are dotted borders at each position of 4×4 and 8×8 feature maps. These fixed-size borders are default boxes, and their scale parameters are designed by the different feature maps scales. For example, assuming that we need M feature maps to predict, the scale parameters of the default box are as follows:

$$S_k = S_{\min} + \left(\frac{S_{\max} - S_{\min}}{m-1}\right) \bullet (k-1), k \in [1, m] \tag{22}$$

where $S_{\min}$ is the bottom scale, and $S_{\max}$ is top scale. The length-width radio of feature default can be expressed as: $a_r \in \{1, 2, 3, 1/2, 1/3\}$. So the feature default box length is $W_k^a = S_k \sqrt{a_r}$ and the width is $H_k^a = S_k / \sqrt{a_r}$.

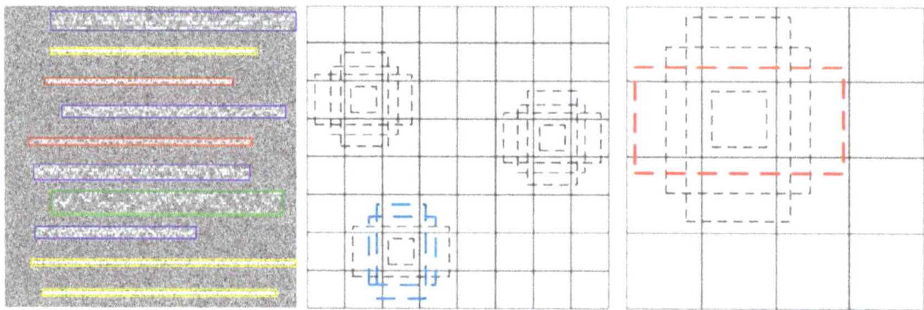

Figure 6. The Design of the default box.

Part 3: Detection and Recognition. In this part, we can predict the target category and location. We add a set of convolution kernels behind several different scales feature maps. Using these convolution kernels, we can get a fixed set of test results. For an $m \times n \times p$ feature maps, a small convolution kernel with $3 \times 3 \times p$ size is used as the fundamental prediction element. Finally, the classification probability of each feature default box and the offsets are obtained.

Part 4: Non-maximum suppression. In the last part, we use non-maximum suppression to select the best prediction results. For the feature default boxes that are matched by each real target border, we calculate their intersection-parallel ratios. The expression is shown as follow

$$IoU = (A \cap B)/(A \cup B) \tag{23}$$

where A and B are two borders. We will select the feature default box whose IoU are greater than 0.5 as best results, and then obtains the highest confidence degree feature default box by non-maximum suppression.

In offline train stage, the whole objective optimal function of the SSD networks includes two parts: confidence loss and location loss. The expression is shown as follow

$$L(x,c,l,g) = \frac{1}{N}\left(L_{conf}(x,c) + \alpha L_{loc}(x,l,g)\right) \tag{24}$$

where x is used to indicate whether the feature default box is a target or not. N is the number of the feature default boxes that are matched to real target borders. Parameter α is used to adjust the radio between L_{conf} and L_{loc}, default $\alpha = 1$. L_{conf} is softmax loss function. L_{loc} is used to measure the performance of the boundary box prediction, and in our initial research, we use the typical $smooth_{L1}$ function to calculate

$$L_{conf}(x,c) = -\sum_{i \in Pos}^{N} x_{ij}^{p} \log\left(\hat{c}_i^p\right) - \sum_{i \in Neg} \log\left(\hat{c}_i^0\right), \hat{c}_i^p = \frac{\exp\left(c_i^p\right)}{\sum_p c_i^p} \tag{25}$$

$$L_{loc}(x,l,g) = \sum_{i \in Pos}^{N} \sum_{m \in \{cx,cy,w,h\}} x_{ij}^{k} smooth_{L1}\left(l_i^m - \hat{g}_j^m\right) \tag{26}$$

where Pos and Neg represent all positive and negative borders, respectively. c_i^p represents the confidence degree for pth feature default matching ith target. l_i^m represents the prediction bias of the i-th feature default box. (cx, cy) is the box center coordinates and (w, h) is the box width and height. $\hat{g}_j^m$ represents the deviation between the real target border g_j^m and the default box d_i^m. $\hat{g}_j^m$ is calculated as follow:$\hat{g}_j^{cx} = \left(g_j^{cx} - d_i^{cx}\right)/d_i^w$, $\hat{g}_j^{cy} = \left(g_j^{cy} - d_i^{cy}\right)/d_i^h$, $\hat{g}_j^w = \log\left(g_j^w/d_i^w\right)$, $\hat{g}_j^h = \log\left(g_j^h/d_i^h\right)$.

3.2. Multi-Inputs CNNs for Modulation Recognition

For signal modulation recognition task, the modulation set is {BPSK, QPSK, OQPSK, 8PSK, 16QAM, 16APSK, 32APSK, 64QAM}, because they are all belonging to amplitude-phase modulation and we cannot distinguish each other from time-frequency characteristic in SSD network. Hence, we use the eye diagram and vector diagram as the model inputs. The multi-inputs CNNs model is shown in Figure 7. The initial size of the samples is 128×128, and we use softmax as the output layer's active function and relu as all other network layers' active function.

The signal features extraction can be divided into three stages. On the first stage, we use 7×7 convolution kernels to convolute IQ eye diagram and vector diagram, respectively. To ensure the dynamic range unification of the feature maps, we apply the batch normalization (BN) [49] on first layer network outputs. We perform the max pooling operation on the BN feature maps. Then we connect the feature maps from IQ eye diagram inputs.

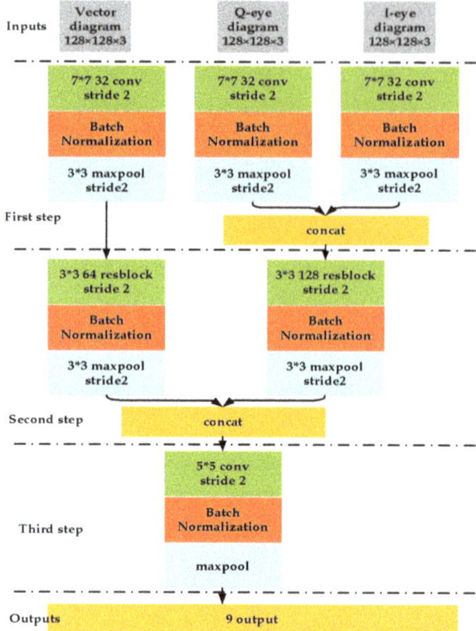

Figure 7. The network model for signal modulation recognition.

On the second signal feature extraction stage, we adopt the residual network structure to avoid the degradation caused by the network over-depth. The basic structure of ResNet [50] is shown in Figure 8. After the second feature extraction stage, each input feature maps are connected. After the batch normalization in the third stage, we directly process the feature maps by global maximum sampling to reduce the network parameters.

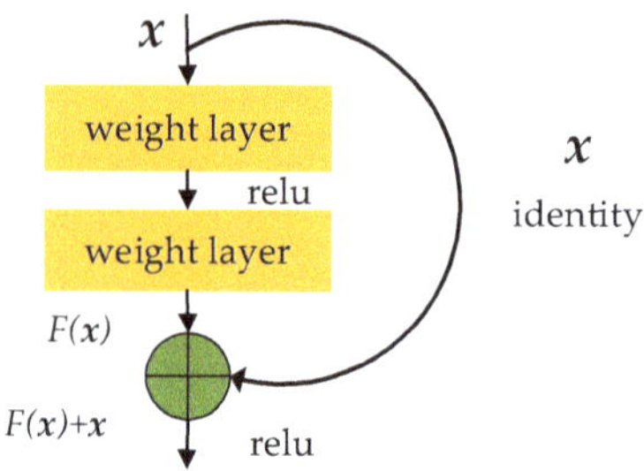

Figure 8. The basic structure of ResNet.

In the processing of network optimization, we adopt Adam algorithm [51] to solve the network parameters optimal solution. The categorical cross-entropy error is chosen as the loss function, which is represented as:

$$J_1(\boldsymbol{w},\boldsymbol{b};\boldsymbol{x}_1,\boldsymbol{x}_2,\boldsymbol{x}_3,\boldsymbol{y}) = -\sum_{i}^{N} (\boldsymbol{y}_i)^{\mathrm{T}} \log(f_1(\boldsymbol{x}_{1,i},\boldsymbol{x}_{2,i},\boldsymbol{x}_{3,i};\boldsymbol{w},\boldsymbol{b})) + \lambda_1 \sum \|\boldsymbol{w}\|^2 \tag{27}$$

3.3. The Description for Deep Learning Framework

According to above introduction of the signal detection network and the modulation recognition network, we describe the use of our DL framework. Figure 9 presents the system model. The steps of the model used are as follows:

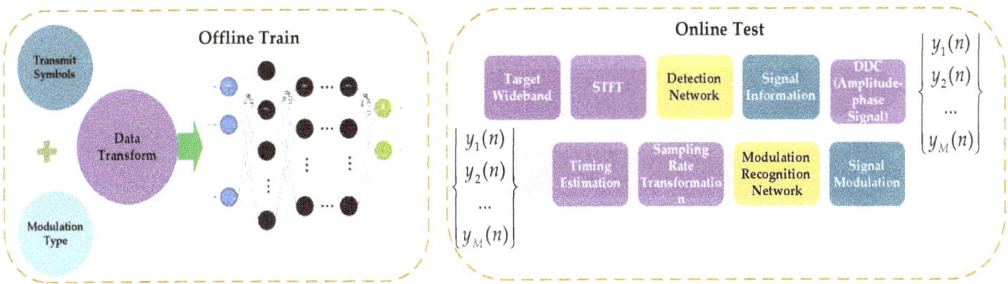

Figure 9. System model.

Step 1: We construct the signal detection network and modulation recognition network and train each network with their appropriate samples.

Step2: In the online testing phase, we perform STFT for the wideband signals, and use the trained SSD networks to detect the signals in the time-frequency spectrum. In this step, we can obtain the center frequency and start-stop time of each signal. And for MFSK signal, we can get its modulation format.

Step 3: For the amplitude-phase modulation signal, we can get the central frequency and start-stop time in step 2. With this knowledge, we filter the target signal and use the envelope spectrum to estimate the signal symbol rate. Then we down convert the signal and perform the matched filter by using the estimated symbol rate.

Step 4: If the timing deviation exists in the target signal, it is necessary to extract the sample value at the optimum sampling position for signal eye diagram and vector diagram. We use the non-data-aided timing estimation algorithm in [52]. The specific expression is as follows, where L_0 is the length of the signal symbols, T is the sampling period and N is the oversampling number:

$$\hat{\tau} = \arg\left\{\sum_{k=0}^{NL_0-1}\left|s\left(\frac{kT}{N}\right)\right|^2 e^{-j2\pi k/N}\right\} \tag{28}$$

Step 5: We alter the target signal sampling rate, and obtain the baseband signal with a maximum delay 32 sampling period. Moreover, we generate the eye-diagram and vector diagram with the processed signal.

Step 6: We use the trained modulation recognition network to identify the signal by its eye diagram and vector diagram. Finally, we complete the signal detection and modulation recognition.

4. Results

In Sections 2 and 3, we have discussed the methods which convert complex signal samples into images without noticeable information loss and introduced the structures of our DL framework. Table 1 shows the time complexity of our DL framework on the different process, in which N is the number of signals in the wideband range. It can be seen that our framework has low time complexity due to the evolution of GPUs, which is acceptable for many practical communication systems.

Table 1. The time complexity of the DL framework (ms).

Simple for Signal Detection	Signal Detection	Down Conversion	Simple for Modulation Recognition	Modulation Recognition
28.5	50.5	$10.3 \times N$	20.8	10.6

We also have conducted several experiments to demonstrate the performances of the DL framework for joint signal detection and modulation recognition in wireless communication systems. Our experiments can be divided into two parts: (1) performances on multi-signals detection and (2) performances on modulation recognition. The rest of this section is organized as follows

Multi-signals detection: First, we show some results of our detection network, and explain the reasons for these results. Then, we evaluate the model performances from three aspects: the modulation format, carrier frequency, and start-stop time. We also compare our network with other detection networks.

Signal modulation recognition: We evaluate our model recognition performances on each modulated signal. We also discuss the network performances when the frequency offset exists. Meanwhile, we compare our method with traditional methods and other DL based methods. Finally, we discuss the influence of symbol and eye numbers and compare the performance between signal-input networks and multi-inputs networks.

4.1. Performance on Signal Detection

For multi-signal detection, we need to know each signal carrier frequency, start-stop time and modulation format. Figure 10 shows some detection results from our model. From Figure 10a,b, it is indicated that our model is beneficial for multi-signals detection, and it can accurately estimate the relevant information about each signal. Moreover, our model has very a promising application prospect in engineering because it has a good visualization effect.

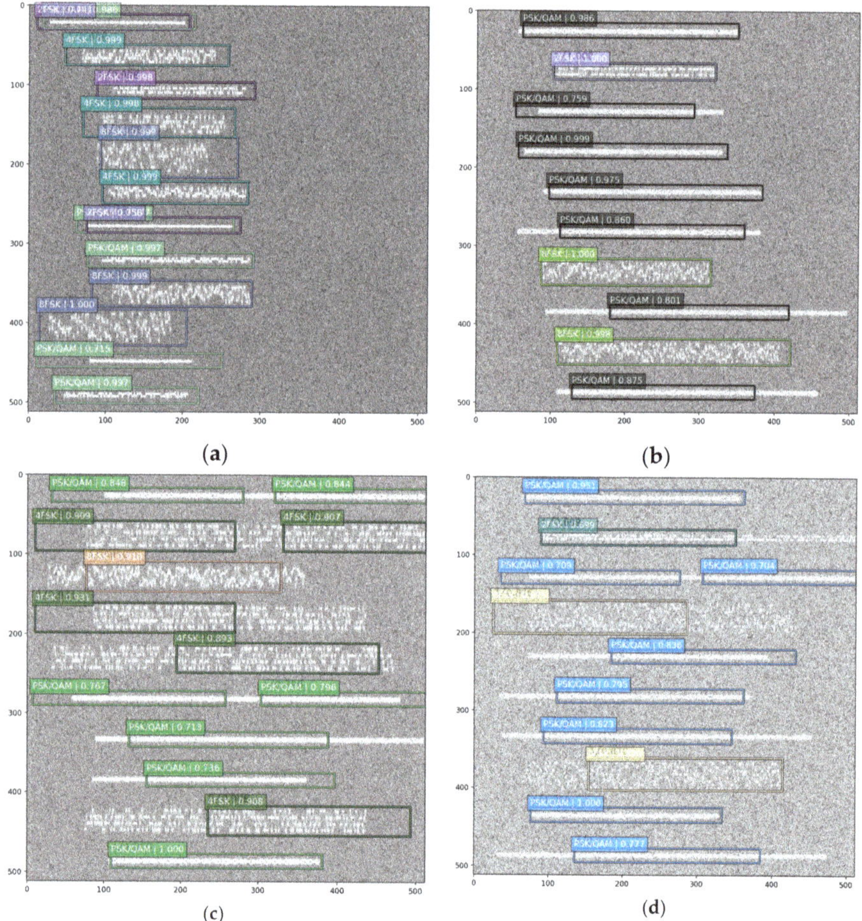

Figure 10. *Cont.*

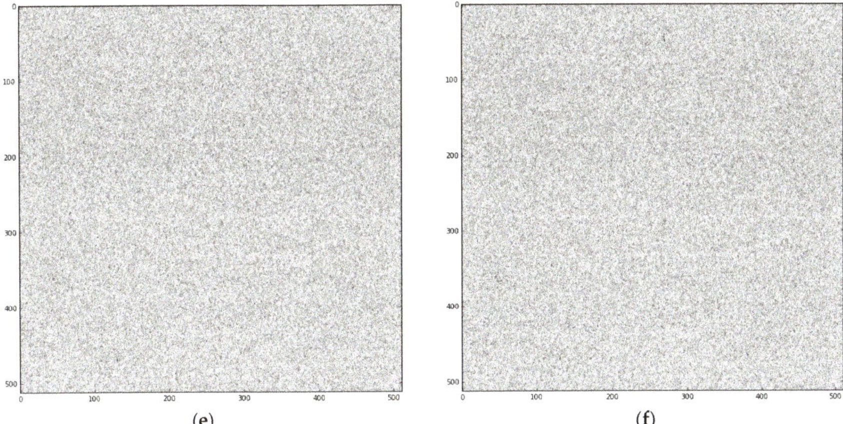

Figure 10. The SSD networks detection results. (**a**) ideal result 1; (**b**) ideal result 2; (**c**) not perfect result 1; (**d**) not perfect result 2; (**e**) no signal result 1; (**f**) no signal result 2.

To some extent, our model is not perfect yet, and there are still some aspects that need to be improved. From Figure 10c,d, we can learn that once the signal length is large, the estimation of the signal start-stop time is not precise, while the estimation of the carrier frequency is precise. The cause of this phenomenon may be that the time-frequency spectrum has large deformation and extreme length-with radio, while the natural image is not. Therefore, we need to further optimize the default box in the SSD networks. Figure 10e,f show the network performance when there is no signal exists. It can be observed that the model does not produce a false alarm, which is useful in engineering.

Figure 11 shows our model detection precision versus different SNRs. We choose the mean Average Precision (mAP) as the performance index of the model. To calculate the mAP, we need to calculate precision and recall. For calculating precision and recall, we need to identify True Positive (TP), False Positive (FP), True Negative (TN), and False Negative (FN). Recall is defined as the proportion of all positive examples ranked above a given rank. Precision is the proportion of all examples above that rank which are from positive. The Average Precision (AP) summarizes the shape of the precision/recall curve. Hence, the mAP is the mean of all the AP values across all classes as measured above. They can be calculated as follows

$$\text{Precision} = \frac{TP}{TP + FP}, \text{Recall} = \frac{TP}{TP + FN} \tag{29}$$

$$AP = \int_0^1 \mathrm{P}(r)dr \tag{30}$$

$$mAP = \frac{\sum\limits_{\text{num\_classes}} AP_i}{\text{num\_classes}} \tag{31}$$

It can be deduced that with the increase of the SNR, the mAP value of the SSD network is increasing. When the SNR is 5 dB, the mAP value can reach 90% in *IoU* is 0.5. Different *IoU* thresholds can lead to different results. Although the increase of the threshold can obtain more reliable signal carrier frequency and start-stop time, it sacrifices the precision of signal detection. Besides, we can adopt some traditional methods to further estimate these signal parameters. Finally, we choose 0.5 as the threshold of *IoU*.

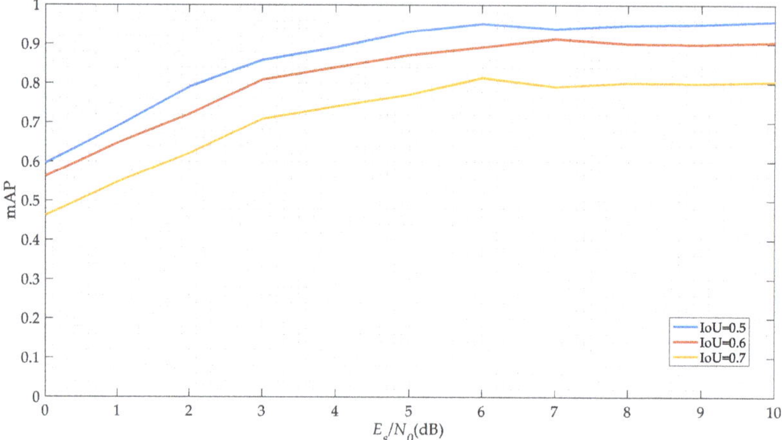

Figure 11. The SSD networks performances.

Once we detected the signals, we need to evaluate the precision of the estimated parameters. We use the normalized offset of the estimated and the actual parameters as the criterion of measurement. They can be presented as follows:

$$\Delta f = \frac{|f_{pre} - f_{real}|}{R} \tag{32}$$

$$\Delta t_{start} = \frac{|t_{pre\_start} - t_{true\_start}|}{T}, \Delta t_{stop} = \frac{|t_{pre\_stop} - t_{true\_stop}|}{T} \tag{33}$$

where f_{pre} is the predicted value of the carrier frequency, f_{real} is the actual value of the carrier frequency, R is the symbol rate, t_{pre_start} and t_{pre_stop} are the predict values of the start and the stop time, t_{true_start} and t_{true_stop} are the actual values of the start and the stop time, and T is the signal duration. Table 2 shows the carrier frequency and the start-stop time precision when the signal is detected. It can be seen that the precision of the carrier frequency is higher than start and stop time. These phenomena are consistent with Figure 10c,d. And in future research, we need to combine the prior information of the signal to design the default boxes and the networks.

Table 2. The offset in the estimation of the various parameters.

Offset	Carrier Frequency	Start Time	Stop Time
0 dB	2.0%	35.2%	38.6%
5 dB	1.3%	22.3%	21.4%
10 dB	0.5%	9.6%	8.7%

We also compare our model performances with the RCNN networks and the Fast RCNN. From Figure 12a, we can see that the mAP of the Fast RCNN and the RCNN is higher than the SSD networks, but the improvement is not significant. And from Figure 12b, we can infer that the SSD networks has considerable advantages in processing speed compared with the RCNN and Fast RCNN. Our model processing speed can reach 0.05 s for each time-frequency spectrum, and such a computational complexity is acceptable for many practical communications systems.

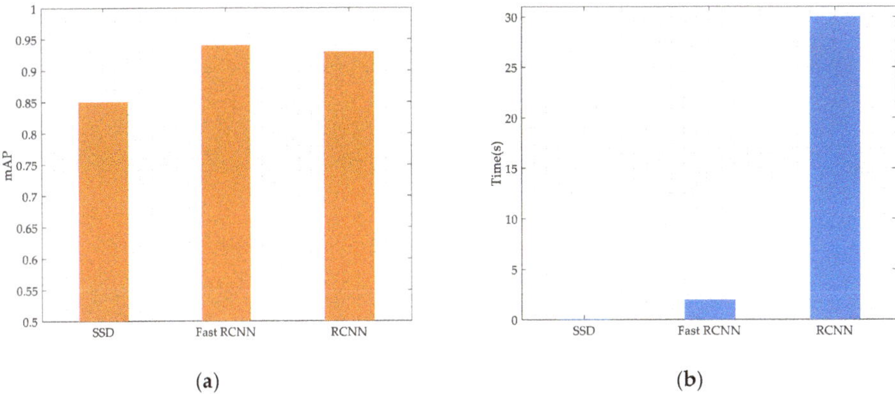

Figure 12. Different networks performances. (**a**) performances for mAP; (**b**) performances for time.

4.2. Performance on Modulation Recognition

For signal modulation recognition, we set a series of experiments to test the network performances. Figure 13a shows the recognition performances of each modulated signal under the different SNR. It can be seen that the algorithm can still achieve better performance when the SNR is very low. Because its modulation complexity, the performance of 64QAM signal is worse than other signals, but it still can achieve 94% accuracy at 7 dB. For BPSK and OQPSK signals, they have distinct visual characteristics from other modulated signals in the eye diagram and the vector diagram, which recognition accuracy can reach 100% even in 0 dB. And it is also obvious that the recognition performance of circular modulation signals {8PSK, 16APSK, 32APSK} is better than QAM modulation signal. To understand the results better, the confusion matrices in different SNR levels are presented in Figure 13b–d. It can be seen that the network shows excellent performance in discriminating BPSK, QPSK, OQPSK, 8PSK, and 16APSK. Moreover, in our experiments, it can be seen that 16QAM is more likely confused with 64QAM, while 16APSK is more likely confused with 64QAM.

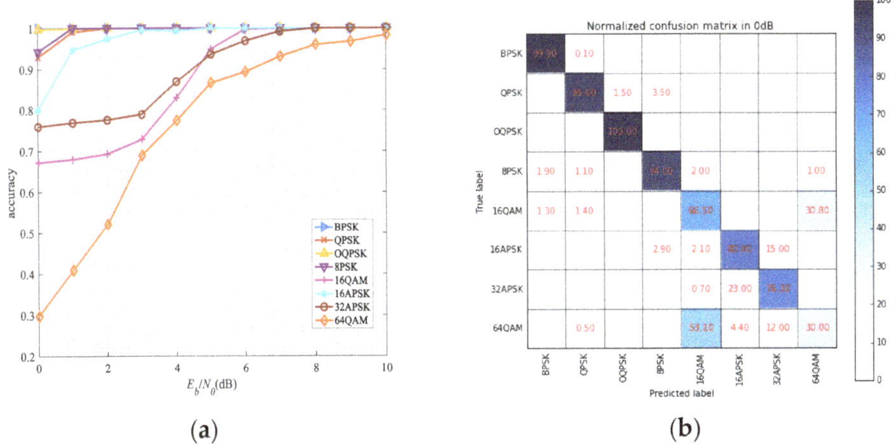

Figure 13. *Cont.*

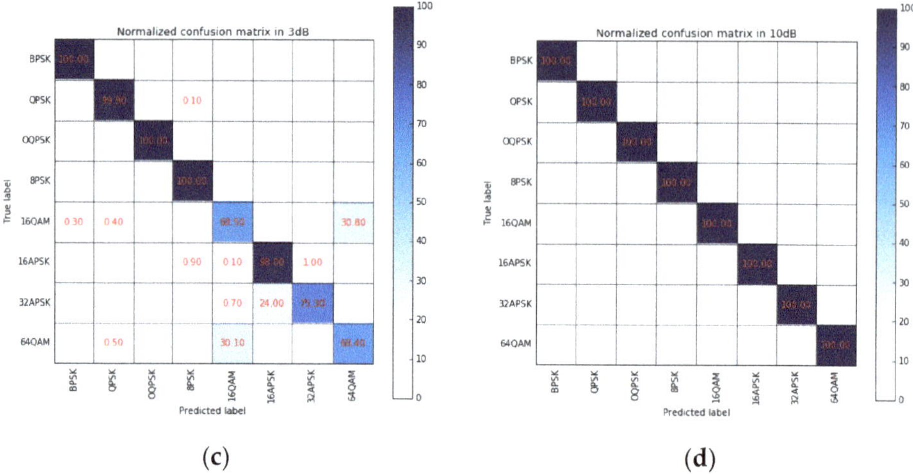

Figure 13. The performance of each modulation (**a**) classification accuracy for each modulation versus SNR; (**b**) normalized confusion matrix in 0 dB; (**c**) normalized confusion matrix in 3 dB; (**d**) normalized confusion matrix in 10 dB

For accuracy comparison, we consider four different modulation classification algorithms.

(1) Cumulant: A traditional signal processing algorithm using the fourth-order cumulant C40 as the classification statistics [19].
(2) SVM-7: An ML-based algorithm using the SVM with seven features, including three fourth-order cumulants C40, C41, and C42 and four sixth-order cumulants C60, C61, C62, and C63 [20].
(3) CNNs for IQ waveform: A DL-based algorithm using the CNNs with the signal IQ waveform [4].
(4) CNNs for constellation: An DL-based algorithm using the CNNs with the signal constellation [35].

Figure 14 presents the average classification accuracy of five algorithms versus SNR. The average accuracy is obtained by averaging the classification performance of eight modulation categories. The performance results of our algorithm outperform all other algorithms.

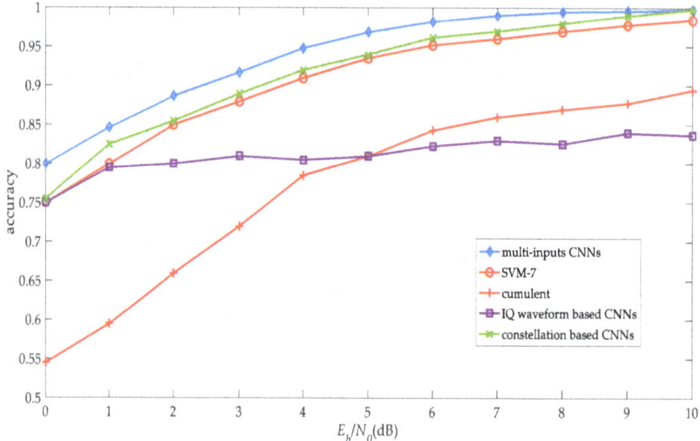

Figure 14. Different methods performance versus SNR.

Considering the error of the carrier frequency estimation by SSD networks and FFT in practice, we research the network recognition performance in different frequency offsets. We set a series of frequency offset for signals, and the result is shown in Figure 15. It can be seen that the recognition accuracy of the signals with a frequency offset is lower than those without frequency offset. When the signals have a large frequency offset, the network is no longer suitable. We also collect some signals from a real satellite communication system, and the real-time wireless channel is performed in the received signals. And then, we use a signal playback device, a DSP card, and PCs to simulate signal reception process. From Figure 15a, we can obtain that the recognition accuracy on real data is lower on simulated data at same SNR level. It may be due to the training data, which not consider the actual channel environment clearly. But the recognition accuracy can still reach 90% when the SNR is 4 dB. And for further research, we will make full use of the real signal to make our model more robust.

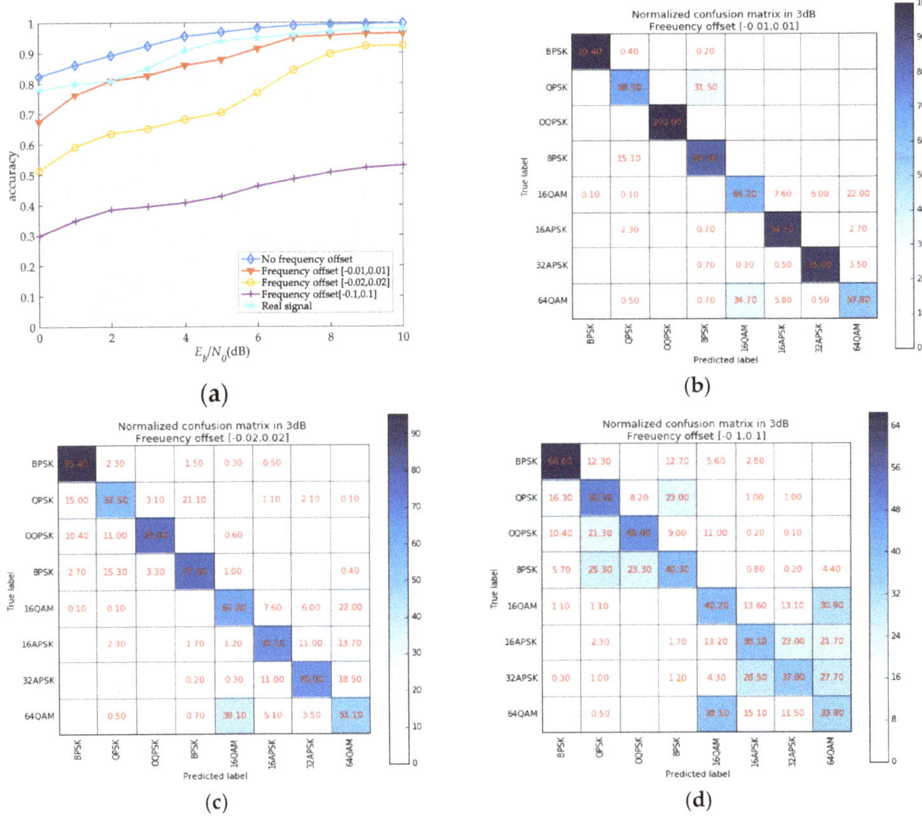

Figure 15. The network performance on different frequency offsets range. (**a**) classification accuracy for different frequency offset versus SNR; (**b**) normalized confusion matrix in 3 dB when the frequency offset is [−0.01, 0.01]; (**c**) normalized confusion matrix in 3dB when the frequency offset is [−0.02, 0.02]; (**d**) normalized confusion matrix in 10 dB when the frequency offset is [−0.1, 0.1].

We also consider the influence of the symbol numbers and the eye number in eye diagram on the network performance. We obtain the best parameter settings of samples by grid search. The symbol number is set as 200, 400, 800, and 1000, respectively, while the eye number is set as 2, 3, 4, and 5. The results are shown in Figure 16. It can be seen that theses parameters do affect network performance. With the increase of symbol number and eye number, the overall accuracy of the model is gradually increasing. But we also can see that when the symbol number is 1000 and the eye number is 5, the improvement of performance is not obvious. Therefore, we finally choose 800 symbols and 4 eye numbers to generate the eye diagram and the vector diagram.

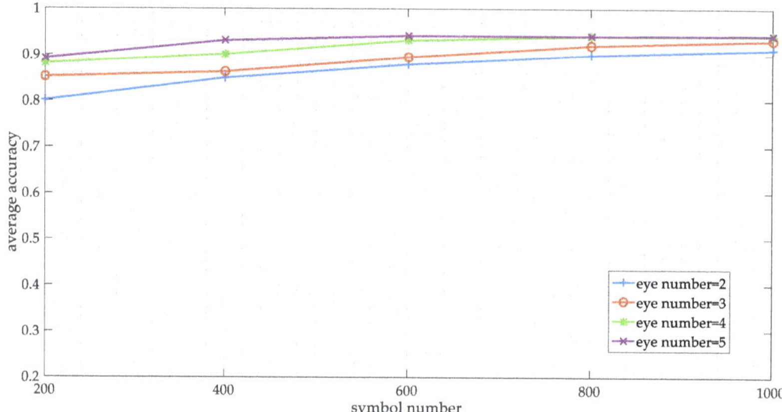

Figure 16. The network performance on the different sample parameters.

Finally, we compare the performance of the single input network with the multi-inputs network in this work. The results are shown in Figure 17. The modulation recognition algorithm based on a single eye diagram has poor performance. The performance of the I-eye diagram is lower than that of the Q-eye diagram, which may be due to the setting of the initial phase in the same modulation format. And the performance of the vector diagram based method is also inferior to our method, since it does not make full use of the signal waveform information.

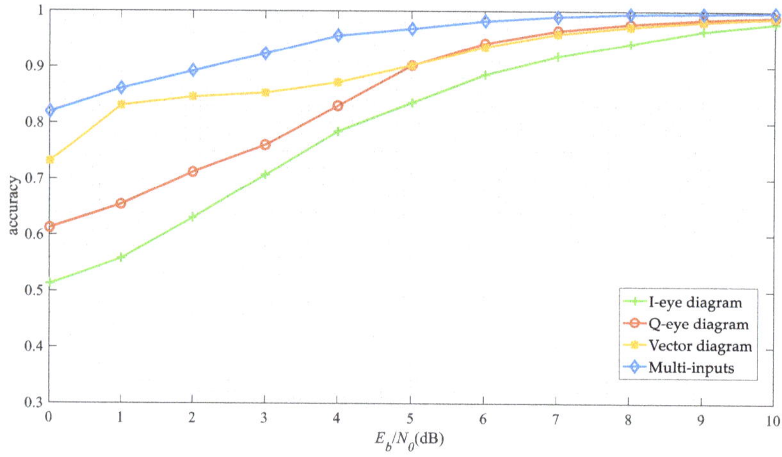

Figure 17. The network performance on the different input model.

5. Conclusions and Discussions

In our research, we have demonstrated our initial efforts to establish a DL framework for multi-signals detection and modulation classification problem. In our method, the time-frequency spectrums are exploited for multi-signals detection task, while the eye-diagrams and vector diagrams are exploited for the modulation classification task. The simulation results prove that DL technologies have the ability to solve the problems in the communication field and have higher performance than other methods.

However, in the future, we will do more rigorous analysis and more comprehensive experiments. Besides, for practical use, we will collect the samples generated from the real channels, and then retrain or fine-tune the model for better performance.

Author Contributions: Conceptualization, X.Z. and H.P.; methodology, X.Z.; software, X.Z.; validation, X.Z., H.P. and X.Q.; formal analysis, G.L.; investigation, X.Z.; resources, G.L.; data curation, X.Z.; writing—original draft preparation, X.Q.; writing—review and editing, G.L.; visualization, S.Y.; supervision, X.Z.; project administration, H.P.; funding acquisition, H.P.

Funding: This research was funded by the National Natural Science Foundation of China (No. 61401511) and the National Natural Science Foundation of China (No. U1736107).

Conflicts of Interest: The authors declare no conflict of interest.

References

1. Mitola, J.; Maguire, G.Q. Cognitive radio: Making software radios more personal. *IEEE Pers. Commun.* **1999**, *6*, 13–18. [CrossRef]
2. Axell, E.; Leus, G.; Larsson, E.G.; Poor, H.V. Spectrum Sensing for Cognitive Radio: State-of-the-Art and Recent Advances. *IEEE Signal. Process. Mag.* **2012**, *29*, 101–116. [CrossRef]
3. Haykin, S. Cognitive radio: Brain-empowered wireless communications. *IEEE J. Sel. Areas Commun.* **2005**, *2*, 201–220. [CrossRef]
4. O'Shea, T.J.; Corgan, J.; Clancy, T.C. Convolutional Radio Modulation Recognition Networks. In Proceedings of the International Conference on Engineering Applications of Neural Networks, Aberdeen, UK, 2–5 September 2016; pp. 213–226.
5. Wang, Y.; Liu, M.; Yang, J.; Gui, G. Data-Driven Deep Learning for Automatic Modulation Recognition in Cognitive Radios. *IEEE Trans. Veh. Technol.* **2019**, *68*, 4074–4077. [CrossRef]
6. Wu, H.C.; Saquib, M.; Yun, Z. Novel Automatic Modulation Classification Using Cumulant Features for Communications via Multipath Channels. *IEEE Trans. Wirel. Commun.* **2008**, *7*, 3098–3105.
7. Fu, J.; Zhao, C.; Li, B.; Peng, X. Deep learning based digital signal modulation recognition. In Proceedings of the 2016 IEEE International Conference on Electronic Information and Communication Technology (ICEICT), Hohhot, China, 20–22 August 2016; pp. 955–964.
8. Salt, J.E.; Nguyen, H.H. Performance prediction for energy detection of unknown signals. *IEEE Trans. Veh. Technol.* **2008**, *57*, 3900–3904. [CrossRef]
9. Tadaion, A.A.; Derakhtian, M.; Gazor, S.; Nayebi, M.M.; Aref, M.R. Signal activity detection of phase-shift keying signals. *IEEE Trans. Commun.* **2006**, *54*, 1439–1445. [CrossRef]
10. Lehtomaki, J.J.; Vartiainen, J.; Juntti, M.; Saarnisaari, H. Analysis of the LAD methods. *IEEE Signal. Process. Lett.* **2008**, *15*, 237–240. [CrossRef]
11. Lehtomaki, J.J.; Vartiainen, J.; Juntti, M.; Saarnisaari, H. CFAR outlier detection with forward methods. *IEEE Trans. Signal. Process.* **2007**, *55*, 4702–4706. [CrossRef]
12. Macleod, M.D. Nonlinear recursive smoothing filters and their use for noise floor estimation. *Electron. Lettc.* **1992**, *28*, 1952–1953. [CrossRef]
13. Salembier, P.; Liesegang, S.; Lopez-Martinez, C. Ship Detection in SAR Images Based on Maxtree Representation and Graph Signal Processing. *IEEE Trans. Geosci. Remote Sens.* **2019**, *57*, 2709–2724. [CrossRef]
14. Bao, D.; De Vito, L.; Rapuano, S. A histogram-based segmentation method for wideband spectrum sensing in cognitive radios. *IEEE Trans. Instrum. Meas.* **2013**, *62*, 1900–1908. [CrossRef]
15. Bao, D.; De Vito, L.; Rapuano, S. Spectrum segmentation for wideband sensing of radio signals. In Proceedings of the 2011 IEEE International Workshop on Measurements and Networking, Anacapri, Italy, 10–11 October 2011; pp. 47–52.
16. Koley, S.; Mirza, V.; Islam, S. Gradient-Based Real-Time Spectrum Sensing at Low SNR. *IEEE Commun. Lett.* **2015**, *19*, 391–394. [CrossRef]

17. Mallat, S.; Zhong, S. Characterization of Signals from Multiscale Edges. *IEEE Trans. Pattern. Anal. Mach. Intell.* **1992**, *14*, 710–732. [CrossRef]
18. Nandi, A.K.; Azzouz, E.E. Algorithms for automatic modulation recognition of communication signals. *IEEE Trans. Commun.* **1998**, *4*, 431–436. [CrossRef]
19. Xie, L.J.; Wan, Q.; Swami, C.A.; Sadler, B.M. Hierarchical digital modulation classification using cumulants. *IEEE Trans. Commun.* **2000**, *48*, 416–429.
20. Aslam, M.W.; Zhu, Z.; Nandi, A.K. Automatic modulation classification using combination of genetic programming and KNN. *IEEE Trans. Wirel. Commun.* **2012**, *11*, 2742–2750.
21. Xie, L.; Wan, Q. Cyclic Feature-Based Modulation Recognition Using Compressive Sensing. *IEEE Wirel. Commun. Lett.* **2017**, *6*, 402–405. [CrossRef]
22. Shermeh, A.E.; Ghazalian, R. Recognition of communication signal types using genetic algorithm and support vector machines based on the higher order statistics. *Digit. Signal. Process.* **2010**, *20*, 1748–1757. [CrossRef]
23. LeCun, Y.; Bengio, Y.; Hinton, G. Deep learning. *Nature* **2015**, *521*, 436–444. [CrossRef]
24. Hochreiter, S.; Schmidhuber, J. Long short-term memory. *Neural Comput.* **1997**, *9*, 1735–1780. [CrossRef] [PubMed]
25. Dolan, R.; DeSouza, G. GPU-based simulation of cellular neural networks for image processing. In Proceedings of the 2009 International Joint Conference on Neural Networks, Atlanta, GA, USA, 14–19 June 2009.
26. Jalil, B.; Leone, G.R.; Martinelli, M.; Moroni, D.; Pascali, M.A.; Berton, A. Fault Detection in Power Equipment via an Unmanned Aerial System Using Multi Modal Data. *Sensors* **2019**, *19*, 3014. [CrossRef] [PubMed]
27. Montavon, G.; Samek, W.; Müller, K.R. Methods for interpreting and understanding deep neural networks. *Digit. Signal. Process.* **2018**, *73*, 1–15. [CrossRef]
28. Reitter, D.; Moore, J.D. Alignment and task success in spoken dialogue. *J. Mem. Lang.* **2014**, *76*, 29–46. [CrossRef]
29. Yuan, Y.; Sun, Z.; Wei, Z.; Jia, K. DeepMorse: A Deep Convolutional Learning Method for Blind Morse Signal Detection in Wideband Wireless Spectrum. *IEEE Access* **2019**, *7*, 80577–80587. [CrossRef]
30. Ke, D.; Huang, Z.; Wang, X.; Li, X. Blind Detection Techniques for Non-Cooperative Communication Signals Based on Deep Learning. *IEEE Access* **2019**, *7*, 89218–89225. [CrossRef]
31. Mendis, G.J.; Wei, J.; Madanayake, A. Deep Learning based Radio-Signal Identification with Hardware Design. *IEEE Trans. Aerosp. Electron. Syst.* **2019**. [CrossRef]
32. Ali, A.; Yangyu, F. Unsupervised feature learning and automatic modulation classification using deep learning model. *Phys. Commun.* **2017**, *25*, 75–84. [CrossRef]
33. Meng, F.; Chen, P.; Wu, L.; Wang, X. Automatic Modulation Classification: A Deep Learning Enabled Approach. *IEEE Trans. Veh. Technol.* **2018**, *67*, 10760–10772. [CrossRef]
34. Zheng, S.; Qi, P.; Chen, S.; Yang, X. Fusion Methods for CNN-Based Automatic Modulation Classification. *IEEE Access.* **2019**, *7*, 66496–66504. [CrossRef]
35. Peng, S.L.; Jiang, H.Y.; Wang, H.X. Modulation Classification Based on Signal Constellation Diagrams and Deep Learning. *IEEE Trans. Neural. Netw. Learn. Syst.* **2019**, *3*, 718–727. [CrossRef] [PubMed]
36. Tang, B.; Tu, Y.; Zhang, Z.; Lin, Y. Digital Signal Modulation Classification with Data Augmentation Using Generative Adversarial Nets in Cognitive Radio Networks. *IEEE Access.* **2018**, *6*, 15713–15722. [CrossRef]
37. Morello, A.; Mignone, V. DVB-S2: The second generation standard for satellite broad-band services. *Proc. IEEE* **2006**, *94*, 210–227. [CrossRef]
38. Griffin, D.; Lim, J. Signal estimation from modified short-time Fourier transform. *IEEE Trans. Acoust.* **1984**, *2*, 236–243. [CrossRef]
39. Lecun, Y.; Bottou, L.; Bengio, Y.; Haffner, P. Gradient-based learning applied to document recognition. *Proc. IEEE* **1998**, *11*, 2278–2324. [CrossRef]
40. Redmon, J.; Divvala, S.; Girshick, R.; Farhadi, A. You Only Look Once: Unified, Real-Time Object Detection. *arXiv* **2015**, arXiv:1506.02640.
41. Redmon, J.; Farhadi, A. YOLO9000: Better, faster, stronger. *arXiv* **2016**, arXiv:1612.08242.
42. Redmon, J.; Farhadi, A. YOLOv3: An Incremental Improvement. *arXiv* **2018**, arXiv:1804.02767.
43. Liu, W.; Anguelov, D.; Erhan, D.; Szegedy, C.; Reed, S.; Fu, C.Y.; Berg, A.C. SSD: Single Shot MultiBox Detector. In Proceedings of the 14th European Conference ECCV 2016, Amsterdam, The Netherlands, 11–14 October 2016; pp. 21–37.

44. Fu, C.-Y.; Liu, W.; Ranga, A.; Tyagi, A.; Berg, A.C. DSSD: Deconvolutional Single Shot Detector. *arXiv* **2017**, arXiv:1701.06659.
45. Girshick, R.; Donahue, J.; Darrell, T.; Malik, J. Rich feature hierarchies for accurate object detection and semantic segmentation. In Proceedings of the 2014 IEEE Conference on Computer Vision and Pattern Recognition, Columbus, OH, USA, 23–28 June 2014; pp. 580–587.
46. Girshick, R. Fast R-CNN. In Proceedings of the IEEE 2015 International Conference on Computer Vision, Santiago, Chile, 7–13 December 2015; pp. 1440–1448.
47. Ren, S.; He, K.; Girshick, R.; Sun, J. Faster R-CNN: Towards Real-Time Object Detection with Region Proposal Networks. *IEEE Trans. Pattern Anal. Mach. Intell.* **2017**, *39*, 1137–1149. [CrossRef]
48. Neubeck, A.; Van Gool, L. Efficient non-maximum suppression. In Proceedings of the 18th International Conference on Pattern Recognition (ICPR'06), Hong Kong, China, 20–24 August 2006; pp. 850–855.
49. He, K.; Zhang, X.; Ren, S.; Sun, J. Deep residual learning for image recognition. In Proceedings of the IEEE International Conference on Computer Vision and Pattern Recognition (CVPR), Las Vegas, NV, USA, 26 June–1 July 2016; pp. 770–778.
50. Ioffe, S.; Szegedy, C. Batch normalization: Accelerating deep network training by reducing internal covariate shift. In Proceedings of the 32nd International Conference on International Conference on Machine Learning, Lille, France, 6–11 July 2015; pp. 448–456.
51. Kingma, D.P.; Ba, J. Adam: A method for stochastic optimization. *arXiv* **2014**, arXiv:1412.6980.
52. Morelli, M.; D'Andrea, A.N.; Mengali, U. Feedforward ML-based timing estimation with PSK signals. *IEEE Commun. Lett.* **1997**, *1*, 80–82. [CrossRef]

Article

A Neuron-Based Kalman Filter with Nonlinear Autoregressive Model

Yu-ting Bai [1,2], Xiao-yi Wang [1,2,*], Xue-bo Jin [1,2,*], Zhi-yao Zhao [1,2] and Bai-hai Zhang [3]

1 School of Computer and Information Engineering, Beijing Technology and Business University, Beijing 100048, China; baiyuting@btbu.edu.cn (Y.-t.B.); zhaozy@btbu.edu.cn (Z.-y.Z.)
2 Beijing Key Laboratory of Big Data Technology for Food Safety, Beijing Technology and Business University, Beijing 100048, China
3 School of Automation, Beijing Institute of Technology, Beijing 100811, China; smczhang@bit.edu.cn
* Correspondence: wangxy@btbu.edu.cn (X.-y.W.); jinxuebo@btbu.edu.cn (X.-b.J.)

Received: 3 December 2019; Accepted: 2 January 2020; Published: 5 January 2020

Abstract: The control effect of various intelligent terminals is affected by the data sensing precision. The filtering method has been the typical soft computing method used to promote the sensing level. Due to the difficult recognition of the practical system and the empirical parameter estimation in the traditional Kalman filter, a neuron-based Kalman filter was proposed in the paper. Firstly, the framework of the improved Kalman filter was designed, in which the neuro units were introduced. Secondly, the functions of the neuro units were excavated with the nonlinear autoregressive model. The neuro units optimized the filtering process to reduce the effect of the unpractical system model and hypothetical parameters. Thirdly, the adaptive filtering algorithm was proposed based on the new Kalman filter. Finally, the filter was verified with the simulation signals and practical measurements. The results proved that the filter was effective in noise elimination within the soft computing solution.

Keywords: kalman filter; nonlinear autoregressive; neural network; noise filtering

---

1. Introduction

In the typical control systems, the measurement is the primary component that senses the system status and provides the input information. The measurement accuracy influences the control effect directly. Especially in intelligent terminals, such as industrial robots, unmanned aerial vehicles and unmanned vehicles, various sensors are implemented to measure the motion state and working condition. The sensors are expected to be of high precision. However, the precision is limited by the manufacturing technique when the intelligent terminals are with small size and low cost relying on the micro-electromechanical sensors [1,2]. For the complicated noises, it is essential to filter the noises. Then the motion state and condition can be estimated accurately for the target tracking and control.

In the noise filtering and state estimation field, many methods have been studied, such as the wavelet filter [3], time-frequency peak filtering [4], empirical mode decomposition [5] and the Kalman filter [6]. These filtering and estimation algorithms are often based on the mathematical models and established using the iterative schemes [7–9] or recursive schemes [10,11]. Some filtering-based estimation algorithms use input-output representations [12,13], and others use state-space models [14–16]. Among these methods, Kalman filter is a state estimation algorithm based on the state-space model. It introduces the state space into the stochastic estimation theory and obtains the optimal estimation without requiring a vast amount of historical data. However, it is obvious that the Kalman filter depends highly on some assumptions that the system model is linear, the process and measurement noises are standard Gaussian, and their covariance matrixes are all known. When these assumptions are seriously limited in reality, two categories of methods have been explored. On the one hand, the adaptive Kalman filter (AKF) was proposed focusing on the parameter adjustment to approximate the filtering process to the practical system. Then common AKF includes innovation-based adaptive estimation (IAE) [17], multiple model adaptive estimation (MMAE) [18] and adaptive fading Kalman filter (AFKF) [19]. On the other hand, some methods focus on nonlinear systems, such as extended Kalman filter (EKF) [20], unscented Kalman filter (UKF) [21], noise-robust filter [22] and other estimation methods [23,24]. The two categories of the methods try to describe and represent the system features with the variable approximation. The filters will be efficient if the alternative expression of the model and parameter is similar to the system dynamic characteristic.

The methods above mainly extract and represent the system characteristics with the existing information. While the characteristic extraction is the specialty of machine learning, the artificial neural network (ANN) has been introduced in noise filtering and state estimation. The leading solution is the distributed mode in which Kalman filter and ANN are applied separately in sequential order [25–27]. In the mode, the neural network mainly preprocesses or reprocesses the data before or after the filtering process. However, the inner relation in the Kalman filter has not been explored deeply with ANN. Then it becomes an issue on how to extract the relationship of parameters in the Kalman filter and optimize the filtering results with the limited existing information.

Because of the advantages in Kalman filter and the neural network, a new neuron-based Kalman filter is built in this paper. It mainly enhances the filtering process with the existing information. The potential numerical relation of the intermediate variables in the Kalman filter is explored with the feature extraction and nonlinear fitting ability of the neural network. In the paper, neurocomputing is integrated with the inner components of the Kalman filter. The nonlinear autoregressive model is introduced and constructed to predict and modify the critical intermediate variables in the Kalman filter. The simulation and practical experiments have verified the precision and feasibility of the proposed filter.

This paper is organized as follows: Section 2 introduces the underlying theory and related works on noise filtering. Section 3 presents the main proposed filter with the framework and network design. The simulation and experiment are designed and conducted in Section 4. The results and work are discussed in Sections 5 and 6 finally concludes the paper.

2. Related Work

As the typical filtering method, the Kalman filter is selected as the basic framework in this paper. The basic theory and developments of the Kalman filter are introduced firstly. Then the related work is presented on the integration of the filter and neural networks.

2.1. Kalman Filter and Its Improvement

Because of its clearness and convenience in computer calculation, the Kalman filter has been the classical method in the filtering and estimation of Gaussian stochastic systems [28,29]. It is applied widely in target tracking [30], integrated navigation [31], communication signal processing [32], etc. Kalman filter introduces the state space description in the time domain, in which the estimated signal is set as the output of the stochastic linear system in the action of white noise. Kalman filter is appropriate for the stationary process and the non-stationary Markov sequence.

For the detailed analysis in the paper, the main algorithm of the Kalman filter is presented here. The discrete model can be expressed as:

$$x(k+1) = A(k)x(k) + w(k) \tag{1}$$

$$z(k) = C(k)x(k) + v(k) \tag{2}$$

where $x(k)$ is the to-be-estimated variable or state variable, $z(k)$ is the measurement value from sensors, A is the state transition matrix or process matrix, C is the measurement matrix, $w(k)$ is the process noise, $v(k)$ is the measurement noise. The concrete Kalman filter algorithm is shown as follows:

(1) State estimation updating:

$$\hat{x}(k|k) = \hat{x}(k|k-1) + K(k)[z(k) - C(k)\hat{x}(k|k-1)] \quad (3)$$

(2) One step forward prediction:

$$\hat{x}(k|k-1) = A(k-1)\hat{x}(k-1|k-1) \quad (4)$$

(3) Filtering gain calculation:

$$K(k) = P(k|k-1)C^T(k)[C^T(k)P(k|k-1)C(k) + R(k)] \quad (5)$$

(4) The variance of the state estimation calculation:

$$P(k|k-1) = A(k-1)P(k-1|k-1)A^T(k-1) + Q(k-1) \quad (6)$$

$$P(k|k) = [I - K(k)C(k)]P(k|k-1) \quad (7)$$

where $\hat{x}(k|k)$ is the posterior estimation, $\hat{x}(k|k-1)$ is the prior estimation which is also called the prediction, K is the filtering gain, P is the variance of the state estimation, Q is the variance of the process noise, R is the variance of the measurement noise.

There are assumptions in Kalman filter, namely that the process and measurement noises are standard Gaussian noises, and their covariance matrixes are all known. The assumptions deviate from the real systems. Then many studies have been carried out to improve Kalman filter from different solutions.

Some improvements were proposed for the nonlinear system, and the typical methods include EKF [20] and UKF [21]. In EKF, the Taylor expansion of the nonlinear function is truncated with the first-order linearization, and other higher-order terms are ignored. Then the nonlinear problem can be transformed into the linearity, which is suitable for the Kalman filter. In UKF, the prediction measurement values are represented with the sampling points, and the unscented transformation is used to deal with the nonlinear transfer of mean and covariance. EKF and UKF have been improved, as well as the integration with other methods [33–35].

Aside from nonlinear system methods, AKF methods have been studied to solve problems where mainly the settled and experiential parameters are given. The representative IAE [17], MMAE [18], and AFKF [19] are proposed based on the thought that the model parameter and noise statistics are modified with the observation and judgment during the filtering process. From a literature search, it was seen that some improvements in AKF [36–38] were presented recently. In the latest work [38], the colored noise is analyzed with the adaptive parameter. The second-order adaptive statistical model and Yule-Walker algorithm are used to recognize and filter the noises. The work is one of the latest representative improvements of AKF, and it can be set as a contrast in the experimental research.

The two categories of methods above, nonlinear and adaptive filters, mainly improve the filtering performance from the approximate system modeling and parameter adjustment. They are conducted based on inherent mathematics and statistic derivation. They provide an effective solution to promote the Kalman filter in the system mechanism analysis idea. The filtering and prediction are based on the mathematical models by assuming that the model parameters are known or estimated using some parameter identification methods, including the iterative algorithms [39–41], the particle-based algorithms [42–44] and the recursive algorithms [45–48].

2.2. Filter with Neural Network

The methods in Section 2.1 improve the Kalman filter by modifying the system model and parameters based on the mathematic mechanism. The idea can be carried out with another data-driven solution. For the filtering parameter adjustment, the core task is to find and express the relation between parameters and process data, which meets the ability of neural networks. ANN has caused great concerns again with the trends of deep learning and artificial intelligence. ANN can fit the nonlinear model with excellent performance. It can solve the nonlinear and time-varying problems without a concrete internal mechanism model. For the problematic modeling of process and noise in the Kalman filter, ANN can be considered as a helpful tool to reconstitute the unknown elements in the filter. Scholars have made some efforts to explore the integrations of ANN and Kalman filter. The related research can be divided into two categories, including the distributed and crossed integration.

2.2.1. Distributed Integration of Kalman filter and ANN

For the distributed integration, Kalman filter and ANN are applied separately in sequential order. Liu et al. [25] smoothed the measurement value with the Kalman filter, and the filtered results were set as the input of the backpropagation neural network (BPNN). Hu et al. [49] estimated the target location with Kalman filter and the estimation was imported into BPNN to classify the targets. Liu et al. [50] utilized Kalman filter and fuzzy neural network (FNN) in a multi-source data fusion framework of an adaptive control system, in which data was processed firstly with Kalman filter, and the filtered results were set as the input of FNN. Others [26,27,51] used a Kalman filter and ANN in reverse order, in which ANN is constructed before the Kalman filter. Leandro et al. [26,27] built up BPNN to predict a variable, which is an important state variable in the Kalman filter. Cui et al. [51] proposed a radial basis function neural network (RBF) to train the GPS signals, and the RBF output is the input of adaptive Kalman filter, aiming at improving the processing precision.

2.2.2. Crossed Integration of Kalman Filter and ANN

Different from the methods above, Kalman filter and ANN are combined in a tight pattern, in which ANN is applied during the internal filtering procedure. Shang et al. [52] predicted the model error in the filter with FNN, and the error level was considered to confirm the measurement noise covariance, which was set as 0 or infinity. Li et al. [53] thought that the gain of EKF was usually modified with the erroneous measurement, which reduced the gain precision. They used BPNN to train the gain with the input of measurement, estimation, and error, and then the precision was increased. Deep neural networks [54] have been studied recently. Pei et al. [55] combined a deep neural network with the Kalman filter in the emotion recognition of the image and audio. The features extracted by the deep neural network were input into the switching Kalman filter to obtain the final estimation results.

In the research reported in the literature, more works are conducted in the first separately distributed mode, in which Kalman filter and ANN process the data respectively. The mode does not adjust the inner parameters of the Kalman filter. The works of tightly crossed integration are relatively few. They can be improved with the relation exploration in filter parameters with ANN. Besides, the category and structure of the neural network can be studied to meet the demand for filtering calculation procedures.

3. Neuron-Based Kalman Filter

3.1. *Framework of Neuron-Based Kalman Filter*

Kalman filter provides a feasible framework to filter the noises and estimate the system state. The components of the Kalman filter can be divided into two categories, namely the models and intermediate variables. The models describe the system dynamic and the measurement process, including the system process equation and measurement equation. The intermediate variables influence the filtering results seriously, but they are difficult to determine in practice. In this paper, the effect of the intermediate variables on the filtering is explored with neural networks. The neurons are integrated into the Kalman filter, and the neurons can help to optimize the filtering process with the limited existing information.

The main influence factors of Kalman filter include the process equation, process noise, and measurement noise, expressed as the matrix A, Q and R. Notably, the noise variances are the critical intermediate variables that affect the estimation results. The effect of noise variances is expressed in the filtering gain K, and the filtering gain determines the estimation result as an important weight. In the optimizing thought with the neurons, the influence relation of the filtering results and the variables should be explored. Then the framework of the integrated Kalman filter is designed firstly, shown in Figure 1, and the concrete design ideas will be interpreted later.

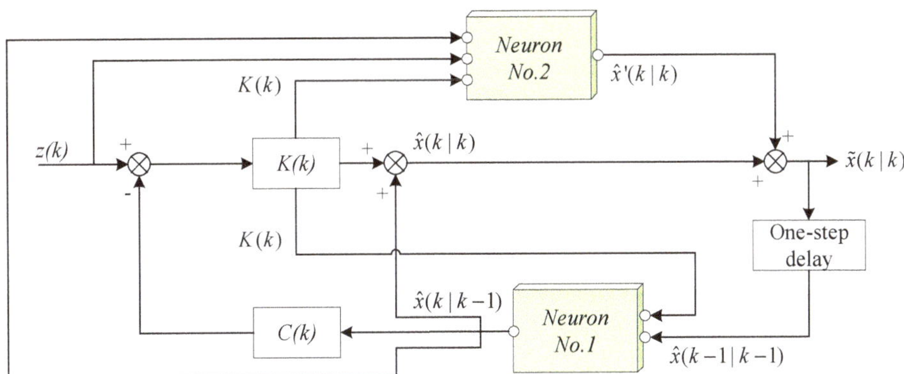

Figure 1. Framework structure of the neuron-based Kalman filter.

Considering the prediction process of the Kalman filter in Equation (4), the estimation result is affected by the precision of the process equation. The process equation describes the system change along the time, but the common simplified equation is difficult to model the actual system. In the view of data-driven, the core of the process equation is the time series relation of the system change. And the time series can be modeled well with neural computing. Then a neuro unit is introduced to model the process in a black-box thought. The first neuro unit can be expressed as:

$$\hat{x}(k) = f_1(K(k), K(k-1), K(k-2), \cdots, \hat{x}(k-1), \hat{x}(k-2), \cdots) \tag{8}$$

The inputs in the first neuro unit are the filtering gain K series and the state estimation value $\hat{x}$ at the previous time points, represented with $k-1, k-2, \cdots$. The output is the estimation $\hat{x}$ at the time k. The key to the model is the fitting function f_1 which aims at the excavation of the time series features in the estimated variables and intermediate filtering variables. As a vital variable, the filtering gain K is obtained from the variance of process and measurement noise Q and R, and it can represent the noise features to some extent. Therefore, the filtering gain is set as an input to transmit the noise features to the system process model. The output estimation value is related to the state at the previous time and the filtering gain. Then the new estimation can be regarded as a more accurate predictive value of the system, and it replaces the initial prediction value in Equation (4) to continue the computing process of the Kalman filter.

For neural computing, it needs training with the existing data. In the training of the first neuro unit, the filtering process data and the final estimation result are collected as the training set. In detail, the series data of the filtering gain from time step $(k-m)$ to k and the estimation value from $(k-m)$ to $(k-1)$ are set as the training input data, where m is the prediction length set in the neural computing. The estimation value at k is set as the training output. In fact, the previous one-step prediction value is optimized with the final estimation value in the network. The fitting function f_1 can be obtained with the training data and the learning algorithm which will be discussed in Section 3.2.

Considering the final estimation process of Kalman filter in Equation (3), the estimation result is determined by two parts, of which one is the prediction value, and the other one is the measurement residual error. The second neuro unit is built to discover the mapping relation between the two parts and the final estimation result, expressed as:

$$\hat{x}'(k) = f_2(K(k), K(k-1), \cdots, z(k), z(k-1), \cdots, \hat{x}(k), \hat{x}(k-1), \cdots) \tag{9}$$

The inputs of the second neuro unit are the measurement value z, the prediction value $\hat{x}$ and the filtering gain K at the previous time points, represented with $k-1, k-2, \cdots$ The output is the final estimation value $\hat{x}'$ at the time k. The final estimation synthesizes the measurement and the filtering intermediate variables such as the noise variance matrix (reflected by the filtering gain) and the prediction variables. The function f_2 is trained to realize the synthetization. In the training, the values of K, z, $\hat{x}$ from $(k-m)$ to k are set as the input, and the final estimation value $\hat{x}'$ at k is set as the output.

It can considered that the estimation via the neuro units is an effective supplement and amendment of the estimation in Kalman filter. Then the new final estimation value can be obtained by synthesizing the two estimation values from the neural computing and Kalman filter:

$$\hat{x}(k|k) = (1-\alpha)\hat{x}(k|k) + \alpha\hat{x}'(k|k) \tag{10}$$

where $\hat{x}(k|k)$ is from Kalman filter, $\hat{x}'(k|k)$ is from the neuro unit. $(1-\alpha)$ and α are the weights of the two estimation values. α is determined by the validation error of the neuro unit, and:

$$\alpha = \frac{n}{\sum_{i=1}^{n} \left| (d_i^v - d_i)/d_i \right|} \tag{11}$$

where d is the validation set during the neuro unit training, d^v is the output of the neuro unit for the validation set, n is the number of data in the validation set.

3.2. Neuro Units Based on Nonlinear Autoregressive Model

In the framework of the neuron-based Kalman filter, the critical components are the neuro units which analyze the intermediate variables to support the final filtering result. Referring to the demand analysis of the two units, the two functions in Equations (8) and (9) should be able to fit the nonlinear relation in multiple variables. Moreover, they should excavate the time-series features in the data. With the two aspects of the demands, the nonlinear autoregressive model with exogenous input (NARX) can be the appropriate solution [56,57]. NARX derives from the time series autoregressive analysis, and it is effective in the reconstitution of the nonlinear systems. The availability of NARX has been proved by various applications [58–60].

NARX belongs to the recurrent neural network. It has a learning efficiency with the better gradient descent. The nonlinear relation between the inputs and outputs in NARX can be expressed as follows:

$$y(t+1) = \phi(y(t), y(t-1), \cdots, y(t-n), u(t+1), u(t), \cdots, u(t-m)) \tag{12}$$

where $y(t+1)$ is the output to be predicted, $y(t)$ to $y(t-n)$ are the historical outputs, $u(t+1)$ to $u(t-m)$ are the related inputs which last to the current moment. ϕ represents the nonlinear relation between inputs and outputs, and it also represents the structure of NARX, shown in Figure 2.

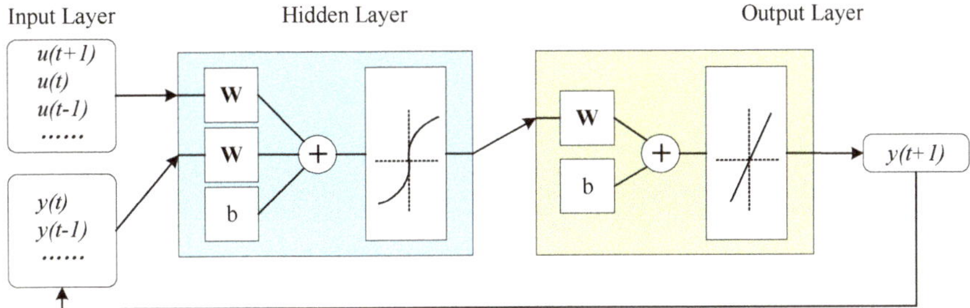

Figure 2. Structure of NARX.

NARX consists of the input layer, hidden layer, and output layer. The transfer function of NARX is similar to the backpropagation neural network, and a one-hidden-layer network is shown as follows:

$$f(\cdot) = g(\sum \omega_h h(\cdot)) \tag{13}$$

$$h(\cdot) = r(\sum \omega_i u_i) \tag{14}$$

where g and r are the activation functions of the output, ω_h is the node weight in the hidden layer, $h(\cdot)$ is the activation function of the hidden layer, ω_i is the weight of all inputs. The vector form of ω_h and ω_i are represented with W in Figure 2.

Based on the primary NARX structure, the two neuro units in the neuron-based Kalman filter can be designed concretely, which are shown in Figure 3.

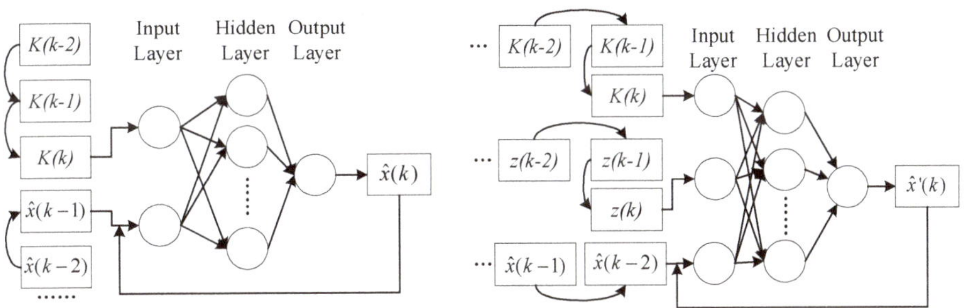

(**a**) First unit to predict the state variable (**b**) Second unit to estimate the final state

Figure 3. Concrete structures of the two neuro units in the proposed Kalman filter.

For the neuron units in Figure 3, the concrete scale can be determined with the traditional empirical mode in the shallow neural network. The number of the hidden layer is set as one according to the number of input and output variables. The number of hidden nodes can be determined with the equations, such as $n = \log_2 p$, $n = \sqrt{p+q} + a$, where n is the number of hidden nodes, p is the number of input nodes, q is the number of output nodes, and a is a constant between 1 and 10. Besides, the number of hidden nodes can be adjusted in the network training, following the network performance.

Based on the static construction of the neural network, the appropriate training method should be selected to obtain favorable dynamic performance. The gradient descent method is the core solution in neural network training. Some improved algorithms have been proposed. Levenberg-

Marquardt (L–M) [61,62] is a rapid training method that combines the basic gradient descent method and Gauss-Newton method. Its error target function is:

$$E(w) = \frac{1}{2}\sum_{i=1}^{p} ||Y_i - Y_i'||^2 = \frac{1}{2}\sum_{i=1}^{p} e_i^2(w) \quad (15)$$

where Y_i is the expected output, Y_i' is the actual output, $e_i(w)$ is the error, p is the number of samples, w is the vector consisting of network weights and threshold values.

The k-th iterative vector of weights and threshold values is w^k, and the new vector is:

$$w^{k+1} = w^k + \Delta w \quad (16)$$

and the increment in L–M is calculated as:

$$\Delta w = \left[J^T(w)J(w) + \mu I\right]^{-1} J^T(w)e(w) \quad (17)$$

where I is the unit matrix, μ is the learning rate, $J(w)$ is the Jacobian matrix, and:

$$J(w) = \begin{bmatrix} \frac{\partial e_1(w)}{\partial w_1} & \frac{\partial e_1(w)}{\partial w_1} & \cdots & \frac{\partial e_1(w)}{\partial w_1} \\ \frac{\partial e_1(w)}{\partial w_1} & \frac{\partial e_1(w)}{\partial w_1} & \cdots & \frac{\partial e_1(w)}{\partial w_1} \\ \vdots & \vdots & \ddots & \vdots \\ \frac{\partial e_1(w)}{\partial w_1} & \frac{\partial e_1(w)}{\partial w_1} & \cdots & \frac{\partial e_1(w)}{\partial w_1} \end{bmatrix} \quad (18)$$

With the training method, the neuro units can be built with the anticipative functions. Then the improved Kalman filter can be established with the functional neuro units.

3.3. Adaptive Filtering Algorithm

In the framework of neuron-based Kalman filter, two neuron units are introduced into the basic consistent of Kalman filter. The input, output, and inner structure of the neuro unit are designed to improve the filtering. Finally, the adaptive filtering algorithm based on the improved Kalman filter is proposed here, in which the neuro units are trained firstly and applied to the filter. The flow of the algorithm is presented in Figure 4.

As shown in Figure 4, the algorithm consists of two parts, namely the training process on the left and the filtering process on the right. The concreter flow of the algorithm is as follows:

(A) Training process

(1) The system and measurement equations are established according to the object. The parameters in the Kalman filter can be initialized with empirical values.

(2) The primary calculation of the Kalman filter is conducted iteratively following Equations (1)–(7). The measurement vectors are imported into the filter along with time. The intermediate and final values are recorded, including the one-step prediction value, Kalman gain, measurement, estimation result, etc. The recorded values are all labeled with a time stamp. Meanwhile, the iterative steps should be no less than about 150 for the following neuro unit training. The number of sample steps may be adjusted according to the complexity of signals.

(3) With the filtering values in step 2, they are marked with the step number to form the time series sets. Then the prediction value and filter gain are imported into the first neuro unit. The prediction value, filter gain, and measurement are imported into the second neuro unit. The estimation result is set as the reference output of the two units.

(4) The neuro units are trained with the learning method L–M in Section 3.2. The trained neuro units are obtained when the preset iteration conditions are met, including the numbers of iteration or the convergence error.

(B) Filtering process with trained neuro units

(5) Based on the model equations and the initialized parameters in Kalman filter, the initial variable and filter gain are imported into the first neuro unit, and the prediction value is outputted and set as the basis of prediction error.

(6) The filter gain is updated and used to calculate the estimation value with the measurement. Meanwhile, the prediction value, filter gain, and measurement are imported into the second neuro unit to obtain another estimation value.

(7) The two estimation values are fused following Equation (10).

(8) The step moves forward to conduct steps (5)–(7) iteratively. In the iteration, the measurement vectors are calculated along with time.

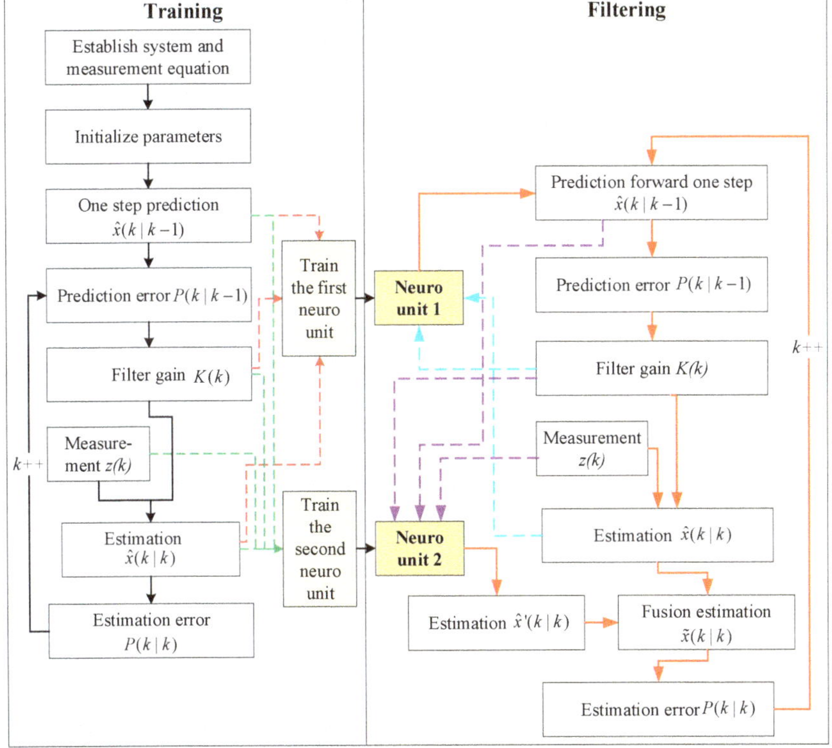

Figure 4. Algorithm flow of the adaptive filtering with the neuron-based Kalman filter.

4. Experiment and Result

The simulation and practice experiments are conducted to test the filter proposed. In the simulation, different noises are generated to simulate the complex noises in the sensors. In the experiment, the wheeled robot path is measured with low-cost GPS. All the computing runs with MATLAB 2017a on a PC equipped with an Intel Core i5-6200U CPU@2.30 GHz and 8 GB RAM. The experiment setting and results are presented in this section.

4.1. Simulation and Result

The common noises in sensors are the white noise and color noise. The signals with the two kinds of noises are generated in the simulation. The two sets of the signals are:

$$x_1(k) = \sigma(k)f(k) \tag{19}$$

$$x_2(k) = G(z^{-1})f(k) \tag{20}$$

where $f(k)$ is the Gaussian white noise. $\sigma(k)$ is the standard deviation of $f(k)$, and $\sigma(k) = (L+k)/L$, L is the number of signal samples, and k is the sample number. $G(z^{-1})$ is the transfer function of a system which can be second order or third order to simulate the noise change process.

The first set is the approximately linear noises, and the second one is the sinusoidal noises, corresponding to the white noise and color noise, respectively. In the simulation, the sampling interval is 0.02 s. The numbers of signal samples are all 2000. That is, the sampling time is the 40 s. The simulation signals are shown in Figure 5, in which the true values of x_1 and x_2 are 0.

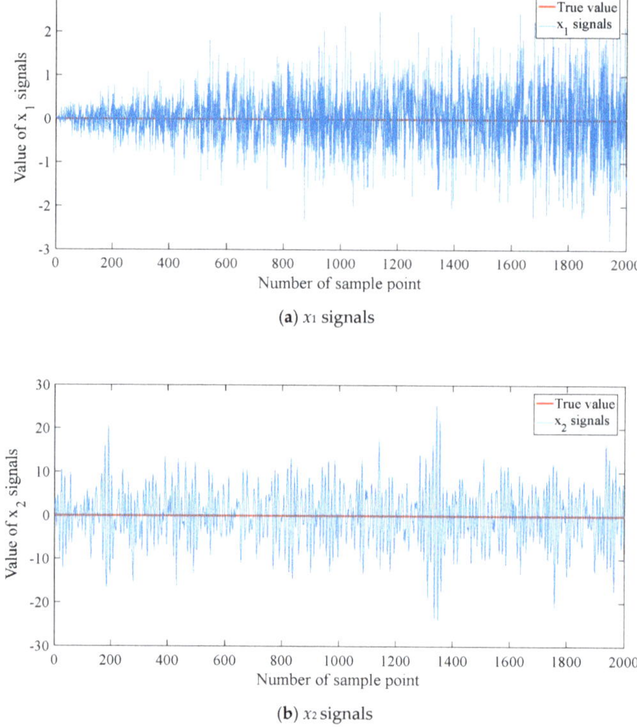

(**a**) x_1 signals

(**b**) x_2 signals

Figure 5. Simulation signals with different noises.

In the filtering of the simulation signals, the system model is established with the classical Jerk model, which also can be replaced with other motion models such as constant velocity, constant acceleration, Singer, interacting multiple model algorithms, and so on. For the Kalman filter, the initial state estimate x_0 and covariance P_0 are assumed to be x_0 = $[0\ 0\ 0]^T$ and P_0 = 1000*$eye(4)$.

Because the neuro unit needs the training, the first 70% of the data are set as the training data, and the rest is used to test the filtering result. In the setting of the two neuro units, the number of hidden nodes are set as 3 and 6, respectively. Other settings are also tested to obtain the optimal performance. The training results of NARX are shown in Figure 6.

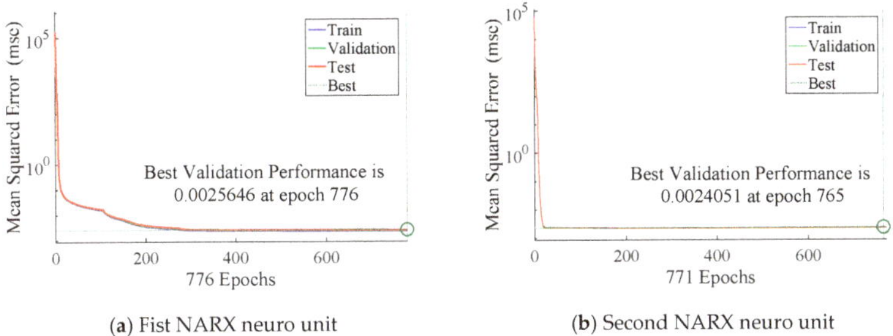

(a) Fist NARX neuro unit (b) Second NARX neuro unit

Figure 6. Training results of NARX in integrated filter.

Based on the trained neuro units, the data are imported into the proposed filter to estimate the variable values. For verifying the estimation performance, the traditional Kalman filter is set as a contrast, abbreviated as KF. Moreover, the proposed filter can be regarded as a solution to adaptive filtering. Then one of the latest improvements of AKF in [38] is also set as the contrast, abbreviated as IAKF. The proposed filter in this paper is abbreviated as NKF. The filtering results are shown in Figure 7. For the quantitative evaluation of errors, the mean absolute error (MAE) and root-mean-square error (RMSE) are calculated and listed in Table 1.

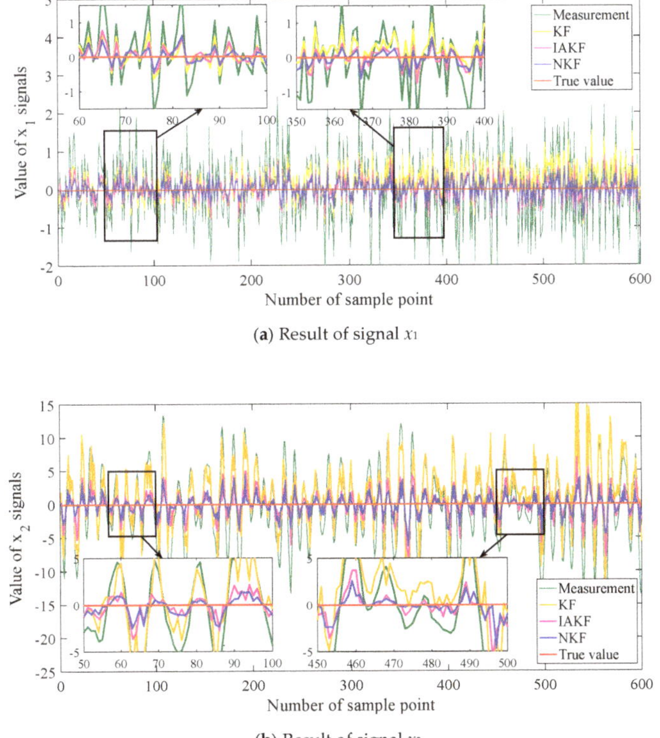

(a) Result of signal x_1

(b) Result of signal x_2

Figure 7. Filtering results of simulation signals.

Table 1. Evaluation of filtering errors.

		KF	**IAKF**	**NKF**
Signal x_1	MAE	0.3692	0.2550	0.2004
	RMSE	0.4577	0.3170	0.2507
Signal x_2	MAE	3.3379	1.2678	1.0294
	RMSE	4.4763	1.8295	1.4429

For the first set of signals which are of the white noise, the results of the three methods are relatively similar in the curve graph. NKF performs better slightly than the others. The trend is evident in the error evaluation criteria. MAE represents the mean level of errors. MAE of NKF declines 45.72% of KF and 21.41% of IAKF. RMSE shows the fluctuation degree of errors. RMSE of NKF decreases 45.23% of KF and 20.91% of IAKF.

For the second set, the filtering results show the distinguishable trends. The results of KF fluctuate sharply and become diverging in the latter period. IAKF and NKF can trace the signals more closely, and NKF is more effective in the intuitionistic graph. MAE of NKF has been reduced by 69.16% of KF, and 18.80% of IAKF. The decreasing percentage of RMSE reaches 67.77% and 21.13% for NKF to KF and IAKF. The error reduction of the second set is larger than the first set.

4.2. Practical Experiment and Result

Except for the simulation, a practical experiment is also conducted to verify the proposed method. A trajectory of the wheeled robot (shown in Figure 8) is measured on the playground, and the presupposed trajectory is presented in Figure 9. The robot started from the top right corner and ended at the same point. A low-cost GPS receiver is used to obtain the location information, including the longitude and latitude. The relative coordinates are transformed from the longitude and latitude:

$$d = 111.12 \cdot \cos \frac{1}{\sin\phi_{t-1}\sin\phi_t + \cos\phi_{t-1}\cos\phi_t(\lambda_t - \lambda_{t-1})} \tag{21}$$

where d is the displacement, ϕ is the latitude and λ is the longitude. The displacement can be decomposed into the coordinates on a plane. The measurements and true trajectory in the relative coordinates are shown in Figure 10.

Figure 8. Wheeled robot used to measure the trajectory, and the robot is developed and assembled by laboratory of system engineering in Beijing Institute of Technology, Beijing, China.

Figure 9. Presupposed trajectory in the practical experiment.

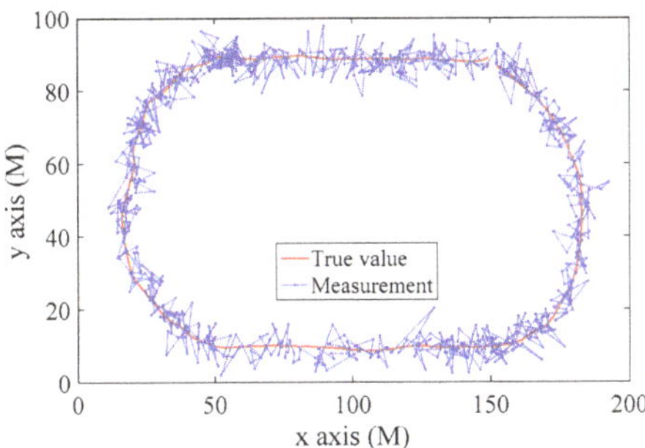

Figure 10. Relative coordinates transformed from the practical trajectory measurements.

In this part, the data of the whole trajectory is filtered firstly. Then a segment of the trajectory in another measurement is tested again.

4.2.1. Result of the Whole Trajectory

Similar to the simulation, the traditional Kalman filter and improvement of AKF in [38] are set as the contrast methods. The filtering results are shown in Figure 11, including the distance in the x-axis, y-axis, and x-y plane. The absolute errors are shown in Figure 12, and the evaluation indexes are in Table 2.

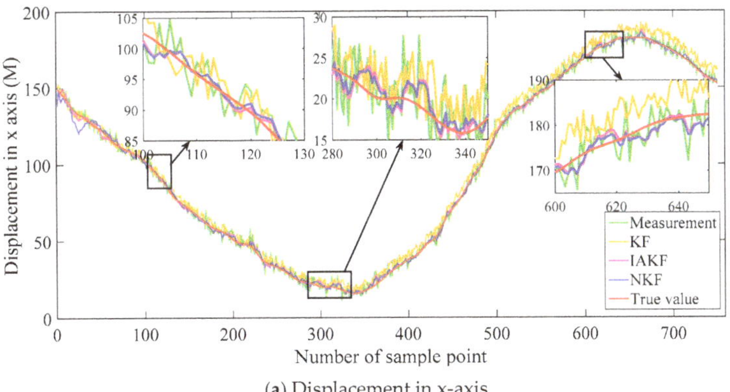

(**a**) Displacement in x-axis

Figure 11. *Cont.*

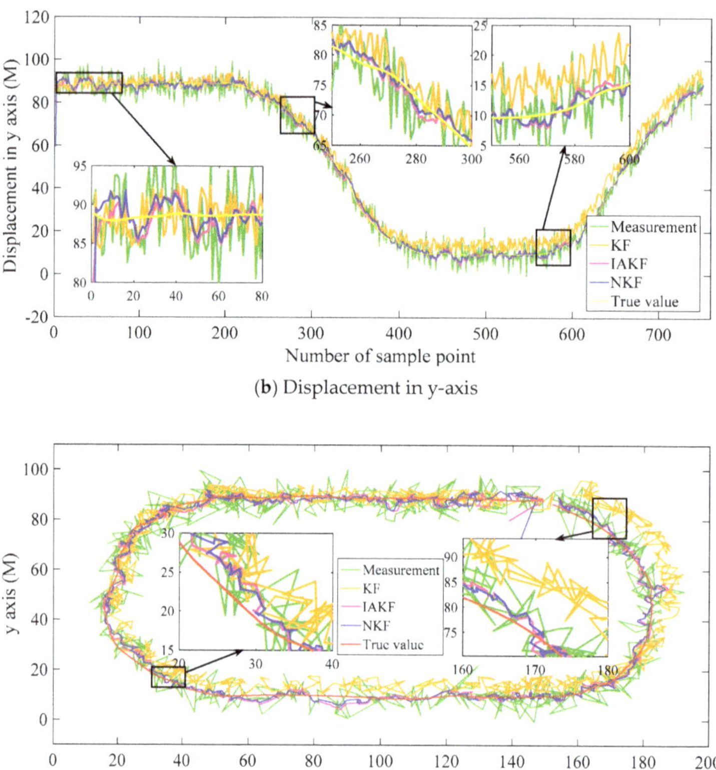

(**b**) Displacement in y-axis

(**c**) Whole trajectory in the x-y plane

Figure 11. Filtering results of the whole trajectory.

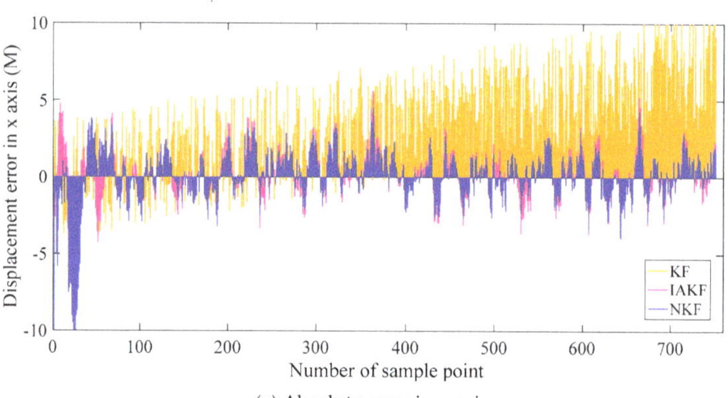

(**a**) Absolute error in x-axis

Figure 12. *Cont.*

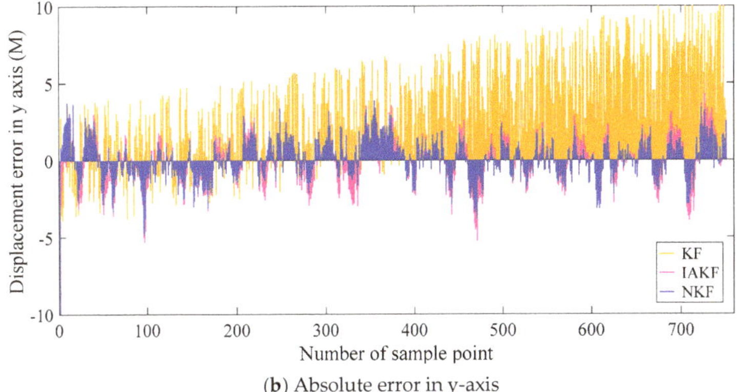

(**b**) Absolute error in y-axis

Figure 12. Absolute errors of displacement in x and y axes.

The results in Figure 11 show intuitively the small differences in the filtering results, and it is because the difference is in the order of magnitudes less than 1 m. The differences are more evident in the absolute errors in Figure 12. The general trends of the filtering results with the three methods are similar to the pattern of the second set signal in the simulation. It is because that the noises in the real sensors mainly are the color noises instead of the white noises. The basic Kalman filter performs badly in the practical system. Its results are unsteady and diverging along with the time. NKF perform better than others in various time periods besides the beginning. The neuro unit based on NARX reaches stable performance after the drastic fluctuation, which usually occurs at the beginning. Therefore, NKF is superior to the Kalman filter on the whole without the initial period. The performance can be analyzed quantitatively with the error criteria in Table 2.

Table 2. Evaluation of filtering errors (whole trajectory).

		KF	IAKF	NKF
x axis	MAE	3.8730	1.3117	1.3048
	RMSE	4.6732	1.9079	1.6594
y axis	MAE	3.7327	1.3184	1.1651
	RMSE	4.5560	1.7578	1.6430

The general trend is consistent in the x-axis and y-axis for the three indexes. There is a conspicuous promotion in MAE for NKF and the RMSE of NKF declines about 6% with the Kalman filter and 8% with IAKF.

4.2.2. Result of Segment in the Trajectory

To test the proposed filter with more data, a segment of the whole trajectory is selected from one of the multiple measurements, which is not in the same measurement with the whole trajectory above.

The data from 180 s to 450 s are selected, including the displacements in x-axis and y-axis. The contrast methods are the same as the experiment above. The filtering results are shown in Figure 13, and the evaluation indexes of errors are listed in Table 3.

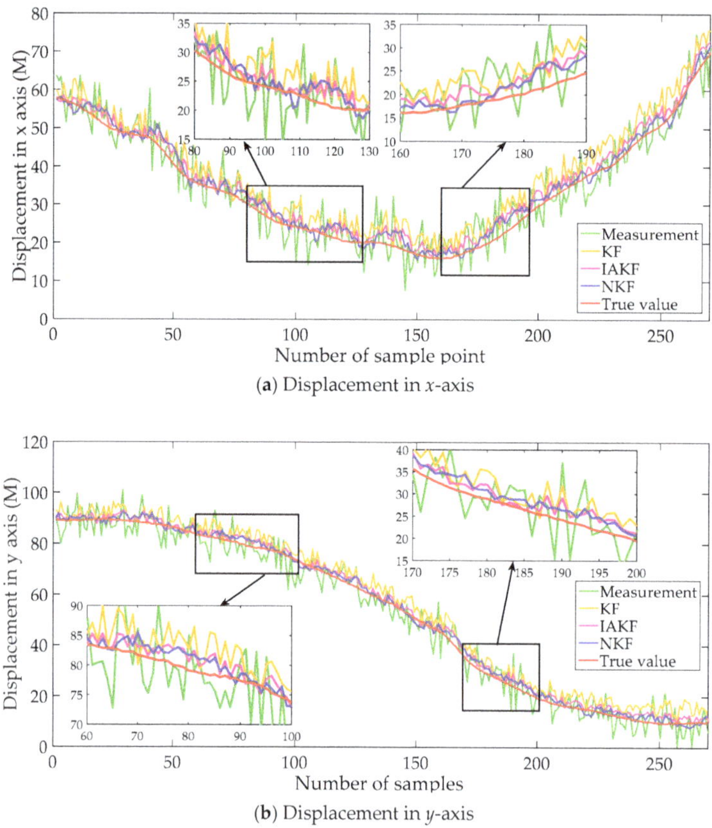

(**a**) Displacement in x-axis

(**b**) Displacement in y-axis

Figure 13. Filtering results of the selected segment of the trajectory.

Table 3. Evaluation of filtering errors (segment of the trajectory).

		KF	IAKF	NKF
x axis	MAE	4.0157	2.0905	1.7159
	RMSE	4.7610	2.5017	2.0879
y axis	MAE	3.8769	1.7897	1.5230
	RMSE	4.5707	2.1330	1.8024

It can be found from Figure 13 that the general trend is similar to the results of the whole trajectory in Figure 11. The results of NKF and IAKF are intuitively approximately the same. From careful and detailed identification in the magnifying subfigure, the proposed NKF can filter the noises closer to the true value than other methods. In the results of the x-axis and y-axis, the basic KF performs badly with the obvious deviation.

The filtering performance can be evaluated more accurately with Table 3. For results in the x-axis, MAE of NKF is 32.73% of KF, 82.01% of IAKF. RMSE of NKF is similar to IAKF, but relatively lower than KF. The trend of results in the y-axis is consistent with the ones in x-axis. Compared with the error indicators in Table 2, the errors of the segment are larger than the whole trajectory, which may be due to the fewer data to train the neural networks.

5. Discussion

In this paper, a novel Kalman filter is designed by introducing neural computing. Simulations and experiments are carried out, and the results are presented and described briefly. Following the results, the methods are discussed in this section.

From the filtering results of simulation signals and practical measurements, it can be proved that the proposed filter can eliminate the noises to the anticipated degree. It performs distinctly better than the traditional Kalman filter, especially for complex noises. Besides, the proposed filter can achieve the latest improvements of AKF. The core thought of the proposed filter is to obtain more knowledge from the existing limited data during the filtering procedure. The process variables in the filtering are reutilized with the neural units, while the reutilization in AKF [36–38] is conducted with statistical methods. Therefore, the proposed filter can be regarded as a new exploration of parameter adjusting, which is similar to the essential thought of AKF.

In the proposed filter, the neuro unit is built based on the nonlinear autoregressive model. The neuro unit specializes in the nonlinear time-series feature extraction with a small-scale structure. Although many more networks have been proposed, it should be conservative in the selection of networks. The complex network may destroy the efficient and straightforward features of the Kalman filter. Besides, the complex network may be not suitable for the terminal applications without the high-performance processor.

Except for the intuitive estimation results, the computational complexity can be analyzed for the proposed and contrast method. According to the basic evaluation method of computational complexity, the complexity of KF, IAKF, and NKF is $O(n^2)$, $O(n^2)$ and $O(n^3)$, respectively, where n is the number of state variables. The complexity of NKF increases because dual matrix multiplication is introduced by neuro units. The operation time is also recorded in the experiment, shown in Table 4.

Table 4. Operation time of different methods in simulation and experiment (time unit: s).

	Simulation		Practical Experiment (Whole Trajectory)	
	Signal x_1	Signal x_2	x Axis	y Axis
KF	1.23	1.45	2.15	2.09
IAKF	1.37	1.73	2.32	2.43
NKF	1.27	1.79	2.41	2.24

For computing time, the methods distinct slightly, although the complexity of NKF is higher. However, an important fact that cannot be ignored is the training of the neuro units. The time above is the test procedure, while NKF needs prior training. The training time is between 3 s and 7 s, according to the preset convergence conditions. In this paper, the training requires historical data in the offline mode. The filtering is conducted after the training, which reduces the real-time performance. It is the challenge how to realize online learning along with the filtering process, which can be studied in the future.

The neurons in the proposed method work well from the experiment results. Although a good filtering performance has been obtained, the inherent mechanism of the proposed method is actually not completely clear. Hence the theoretical analysis should be conducted, and the effect of the neural network on the filter should be deduced in the view of numerical analysis in the future.

In the proposed method, ANN inspired us to optimize the intermediate factors and calculating process in the Kalman filter with the black-box thought. It mainly solves the problem of modeling and parameter adjusting of the traditional filter. It can be a useful tool in the target tracking, trajectory estimation, and pedestrian navigation, especially in the situations of inexperienced modeling of complex systems and the parameter settings.

6. Conclusions

For the intelligent terminals and objects in the internet of things, it has been the vital task to sense the environment and self-status accurately. An improved Kalman filter is proposed with neural computing for accurate sensing. Kalman filter provides a favorable framework in which the system model can be replaced according to the concrete applications. The neuro unit based on NARX is a powerful tool to examine nonlinear and time-series relations. The proposed filter focuses on the data change features and tries to lower the impact of model analysis. In future work, the stability and rapidity of neural computing should be studied deeply. The neuron-based Kalman filter can develop more fully with smarter and faster online neural computing. Moreover, the theoretical derivation should be carried out to support the neuro-based filter. The proposed method can combine other identification approaches [63–66] to study the modeling and filtering problems of other dynamic time series and stochastic systems with colored noises [67–70], and can be applied to other fields [71–74], such as signal modeling and control systems [75–79] studied in other literature [80].

Author Contributions: Conceptualization, Y.-t.B. and X.-b.J.; methodology, Y.-t.B.; writing—original draft preparation, Y.-t.B.; project administration, X.-y.W.; funding acquisition, Z.-y.Z.; supervision, B.-h.Z. All authors have read and agreed to the published version of the manuscript.

Funding: This research was funded by National Key Research and Development Program of China No. 2017YFC1600605, National Natural Science Foundation of China No. 61903008, 61673002, Young Teacher Research Foundation Project of BTBU No. QNJJ2020-26.

Conflicts of Interest: The authors declare no conflict of interest.

References

1. Mohd-Yasin, F.; Nagel, D.J.; Korman, C.E. Noise in MEMS. *Meas. Sci. Technol.* **2009**, *21*, 012001. [CrossRef]
2. Shiau, J.K.; Huang, C.X.; Chang, M.Y. Noise characteristics of MEMS gyro's null drift and temperature compensation. *J. Appl. Sci. Eng.* **2012**, *15*, 239–246.
3. Jiang, Z.; Ni, M.; Lu, Q.; Liu, Z.; Zhao, Y. Wavelet filter: Pure-intensity spatial filters that implement wavelet transforms. *Appl. Opt.* **1996**, *35*, 5758–5760. [CrossRef] [PubMed]
4. Yu, P.; Li, Y.; Lin, H.; Wu, N. Seismic random noise removal by delay-compensation time-frequency peak filtering. *J. Geophys. Eng.* **2017**, *14*, 691–697. [CrossRef]
5. Boudraa, A.O.; Cexus, J.C.; Benramdane, S.; Beghdadi, A. Noise filtering using empirical mode decomposition. In Proceedings of the 9th International Symposium on Signal Processing and Its Applications, Sharjah, UAE, 12–15 February 2007; pp. 1–4.
6. Harvey, A.C. *Forecasting, Structural Time Series Models and the Kalman Filter*; Cambridge University Press: Cambridge, UK, 1990.
7. Wang, Y.J.; Ding, F.; Wu, M.H. Recursive parameter estimation algorithm for multivariate output-error systems. *J. Frankl. Inst.* **2018**, *355*, 5163–5181. [CrossRef]
8. Ding, F.; Zhang, X.; Xu, L. The innovation algorithms for multivariable state-space models. *Int. J. Adapt. Control Signal Process.* **2019**, *33*, 1601–1608. [CrossRef]
9. Pan, J.; Jiang, X.; Wan, X.K.; Ding, W. A filtering based multi-innovation extended stochastic gradient algorithm for multivariable control systems. *Int. J. Control. Syst.* **2017**, *15*, 1189–1197. [CrossRef]
10. Ding, F. Two-stage least squares based iterative estimation algorithm for CARARMA system modeling. *Appl. Math. Model.* **2013**, *37*, 4798–4808. [CrossRef]
11. Ding, F. Decomposition based fast least squares algorithm for output error systems. *Signal Process.* **2013**, *93*, 1235–1242. [CrossRef]
12. Li, M.H.; Liu, X.M.; Ding, F. The filtering-based maximum likelihood iterative estimation algorithms for a special class of nonlinear systems with autoregressive moving average noise using the hierarchical identification principle. *Int. J. Adapt. Control Signal Process.* **2019**, *33*, 1189–1211. [CrossRef]
13. Liu, L.J.; Ding, F.; Xu, L.; Pan, J.; Alsaedi, A.; Hayat, T. Maximum likelihood recursive identification for the multivariate equation-error autoregressive moving average systems using the data filtering. *IEEE Access* **2019**, *7*, 41154–41163. [CrossRef]

14. Zhang, X.; Ding, F.; Xu, L.; Yang, E. State filtering-based least squares parameter estimation for bilinear systems using the hierarchical identification principle. *IET Control Theory Appl.* **2018**, *12*, 1704–1713. [CrossRef]
15. Gu, Y.; Ding, F.; Li, J.H. States based iterative parameter estimation for a state space model with multi-state delays using decomposition. *Signal Process.* **2015**, *106*, 294–300. [CrossRef]
16. Liu, Y.J.; Ding, F.; Shi, Y. An efficient hierarchical identification method for general dual-rate sampled-data systems. *Automatica* **2014**, *50*, 962–970. [CrossRef]
17. Mehra, R.K. On the identification of variances and adaptive Kalman filtering. *IEEE Trans. Autom. Control.* **1970**, *15*, 175–184. [CrossRef]
18. Mohamed, A.H.; Schwarz, K.P. Adaptive Kalman filtering for INS/GPS. *J. Geod.* **1999**, *73*, 193–203. [CrossRef]
19. Rutan, S.C. Adaptive Kalman filtering. *Anal. Chem.* **1991**, *63*, 687–689. [CrossRef]
20. Julier, S.J.; Uhlmann, J.K. New extension of the Kalman filter to nonlinear systems. In Proceedings of the Signal Processing, Sensor Fusion, and Target Recognition VI, Orlando, FL, USA, 28 July 1997; Volume 3068, pp. 182–193.
21. Xiong, K.; Zhang, H.; Chan, C. Performance evaluation of UKF-based nonlinear filtering. *Automatica* **2006**, *42*, 261–270. [CrossRef]
22. Chen, B.; Liu, X.; Zhao, H.; Principe, J.C. Maximum correntropy Kalman filter. *Automatica* **2017**, *76*, 70–77. [CrossRef]
23. Zhang, X.; Xu, L.; Ding, F.; Hayat, T. Combined state and parameter estimation for a bilinear state space system with moving average noise. *J. Frankl. Inst.* **2018**, *355*, 3079–3103. [CrossRef]
24. Zhang, X.; Ding, F.; Xu, L.; Yang, E. Highly computationally efficient state filter based on the delta operator. *Int. J. Adapt. Control Signal Process.* **2019**, *33*, 875–889. [CrossRef]
25. Liu, M.; Tian, Z.; Qi, H.; Zhang, C.; Liu, X. Cooperative fusion model based on Kalman-BP neural network for suspended sediment concentration measurement. *J. Basic Sci. Eng.* **2016**, *5*, 970–977. (In Chinese)
26. Leandro, V.M.; Boada, B.L.; Boada, M.J.L.; Gauchía, A.; Díaz, A. A sensor fusion method based on an integrated neural network and Kalman filter for vehicle roll angle estimation. *Sensors* **2016**, *16*, 1400.
27. Leandro, V.M.; Boada, B.L.; Boada, M.J.L.; Gauchía, A.; Díaz, A. Sensor Fusion based on an integrated neural network and probability density function (PDF) dual Kalman filter for on-line estimation of vehicle parameters and states. *Sensors* **2017**, *17*, 987.
28. Sinopoli, B.; Schenato, L.; Franceschetti, M.; Poolla, K.; Jordan, M.I.; Sastry, S.S. Kalman filtering with intermittent observations. *IEEE Trans. Autom. Control* **2004**, *49*, 1453–1464. [CrossRef]
29. Li, S.E.; Li, G.; Yu, J.; Cheng, B.; Wang, J.; Li, K. Kalman filter-based tracking of moving objects using linear ultrasonic sensor array for road vehicles. *Mech. Syst. Signal Process.* **2018**, *98*, 173–189. [CrossRef]
30. Khan, M.W.; Salman, N.; Ali, A.; Khan, A.M.; Kemp, A.H. A comparative study of target tracking with Kalman filter, extended Kalman filter and particle filter using received signal strength measurements. In Proceedings of the IEEE International Conference on Emerging Technologies, Peshawar, Pakistan, 19–20 December 2015; pp. 1–6.
31. Chang, L.; Li, K.; Hu, B. Huber's M-estimation-based process uncertainty robust filter for integrated INS/GPS. *IEEE Sens. J.* **2015**, *15*, 3367–3374. [CrossRef]
32. Durantin, G.; Scannella, S.; Gateau, T.; Delorme, A.; Dehais1, F. Processing functional near infrared spectroscopy signal with a Kalman filter to assess working memory during simulated flight. *Front. Hum. Neurosci.* **2016**, *9*, 707. [CrossRef]
33. Mou, Z.; Sui, L. Improvement of UKF algorithm and robustness study. In Proceedings of the 2009 IEEE International Workshop on Intelligent Systems and Applications, Wuhan, China, 23–24 May 2009; pp. 1–4.
34. Huang, Y.; Zhang, Y.; Li, N.; Chambers, J. Robust Student'st based nonlinear filter and smoother. *IEEE Trans. Aerosp. Electron. Syst.* **2016**, *52*, 2586–2596. [CrossRef]
35. Zhang, X.; Ding, F.; Yang, E. State estimation for bilinear systems through minimizing the covariance matrix of the state estimation errors. *Int. J. Adapt. Control Signal Process.* **2019**, *33*, 1157–1173. [CrossRef]
36. Zhou, Q.; Zhang, H.; Wang, Y. A redundant measurement adaptive Kalman filter algorithm. *Acta Aeronaut. Astronaut. Sin.* **2015**, *36*, 1596–1605.
37. Qian, X.; Yong, Y. Fast, accurate, and robust frequency offset estimation based on modified adaptive Kalman filter in coherent optical communication system. *Opt. Eng.* **2017**, *56*, 096109.

38. Yi, S.; Jin, X.; Su, T.; Tang, Z.; Wang, F.; Xiang, N.; Kong, J. Online denoising based on the second-order adaptive statistics model. *Sensors* **2017**, *17*, 1668. [CrossRef] [PubMed]
39. Ding, F.; Pan, J.; Alsaedi, A.; Hayat, T. Gradient-based iterative parameter estimation algorithms for dynamical systems from observation data. *Mathematics* **2019**, *7*, 428. [CrossRef]
40. Ding, F.; Lv, L.; Pan, J.; Wan, X.; Jin, X. Two-stage gradient-based iterative estimation methods for controlled autoregressive systems using the measurement data. *Int. J. Control Autom. Syst.* **2020**, *18*, 1–11. [CrossRef]
41. Xu, L.; Ding, F. Iterative parameter estimation for signal models based on measured data. *Circuits Syst. Signal Process.* **2018**, *37*, 3046–3069. [CrossRef]
42. Ding, J.; Chen, J.; Lin, J.X.; Wan, L.J. Particle filtering based parameter estimation for systems with output-error type model structures. *J. Frankl. Inst.* **2019**, *356*, 5521–5540. [CrossRef]
43. Ding, J.; Chen, J.Z.; Lin, J.X.; Jiang, G.P. Particle filtering-based recursive identification for controlled auto-regressive systems with quantised output. *IET Control Theory Appl.* **2019**, *13*, 2181–2187. [CrossRef]
44. Ding, F.; Xu, L.; Meng, D.D.; Jin, X.; Alsaedi, A.; Hayate, T. Gradient estimation algorithms for the parameter identification of bilinear systems using the auxiliary model. *J. Comput. Appl. Math.* **2020**, *369*, 112575. [CrossRef]
45. Cui, T.; Ding, F.; Jin, X.B.; Alsaedi, A.; Hayat, T. Joint multi-innovation recursive extended least squares parameter and state estimation for a class of state-space systems. *Int. J. Control Autom. Syst.* **2020**, *18*, 1–13. [CrossRef]
46. Ding, F. Coupled-least-squares identification for multivariable systems. *IET Control Theory Appl.* **2013**, *7*, 68–79. [CrossRef]
47. Xu, L.; Xiong, W.L.; Alsaedi, A.; Hayat, T. Hierarchical parameter estimation for the frequency response based on the dynamical window data. *Int. J. Control Autom. Syst.* **2018**, *16*, 1756–1764. [CrossRef]
48. Ding, F. Hierarchical multi-innovation stochastic gradient algorithm for Hammerstein nonlinear system modeling. *Appl. Math. Model.* **2013**, *37*, 1694–1704. [CrossRef]
49. Hu, Y.; Li, L. The application of Kalman filtering-BP neural network in autonomous positioning of end-effector. *J. Beijing Univ. Posts Telecommun.* **2016**, *39*, 110–115. (In Chinese)
50. Liu, J.; Cheng, K.; Zeng, J. A novel multi-sensors fusion framework based on Kalman Filter and neural network for AFS application. *Trans. Inst. Meas. Control* **2015**, *37*, 1049–1059. [CrossRef]
51. Cui, L.; Gao, S.; Jia, H.; Chu, H.; Jiang, R. Application of neural network aided Kalman filtering to SINS/GPS. *Opt. Precis. Eng.* **2014**, *22*, 1304–1311. (In Chinese)
52. Shang, Y.; Zhang, C.; Cui, N.; Zhang, Q. State of charge estimation for lithium-ion batteries based on extended Kalman filter optimized by fuzzy neural network. *Control Theory Appl.* **2016**, *33*, 212–220. (In Chinese)
53. Li, S.; Ma, W.; Liu, J.; Chen, H. A Kalman gain modify algorithm based on BP neural network. In Proceedings of the International Symposium on Communications and Information Technologies, Qingdao, China, 26–28 September 2016; pp. 453–456.
54. Zheng, Y.Y.; Kong, J.L.; Jin, X.B.; Wang, X.Y.; Su, T.l.; Wang, J.L. Probability fusion decision framework of multiple deep neural networks for fine-grained visual classification. *IEEE Access* **2019**, *7*, 122740–122757. [CrossRef]
55. Pei, E.; Xia, X.; Yang, L.; Jiang, D.; Sahli, H. Deep neural network and switching Kalman filter based continuous affect recognition. In Proceedings of the IEEE International Conference on Multimedia & Expo Workshops, Seattle, WA, USA, 11–15 July 2016; pp. 1–6.
56. Menezes, J.M.P., Jr.; Barreto, G.A. Long-term time series prediction with the NARX network: An empirical evaluation. *Neurocomputing* **2008**, *71*, 3335–3343. [CrossRef]
57. Goudarzi, S.; Jafari, S.; Moradi, M.H.; Sprott, J.C. NARX prediction of some rare chaotic flows: Recurrent fuzzy functions approach. *Phys. Lett. A* **2016**, *380*, 696–706. [CrossRef]
58. Ouyang, H. Nonlinear autoregressive neural networks with external inputs for forecasting of typhoon inundation level. *Environ. Monit. Assess.* **2017**, *189*, 376. [CrossRef] [PubMed]
59. Bai, Y.; Jin, X.; Wang, X.; Su, T.; Kong, J.; Lu, Y. Compound autoregressive network for prediction of multivariate time series. *Complexity* **2019**, *2019*, 9107167. [CrossRef]
60. Bai, Y.; Wang, X.; Sun, Q.; Jin, X.B.; Wang, X.K.; Su, T.L.; Kong, J.L. Spatio-temporal prediction for the monitoring-blind area of industrial atmosphere based on the fusion network. *Int. J. Environ. Res. Public Health* **2019**, *16*, 3788. [CrossRef] [PubMed]

61. Lourakis, M.I.A. A brief description of the Levenberg-Marquardt algorithm implemented by levmar. *Found. Res. Technol.* **2005**, *4*, 1–6.
62. Wilamowski, B.M.; Yu, H. Improved computation for Levenberg-Marquardt training. *IEEE Trans. Neural Netw.* **2010**, *21*, 930–937. [CrossRef]
63. Ma, H.; Pan, J.; Ding, F.; Xu, L.; Ding, W. Partially-coupled least squares based iterative parameter estimation for multi-variable output-error-like autoregressive moving average systems. *IET Control Theory Appl.* **2019**, *13*, 3040–3051. [CrossRef]
64. Liu, S.Y.; Ding, F.; Xu, L.; Hayat, T. Hierarchical principle-based iterative parameter estimation algorithm for dual-frequency signals. *Circuits Syst. Signal Process.* **2019**, *38*, 3251–3268. [CrossRef]
65. Zheng, Y.Y.; Kong, J.L.; Jin, X.B.; Wang, X.Y.; Su, T.L.; Zuo, M. CropDeep: The crop vision dataset for deep-learning-based classification and detection in precision agriculture. *Sensors* **2019**, *19*, 1058. [CrossRef]
66. Xu, L. Application of the Newton iteration algorithm to the parameter estimation for dynamical systems. *J. Comput. Appl. Math.* **2015**, *288*, 33–43. [CrossRef]
67. Xu, L. The damping iterative parameter identification method for dynamical systems based on the sine signal measurement. *Signal Process.* **2016**, *120*, 660–667. [CrossRef]
68. Ding, F.; Wang, F.F.; Xu, L.; Wu, M.H. Decomposition based least squares iterative identification algorithm for multivariate pseudo-linear ARMA systems using the data filtering. *J. Frankl. Inst.* **2017**, *354*, 1321–1339. [CrossRef]
69. Ding, F.; Liu, G.; Liu, X.P. Partially coupled stochastic gradient identification methods for non-uniformly sampled systems. *IEEE Trans. Autom. Control.* **2010**, *55*, 1976–1981. [CrossRef]
70. Ding, J.; Ding, F.; Liu, X.P.; Liu, G. Hierarchical least squares identification for linear SISO systems with dual-rate sampled-data. *IEEE Trans. Autom. Control.* **2011**, *56*, 2677–2683. [CrossRef]
71. Xu, L.; Chen, L.; Xiong, W.L. Parameter estimation and controller design for dynamic systems from the step responses based on the Newton iteration. *Nonlinear Dyn.* **2015**, *79*, 2155–2163. [CrossRef]
72. Xu, L. The parameter estimation algorithms based on the dynamical response measurement data. *Adv. Mech. Eng.* **2017**, *9*, 1–12. [CrossRef]
73. Wang, Y.J.; Ding, F. Novel data filtering based parameter identification for multiple-input multiple-output systems using the auxiliary model. *Automatica* **2016**, *71*, 308–313. [CrossRef]
74. Ding, F.; Liu, Y.J.; Bao, B. Gradient based and least squares based iterative estimation algorithms for multi-input multi-output systems. *Proc. Inst. Mech. Eng. Part I J. Syst. Control Eng.* **2012**, *226*, 43–55. [CrossRef]
75. Xu, L.; Ding, F.; Gu, Y.; Alsaedi, A.; Hayat, T. A multi-innovation state and parameter estimation algorithm for a state space system with d-step state-delay. *Signal Process.* **2017**, *140*, 97–103. [CrossRef]
76. Ma, H.; Pan, J.; Lv, L.; Xu, G.; Ding, F.; Alsaedi, A.; Hayat, T. Recursive algorithms for multivariable output-error-like ARMA systems. *Mathematics* **2019**, *7*, 558. [CrossRef]
77. Ma, J.X.; Xiong, W.L.; Chen, J.; Feng, D. Hierarchical identification for multivariate Hammerstein systems by using the modified Kalman filter. *IET Control Theory Appl.* **2017**, *11*, 857–869. [CrossRef]
78. Ding, F.; Liu, X.G.; Chu, J. Gradient-based and least-squares-based iterative algorithms for Hammerstein systems using the hierarchical identification principle. *IET Control Theory Appl.* **2013**, *7*, 176–184. [CrossRef]
79. Wan, L.J.; Ding, F. Decomposition- and gradient-based iterative identification algorithms for multivariable systems using the multi-innovation theory. *Circuits Syst. Signal Process.* **2019**, *38*, 2971–2991. [CrossRef]
80. Jin, X.; Yang, N.; Wang, X.; Bai, Y.; Su, T.; Kong, J. Integrated predictor based on decomposition mechanism for PM2.5 long-term prediction. *Appl. Sci.* **2019**, *9*, 4533. [CrossRef]

Article

Laplacian Scores-Based Feature Reduction in IoT Systems for Agricultural Monitoring and Decision-Making Support

Giorgos Tsapparellas [1], Nanlin Jin [2,*], Xuewu Dai [3] and Gerhard Fehringer [2]

1 Department of Maritime and Mechanical Engineering, Liverpool John Moores University, Liverpool L3 3AF, UK; g.tsapparellas@ljmu.ac.uk
2 Department of Computer and Information Sciences, Northumbria University, Newcastle upon Tyne NE1 8ST, UK; gerhard.fehringer@northumbria.ac.uk
3 Department of Mathematics, Physics and Electrical Engineering, Northumbria University, Newcastle upon Tyne NE1 8ST, UK; xuewu.dai@northumbria.ac.uk
* Correspondence: nanlin.jin@northumbria.ac.uk

Received: 31 July 2020; Accepted: 2 September 2020; Published: 8 September 2020

Abstract: Internet of things (IoT) systems generate a large volume of data all the time. How to choose and transfer which data are essential for decision-making is a challenge. This is especially important for low-cost and low-power designs, for example Long-Range Wide-Area Network (LoRaWan)-based IoT systems, where data volume and frequency are constrained by the protocols. This paper presents an unsupervised learning approach using Laplacian scores to discover which types of sensors can be reduced, without compromising the decision-making. Here, a type of sensor is a feature. An IoT system is designed and implemented for a plant-monitoring scenario. We have collected data and carried out the Laplacian scores. The analytical results help choose the most important feature. A comparative study has shown that using fewer types of sensors, the accuracy of decision-making remains at a satisfactory level.

Keywords: Laplacian scores; data reduction; sensors; Internet of Things (IoT); LoRaWAN

1. Introduction

The Internet of Things (IoT) interconnects and embeds objects, machines and devices, forming a highly distributed network of device broadcasting with humans and other devices [1]. Recent application areas progressing within the IoT sector include smart cities, agriculture, building, healthcare, and shopping [2–4]. This paper proposes an open-source [5] and low-cost Long-Range Wide-Area Network (LoRaWAN) solution for strawberry-plant monitoring.

Conducting data-mining on raw IoT data will help to reduce the cost of powering sensors, the amount of packet transmission, latency, and response delay [6]. Moreover, the discovery of information from raw data improves system performance. Data-mining generates knowledge models from data received to support decision-making. Common methods include data compression, data-mining, and data reduction [6].

Recent research introduces a range of methods, including compression [7] and reconstruction [8], aggregation [9], redundancy removal, and reduction of the number of sensors [6] using time-discounted histogram encoding. To replace multiple sensors that send appliance energy usage in households, smart phone data is used as the only data source to estimate user activities [10]. A lightweight monitoring framework has been developed to cope with limited processing capabilities. It adapts the amount of data disseminated through the network over time [11]. Another framework transmits updates when the sensor readings are detected to be unusual, and have triggered dissemination [12], adapting the monitoring sensing intensity and dynamically adjusting the data volume payload.

Reducing the amount of data to be analyzed in IoT systems can be done either offline or online. The offline analysis is to collect data as much as possible during trials, to conduct offline analysis, and discover patterns. During the real-time operation, data will be checked against such learned patterns while running lightweight data analysis programs, for example signature-based network-intrusion-detection systems [13]. The online analysis operates data reduction in real time, to calculate the difference from normal behaviors, for example anomaly-based detection network-intrusion-detection systems [11,12,14]. The offline approach might out-perform the online approach in finding the previously trained events or situations, but the second approach would be better if unknown situations happens during real-time operation. This paper focuses on the offline approach. Here, the feature selection will not serve as pre-processing techniques for data-mining only, but also determine the sources of data to be collected in deployment and operation.

The rest of this paper is structured as follows: Section 2 showcases Related Work, Section 3 shows the motivation and analysis behind the Data reduction in IoT monitoring, Section 4 presents the Problem and System Architecture, including the design and implementation of the proposed LoRaWAN-based IoT system for strawberry-plant monitoring, Section 5 shows the experimental results and analysis including experimental set-up and sensors calibration, traffic analysis, data visualization, feature selection and evaluation, and example in decision support, and Section 6 provides conclusions with any future research directions.

2. Related Work

2.1. Usage of Sensors in Agriculture

The usage of sensors and actuators has been replacing the traditional human-intensive ways of monitoring in agriculture [15]. Sensors can measure environmental parameters and convert them into meaningful signals [16], for example, water resource monitoring for irrigation [17]. It is reported that in 2000, there were approximately 525 million farms on record across the globe, but none connected to the IoT. However, by 2025 for the same base of 525 million farms, it is expected for there to be around 600 million sensors installed, connected and in use in these farms [18]. The technological advancement as well as size abatement of devices make employment of sensors feasible for agriculture applications [16].

2.2. IoT LPWAN Communication Protocols: LoRa and LoRaWAN

Low-power wide-area (LPWAN) communication protocols are designed for low-power consumption, suitable for applications which demand limited efforts for maintenance. One of the protocols, LoRaWAN, has been introduced by the LoRa Alliance organization as the protocol for low-power and wide-area coverage [19]. LoRaWAN, which stands for long-range wide-area network, defines the communication protocol and the system architecture for the network [20].

By definition, LoRa is the physical layer or the wireless modulation used to create long communication links. In terms of the LoRa functionality, an end-device communicates to a gateway which is employing LoRa with LoRaWAN. To be more specific, a LoRa gateway passes raw LoRaWAN packets from the end-devices to a network server [19]. Major advantages of LoRa are its low-power consumption, long-range capability, security and relatively easily expandable network. However, LoRa advantages have their trade-offs: for example, the time delay for the data to be stored in the cloud after being obtained, and the final data usage or display [21]. Therefore, it might not be the ideal choice for those applications requiring immediate responses or high-resolution data.

However for low-cost and low-power IoT systems, the data transmission is constrained. Therefore, how to reduce the volume of data to be sent from a LoRa node to a network server, while still enabling data-driven decision, is a challenge.

2.3. Feature Reduction

To reduce the number of features to be used, the main data-mining methods include: feature selection, which selects a subset of the original feature set; and feature extraction, which creates a set of new features by combining original features. The choice of selecting features are problem-dependent, but the resulting subset features should remain a faithful, perhaps simplified representation to the original data set and preserve the intrinsic knowledge accurately. This paper focuses on feature selection.

Feature selection methods were used to identify the set of features which brings high accuracy to detect cyber-attacks [22]. It has been found that features have discriminatory contribution to classification accuracy in identifying attacks. Some features are redundant, irrelevant, partially relevant to the learning target and some even reduce accuracy, for example noise.

In addition, feature construction or feature transformation can create new features or transform existing features into a new set of features, smaller than the original set [23]. This method requires decent domain related knowledge, for example the understanding of energy usage patterns as shown in [23]. Principal component analysis (PCA) also summarizes data into fewer dimensions by projecting it onto an orthogonal basis.

Deep learning has demonstrated high performance in terms of accuracy [24]. However in the setting of real-time operation in IoT, response time is one of key requirements and edge devices or even gateways have limited computational resources to use the computationally demanding method deep learning, especially for large scale of IoT systems. In addition, the results from deep learning is difficult to be interpreted. This method is especially unpractical when human involves in analysis, monitoring, decision-making and control.

3. Data Reduction in IoT Monitoring

To illustrate the feature reduction, we provide a sample scenario in a plant-monitoring context. For example, some sensors can be used: temperature $w(t)$, humidity $h(t)$, lighting $b(t)$ and soil moist $s(t)$. The decision-making for a specific action can be represented as a function $f : \Re^{4k} \rightarrow \Re$ as follows:

$$d(t) = f(\mathbf{w}(t), \mathbf{h}(t), \mathbf{b}(t), \mathbf{s}(t), \theta) \quad (1)$$

where $d(t) \in \Re$ is the decision variable representing the action to be taken. For example, $d(t) = 1$ means watering and $d(t) = 0$ means no watering. And $\mathbf{w}(t) \in \Re^k$, $\mathbf{h}(t) \in \Re^k$, $\mathbf{b}(t) \in \Re^k$ and $\mathbf{s}(t) \in \Re^k$ are data vectors for temperature, humidity, lighting and soil moist for the last k samples until time t, For example, $\mathbf{w}(t) = [w(t-k+1), w(t-k+2), \cdots, w(t-1), w(t)]^T$ is the last k samples of the temperature at time t. k is referred to as the sampling window.

The research question is how to make the correct decision with less data. More specifically, the data reduction problem can be stated as follows: Are all these four types of data needed to make the decision? Would it be possible to just use three type of data and which three type of data should be selected to make the decision?

3.1. Feature Selection Using Laplacian Score

Carrying out data analysis on many features is always computationally expensive. Its computational complexity increases while the dimensions or the number of features increase. Therefore, to select the most important features becomes necessary, especially in source-limited situations.

There is a rich range of dimensionality reduction methods. Some are suitable for classification, for example, to rank features using neighborhood component analysis, to rank features using minimum redundancy maximum relevance algorithm, to estimate predictor importance for classification tree. Some are suitable for regression, to select those independent variables which have the best relation to the predictor, i.e., the dependent variable, for example, to rank features using F-test. This method will be useful if the dependent variable is known and its data is collected. In our IoT system, it has a set of sensors for monitoring, but its predicting variable is unknown. Therefore we will need to consider feature selection for unsupervised learning. For unsupervised learning, Laplacian scores have been used to rank features.

Laplacian score was designed to select features in unsupervised learning [25]. Feature selection in unsupervised learning is more difficult than supervised learning, due to lacking of class labels to guide search. Laplacian score was introduced as a filter method to evaluate a feature by "its power of locality preserving", using local neighborhood relationships between data points [25].

For feature selection in supervised learning, Laplacian score has been used for multi-label classification, to measure feature relevance [26] to be used together with manifold learning which is non-linear dimensionality reduction [27]. For feature selection in unsupervised learning, Laplacian score concept has been used to produce pseudo class labels [28], in clustering [29], and to rank multi-cluster structure [30].

3.2. Laplacian Scores to Rank Features for Unsupervised Learning

To reduce the volume of data for specific tasks, class labels are normally available for supervised learning. However in many applications, feature reduction is needed for general usage, not limited to a specific task. This falls into unsupervised learning. Laplacian scores can rank features and users can select important features from the resulting rank [25] for the situations where no class label is available.

The similarity $S_{i,j}$ is defined as:

$$S_{i,j} = exp(-(\frac{D_{i,j}}{\delta})) \tag{2}$$

where δ is a scale factor and $D_{i,j}$ is the distance of two data points i and j in a local neighborhood. The ith element, Dg, of the Degree matrix D is defined as

$$D_g(i,i) = \sum_{j=1}^{n} S_{i,j} \tag{3}$$

The Laplacian matrix is defined as the difference between the degree matrix D_g and the similarity matrix S:

$$L = D_g - S \tag{4}$$

Alternatively, the feature selection results agree with to minimize the value:

$$\frac{\sum_{i,j}(x_{ir} - x_{jr})^2 \times S_{i,j}}{Var(x_r)} \tag{5}$$

where r is the rth feature, x_{ir} is the ith observation of the rth feature. This means that features with large variance is preferred.

In the next section, a simple IoT system is designed to install four sensor measurements (temp, humidity, lighting, soil moisture) to monitor an environment. Then our planned feature reduction will be tested IoT systems. We will select more important features from the aforementioned four and evaluate whether the reduced dataset can achieve comparative performance with the full dataset.

4. Problem Definition and System Architecture

This section starts with the design of the IoT system architecture, followed by five building blocks and their choices of hardware/software for implementation. The gathered data of the real-world plant-monitoring IoT system is then used to test the proposed data-mining method.

4.1. System Architecture

The proposed system will be able to (1) collect data from sensors to monitoring agriculture related variables; (2) transmit such data to the gateway; (3) facilitate the gateway to send data to the cloud server; (4) enable data to be displayed at mobile APP or a client service.

The overall system design is illustrated in Figure 1. Starting from the left, sensors/actuators for monitoring, such as temperature, humidity, light intensity and soil moisture are attached to a low-cost development platform. This platform consists of both a FRDM-K64F ARM mbed evaluation board (as the base) and a SX1272MB2xAS LoRa radio shield, to be explained later in this section. The main function of this platform is to transmit sensor data to a gateway. This cluster of physically connected devices is named "LoRa Node" in this paper. The LoRa Node sits next to the test site, for example, a plant.

The LoRa Node is transmitting data to a Gateway, using LoRa wireless communication. This wireless communication will be explained in Section 4.4. The Gateway is responsible for establishing an IP communication with, and sending data to an IoT Cloud Server. The Cloud Server sends data and its visualization to the end-user(s) through web and mobile dashboards. The following sections will explain the main building blocks in details.

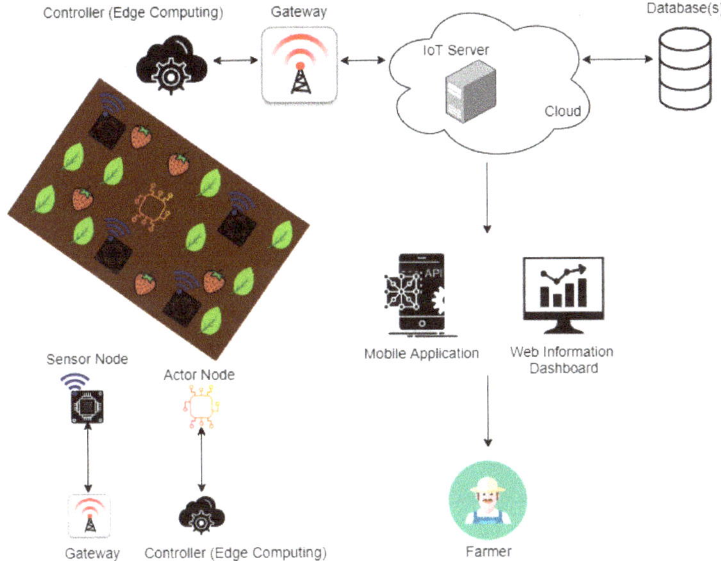

Figure 1. Overview of IoT system for strawberry-plant monitoring using LoRaWAN.

4.2. IoT Platform Development

This platform consists of both a FRDM-K64F ARM mbed evaluation board (as the base) and a SX1272MB2xAS LoRa radio shield, to be explained later in this section. The main function of this platform is to transmit sensor data to a gateway. This cluster of physically connected devices is named "LoRa Node" in this paper. The LoRa Node sits next to the test site, for example, a plant.

The LoRa Node is transmitting data to a Gateway, using LoRa wireless communication. This wireless communication will be explained in Section 4.4. The Gateway is responsible for establishing an IP communication with, and sending data to an IoT Cloud Server. The Cloud Server sends data and its visualization to the end-user(s) through web and mobile dashboards. The following sections will explain the main building blocks in detail.

4.2.1. Sensors

There is a rich range of sensors available in the market. The sensors chosen here are examples.

Soil Moisture Sensor

A soil moisture sensor detects the moisture of soil based on soil resistance measurement. In other words, sensor output value will decrease once soil moisture deficits. The output signal from the sensor is an analog value [31]. Notice that its measurements can be converted to a specific unit (e.g., voltage extraction) by employing FRDM–K64F ARM mbed board's 16-bit ADC converter for meaningful data. The soil resistance measurement is in a range of 0 to 5 Volts soil moisture excitation. For instance, the soil resistance measurement can be calculated using the analog value as:

$$moistureVoltage = moistureAnalog * (5.0/65,536.0) \tag{6}$$

Temperature and Humidity Sensor

The chosen temperature and humidity sensor provides both temperature and humidity measurements as a pre-calibrated digital output using a negative temperature coefficient thermistor and a capacitive sensor element, accordingly [32]. Its detailed characteristics can be viewed through Table 1. At the beginning, the temperature and humidity sensor starts running the active mode from the low-power consumption mode once MCU sends a trigger signal. As a result, 40-bit data is collected back by the MCU consisting of 16-bit humidity data, 16-bit temperature data and 8-bit checksum number.

Table 1. Temperature and Humidity Sensor Main Characteristics.

Grove Temperature and Humidity Sensor	
VCC	3.3–5 Volts
Measuring Range: Temperature	0–50 °C
Measuring Range: Humidity	20–90%
Sensitivity: Humidity	1%
Sensitivity: Temperature	1 °C

Light-Intensity Sensor

A light-intensity sensor exposes the intensity of light based on the resistance value of a photo-resistor (for the device chosen, GL5528 photo-resistor (Seeed, Shenzhen, China)). In particular, the resistance of a photo-resistor increases when the intensity of light decreases. The output signal is an analog value [33]. The measurements can be converted to a specific unit (e.g., voltage extraction) by deploying FRDM–K64F ARM mbed board's 16-bit ADC converter for meaningful data gathering. For example, the following calculation:

$$lightVoltage = lightAnalog \ * \ (5.0/65,536.0) \tag{7}$$

can be considered to be a 0 to 5 Volts light-intensity excitation.

4.2.2. Lora Node Platform

As shown in Figure 1, a development platform attaches sensors and a transceiver send such data to a gateway.

FRDM–K64F ARM Mbed Board

FRDM–K64F ARM mbed board (ARM mbed, Cambridge, UK) is an ultra-low-cost development platform designed by NXP in collaboration with ARM mbed [34]. FRDM–K64F ARM mbed board will be the base device of LoRa Node along with SX1272MB2xAS LoRa shield and temperature, humidity, light intensity and soil moisture sensors. The sensors are physically attached on it. The specification of a FRDM–K64F ARM mbed board is in Table 2.

Table 2. FRDM–K64F ARM mbed board Main Hardware Specifications.

FRDM–K64F ARM Mbed Board	
MCU	Kinetis MK64FN1M0VLL12 (ARM Cortex-M4)
Flash	1024 KB
RAM	256 KB
CPU max. frequency	120 MHz

SX1272MB2xAS Semtech LoRa Shield

A SX1272MB2xAS Semtech LoRa shield (ARM mbed, Cambridge, UK) contains a SX1272 transceiver which features a spread communication using LoRa modulation over either 868 MHz or 915 MHz frequency [35]. For this particular product, the SX1272MB2xAS Semtech LoRa shield is attached to the base device FRDM–K64F ARM mbed board, constructing the desired LoRa node. The SX1272MB2xAS Semtech LoRa shield provides a reliable transmitting sensor measurement directly to a Gateway. The specification of the SX1272MB2xAS Semtech LoRa shield is in Table 3.

Table 3. SX1272MB2xAS Semtech LoRa shield Main Hardware Specifications.

SX1272MB2xAS Semtech LoRa Shield	
Transceiver	SX1272
Frequency Ranges	868 MHz and 915 MHz
Link Budget	157dB max.
Sensitivity	down to –137 dBm
Bit-Rate	300 kbps
Dynamic Range RSSI	127 dB

4.2.3. Gateway

A Dragino LG01–P LoRa Gateway (Dragino, China) is a single-channel gateway that bridges the data gathered from the LoRa node (s) to the dedicated cloud service using either Wi-Fi, Ethernet, 3G or 4G cellular [36]. The specification of a Dragino LG01–P LoRa Gateway is in Table 4.

Table 4. Dragino LG01–P LoRa Gateway Main Hardware Specifications.

Dragino LG01–P LoRa Gateway	
Processor	400 MHz
MCU	ATMega328P
Flash	32 KB
Link Budget	168dB max.
Dynamic Range RSSI	127 dB
Bit-Rate	up to 300 kbps
RJ45 Ports	2 (WAN and LAN)
Wi-Fi	IEEE 802.11 b/g/n
Power Input	12V DC

4.2.4. Cloud Server

The "Things Network" Cloud Server is an open-source decentralized network service enabling devices (such as a LoRa Node) as well as Gateways (such as Dragino LG01-P LoRa Gateway) to be connected to it [37]. The Things Network is an open community with more than 3000 Gateways up and running, and 35,000 registered members. The goal of The Things Network is building a distributed IoT data infrastructure by creating sufficient data connectivity through LoRaWAN technology [37].

Certainly, there are various alternative options of Cloud Servers, such as the Mbed Cloud and the IBM Watson. Here the Things Network Cloud Server is chosen for its open-source providence and its concentration to the LoRaWAN technology. This aligns with this research which applies LoRaWAN into monitoring agriculture.

4.2.5. Data Visualization and Client-Side Application

The "All Things Talk" application platform is chosen as it provides open-source data visualization through either web or mobile dashboards using an in-house HTTP API [38]. Some core features of All Things Talk API are real-time data gathering and instant notifications through either Web/Mobile dashboards or registered e-mail. Finally, All Things Talk API's end-user(s) has/have the privilege of viewing, processing and downloading any historical measurements for data analysis purposes.

4.3. *Software Development*

Lora Node

Software architecture of LoRa Node can be observed in Algorithm 1. Functions, events and possible errors are illustrated. At first instance, setUp() function represents a local function call intended to initialize ARM mbed operating system environment as well as SX1272 Radio's and IBM's LMiC libraries configuration aspects. As a result, LMIC_setSession() application callback can then be implemented for acquiring an activation by personalization session. For a successful session establishment, static constants such as Network ID, Device Address, Network Session Key and Application Session Key extracted from The Things Network Cloud Server should be employed. After LMIC_setSession() callback, LoRa stack should output either EV_JOINED or EV_JOINED_FAILED event, indicating successful or unsuccessful join to the network service.

Then, getTemperatureHumidity() local function call is core for gathering related measurement parameters. Beyond that, DHT11 library which is intended to be used for temperature and humidity sensor's implementation provides various error enumerations. Specifically, error enumerations of temperature and humidity sensors are ERROR_NONE, BUS_BUSY, ERROR_NOT_PRESENT, ERROR_ACK_TOO_LONG, ERROR_SYNC_TIMEOUT, ERROR_DATA_TIMEOUT, ERROR_CHECKSUM and ERROR_NO_PATIENCE in sequence.

Additionally, both light-intensity and soil moisture measurement parameters are collected through getLightIntensity() and getSoilMoisture() local function calls, respectively.

Moving to data transmission, a time-triggered local function call should be initialized for sending desired LoRa packet to the Gateway in a context of set time interval. As with LMIC_setSession() application callback, events such as EV_TXCOMPLETE, EV_LOST_TSYNC or EV_LINK_DEAD should be outputted from transmit() function call indicating whether LoRa packet had successfully be transmitted to the connected Gateway. Apart from that, IBM's LMiC library provides a _setTimedCallback() application callback which settles the program down until set time interval is being triggered signaling the next LoRa packet transmission.

Following embedded systems good principles, reset button deployment is giving the opportunity of completely resetting the LoRa Node, manually, at any time.

Finally, yet importantly, software architecture of LoRa Node has been implemented in a sequential form, avoiding any unnecessary computational complexity which could result in poor performance and increased power-consumption.

Algorithm 1: LoRa Node algorithmic software architecture.

```
TIMEINTERVAL ⇐ 300                          ▷ 5 min. transmission rate;
setUp();
LMICsetSession();
if LMICsetSession() is EV JOINED then
  | EV JOINED;
else
  | EV JOIN FAILED;
end
while EV JOINED and TIME INTERVAL is 300 do
  | getTemperatureHumidity();
  | getLightIntensity();
  | getSoilMoisture();
  | transmit();
  | os setTimedCallback();
end
```

4.4. Network Architecture

This section provides the implementation of network architecture.

4.4.1. Gateway

The LoRa Gateway's block architecture is given in Figure 2. This Gateway can handle LoRa packets coming from the LoRa Node using the SX1276/78 LoRa wireless module which is attached on ATMega328P micro-controller. Then the Arduino environment communicates and passes LoRa packets to the Dragino HE AR9331 Linux module by employing a bridge library.

The Linux environment of Dragino's LG01–P LoRa Gateway provides three different options for bridging the LoRa wireless network to an IP network for the successful transmission of LoRa packets to a Cloud Server: namely 802.11 b/g/n Wi-Fi, Ethernet (LAN and WAN RJ45 communications), and 3G/4G module. Please note that the chosen Dragino LG01–P LoRa Gateway does not include an internal 3G/4G module. As a result, cellular communications cannot be applied to our IoT system for agriculture.

The gateway is configured in a way that acts as the "middle station" between the LoRa Node and the IoT Cloud Server.

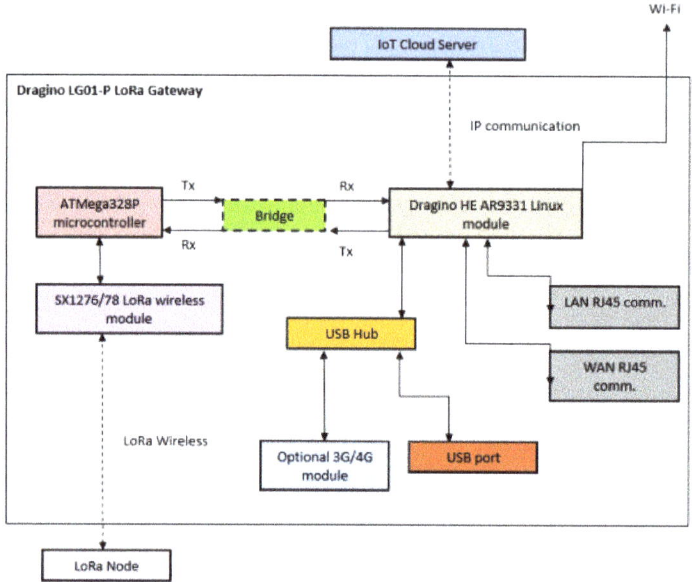

Figure 2. Block diagram of Dragino LG01–P LoRa Gateway architecture [36].

4.4.2. Cloud Server

The block architecture of the "Things Network" Cloud Server is illustrated in Figure 3. Its open-sourced elements such as packet forwarder, router, broker, handler, network and discovery servers enable the employment of the LoRaWAN standard for IoT systems [39].

The main functionality of this cloud server includes: first, this cloud server forwards LoRa messages using a remotely configurable and secure packet forwarder [39]. Then, the router micro-service is liable for identifying a broker to forward the LoRa message [39]. When it comes to the handling procedure, a micro-service handler is reliable to encrypt as well as decrypt the play-load and therefore publishes it to the desired Application Manager API through a suitable integration [39]. Please note that the integration functionality bridges The Things Network Cloud Server with the IoT applications to support data visualization, analysis and storage [39].

In this infrastructure, both Discovery and Network servers are being employed. The discovery server keeps track of network's components such as router, broker and handler. On the other hand, the network server monitors device states as well as device registries [39].

This cloud server is designed and implemented in a distributed and scalable way by allowing high-performance, high-availability and end-to-end security [39]. In addition, stack components such as gateway software, device libraries, cloud routing services and integration are being covered [39].

4.4.3. Client/User Interface API

We have chosen "All Things Talk API" to support clients and User Interface. Its architecture is illustrated in Figure 4. Entities such as applications, notifications, connectivity and data management, together build up an application manager API. This offers end users the opportunities to visualize, store and process gathered sensor data (measurements).

The All Things Talk API offers the choices of joining a device through either WAN, LPWAN or Gateway connections. In particular, the LoRa Node of our IoT system for agriculture, is integrated through the Low-Power Wide-Area Network (LPWAN) connection with The Things Network Cloud Server.

In the IoT system for agriculture, the measurement parameters such as temperature, humidity, light intensity and soil moisture will be displayed on the client side. In addition, a virtual watchdog has been initialized for monitoring any potential warnings or errors.

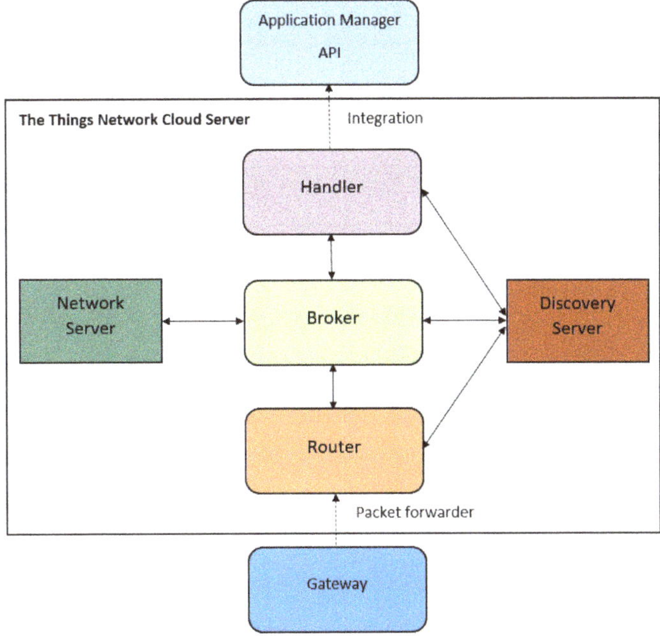

Figure 3. Block diagram of The Things Network Cloud Server architecture [39].

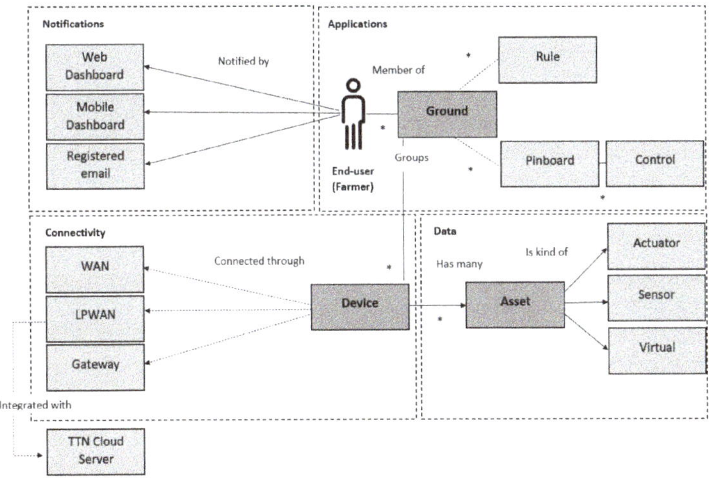

Figure 4. Domain diagram of All Things Talk API architecture [40].

5. Experimental Results, Analysis and Discussion

5.1. Experimental Set-Up and Sensors Calibration

To measure the functionality and performance of the proposed LoRaWAN empowered IoT architecture and implementation for agriculture, a testbed has been setup. An indoor greenhouse is used for this purpose, as seen in Figure 5a.

The hardware used are the Dragino LG01-P LoRa Gateway, FRDM–K64F ARM mbed board, LoRa shield, light-intensity sensor, soil moisture sensor, temperature and humidity sensor. A strawberry plant in this greenhouse is used for the tests which could be assumed to be representative of a plot in the greenhouse.

Sensors are attached to FRDM–K64F ARM mbed board and Semtech SX1272MB2xAS LoRa shield as seen in Figure 5b. The temperature and humidity sensors are connected through the D6 digital input port of the LoRa shield, while the soil moisture sensors are attached by employing the A3 analog input port. Similarly, the light-intensity sensor is used through A1 analog input port of the LoRa shield.

(**a**) Overall view of strawberry greenhouse.

(**b**) LoRa Node, sensors and strawberry plant.

Figure 5. Greenhouse and LoRa node monitoring strawberry-plant growth.

The Gateway is placed approximately 100 m away, due to the size of the greenhouse, from the above connected devices. The required Internet connection of Dragino LG01-P LoRa Gateway is established by deploying the WAN port of the device connected to an Ethernet admission. After that, the soil moisture sensor is placed inside the soil surrounding the strawberry plant, while the temperature, humidity and light-intensity sensors settle nearby, as seen in Figure 5b.

Data is flashed into the FRDM–K64F ARM mbed evaluation board's micro-controller through the Mbed online compiler. The IoT system runs as an autonomous time-triggered program based on set transmit interval. Once data is collected, it will be sent to the cloud server, i.e., "The Things Network" and consecutively to the client interface API, i.e., "All Things Talk API".

Before powering-up the whole IoT system, where compulsory, calibration tests have been conducted to measure the accuracy and stability of the sensor readings. For example, the temperature and humidity sensor is pre- calibrated with minimal sensitivity levels of humidity 1% RH and temperature 1 °C (see Table 1). On the other hand, for the soil moisture sensor, calibration has been conducted for three different levels of moisture; (A) sensor in dry soil, (B) sensor in humid soil and (C) sensor in water. Similarly, for the light-intensity sensor, calibration has been deployed for two different levels of light; (A) HIGH when sensor in daylight and (B) LOW when sensor in dark. The results for soil moisture and light-intensity sensors during calibration test are shown in

Figure 6. Data gathered from temperature and humidity sensor is also being visualized for a more comprehensive review.

After the sensor calibration test, the real-environment test has been deployed. *Test 1 (Real-condition)* was set to transmit all sensor data at the interval of 300,000 milliseconds, which is 5 min.

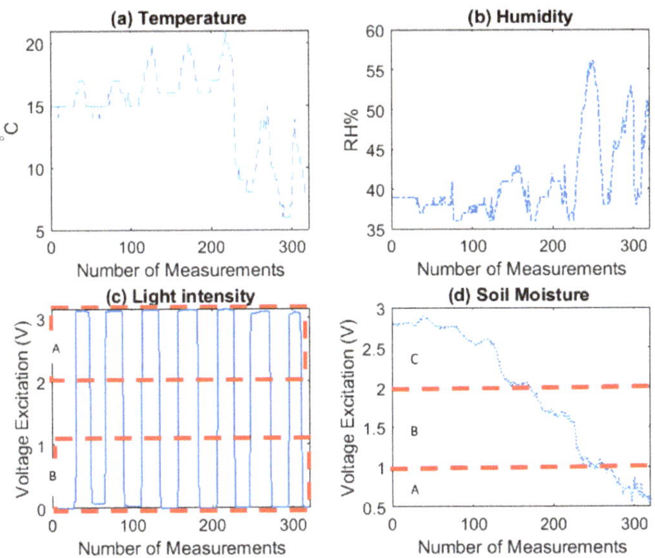

Figure 6. Visualization of sensors calibration test data. x axis is time (Number of Measurements). y axis represents the sensor readings. (**a**) Temp unit is °C, (**b**) Hum unit is % RH, (**c**) LightInt unit is Volts and (**d**) SoilMoist unit is Volts. Soil moisture calibrated against three different levels; (A) sensor in dry soil, (B) sensor in humid soil and (C) sensor in water. For the light-intensity sensor, calibration has been deployed for two different levels of light; (A) HIGH when sensor in daylight and (B) LOW when sensor in dark.

5.2. Traffic Analysis

As seen in Tables 5 and 6, Sensors Calibration Test is executed 4% data transmission loss, while Test 1 (Real-condition) is executed with 12% data transmission loss. It is clear that a higher number of measurements causes a higher data transmission loss.

Table 5. Sensors Calibration Test Traffic Analysis.

Sensors Calibration Test Traffic Analysis (321 Num. of Measurements).	
LoRa packets to send	336
LoRa packets to arrive	321
LoRa packets lost	15
LoRa packet loss percentage	4%

Table 6. Test 1 (Real-condition) Traffic Analysis.

Test 1 (Real-Condition) Traffic Analysis (1776 Num. of Measurements).	
LoRa packets to send	2016
LoRa packets to arrive	1776
LoRa packets lost	240
LoRa packet loss percentage	12%

5.3. Data Visualization

Figure 7 visualizes the sensing reading of temperature (legend: Temp), humidity (legend: Hum), light intensity (legend:LightInt) and soil moisture (legend: SoilMoist), collected in the Test 1 (Real-condition), in total of 1776 observations. For decision-making purposes, three different watering events have been tested and can be observed; (A) Strawberry plant not watered, (B) Strawberry plant in humid soil and (C) Strawberry plant watered.

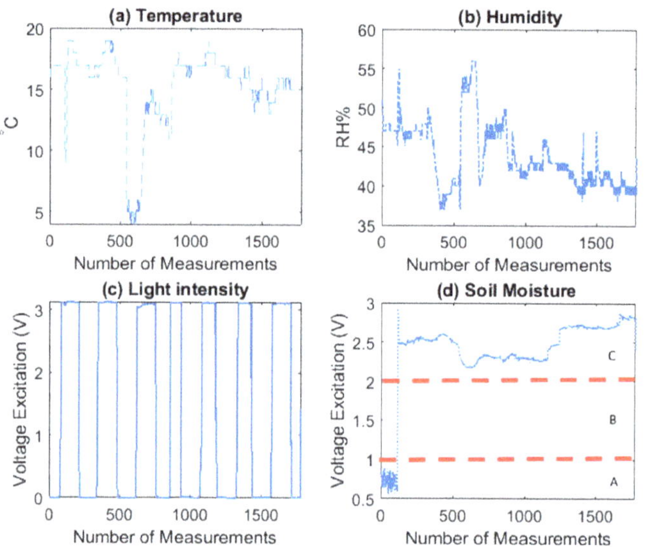

Figure 7. Visualization of Test 1 (Real-condition) data. x axis is time (Number of Measurements). y axis represents the sensor readings. (**a**) Temp unit is °C, (**b**) Hum unit is % RH, (**c**) LightInt unit is Volts and (**d**) SoilMoist unit is Volts. For decision-making purposes, three different watering events have been tested and can be observed; (A) Strawberry plant not watered, (B) Strawberry plant in humid soil and (C) Strawberry plant watered.

Correlation Coefficients

Correlation coefficients are used to measure the dependence of the readings between two sensors X and Y. The Pearson correlation coefficient is defined as:

$$\rho(X,Y) = \frac{cov(X,Y)}{\sigma_X \sigma_Y},$$

where $cov(X,Y)$ is the covariance of X and Y, and σ_X and σ_Y are the standard deviation of X and Y, respectively. The values of the coefficients can range from -1 to 1. Value -1 represents a directly negative correlation, 0 represents no correlation, and 1 represents a directly positive correlation.

Table 7 lists the $\rho(X, Y)$ values for each pairwise variable combinations of temperature, humidity, light intensity and soil moisture, shorted as Temp, Hum, LightInt and SoilMoist respectively. It shows that Temp and Hum has a strong negative linear relationship, Temp and SoilMoist has a moderate positive linear relationship, and Hum and SoilMoist has a moderate negative linear relationship.

These findings are consistent with domain knowledge in agriculture: relative humidity relies on both pressure and temperature. At a lower temperature, less water vapor is needed to reach a high level of humidity. However, at a higher temperature, a higher water vapor is needed to obtain a high level of relative humidity.

Table 7. Correlation coefficients ρ values.

ρ	Temp	Hum	LightInt	SoilMoist
Temp	1	−0.8381	0.34	0.6573
Hum		1	−0.2273	−0.685
LightInt			1	0.0148
SoilMoist				1

5.4. Feature Selection and Evaluation

Data generating in this IoT system comes from four sensors. They measure temperature, humidity, light intensity and soil moisture. In the dataset, one feature contains the readings from one sensor. Data of each feature is being generated by the according sensor node. Laplacian scores are calculated to measure the important of features.

Laplacian scores here are for unsupervised learning. To further evaluate it, we test the result on the following example, as an application in future decision support.

Example in Decision Support

The outputs from the unsupervised method Laplacian scores can be used to for decision-making. For example, an expert labeled the data collected and decided when watering is needed. We compare the classification outcomes of using the selected features from using Laplacian scores and of using the all sensor data. Please note that the class label is only for one action here, while Laplacian scores is generated without class labels for general purpose.

Classifiers' accuracy and performance measured using data inputs with 5 min transmission rate and last 2 h average. In both cases, classification conducted using the 4 and 3 most important features based on their scores.

The accuracy and performance of resulting classifiers using data inputs with 5 min transmission rate is shown in Table 8 for the 4 and 3 most important features, respectively.

Table 8. The accuracy and performance of resulting classifiers using data inputs with 5 min transmission rate for the 4 and 3 most important features.

Features	Correctly Classified Instances	Incorrectly Classified Instances
Hum, Temp, Light, Soil	1776 (100%)	0 (0%)
Hum, Temp, Light	1680 (94.5946%)	96 (5.4054%)

The accuracy and performance of resulting classifiers using data inputs with last 2 h average is shown in Table 9 for the 4 and 3 most important features, respectively.

Overall, the classification results showed that the decision of watering or not a plant can be made using a reduced number of sensors. With 5 min transmission rate, the accuracy for decision-making achieved 95% when the least important feature has been removed. With the last 2 h average data set, the accuracy for decision-making achieved 97% when reducing the features to 3.

Table 9. The accuracy and performance of resulting classifiers using data inputs with last 2 h average for the 4 and 3 most important features.

Features	Correctly Classified Instances	Incorrectly Classified Instances
Hum, Temp, Light, Soil	1752 (99.943%)	1 (0.057%)
Hum, Temp, Light	1698 (96.8625%)	55 (3.1375%)

Often the acceptable level of accuracy is user defined, depending the nature of the subject or scenarios [41]. In this case, the accuracy reduces from nearly 100% to 97% and 95%, which means the error is within 5%. In statistics, when the type of error rate is within 5%, which is acceptable to have a 5% probability of incorrectly rejecting the true null hypothesis [41]. In addition, it is a common practice.

This approach can be promising for a large-scale deployment. The sum of a large amount of data from the least important sensor(s) might be reduced, if using appropriate data-mining methods to select sensors which are more important to the chosen decision-making.

6. Conclusions

This paper addresses the open challenge of feature reduction in IoT systems for agricultural plant-monitoring and decision-making support. Our data reduction approach is unsupervised learning using Laplacian scores. This approach is especially useful when class labels are unavailable. Using similarity and difference, features are ranked, so that users can select the most important features, rather than the whole feature set. Giving high resolutions of some features in real-world IoT applications, this will help reduce the volume of data to be transmitted. To evaluate our proposal, a real-world strawberry-plant monitoring IoT system has been implemented, calibrated and tested, measuring real-condition parameters such as temperature, relative humidity, soil moisture and light intensity. Our research has demonstrated that the proposed feature reduction can significantly reduce the volume data required to be transferred from the LoRa Node (edge device) to the network, while keeping the IoT system functioning at high accuracy levels. Moreover, the proposed IoT system has been tested on a specific decision-making support task (to water or not to water). The experimental results clearly show that the accuracy of decision-making on the reduced data decreases at an acceptable level (only 3–5%). The proposed research can potentially be used and provide insights for a rich range of decision-making tasks related to agricultural monitoring which can release the burden of data volume off the IoT systems.

In the future, this work can be expanded to another decision-making task except for watering a plant. For instance, if a greenhouse includes cooling fans, the event of turning them on/off could be controlled through an IoT system, similarly to what is proposed above. Strawberry and any other plants are very sensitive to very high/low level values of temperature or relative humidity so this could prevent them from being destroyed. Moreover, farmers can take advantage of this decision-making support to become more efficient on the usage of cooling fans, preventing high amount electricity bills. This decision-making scenario is planned to be conducted in the future when a greenhouse with such cooler fans is identified.

Author Contributions: Conceptualization, G.T., X.D., G.F. and N.J.; Data curation, G.T. and N.J.; Formal analysis, G.T., X.D. and N.J.; Funding acquisition, G.F. and N.J.; Investigation, G.T., X.D. and N.J.; Methodology, G.T., X.D., G.F. and N.J.; Project administration, X.D., G.F. and N.J.; Resources, G.F. and N.J.; Software, G.T. and N.J.; Supervision, X.D., G.F. and N.J.; Validation, G.T., X.D., G.F. and N.J.; Visualization, G.T. and N.J.; Writing–original draft, G.T., X.D., G.F. and N.J.; Writing–review & editing, G.T., X.D., G.F. and N.J. All authors have read and agreed to the published version of the manuscript.

Funding: Nanlin Jin is partly funded by Landslide Mitigation Informatics (LIMIT): Effective decision-making for complex landslide geo-hazards provided by NERC (NE/T005653/1) for this research.

Acknowledgments: The authors would like to thank Alun Moon and David Kendall of Northumbria University at Newcastle for the support throughout the project. The authors would also like to thank K & M Yiannoukkou Strawberry Production for the providence of their greenhouse premises for real-world plant-monitoring, data gathering and testing.

Conflicts of Interest: The authors declare no conflict of interest.

References

1. Feng, X.; Laurence, Y.; Lizhe, W.; Alexey, V. Internet of Things. *Int. J. Commun. Syst.* **2012**, *25*, 1101–1102.
2. Li, R.; Song, T.; Capurso, N.; Yu, J.; Couture, J.; Cheng, X. IoT Applications on Secure Smart Shopping System. *IEEE Internet Things J.* **2017**, *4*, 1945–1954. [CrossRef]
3. Venkatesh, J.; Aksanli, B.; Chan, C.S.; Akyurek, A.S.; Rosing, T.S. Modular and Personalized Smart Health Application Design in a Smart City Environment. *IEEE Internet Things J.* **2018**, *5*, 614–623. [CrossRef]
4. Zhang, H.; Li, J.; Wen, B.; Xun, Y.; Liu, J. Connecting Intelligent Things in Smart Hospitals Using NB-IoT. *IEEE Internet Things J.* **2018**, *5*, 1550–1560. [CrossRef]
5. Tsappparellas, G. GitHub–LoRaWAN_mbed_lmic_agriculture_app. 2018. Available online: https://github.com/GTsapparellas/LoRaWAN_mbed_lmic_agriculture_app (accessed on 11 April 2018).
6. Gaura, E.I.; Brusey, J.; Allen, M.; Wilkins, R.; Goldsmith, D.; Rednic, R. Edge Mining the Internet of Things. *IEEE Sens. J.* **2013**, *13*, 3816–3825. [CrossRef]
7. Zhang, X.; Ma, Y.; Qi, H.; Gao, Y.; Xie, Z.; Xie, Z.; Zhang, M.; Wang, X.; Wei, G.; Li, Z. Distributed Compressive Sensing Augmented Wideband Spectrum Sharing for Cognitive IoT. *IEEE Internet Things J.* **2018**, *5*, 3234–3245. [CrossRef]
8. Rani, M.; Dhok, S.; Deshmukh, R. A Machine Condition Monitoring Framework Using Compressed Signal Processing. *Sensors* **2020**, *20*, 319. [CrossRef] [PubMed]
9. Wen, D.; Zhu, G.; Huang, K. Reduced-Dimension Design of MIMO Over-the-Air Computing for Data Aggregation in Clustered IoT Networks. *IEEE Trans. Wirel. Commun.* **2019**, *18*, 5255–5268. [CrossRef]
10. Englert, F.; Diaconita, I.; Reinhardt, A.; Alhamoud, A.; Meister, R.; Backert, L.; Steinmetz, R. Reduce the Number of Sensors: Sensing Acoustic Emissions to Estimate Appliance Energy Usage. In Proceedings of the 5th ACM Workshop on Embedded Systems For Energy-Efficient Buildings, Italy, Rome, 11 November 2013; pp. 1–8.
11. Trihinas, D.; Pallis, G.; Dikaiakos, M. Low-Cost Adaptive Monitoring Techniques for the Internet of Things. *IEEE Trans. Serv. Comput.* **2018**. [CrossRef]
12. Trihinas, D.; Pallis, G.; Dikaiakos, M.D. ADMin: Adaptive monitoring dissemination for the Internet of Things. In Proceedings of the IEEE Conference on Computer Communications, Atlanta, GA, USA, 1–4 May 2017; pp. 1–9.
13. Bhuyan, M.H.; Bhattacharyya, D.K.; Kalita, J.K. Network Anomaly Detection: Methods, Systems and Tools. *IEEE Commun. Surv. Tutor.* **2014**, *16*, 303–336. [CrossRef]
14. Pajouh, H.H.; Javidan, R.; Khayami, R.; Dehghantanha, A.; Choo, K.R. A Two-Layer Dimension Reduction and Two-Tier Classification Model for Anomaly-Based Intrusion Detection in IoT Backbone Networks. *IEEE Trans. Emerg. Top. Comput.* **2019**, *7*, 314–323. [CrossRef]
15. Prathibha, S.R.; Hongal, A.; Jyothi, M.P. IOT Based Monitoring System in Smart Agriculture. In Proceedings of the 2017 International Conference on Recent Advances in Electronics and Communication Technology (ICRAECT), Bangalore, India, 16–17 March 2017; pp. 81–84. [CrossRef]
16. Aqeel-Ur-Rehman.; Abbasi, A.Z.; Islam, N.; Shaikh, Z.A. A Review of Wireless Sensors and Networks' Applications in Agriculture. *Comput. Stand. Interfaces* **2014**, *36*, 263–265. [CrossRef]
17. Zhao, W.; Lin, S.; Han, J.; Xu, R.; Hou, L. Design and Implementation of Smart Irrigation System Based on LoRa. In Proceedings of the 2017 IEEE Globecom Workshops (GC Wkshps), Singapore, 4–8 December 2017; pp. 1–6.
18. MoboDexter. IoT Solutions for Agriculture. 2018. Available online: https://www.mobodexter.com/wp-content/uploads/2018/07/Whitepaper_on_IOT_Solution_for_Agriculture.pdf (accessed on 10 August 2018).
19. Jawad, H.M.; Nordin, R.; Gharghan, S.K.; Jawad, A.M.; Ismail, M. Energy-Efficient Wireless Sensor Networks for Precision Agriculture: A Review. *Sensors* **2017**, *17*, 1781. [CrossRef] [PubMed]
20. LoRaAlliance. LoRaWAN–What is It? A technical overview of LoRa and LoRaWAN. 2015. Available online: https://docs.wixstatic.com/ugd/eccc1a_ed71ea1cd969417493c74e4a13c55685.pdf (accessed on 10 October 2017).

21. Stoces, M.; Vanek, J.; Masner, J.; Pavlik, J. Internet of Things (IoT) in Agriculture–Selected Aspects. *Agris-Line Pap. Econ. Inform.* **2016**, *8*, 83–88. [CrossRef]
22. Bahşi, H.; Nõmm, S.; La Torre, F.B. Dimensionality Reduction for Machine Learning Based IoT Botnet Detection. In Proceedings of the 2018 15th International Conference on Control, Automation, Robotics and Vision (ICARCV), Singapore, 18–21 November 2018; pp. 1857–1862.
23. Al-Otaibi, R.; Jin, N.; Wilcox, T.; Flach, P. Feature Construction and Calibration for Clustering Daily Load Curves from Smart Meter Data. *IEEE Trans. Ind. Inform.* **2016**, *12*, 1–10. [CrossRef]
24. Meidan, Y.; Bohadana, M.; Mathov, Y.; Mirsky, Y.; Shabtai, A.; Breitenbacher, D.; Elovici, Y. N-BaIoT—Network-Based Detection of IoT Botnet Attacks Using Deep Autoencoders. *IEEE Pervasive Comput.* **2018**, *17*, 12–22. [CrossRef]
25. He, X.; Cai, D.; Niyogi, P. Laplacian Score for Feature Selection. In *Advances in Neural Information Processing Systems 18*; Weiss, Y., Schölkopf, B., Platt, J.C., Eds.; MIT Press: Cambridge, UK, 2006; pp. 507–514.
26. Alalga, A.; Benabdeslem, K.; Taleb, N. Soft-constrained Laplacian score for semi-supervised multi-label feature selection. *Knowl. Inf. Syst.* **2016**, *47*, 75–98. [CrossRef]
27. Huang, R.; Jiang, W.; Sun, G. Manifold-based constraint Laplacian score for multi-label feature selection. *Pattern Recognit. Lett.* **2018**, *112*, 346–352. [CrossRef]
28. Zhang, Y.; Wang, Q.; Gong, D.W.; Song, X.F. Nonnegative Laplacian embedding guided subspace learning for unsupervised feature selection. *Pattern Recognit.* **2019**, *93*, 337–352. [CrossRef]
29. Doan, N.; Azzag, H.; Lebbah, M. Hierarchical Laplacian Score for unsupervised feature selection. In Proceedings of the 2018 International Joint Conference on Neural Networks (IJCNN), Rio de Janeiro, Brazil, 8–13 July 2018; pp. 1–7.
30. Luo, M.; Nie, F.; Chang, X.; Yang, Y.; Hauptmann, A.G.; Zheng, Q. Adaptive Unsupervised Feature Selection With Structure Regularization. *IEEE Trans. Neural Netw. Learn. Syst.* **2018**, *29*, 944–956. [CrossRef] [PubMed]
31. Seeedstudio. Grove–Moisture Sensor. 2018. Available online: http://wiki.seeedstudio.com/Grove-Moisture_Sensor/ (accessed on 10 January 2018).
32. Seeedstudio. Grove–Temperature & Humidity Sensor. 2018. Available online: http://wiki.seeedstudio.com/Grove-TemperatureAndHumidity_Sensor/ (accessed on 10 January 2018).
33. Seeedstudio. Grove–Light Sensor. 2018. Available online: http://wiki.seeedstudio.com/Grove-Light_Sensor/ (accessed on 10 January 2018).
34. Mbed, A. FRDM–K64F. 2018. Available online: https://os.mbed.com/platforms/FRDM-K64F/ (accessed on 10 January 2018).
35. Mbed, A. SX1272MB2xAS/SX1272MB2DAS. 2018. Available online: https://os.mbed.com/components/SX1272MB2xAS/ (accessed on 10 January 2018).
36. Dragino. LG01 LoRa Gateway User Manual. 2018. Available online: http://www.dragino.com/downloads/downloads/UserManual/LG01_LoRa_Gateway_User_Manual.pdf (accessed on 20 January 2018).
37. Network, T.T. The Things Network–Building a Global Internet of Things Network Together. 2018. Available online: https://www.thethingsnetwork.org/ (accessed on 25 January 2018).
38. AllThingsTalk. AllThingsTalk–Make IoT Ideas Happen. 2018. Available online: https://www.allthingstalk.com/ (accessed on 20 February 2018).
39. Stokking, J. The Things Network Architecture. 2017. Available online: https://www.thethingsnetwork.org/article/the-things-network-architecture-1 (accessed on 29 January 2018).
40. AllThingsTalk. Domain Model. 2018. Available online: https://www.allthingstalk.com/faq/domain-model (accessed on 20 February 2018).
41. Lindenmayer, D.; Burgman, M. *Practical Conservation Biology*; CSIRO: Canberra, Australia, 2005.

Article

Mask Gradient Response-Based Threshold Segmentation for Surface Defect Detection of Milled Aluminum Ingot

Ying Liang [1,†], Ke Xu [1,*,†] and Peng Zhou [2,†]

1 Collaborative Innovation Center of Steel Technology, University of Science and Technology Beijing, Beijing 100083, China; liangyinghero@gmail.com

2 Research Institute of Artificial Intelligence, University of Science and Technology Beijing, Beijing 100083, China; zhoupeng@nercar.ustb.edu.cn

* Correspondence: xuke@ustb.edu.cn

† These authors are equally contributed to this work.

Received: 16 July 2020; Accepted: 10 August 2020; Published: 12 August 2020

Abstract: The surface quality of aluminum ingot is crucial for subsequent products, so it is necessary to adaptively detect different types of defects in milled aluminum ingots surfaces. In order to quickly apply the calculations to a real production line, a novel two-stage detection approach is proposed. Firstly, we proposed a novel mask gradient response-based threshold segmentation (MGRTS) in which the mask gradient response is the gradient map after the strong gradient has been eliminated by the binary mask, so that the various defects can be effectively extracted from the mask gradient response map by iterative threshold segmentation. In the region of interest (ROI) extraction, we combine the MGRTS and the Difference of Gaussian (DoG) to effectively improve the detection rate. In the aspect of the defect classification, we train the inception-v3 network with a data augmentation technology and the focal loss in order to overcome the class imbalance problem and improve the classification accuracy. The comparative study shows that the proposed method is efficient and robust for detecting various defects on an aluminum ingot surface with complex milling grain. In addition, it has been applied to the actual production line of an aluminum ingot milling machine, which satisfies the requirement of accuracy and real time very well.

Keywords: surface inspection; aluminum ingot; mask gradient response; Difference of Gaussian; inception-v3

---

1. Introduction

Surface defect detection is a critical step of the metal industry. Since the technologies under development are becoming more and more feasible, and the results are reliable enough for a decision, the optical non-destructive testing (ONDT) has gained more and more attention in this filed. This is mainly due to the development of the used tools: laser, cameras, and those faster computers that are capable of processing large amounts of encrypted data in optical measurements [1]. A review has been provided in [2], which is about the main ONDT technologies, including fiber optics [3], electronic speckle [4], infrared thermography [5], endoscopic, and terahertz technology. The focus of this paper is the digital speckle measurement method because of the use of CCD technology and advanced computer vision technologies. For the high-quality inspection of various types of materials in all kinds of environments, the advanced computer vision technologies have evolved into a mainstream and replaced the conventional manual inspection method, improving on its inefficiency and high labor intensity.

Texture analysis provides a very powerful tool to detect defects in applications for visual inspection, since textures provide valuable information about the features of different materials [6]. In computer vision, texture is broadly divided into two main categories: statistical and structural. As shown in Figure 1, statistical textures are isotropic and do not have easily identifiable primitives. In contrast, structural (or patterned) textures are characterized by a set of repetitive primitives and placement rules, as shown in Figure 2. Both the statistical and structural textures appear as homogeneous (Figure 1a,b and Figure 2a,b) or inhomogeneous (Figure 1c,d and Figure 2c,d). It should be noted that Figures 1d and 2b are respectively quoted from Reference [7] and Reference [8]. As can been seen, the milling surface we deal with features structured homogeneous or inhomogeneous textures (Figure 2a,c,d).

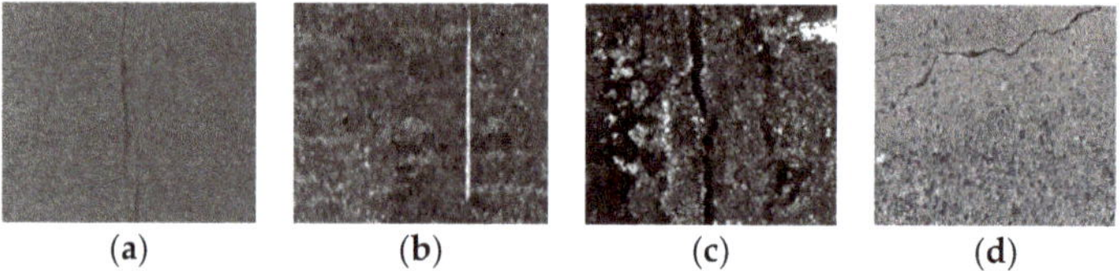

Figure 1. Statistical textures examples. (**a**) Hot-rolled steel strips surface, homogeneous; (**b**) Con-casting slabs surface, homogeneous; (**c**) Con-casting slabs surface, inhomogeneous; (**d**) Bridge deck inhomogeneous.

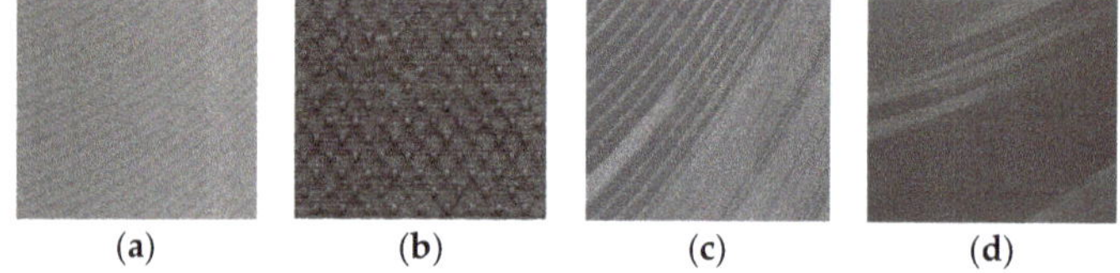

Figure 2. Structural textures examples. (**a**) Milled surface of aluminum ingot, oriented, homogeneous; (**b**) Fabric, isotropic, homogeneous; (**c**,**d**) Milled aluminum ingot surface, inhomogeneous.

In order to enable automatic and non-destructive detection, visual inspection systems have found wide applications in surface detection such as concrete structures [7,9–11] and metal surfaces [12–30]. In the field of concrete structure, there are lots of studies that try to inspect cracks from image analysis [7,9–11]. In the field of metal surfaces, visual inspection systems have been applied in both ferrous metal and nonferrous metal surface detection. For the nonferrous metals, methods to detect the surface defects of various products such as aluminum strips [12–14], aluminum foils [15], and aluminum profiles [16–18] have been well established. About the ferrous metal, the types of steel surfaces studied for defect detection based on vision include slab [14,19,20], plate [21–23], hot strip [24–26], and cold strip [27–29]. The comprehensive survey for typical flat steel products can be found in [30]. In general, the above defect detection techniques can be roughly divided into three categories: statistical, filtering, and machine learning.

The statistical method is to establish a mathematical model using probability theory and mathematical statistics, which can be used to infer, predict, quantitatively analyze, and summarize the spatial distribution data of pixels [31]. Reference [7] presents a multiple features-based cracks detection algorithm of bridge decks. A comprehensive analysis of multiple features (intensity-based, gradient-based, and scale-space) and multiple classifiers (random forests, support vector machines, and adaboost) show a peak classifier performance of 95%. Reference [24] proposed a simple yet robust feature descriptor against noise named the adjacent evaluation completed local binary patterns for hot-rolled steel strip surface defects recognition. Filtering-based methods commonly apply a filter bank to an image to calculate the energy of the filter response. To provide an efficient multi-scale directional representation of different defects, the shearlet transform is introduced in [14]. With the popularity of artificial intelligence in recent years, machine learning has been applied extensively in surface defect detection. Reference [9] used a supervised machine learning method called light gradient boosting machine (LightGBM) to detect cracks from the concrete surface imagery. The features are derived from pixel values and geometric shapes of cracks. In addition, spectral filtering approaches are suitable for the defect detection of uniform textured images composed of basic texture primitives with a high degree of periodicity [32]. Fourier transform (FT) was used in [33] to detect defects in directionally textured surfaces. Nevertheless, the FT-based approaches are inadequate under the circumstances that Fourier frequency components related to the background and defect areas are highly mixed together [34]. Gabor wavelet was used in [30] to extract features of images with periodic texture. Wavelet transform has been successfully applied in defect detection on statistical surfaces such as cold-rolled steel strips [27] and hot-rolled steel strips [35], and it has also been well used for homogeneous patterned surfaces [36]. Navarro et al. [6] present a wavelet reconstruction scheme to detect defects in a wide variety of structural and statistical textures.

Recently, fine-designed deep convolutional neural networks have emerged as powerful tools in a variety of computer vision tasks. Reference [10] proposed an improved You Only Look Once (YOLOv3) with transfer learning, batch renormalization, and focal loss for concrete bridge surface damage detection. The improved single-stage detector achieved a detection accuracy of 80% on a dataset containing a total of 2206 inspection images labeled with four types of concrete damages. Reference [11] proposed a crack detection method based on deep fully convolutional network (FCN) semantic segmentation with the VGG16 backbone on concrete crack images. The FCN network is trained end-to-end on a subset of 500 annotated 227 × 227-pixel crack-labeled images and achieves about 90% in average precision. An end-to-end steel strip defect detection network model was outlined in [28]; this system is based on the symmetric surround saliency map for surface defects detection and deep convolutional neural networks (CNNs) for seven classes of steel strip defects classification. To inspect the defects of a steel surface, Reference [23] presents a new classification priority network (CPN) and a new classification network, multi-group convolutional neural network (MG-CNN).

However, these defect detection methods are primarily used for only crack defects on concrete structures or metal surfaces with non-texture backgrounds. As far as we know, there is no literature on the surface defect detection of aluminum ingots with a milling grain background. The surface of aluminum ingot after milling always has multi-directional and multi-scale grinding texture patterns; sometimes, the distribution of the grinding ridge is uneven. After milling, various surface defects (Figure 3) will appear on the surface of aluminum ingot such as small local defects (Figure 3a), distributed defects with complex texture and fuzzy boundaries (Figure 3b-d), longitudinal linear defects throughout the whole picture (Figure 3e), and large-scale distributed defects with irregular shapes (Figure 3f). In addition, there are many pseudo defects with various patterns on the surface of aluminum ingot, such as aluminum chips (AC), mosquito (Mo) (Figure 3g), and the milling grain (Figure 3h). These factors greatly increase the difficulty of defect detection and recognition. To handle these problems, we propose a detection algorithm of aluminum ingot surface defects combining traditional detection and deep learning classification, which has been applied to the production line of an aluminum ingot milling surface.

The main contributions of the paper are summarized below:

1. In terms of ROI extraction in an aluminum ingot image, we design a novel mask gradient response-based threshold segmentation algorithm to iteratively separate out defects of varying significance. In addition, the combination of the mask gradient response-based threshold segmentation (MGRTS) and Difference of Gaussian (DoG) can effectively improve the detection rate of the above defects.
2. In the classification stage, we use the inception-v3 network structure with focal loss in the training process and data augmentation technologies to overcome the class imbalance problem and realize the accurate identification of various defects.
3. Our method can make full use of central processing unit (CPU) and graphics processing unit (GPU) resources in a workstation or server. Even if the server used in the production line is not configured with GPU, the algorithm can still ensure the realization of rapid defect detection.

4. At the beginning of the project, even without a large number of labeled samples, the algorithm can still deploy and detect the suspicious regions quickly owing to the improved ROI detection algorithm.

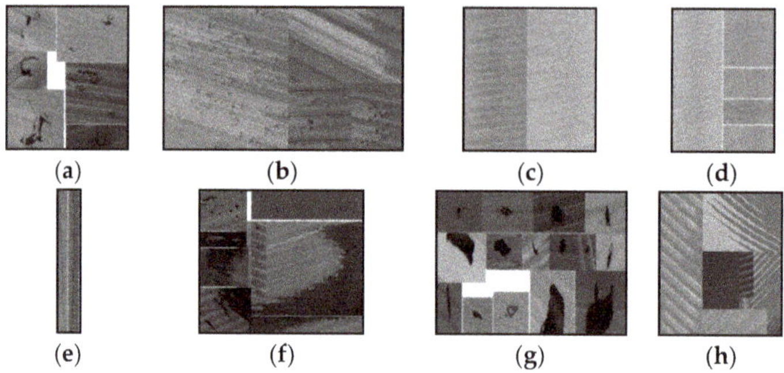

Figure 3. Samples of different defects: (**a**) Slag inclusion (SI); (**b**) Pitted slag inclusion (PSI); (**c**) Adhesion aluminum (AA); (**d**) Scratches (Sc); (**e**) Crack (Cr); (**f**) Oxide film (OF); (**g**) Mosquito (Mo) and aluminum chips (AC); (**h**) Texture background (Tb).

2. Materials and Methods

In this section, the proposed two-stage surface defects detection method will be introduced in detail. Since there is no similar defect database, at the beginning of the project, we need to preliminarily detect the area of interest and collect defect samples. Therefore, we cannot use the end-to-end network which needs a large number of labeled defect images; instead, we design a two-stage target detection method. As shown in Figure 4, the proposed method has two main components: (1) ROI extraction based on the combination of MGRTS, DoG, and similar area merge, and (2) defect ROI classification.

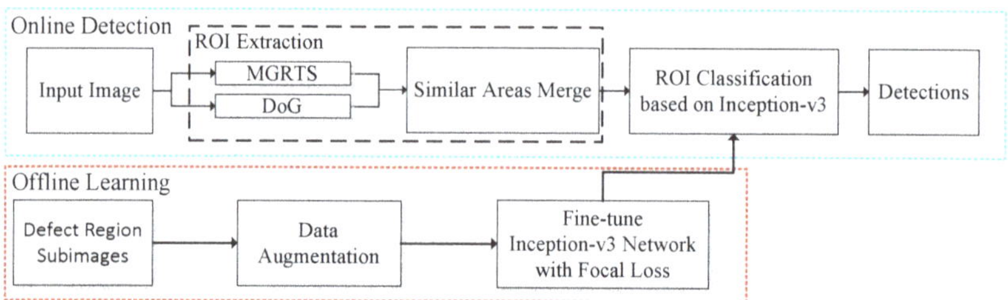

Figure 4. Overview of the defect detection method applied to a real production line. DOG: Difference of Gaussian, MGRTS: mask gradient response-based threshold segmentation, ROI: region of interest.

2.1. ROI Extraction

In the region of interest (ROI) extraction stage, in order to ensure the detection rate of defects, DOG and edge detection with the MGRTS are used to jointly complete the detection of suspicious areas of defects, and special post-processing is adopted to merge similar areas that may be distributed defects. In the proposed method, edge detection with MGRTS can iteratively segment most of the suspicious regions of defects, while the DoG method is mainly used to detect large-scale defects that cannot be completely segmented by MGRTS, and defects that can be missed by the MGRTS when the background texture gradient is strong.

2.1.1. MGRT-Based Iterative Threshold Segmentation

In the MGRTS, the mask gradient response is the gradient map after the strong gradient has been eliminated by the binary mask, so that various defects can be effectively extracted from the mask gradient response map by iterative threshold segmentation. The operation process is as follows.

Firstly, we calculate the horizontal gradient of the original image and get the gradient response map of the Original Gradient (OG). Then, an adaptive threshold segmentation is used to get the binary image of the OG. Next, the binary image is used as a mask to eliminate the strong gradient region on the OG, thus obtaining the mask gradient response. As an iteration, we then repeat the first step on the mask gradient response map. Finally, the binary images obtained by each iteration are combined to obtain the segmentation results of different significant defects.

As shown in Figure 5, the original image (Figure 5a) contains aluminum chips and scratches, and Figure 5b is the gradient response map based on the Sobel operator. By using iterative threshold segmentation guided by mask gradient response maps, the defect areas (Figure 5f) are segmented from the gradient map. In each iteration, adaptive threshold segmentation is realized by Equation (1) and Equation (2).

$$G(x, y) = \frac{1}{2\pi\sigma^2} e^{-\frac{x^2+y^2}{2\sigma^2}} \quad (1)$$

$$f_{bin}(x, y) = \begin{cases} 1, \ f(x, y) > \left[f(x, y) * G(x, y) + \lambda\sigma_g \right] \\ 0, \ f(x, y) \le \left[f(x, y) * G(x, y) + \lambda\sigma_g \right] \end{cases} \quad (2)$$

Equation (1) generates a Gaussian weight matrix of size m × m, where σ is the standard deviation. Equation (2) combines local Gaussian weighted sum and global standard deviation σ_g to adapt to local texture changes, so that the algorithm can better extract details and improve the detection of non-obvious defects. The * denotes the convolution operator, and λ is the weight coefficient.

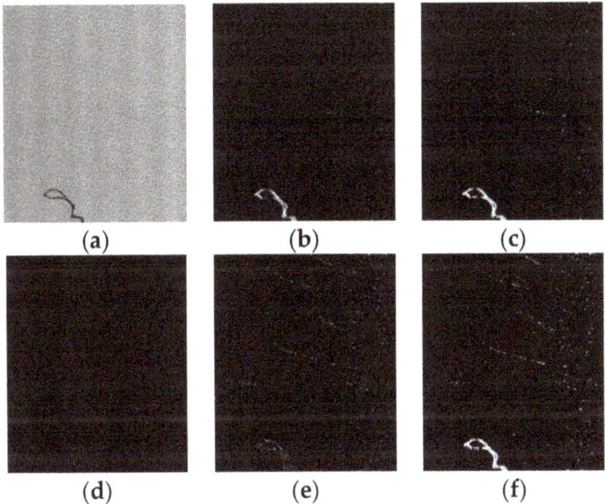

Figure 5. Iterative threshold segmentation of gradient map: (**a**) original image; (**b**) gradient response map (Original Gradient, or OG); (**c**) response map of the first threshold segmentation; (**d**) mask gradient (MG1) after the first threshold segmentation; (**e**) response map of the second threshold segmentation; (**f**) the final segmentation result.

As can be seen from Figure 5, when obvious defects (aluminum chips) and slight defects (scratches) exist at the same time, the slight defects cannot be completely segmented from the gradient map after the first threshold segmentation (Figure 5c). Therefore, we adapt the iterative method and design the termination conditions. Before the second threshold segmentation, the response graph of the first segmentation is reversed to obtain the mask. The mask is applied to the gradient map to eliminate the region with strong gradient value that has been segmented in the first time, and the gradient map (Figure 5d) for the second time is obtained. Figure 5e shows the response map of the second threshold segmentation. Finally, by combining the response maps (Figure 5c, Figure 5e) of the two segmentations, the final segmentation result (Figure 5f) is obtained.

The iteration termination conditions are made up of two parts: the maximum number of iterations and the change degree of Masked Gradient (MG). As long as one condition is satisfied, the iteration will be terminated. The maximum number of iterations is a super parameter N, and the change degree of masked gradient is calculated by Equations (3)–(5):

$$g_i = \text{mean}(MG_i) + \lambda std(MG_i)\,,\ i = 1, 2, \ldots, N \tag{3}$$

$$g_i = \text{mean}(MG_i) + \lambda std(MG_i)\,,\ i = 1, 2, \ldots, N \tag{4}$$

$$Isover = \begin{cases} true, & g_i - g_{i-1} \le \delta\, or\, i = N \\ false, & g_i - g_{i-1} \le \delta\, or\, i < N \end{cases} \tag{5}$$

where mean (MG) calculates the mean value of the masked gradient map, and std (MG) calculates the standard deviation. g_i is used to describe the information distribution of the masked gradient map, and it represents the information change of the gradient map after the i-th iteration. When i is equal to 1, $MGi - 1 = MG0$ is the original gradient OG. λ is the weight mentioned above, and δ is the threshold of change degree. Figure 6 shows the histogram and statistical information of the gradient map in different iterations. Figure 6a shows the histogram distribution and statistical information difference of the Original Gradient (OG) shown in Figure 6b and the Mask Gradient ($MG1$) after the first threshold segmentation shown in Figure 5d, where λ is set to 5 (We first set λ to 3, considering that at the value of 3 sigma, the confidence probability of a normal distribution is 99.7%. To achieve better recall rate and precision, we test the value range from 3 to 3.5 with a step of 0.5. According to the test results, λ can be set within the range of [3.5, 5]. For relatively simple surfaces such as aluminum strip, it can be set to 3.5).

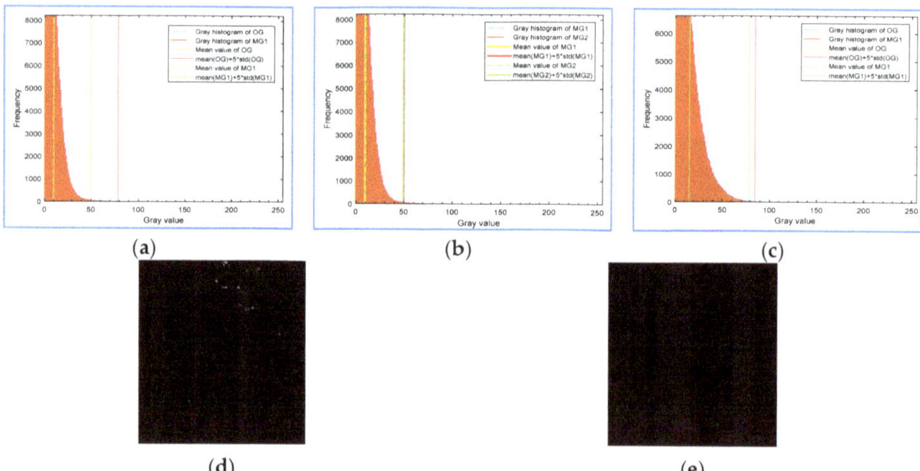

Figure 6. Histogram and statistical information of gradient graph in different iterations: (**a**) the original gradient (OG) and the mask gradient (MG1) after the first threshold segmentation; (**b**) the mask gradients MG1 and MG2 after the second threshold segmentation; (**c**) the OG and the MG1 of another test sample; (**d**) the OG; (**e**) the MG1.

It can be observed that after the first threshold segmentation, the statistical information value g_0 (red solid line in Figure 6a) of *OG* is very different from that (g_1) of *MG*1 (green solid line in Figure 6a), and $g_0 - g_1 = 28.3$, so the second threshold segmentation is needed. Figure 6b shows that after the second threshold segmentation, the statistical information value g_1 of *MG*1 (red solid line) and the statistical information value g_2 of *MG*2 (green solid line) have almost no difference, so the iteration can be ended. Figure 6c shows the distribution of statistical information of another sample after the first iteration of threshold segmentation. Figure 6d,e are the gradient map *OG* and *MG*1, respectively. For this sample, one segmentation is enough, so the distance (5.78) between g_0 and g_1 in Figure 6c provides a reference for the selection of the threshold.

2.1.2. Difference of Gaussians

Difference of Gaussian has been well used in Scale Invariant Feature Transform (SIFT) [37] to identify potential interest points that are invariant to scale and orientation. First, the scale space of an image is defined as a function, $L(x, y, \sigma)$, that is produced from the convolution of a variable-scale Gaussian, $G(x, y, \sigma)$ (defined in Equation (2)) with the input image $f(x, y)$,

$$L(x,y,\sigma) = G(x,y,\sigma) * f(x,y). \tag{6}$$

Then the result image of DoG can be the Difference of Gaussian function convolved with the image, $D(x, y, \sigma)$, which can be computed from the difference of two nearby scales separated by a constant multiplicative factor k,

$$\begin{aligned} D(x,y,\sigma) &= (G(x,y,k\sigma) - G(x,y,\sigma)) * f(x,y) \\ &= L(x,y,k\sigma) - L(x,y,\sigma) \end{aligned}. \tag{7}$$

Considering the time consumption and defect scale, only two scales $\sigma = 0$ (the original image) and $\sigma = 7.1$ (the corresponding window size is 45) are used, and the result of DoG is

$$D(x,y) = G(x,y,7.1) * f(x,y) - f(x,y). \tag{8}$$

The construction of $D(x, y)$ for a surface defect image of aluminum ingot is shown in Figure 7. Figure 7a is the original image with an oxide film defect. Figure 7b is the fuzzy effect image after convolution of the Gaussian function with an original image. The Gaussian window is set as 45 according to the experiment. Figure 7c is the response map of the DoG calculated by Equation (8). In the collected image of an aluminum ingot surface, the gray value of the defect area is lower than the texture background in varying degrees, so this paper uses $G(x,y,7.1) * f(x,y) - f(x,y)$ to reduce the influence of the background. Figure 7d shows the result of $f(x,y) - G(x,y,7.1) * f(x,y)$, which introduces a part of the texture response compared with Figure 7c. Figure 7e is a binary image after segmentation with a fixed threshold, and it will be combined with the result image of MGRTS by a logical OR operator.

2.1.3. Similar Areas Merge

The segmentation results of MGRTS and DoG are merged, and the enclosing rectangle of each defect area is obtained by contour extraction after morphological expansion. In this way, we can locate the bounding box of defects with clear boundaries, but for the distributed defects without clear boundaries, we need to further integrate the similar region, so as to obtain the bounding box of distributed defects more completely. For each defect ROI, the mean value and standard deviation of the original image (src_1) (Figure 8a) and the gradient map (*OG*) (Figure 8b) are calculated respectively, and an information distribution descriptor v with a length of 4 is obtained. Figure 8 shows the examples of a similar region (red box) and dissimilar region (green box) in the original image and gradient image. As shown in Figure 9, for two dissimilar regions, the gray distribution histogram and statistical information (mean, standard deviation) of their original image and gradient image are different to some extent. However, for two similar regions, the difference is very small, as shown in Figure 10.

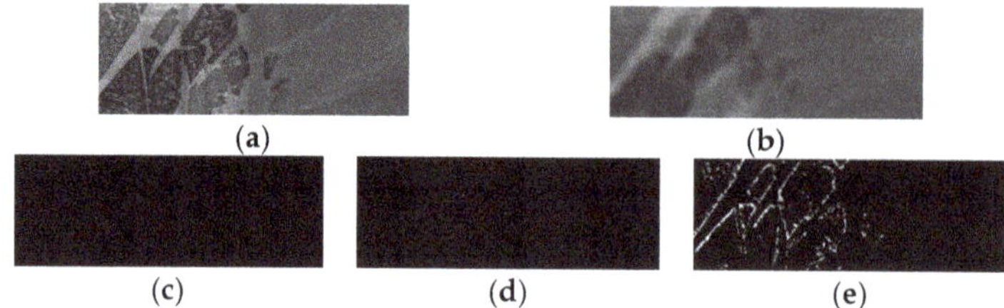

Figure 7. Difference of Gaussian of aluminum ingot image with a large-scale defect: (**a**) the original image; (**b**) the Gaussian blur effect; (**c**) the DoG response map calculated by Equation (8); (**d**) the DoG response map calculated by the opposite of Equation (8); (**e**) the binary image after thresholdsegmentation.

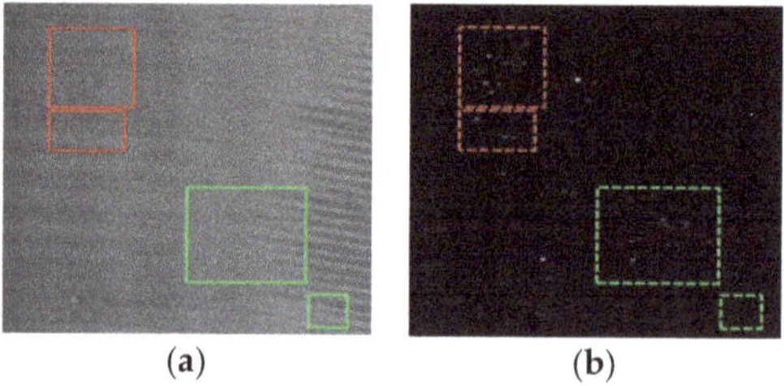

Figure 8. Examples of similar region (red) and dissimilar region (green) in (**a**) the original image and (**b**) the gradient image.

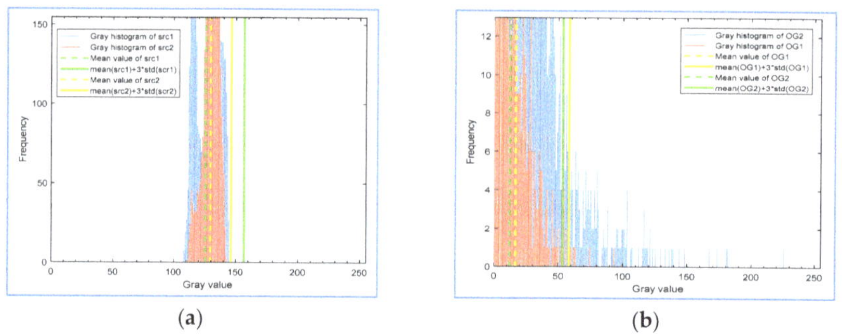

Figure 9. Gray histogram and statistical information of two dissimilar regions: (**a**) original images; (**b**) gradient images.

As shown in Figure 11, when iterating through the extracted candidate regions, we can decide whether to merge them into one window by calculating the spatial distance of two windows and the Euclidean distance of their information distribution descriptors.

If the two windows overlap, or the spatial Euclidean distance d_s (refer to Equation (9)) between their center points (p, q) is very close and less than the threshold δ_s, the information distribution vectors v_1, v_2 will be extracted, and the Euclidean distance dv (refer to Equation (10)) will be calculated.

$$d_s = \sqrt{(p_x - q_x)^2 + (p_y - q_y)^2} \tag{9}$$

$$d_v(v_1, v_2) = \sqrt{\sum_{i=1}^{4} (v_{1i} - v_{2i})^2} \tag{10}$$

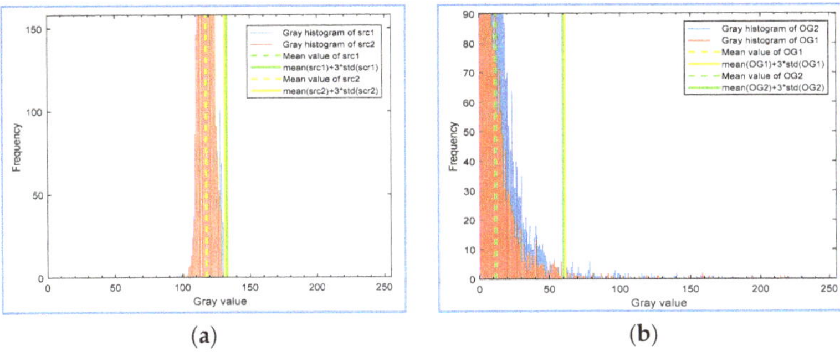

Figure 10. Histogram and statistical information of two similar regions: (**a**) original images; (**b**) gradient images.

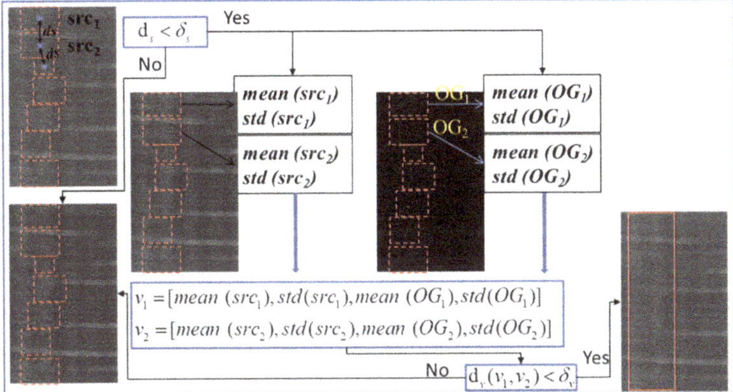

Figure 11. Schematic diagram of similar areas merging.

If the distance is less than the threshold δ_v, the two windows will be merged. After testing the effect of different values on the merge results from similar areas, we set δ_s to 150. As shown in Figure 12, the similar areas merge results are insensitive to the value of δ_s. For the pitted slag inclusion defect in this paper, it is better to set the threshold to 150. It is recommended to set the δ_s higher, as it ensures that similar areas will merge together as much as possible. Thus, defect ROIs can be completely detected, and the reduction of the ROI number will help to improve the speed of subsequent classification.

As for the value of δ_v, we calculate the Euclidean distance of v_1, v_2 extracted from 31,304 pairs of similar areas (Figure 13a) and 13,137 pairs of dissimilar areas (Figure 13b). As shown in Figure 13a, the profile of the histogram (H) is approximately normal distribution. According to the 3 sigma principle of normal distribution and the observation of the two histograms, we set δ_v = 1 (3*std(H) = 0.082).

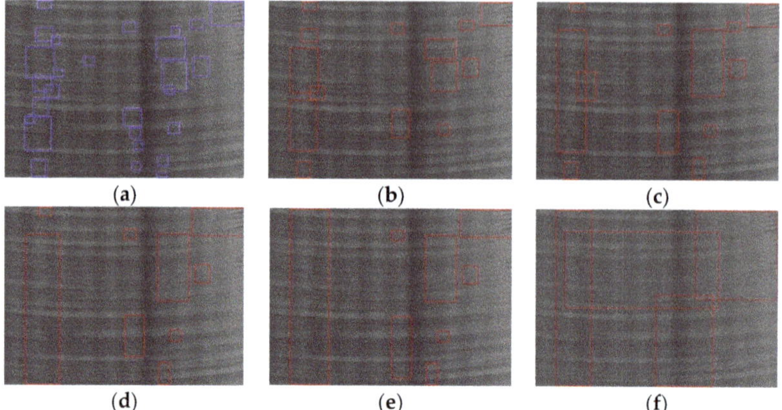

Figure 12. Similar areas merge results with different d_s: (**a**) Before similar areas merge; (**b**) Similar areas merge result when d_s is set to 30 and 40 (d_s = 30, 40); (**c**) d_s = 50; (**d**) d_s = 60,70,80,90; (**e**) d_s = 100,150; (**f**) d_s = 200.

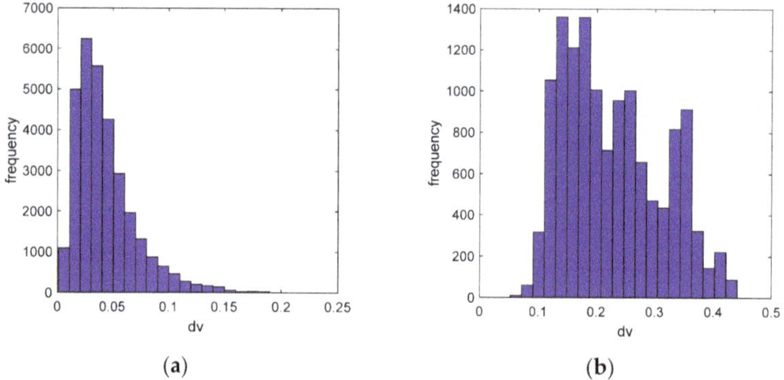

Figure 13. The Euclidian distance (d_v) histogram of similar areas and dissimilar areas: (**a**) Similar areas distance histogram; (**b**) Dissimilar areas distance histogram.

2.2. Defect ROI Classification

In the classification stage, considering the strong feature extraction and representation ability of the CNN network, we use the inception-v3 [38] network structure to realize the accurate identification of various defects with large intra-class variations and high inter-class similarity.

Inception [39] is a popular convolutional neural network model proposed by Google. Its unique and detailed inception block design makes the model increase the depth and width of the network while maintaining the same amount of calculation. The inception-v3 network is the third version. The biggest change of v3 version is to decompose the 7 × 7 convolution kernel into two 1 × 7 and 7 × 1 one-dimensional convolution kernels. In this way, the calculation can be accelerated, and one convolution layer can be divided into two, which can further increase the depth of the network and strengthen the nonlinearity of the network. Since the data set of aluminum ingot surface defects used in this paper is quite different from the data set of ImageNet [40], the method of fine tune is adopted to train the model parameters.

The aluminum ingot defect samples used in this paper are collected from an aluminum ingot production line in China. The data set has the problem that the number of real defect samples is very small, and the number of false defect and texture background samples is very large. In order to overcome the class imbalance problem and improve the accuracy of defect classification, we use data augmentation technology to preprocess the sample set and introduce the focal loss into the loss function.

2.2.1. Data Augmentation

Based on the analysis of the difficult cases in online application, we use basic image transformation such as flipping, contrast enhancement, sharpening, etc. to create a larger data set. Figure 14 shows the original defect images and the corresponding transformed images.

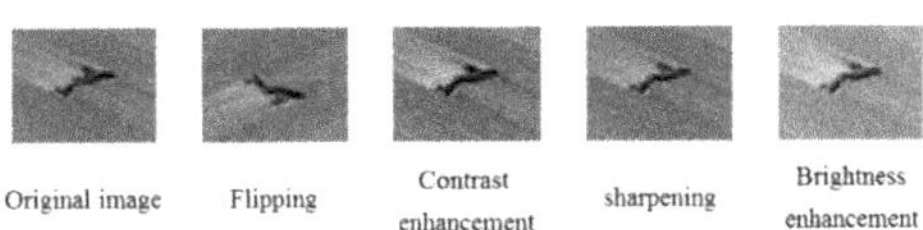

Figure 14. Image transformation of slag inclusion defect.

2.2.2. Focal Loss for Multi-Class

The focal loss [41] was designed by Lin et al. to address the one-stage object detection scenario in which there is an extreme imbalance between foreground and background classes during training. The focal loss for binary classification has been be given by Equation (11),

$$FL(p_t) = -(1-p_t)^{\gamma}\log(p_t), \tag{11}$$

where $p_t \in [0,1]$ is the estimated probability of the model for the class with label y = 1. $(1-p_t)^{\gamma}$ is a modulating factor with a tunable focusing parameter $\gamma \geq 0$ to down-weight easy examples and thus focus training on hard negatives. Similarly, for k-class classification, the formula of focal loss for multi-class (FLM) is as follows,

$$FLM(P_{k\times 1}) = -Y_{k\times 1}(1_{k\times 1} - P_{k\times 1})^{\gamma} Y_{k\times 1}\log(P_{k\times 1}), \tag{12}$$

where $Y_{k\times 1}$ is a one-hot label vector with k elements and $P_{k\times 1}$ is the model's estimated probability vector. The multiplication and logarithm here are all operations at the element level within a vector.

3. Results

The algorithm proposed in this paper is a two-stage target detection algorithm, so corresponding experiments are carried out to analyze and evaluate the performance of the ROI extraction and ROI classification algorithms. Finally, the performance of the whole algorithm is evaluated.

3.1. Evaluation Metric

In the actual production, the impact of defect missing detection is much more serious than that of false detection. Therefore, it is necessary for a surface defect detection system to have a high recall rate for real defects and ensure a high accuracy rate. In the experiments, the precision, recall, and F1-score are used to evaluate the system performance, and the accuracy is used to evaluate the classifier performance. These three metrics are defined as follows,

$$precision = \frac{TP}{TP+FP}, \; recall = \frac{TP}{TP+FN}, \; \text{F1-score} = 2 * \frac{precision * recall}{precision + recall}, \tag{13}$$

$$acc = \frac{TP+TN}{TP+FN+FP+TN}, \tag{14}$$

where TP represents the number of true positives, FP represents the number of false positives, FN represents the number of false negatives, and TN represents the number of true negatives.

3.2. Experimental Analysis of ROI Extraction Algorithm

In the MGRTS, the maximum number of iterations N and the change degree threshold are set to 5 and 6, and the Gaussian weight matrix size m in adaptive threshold segmentation is set to 25. In the DOG, we did experiments to choose the most appropriate Gaussian window size, which is related to the texture scale. We set the window sizes to 25, 35, 45, and 55 respectively to test the effect of DoG. The experiment result (Figure 15) shows that when the window size is 45, the effect is the best; that is, the DoG not only highlights the defect structure, but also suppresses most of the texture background.

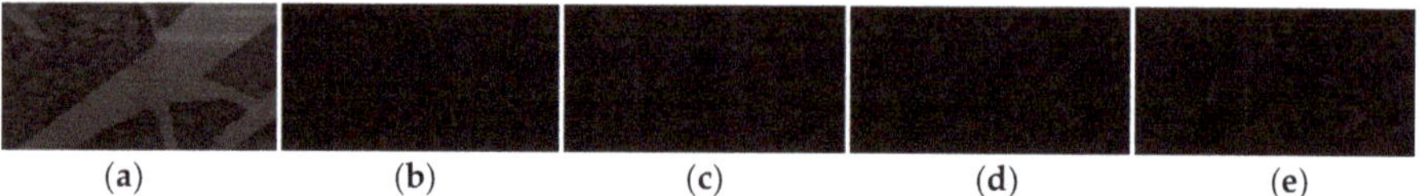

Figure 15. Cropped DoG results with different window sizes: (**a**) Original image; (**b**) DoG image with window size (ws) 25; (**c**) DoG image with ws 35; (**d**) DoG image with ws 45; (**e**) DoG image with ws 55.

We also experimented to test the ROI extraction effectiveness of MGRTS and DoG, and the performance of the algorithm that merges similar areas. Figure 16 shows a few representative results of different defects including oxide film (Figure 16a,b), oil stain (Figure 16c), pitted slag inclusion (Figure 16d), and crack (Figure 16e). The first line of Figure 16 is the binary response map of the MGRTS, the second line is the binary response map of DoG, and the third line is the result of adding the response map of MGRTS and DoG. As shown in Figure 16b, the MGRTS failed to segment the large-scale defect completely, and MGRTS also failed to detect the oil stain in Figure 16c, which is mixed in a dense texture background, but the DoG algorithm makes up for these two disadvantages. The fourth line of Figure 16 shows the enclosing rectangle of each defect area obtained by contour extraction after morphological expansion, and the last line is the final ROI bounding box after merging similar regions. It can be seen that for large-scale defects (Figure 16a,b) and distributed defects without obvious boundaries (Figure 16d), the similar region merging algorithm can integrate local regions to obtain a complete bounding box, which also reduces the number of ROI windows and improves the classification efficiency and accuracy.

Table 1 shows the quantitative evaluation of MGRTS and DoG, and the combination of MGRTS and DoG in terms of recall and precision. We tested on a defective images data set captured in an aluminum ingot milling machine production line. The data set consists of 180 images with the size of 4096 × 1024, including 153 defects such as oxide film, oil stain, crack, slag inclusion, and pitted slag inclusion. It can be seen that the combination of MGRTS and DoG boosts the ROI extraction performance, especially the recall rate, which is more important to the production line.

Table 1. Comparison of MGRTS, DoG, and the combination of both.

Method	Recall	Precision
MGRTS	97.4%	56.2%
DoG	54.9%	93.3%
MGRTS + DoG	99.3%	56.7%

3.3. Experimental Analysis of Defect ROI Classification

In the classification experiment, we used 32,665 defect ROI images of aluminum ingot, among which 10% sample images are selected randomly respectively as validation and training sets; the remaining 80% sample images are used as the training set. The specific number of each type of defect is shown in Table 2. It can be seen from the table that the sample number of each type of defect is extremely imbalanced.

Table 2. Specific of each type of defect used in the classification experiment.

Defects	SI	PSI	Cr	AA	Sc	OF	Mo	Tb	Total
Total	3709	5413	5204	347	1932	6062	902	9096	32,665
Train	2969	4331	4164	249	1546	4850	722	7278	26,139
Validation	370	541	520	34	193	606	90	909	3263
Test	370	541	520	34	193	606	90	909	3263

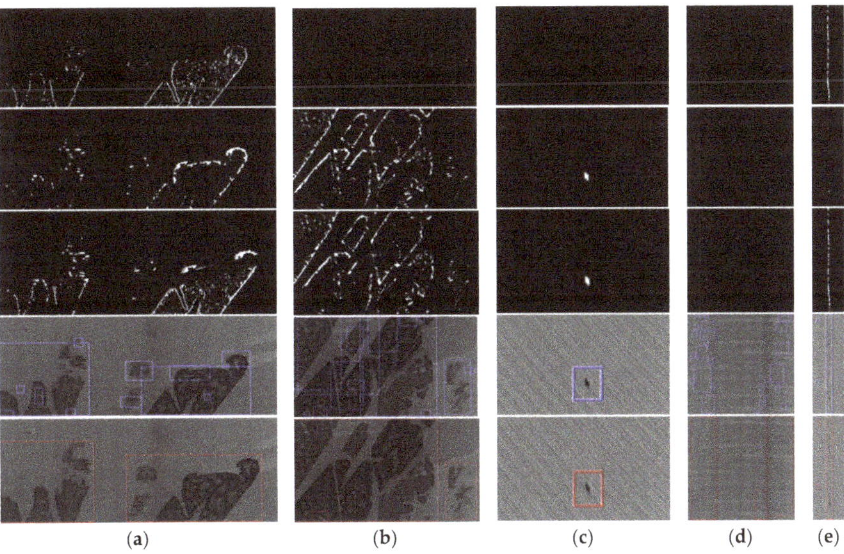

Figure 16. ROI extraction results of different defects: (**a**) OF; (**b**) OF; (**c**) Oil; (**d**) PSI; (**e**) Sc.

In order to verify the ability of focal loss to deal with sample imbalance, we compared the classification effect of using cross entropy loss and using focal loss in an inceptions-v3 network. Figure 17 shows the recall curve of the two methods for each type of defect on the test set. From the green curve in the figure, it can be seen that inception-v3 with focal loss significantly improved the recall rate of adhesion aluminum and mosquito defects with a relatively small number of samples.

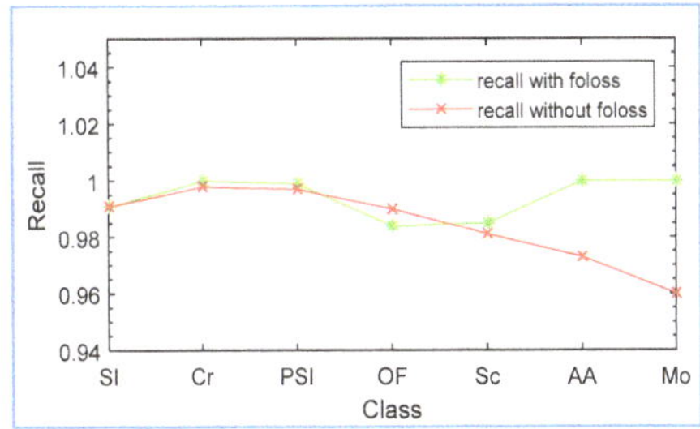

Figure 17. The recall curve of the two methods for each type of defect.

We also compared the improved inceptions-v3 network with the traditional machine learning method proposed in [42]. As described in [42], we also extracted seven features including anisometry, circinal rate, ratio between the width and area, compactness, rectangularity, elongation, and ratio between area and perimeter. Furthermore, the Artificial Neural Networks (ANN) was trained with the features extracted from the aluminum ingot defect images. The classification accuracy comparison is listed in Table 3. The ANN with extracted geometric features failed to recognize the AA and Mo, because the AA defects are similar with PSI, while the Sc and the Mo defects are similar with the SI in geometry. These seven features cannot distinguish them well.

Table 3. Classification accuracy of Artificial Neural Networks (ANN), inceptions-v3, and inceptions-v3 with focal loss.

Defects	SI	PSI	Cr	AA	Sc	OF	Mo	Average
ANN	83.0%	89.0%	84.0%	0.0%	59.0%	42.0%	0.0%	51.0%
inceptions-v3	99.1%	99.9%	100.0%	97.3%	98.1%	99.0%	99.6%	99.0%
inceptions-v3 with focal loss	99.1%	99.7%	99.8%	100.0%	98.5%	98.4%	100.0%	99.4%

3.4. Overall Performance Analysis of the Proposed Algorithm

We test the overall performance of our algorithm using the database of 180 defective images described above. On the premise that the detection resolution meets the needs of the industrial field, we down-sample the image as half of the original image to improve the processing speed.

As a contrast, we also test three one-stage target detection algorithms: YOLOv3 [43], RetinaNet [41], and YOLOv4 [44]. In order to match the network structure, improve the detection accuracy, and reduce the loss of large-scale sampling, we preprocess the original annotation image. First, the original image is down-sampled to half of the original image size, and then the aluminum ingot area image after boundary detection is divided into two parts, and finally, it is normalized to 512 × 512 for network training. Due to the small sample size of the original image, we augment the defective image to three times that of the original; 2/3 of it is used as the training set, and the remaining 1/3 is used as the test set.

Figure 18 shows the detection effect of the four methods for different defects. In order to prevent some small defects from being covered by the bounding boxes, we show the detection results of the four methods on four images. The detection results of YOLOv3, Retina Net, YOLOv4, and our method are shown from left to right in each group of comparison images. YOLOv3 uses multi-scale features to detect objects, and it shows a good ability to identify defects such as large-scale oxide film (Figure 18d), crack (Figure 18e,f), and small-scale slag inclusion (Figure 18c), even though the sample data set used in this paper is small. However, for the distributed pitted slag inclusion (Figure 18a,g) and the scratches (Figure 18b) with low contrast, the effect is poor, especially for the scratch defect, and the recall rate is very low. In contrast, our algorithm can detect scratches and pitted slag inclusion well because of using an iterative threshold segmentation of a masked gradient response map and the merging of similar regions.

Table 4 compares the recall (R), precision (P), and F1-score (F1) of the four methods for each type of defect. At the same time, the reasoning time of each algorithm is also listed. In experiments, the top-1 strategy was used in the statistics of detection results, and no threshold was set for the score. The average recall rate and precision of the algorithm in this paper are over 92.0%, but when influenced by a scratch defect, the average recall rate of YOLOv3 is only 66.1%. Meanwhile, RetinaNet is worse in detecting Cr defects. On the whole, the metrics of YOLOv4 are high, but similar to YOLOv3, the recall rate of scratch defects is low. Our algorithm has a relatively low recall and precision for the defects of pitted slag inclusion. The reason is that the resolution of some defects in the image is low, which affects the accuracy of defect classification. It can be seen from Table 4 that our algorithm achieves the highest F1-score and the shortest inference time.

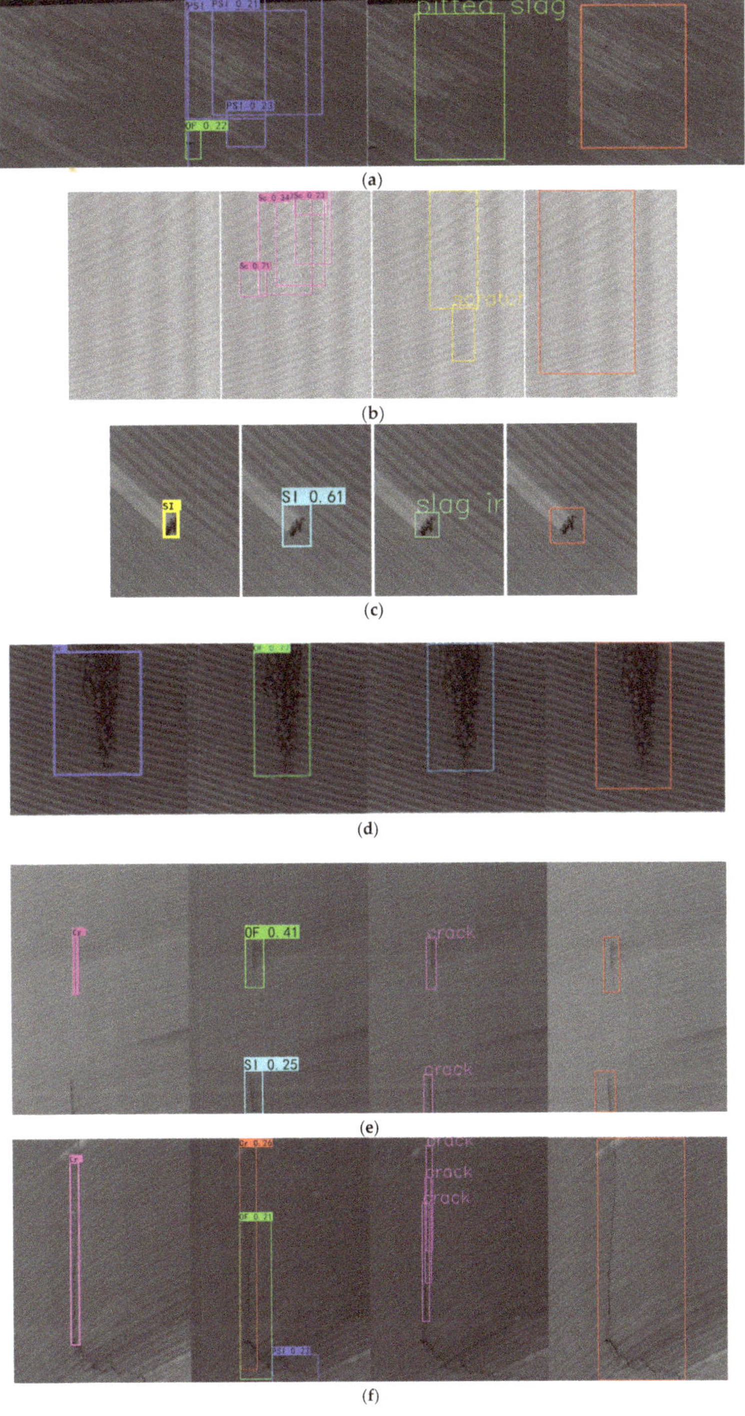

(a)

(b)

(c)

(d)

(e)

(f)

Figure 18. *Cont.*

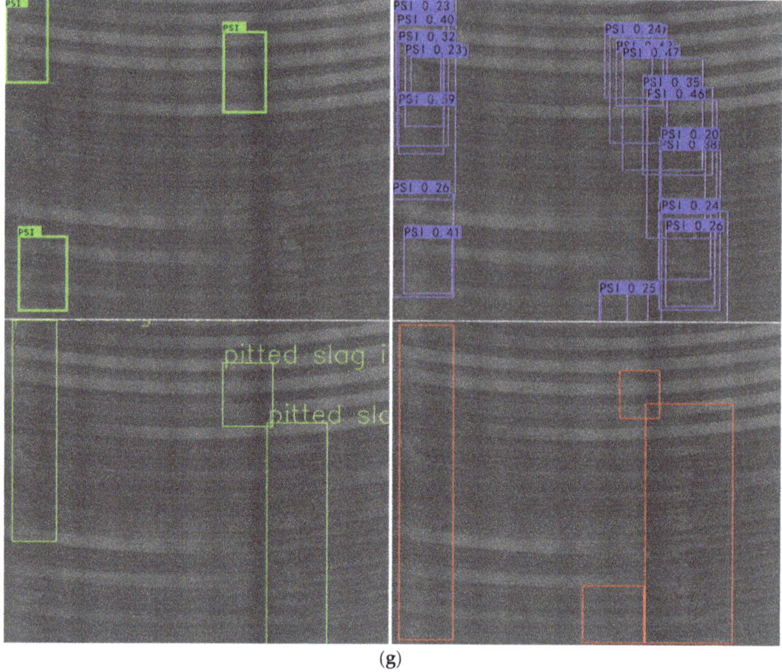

(**g**)

Figure 18. Detection effect of different methods: (**a**) PSI; (**b**) Sc; (**c**) SI; (**d**) OF; (**e**) Cr; (**f**) Cr; (**g**) PSI. The detection results of YOLOv3, RetinaNet, YOLOv4, and our method are shown from left to right in each group of comparison images.

Table 4. Performance comparison of the four methods for each type of defect.

Method	**Our Method**			**YOLOv3**			**Retina Net**			**YOLOv4**		
Metric	**R**	**P** (%)	**F1**	**R**	**P** (%)	**F1**	**R**	**P** (%)	**F1**	**R**	**P** (%)	**F1**
Sc	96.7	93.8	95.2	3.6	66.7	6.8	71.4	93.0	80.8	75.9	100.0	86.3
OF	95.7	98.2	96.9	90.5	93.5	92.0	87.4	86.5	87.0	88.4	97.9	92.9
Cr	98.6	98.6	98.6	88.9	92.3	90.6	15.9	81.3	26.5	100.0	98.8	99.4
SI	94.1	91.9	94.0	76.9	69.0	72.7	57.7	62.5	60.0	84.6	100.0	91.7
PSI	88.6	85.2	86.9	70.3	95.5	81.0	91.9	85.3	88.5	82.8	98.7	90.1
Average	94.7	93.5	94.3	66.1	83.4	68.6	64.8	81.7	68.5	86.3	99.1	92.1
Time (ms) 512 × 512		103			233			304			167	

In order to test the robustness of the algorithm to illumination changes, we enhanced and reduced the brightness of the original image to simulate the change of light source brightness. As shown in Figure 19, the brightness changes of the original image are –40%, –20%, 20%, and 40%, respectively. The detection performance of the algorithm for each defect is basically not affected by illumination. This is due to the following two points: (1) ROI extraction is based on the gradient difference, which is not affected by the overall brightness change of the image. (2) As a result of the data augmentation technology mentioned in Section 2.2.1, ROI classification has a certain robustness to illumination change.

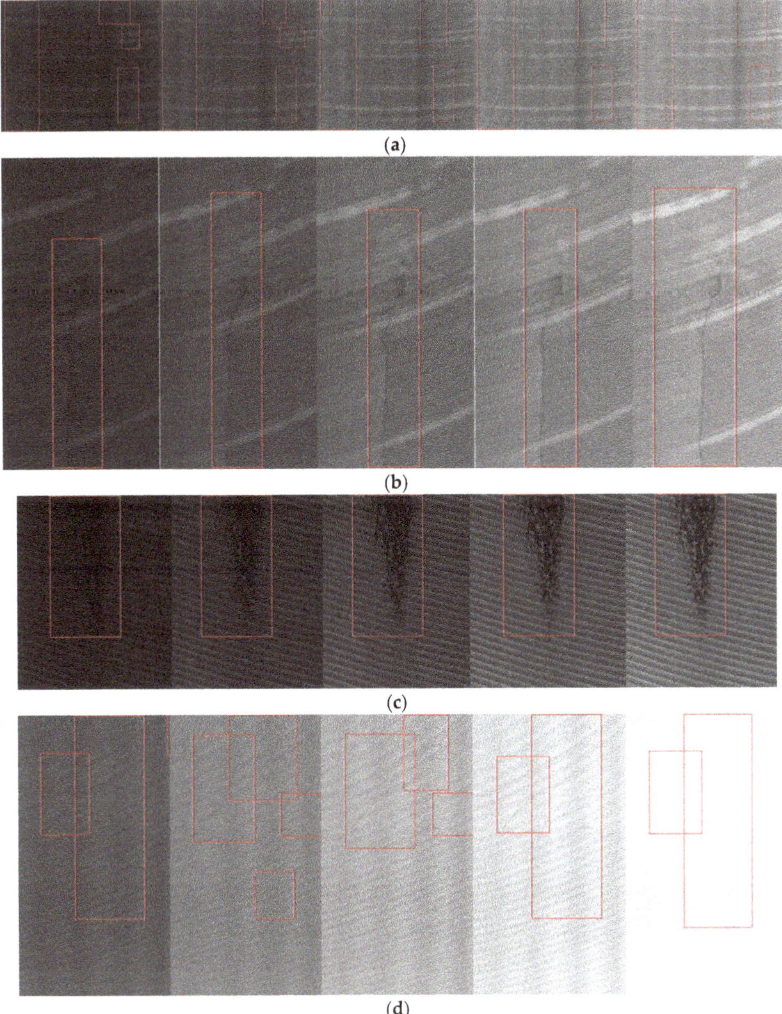

Figure 19. Detection effect of different brightness: (**a**) PSI; (**b**) OF; (**c**) PSI; (**d**) Sc. From left to right, the brightness of the image is reduced by 40%, reduced by 20%, unchanged, increased by 20%, and increased by 40%.

We also analyzed the reason that led to the failure cases (shown in Figure 20) of our method. Figure 20a shows that our method produce false negatives of Sc defect, which are mainly caused by the low contrast and the horizontal distribution similar to the milling grain background. Similarly, some small PSI defects with low contrast are missed in Figure 20c. There were no corresponding samples in the classification network training, so the pitted oil areas are incorrectly detected as PSI defects, as shown in Figure 20b. For the large-scale oxide film shown in Figure 20d, the oxide film coverage area is too large, so that its interior is treated as regular patterns and neglected by the MGRTS +DoG. As a result, only the edge is retained.

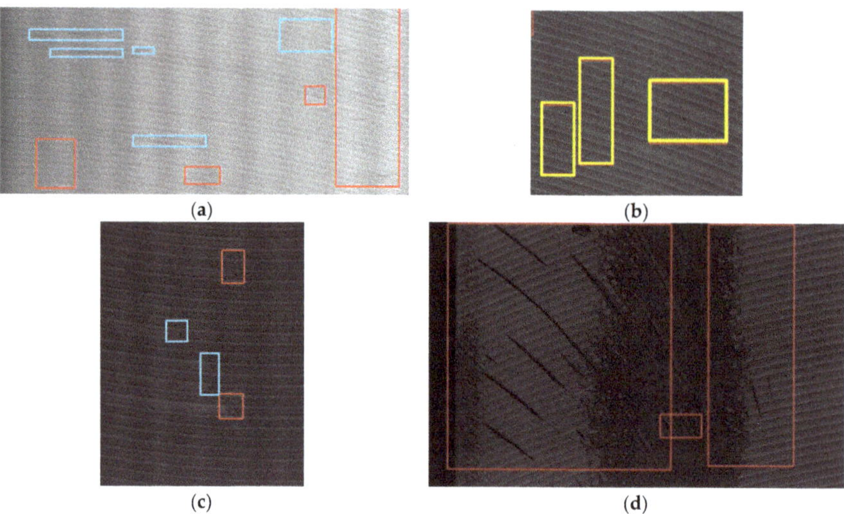

Figure 20. Failure cases of our method: (**a**) False negatives (blue box) of Sc (to facilitate observation, we increased the contrast by 20%); (**b**) False positives (yellow box) of PSI; (**c**) False negatives (blue box) of PSI (to facilitate observation, we increased the contrast by 20%); (**d**) Incomplete detection of OF.

4. Application in Actual Production Line

The algorithm proposed in this paper has been applied to the on-line surface defects inspection system installed at the actual production line of an aluminum ingot milling machine.

4.1. Image Acquisition Devices

Figure 21a shows a concise diagram of the imaging system, and Figure 21b is the corresponding picture of the material object. The image acquisition device includes cameras and a light source. Two line-scan charge-coupled device (CCD) cameras are used to capture images of 4096 × 1024 size of an aluminum ingot surface after milling under the illumination of light source, and the resolution is 0.315 mm/pixels. When the aluminum ingot passes through the acquisition device, the image acquisition program will control the acquisition speed of the camera according to the production speed and store the image. At the same time, the defect detection algorithm starts to process the image, and it alarms in time when defects are found.

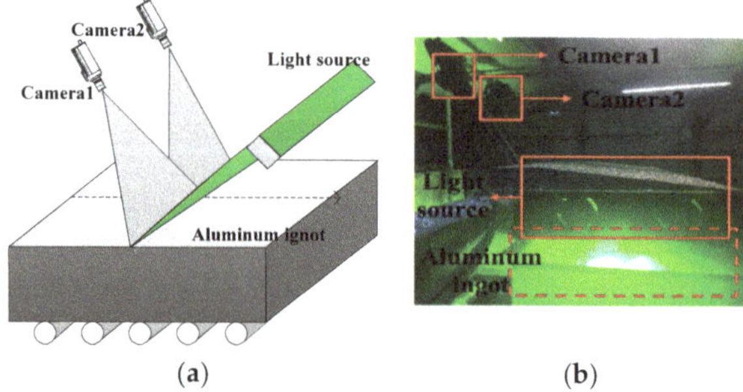

Figure 21. The imaging system: (**a**) a concise diagram and (**b**) picture of the material object.

4.2. Effectiveness of Our Method

For 5 days, we randomly checked the defect detection results of 39 production records and compared them with the real products. The detection rate of the algorithm is over 98.0%, and the accuracy of defect recognition rate is 96.0%. The statistical method of detection rate is as follows: the number of defects detected by the surface inspection system (regardless of defect category) accounts for the percentage of the number of defects on the surface of aluminum ingot.

4.3. Time Efficiency

The aluminum ingot region detection and the ROI extraction of the proposed approach were implemented by using C++ and OpenCV 2.4.6 library in Microsoft Visual Studio 2008, and the defect ROI classification is implemented by using python and Keras. The proposed approach was executed on a workstation with a 2.8 GHz Intel Xeon i5 processor and 16 GB memory, and the workstation is configured with a piece of NVIDIA Tesla k40c. The average time consumption of one image in each step is given in Table 5. Our detection system achieves an average processing speed of approximately 2.43 fps. The production speed of the aluminum ingot milling machine production line is from 3 to 6 m/min, when the actual production speed is 6 m/min, the corresponding camera acquisition speed is approximately 0.31 fps, so our algorithm can meet the real-time requirements.

Table 5. Average time consumption of one image in each step.

Process Stage	Time Consumption (ms)
Aluminum ingot region detection + ROI extraction	272
Defect ROI Classification	140
Total	412

To sum up, by applying our defect detection method to the online surface inspection system, the production is guided by the timely alarm of defects, which has great significance for ensuring product quality and improving production efficiency. In addition, the using effect also proves the promising application of our method in the surface defect detection of aluminum ingot with complex texture background after milling.

5. Conclusions

We proposed a novel two-stage detection approach to adaptively detect different types of defect on the surface of aluminum ingot with a complex milling grain background.

Firstly, the combination of MGRTS, DoG, and the similar region merging for the ROI extraction boosts the detection performance of various defects. Secondly, the data augmentation and the focal loss used in the inception-v3 network fine tuning handled the class imbalance well and improved the classification accuracy. Finally, the experimental results and the application in the actual production line show that when the number of defect ROI samples is large but the number of labeled original image samples is small, the performance of the two-stage defect detection algorithm proposed in this paper is significantly better than that of the one-stage deep learning algorithm. At the same time, it can also meet the real-time requirements.

Our algorithm combines the traditional detection and deep learning classification methods, which has great advantages in field application, because it can not only make full use of CPU and GPU to maximize the processing speed, but it also can be put into use quickly at the beginning of the project in the case of a lack of samples.

In future work, we will continue to collect more samples from the production line, including difficult cases and false defects. Then, we will focus on exploring a multi-scale analysis method and full convolution semantic segmentation network to further improve the detection effect of various defects.

Author Contributions: Methodology, Y.L.; software, Y.L.; validation, P.Z.; formal analysis, K.X.; resources, P.Z.; data curation, K.X.; writing—original draft preparation, Y.L.; writing—review and editing, K.X.; project administration, P.Z.; funding acquisition, K.X. All authors have read and agreed to the published version of the manuscript.

Funding: This research was funded in part by the National Key R&D Program of China under Grant 2018YFB0704304, and in part by the National Natural Science Foundation of China (NSFC) under Grant 51674031 and Grant 51874022.

Conflicts of Interest: The authors declare no conflict of interest.

References

1. Huke, P.; Klattenhoff, R.; von Kopylow, C.; Bergmann, R.B. Novel trends in optical non-destructive testing methods. *J. Europ. Opt. Soc. Rap. Public.* **2013**, *8*, 13043. [CrossRef]
2. Zhu, Y.K.; Tian, G.Y.; Lu, R.S.; Zhang, H. A review of optical NDT technologies. *Sensors* **2011**, *11*, 7773–7798. [CrossRef] [PubMed]
3. Ramakrishnan, M.; Rajan, G.; Semenova, Y.; Farrell, G. Overview of fiber optic sensor technologies for strain/temperature sensing applications in composite materials. *Sensors* **2016**, *16*, 99. [CrossRef] [PubMed]
4. Qiu, Q.W.; Lau, D. A novel approach for near-surface defect detection in FRP-bonded concrete systems using laser reflection and acoustic-laser techniques. *Constr. Build. Mater.* **2017**, *141*, 553–564. [CrossRef]
5. Ciampa, F.; Mahmoodi, P.; Pinto, F.; Meo, M. Recent advances in active infrared thermography for non-destructive testing of aerospace components. *Sensors* **2018**, *18*, 609. [CrossRef]
6. Navarro, P.J.; Fernandezisla, C.; Alcover, P.M.; Suardíazet, J. Defect detection in textures through the use of entropy as a means for automatically selecting the wavelet decomposition level. *Sensors* **2016**, *16*, 1178. [CrossRef]
7. Prasanna, P.; Dana, K.J.; Gucunski, N.; Basily, B.B.; La, H.M.; Lim, R.S.; Parvardeh, H. Automated crack detection on concrete bridges. *IEEE Trans. Autom. Sci. Eng.* **2016**, *13*, 591–599. [CrossRef]
8. Chen, H.; Zhao, H.; Han, D.; Liu, W.; Chen, P.; Liu, K. Structure-aware-based crack defect detection for multicrystalline solar cells. *Measurement* **2020**, *151*, 107170. [CrossRef]
9. Chun, P.J.; Izumi, S.; Yamane, T. Automatic detection method of cracks from concrete surface imagery using two-step light gradient boosting machine. *Comput. Aided Civ. Inf.* **2020**, 1–12. [CrossRef]
10. Zhang, C.; Chang, C.; Jamshidi, M. Concrete bridge surface damage detection using a single-stage detector. *Comput. Aided Civ. Inf.* **2020**, *35*, 389–409. [CrossRef]
11. Dung, C.V.; Anh, L.D. Autonomous concrete crack detection using deep fully convolutional neural network. *Autom. Constr.* **2018**, *99*, 52–58. [CrossRef]
12. Fernandez, C.; Campoy, P.; Platero, C.; Sebastian, J.M.; Aracil, R. On-Line Surface Inspection for Continuous Cast Aluminum Strip. In *Computer Vision for Industry*; Electronic Imaging Device Engineering: Munich, Germany, 1993; pp. 26–37. [CrossRef]
13. Huang, X.Q.; Luo, X.B. A real-time algorithm for aluminum surface defect extraction on non-uniform image from CCD camera. In Proceedings of the International Conference on Machine Learning and Cybernetics(ICMLC), Lanzhou, China, 13–16 July 2014; pp. 556–561. [CrossRef]
14. Xu, K.; Liu, S.H.; Ai, Y.H. Application of shearlet transform to classification of surface defects for metals. *Image Vis. Comput.* **2015**, *35*, 23–30. [CrossRef]
15. Zhai, M.; Shan, F. Applying target maneuver onset detection algorithms to defects detection in aluminum foil. *Signal Process* **2010**, *90*, 2319–2326. [CrossRef]
16. Wei, R.F.; Bi, Y.B. Research on recognition technology of aluminum profile surface defects based on deep learning. *Materials* **2019**, *12*, 1681. [CrossRef] [PubMed]
17. Neuhauser, F.M.; Bachmann, G.; Hora, P. Surface defect classification and detection on extruded aluminum profiles using convolutional neural networks. *Int. J. Mater. Form.* **2019**, *3*, 1–13. [CrossRef]
18. Zhang, D.F.; Song, K.C.; Xu, J.; He, Y.; Yan, Y.H. Unified detection method of aluminium profile surface defects: Common and rare defect categories. *Opt. Lasers Eng.* **2020**, *126*, 105936. [CrossRef]
19. Ai, Y.H.; Xu, K. Feature extraction based on contourlet transform and its application to surface inspection of metals. *Opt. Eng.* **2012**, *51*, 113605. [CrossRef]
20. Ai, Y.H.; Xu, K. Surface detection of continuous casting slabs based on curvelet transform and kernel locality preserving projections. *J. Iron Steel Res.* **2013**, *20*, 83–89. [CrossRef]

21. Ghorai, S.; Mukherjee, A.; Gangadaran, M.; Dutta, P.K. Automatic defect detection on hot-rolled flat steel products. *IEEE Trans. Instrum. Meas.* **2013**, *62*, 612–621. [CrossRef]
22. Tian, S.Y.; Ke, X. An algorithm for surface defect identification of steel plates based on genetic algorithm and extreme learning machine. *Metals* **2017**, *7*, 311. [CrossRef]
23. He, D.; Xu, K.; Zhou, P. Defect detection of hot rolled steels with a new object detection framework called classification priority network. *Comput. Ind. Eng.* **2019**, *128*, 290–297. [CrossRef]
24. Song, K.C.; Yan, Y. A noise robust method based on completed local binary patterns for hot-rolled steel strip surface defects. *Appl. Surf. Sci.* **2013**, *285*, 858–864. [CrossRef]
25. Liu, M.F.; Liu, Y.; Hu, H.J.; Nie, L.Q. Genetic algorithm and mathematical morphology based binarization method for strip steel defect image with non-uniform illumination. *J. Vis. Commun. Image Represent.* **2016**, *37*, 70–77. [CrossRef]
26. Youkachen, S.; Ruchanurucks, M.; Phatrapomnant, T.; Kaneko, H. Defect Segmentation of Hot-rolled Steel Strip Surface by using Convolutional Auto-Encoder and Conventional Image processing. In Proceedings of the International Conference of Information and Communication Technology for Embedded Systems(IC-ICTES), Bangkok, Thailand, 25–27 March 2019; pp. 1–5. [CrossRef]
27. Liu, W.W.; Yan, Y.H. Automated surface defect detection for cold-rolled steel strip based on wavelet anisotropic diffusion method. *Int. J. Ind. Syst. Eng.* **2014**, *17*, 224–239. [CrossRef]
28. Li, Y.; Li, G.Y.; Jiang, M.M. An end-to-end steel strip surface defects recognition system based on convolutional neural networks. *Steel Res. Int.* **2016**, *88*, 176–187. [CrossRef]
29. Liu, K.; Wang, H.Y.; Chen, H.Y.; Qu, E.Q.; Tian, Y.; Sun, H.X. Steel surface defect detection using a new haar-weibull-variance model in unsupervised manner. *IEEE Trans. Instrum. Meas.* **2017**, *66*, 1–12. [CrossRef]
30. Luo, Q.; Fang, X.; Liu, L.; Yang, C.; Sun, Y. Automated visual defect detection for flat steel surface: A survey. *IEEE Trans. Instrum. Meas.* **2020**, *69*, 626–644. [CrossRef]
31. Sun, X.H.; Gun, J.N.; Tang, S.X.; Li, J. Research progress of visual inspection technology of steel products—A review. *Appl. Sci.* **2018**, *8*, 2195. [CrossRef]
32. Kumar, A. Computer-vision-based fabric defect detection: A survey. *IEEE Trans. Ind. Electron.* **2008**, *55*, 348–363. [CrossRef]
33. Tsai, D.; Hsieh, C.Y. Automated surface inspection for directional textures. *Image Vis. Comput.* **1999**, *18*, 49–62. [CrossRef]
34. Asha, V.; Bhajantri, N.U.; Nagabhushan, P. Automatic detection of texture defects using texture-periodicity and Gabor wavelets. *Comput. Netw. Intell. Comput.* **2011**, 548–553. [CrossRef]
35. Wu, X.; Xu, K.; Xu, J. Application of undecimated wavelet transform to surface defect detection of hot rolled steel plates. In Proceedings of the International Congress on Image and Signal Processing, Sanya, Hainan, China, 27–30 May 2008; pp. 528–532. [CrossRef]
36. Ngan, H.Y.; Pang, G.K.; Yung, S.P.; Michael, K.N. Wavelet based methods on patterned fabric defect detection. *Pattern Recognit.* **2005**, *38*, 559–576. [CrossRef]
37. Lowe, D.G. Distinctive image features from scale-invariant keypoints. *Int. J. Comput. Vis.* **2004**, *60*, 91–110. [CrossRef]
38. Szegedy, C.; Vanhoucke, V.; Ioffe, S.; Shlens, J.; Wojna, Z. Rethinking the inception architecture for computer vision. In Proceedings of the Conference on Computer Vision and Pattern Recognition (CVPR), Las Vegas, NV, USA, 27–30 June 2016; pp. 2818–2826. [CrossRef]
39. Szegedy, C.; Liu, W.; Jia, Y.Q.; Sermanet Reed, P.; Anguelov, S.D.; Erhan, D.; Vanhoucke, V.; Rabinovich, A. Going deeper with convolutions. In Proceedings of the Conference on Computer Vision and Pattern Recognition (CVPR), Boston, MA, USA, 7–12 June 2015; pp. 1–9. [CrossRef]
40. Deng, J.; Dong, W.; Socher, R.; Li, L.J.; Li, K.; Li, F.F. Imagenet: A large-scale hierarchical image database. In Proceedings of the Conference on Computer Vision and Pattern Recognition (CVPR), Miami, FL, USA, 20–25 June 2009; pp. 249–255. [CrossRef]
41. Lin, T.Y.; Goyal, P.; Girshick, R.; He, K.M.; Dollár, P. Focal loss for dense object detection. *IEEE Trans. Pattern Anal. Mach. Intell.* **2020**, *42*, 318–327. [CrossRef]
42. Yin, Y.; Tian, G.Y.; Yin, G.F.; Luo, A.M. Defect identification and classification for digital X-ray images. *Appl. Mech. Mater.* **2008**, *10–12*, 543–547. [CrossRef]

43. Joseph, R.; Ali, F. Yolov3: An incremental improvement. *arXiv* **2018**, arXiv:1804.02767.
44. Bochkovskiy, A.; Wang, C.Y.; Liao, H.Y.M. Yolov4: Optimal speed and accuracy of object detection. *arXiv* **2004**, arXiv:2004.10934v1.

Article

Fault Diagnosis of Wind Turbine Gearbox Based on the Optimized LSTM Neural Network with Cosine Loss

Aijun Yin [1,2,*], Yinghua Yan [1,2], Zhiyu Zhang [1,2], Chuan Li [3] and René-Vinicio Sánchez [4]

1 State Key Laboratory of Mechanical Transmissions, Chongqing University, Chongqing 400044, China; 201807021027@cqu.edu.cn (Y.Y.); 20170702004t@cqu.edu.cn (Z.Z.)
2 College of Mechanical Engineering, Chongqing University, Chongqing 400044, China
3 Research Center of System Health Maintenance, Chongqing Technology and Business University, Chongqing 400067, China; chuanli@ctbu.edu.cn
4 Department of Mechanical Engineering, Universidad Politécnica Salesiana, Cuenca 010105, Ecuador; rsanchezl@ups.edu.ec
* Correspondence: aijun.yin@cqu.edu.cn; Tel.: +86-135-0838-4505

Received: 2 March 2020; Accepted: 16 April 2020; Published: 20 April 2020

Abstract: The gearbox is one of the most fragile parts of a wind turbine (WT). Fault diagnosis of the WT gearbox is of great importance to reduce operation and maintenance (O&M) costs and improve cost-effectiveness. At present, intelligent fault diagnosis methods based on long short-term memory (LSTM) networks have been widely adopted. As the traditional softmax loss of an LSTM network usually lacks the power of discrimination, this paper proposes a fault diagnosis method for wind turbine gearboxes based on optimized LSTM neural networks with cosine loss (Cos-LSTM). The loss can be converted from Euclid space to angular space by cosine loss, thus eliminating the effect of signal strength and improve the diagnosis accuracy. The energy sequence features and the wavelet energy entropy of the vibration signals are used to evaluate the Cos-LSTM networks. The effectiveness of the proposed method is verified with the fault vibration data collected on a gearbox fault diagnosis experimental platform. In addition, the Cos-LSTM method is also compared with other classic fault diagnosis techniques. The results demonstrate that the Cos-LSTM has better performance for gearbox fault diagnosis.

Keywords: wind turbine; gearbox fault; cosine loss; long short-term memory network

---

1. Introduction

With the gradual depletion of non-renewable energy and the deteriorating human living environment, wind energy has developed rapidly as one renewable energy source [1]. However, wind turbines (WTs) are mostly installed in remote areas as the main equipment for wind power generation. The harsh operating environment causes frequent failures of key components such as gearboxes and bearings [2]. Therefore, in order to ensure the safe operation of WTs and reduce the operation and maintenance (O&M) costs, it is crucial to study effective fault diagnosis methods for gearboxes [3].

As the vibration and acoustic emission signals are sensitive to the faults of the machine, condition monitoring systems based on vibration [4,5] and acoustic emission [6–8] have been widely used in the field of condition monitoring and fault diagnosis. In order to monitor the health conditions of WTs, the wind energy industry is currently using condition monitoring systems to collect large amounts of real-time data for diagnosing gearbox faults. Since the amount of data collected from gearboxes is increasing, the traditional fault diagnosis method cannot effectively analyze massive data and automatically give accurate diagnosis results [9]. Therefore, intelligent fault diagnosis methods based on artificial intelligence techniques are gaining more attention. Generally, there are two main steps for intelligent fault diagnosis methods: feature extraction and fault classification [10]. Traditional methods such as artificial neural networks (ANN) and support vector machine (SVM) are used to classify faults [11–13]. However, the problem of existing intelligent fault diagnosis methods is that the common machine learning methods rely on well-selected features and have limited ability to learn from complex time-series signals; meanwhile, with these methods it is more difficult to identify faults under variable working conditions, and they have a low classification accuracy. Therefore, a more effective fault identification method is needed. [14–17]. In recent years, deep learning has attracted great attention from various fields due to the powerful ability of feature learning and the superiority of processing massive data. Up to now, deep learning networks have been widely applied in fault diagnosis, such as deep belief networks (DBN) [18], convolutional neural networks (CNN) [19] and recurrent neural networks (RNN) [20]. However, the gearbox has strong time-dependence of faults due to its relatively long operating time [21]. Compared with other deep learning methods, the long short-term memory (LSTM) neural network has great advantages in learning long-term time-dependent characteristics of sequences [22,23].

For the fault diagnosis methods based on LSTM neural networks, the softmax cross entropy is usually used as the loss function of fault classification. However, recent studies found that the traditional softmax loss is insufficient to acquire the discriminating power for classification. To obtain better discriminating performance, Wang et al. [18] proposed a novel loss function called large margin cosine loss (LMCL) for learning the high-resolution depth features used in face recognition. The result shows that the loss function based on cosine distance has a good effect on classification. Therefore, this paper proposes an optimized fault diagnosis method using an LSTM network with cosine loss (Cos-LSTM) to improve the ability of classification. Meanwhile, the energy sequence features and the wavelet energy entropy of the fault vibration data collected on a gearbox fault diagnosis experimental platform are used to validate Cos-LSTM networks. The Cos-LSTM achieves higher accuracy of diagnosis, which is demonstrated through the gear transmission experiments and compared to other fault diagnosis methods.

The rest of the paper is organized as follows. In Section 2, the typical architecture of LSTM and the process of fault diagnosis are briefly introduced. Section 3 details the Cos-LSTM method and the process of gearbox fault diagnosis based on the Cos-LSTM method. The gearbox fault diagnosis experiment and the comparisons of our proposed method and other fault diagnosis methods are presented in Section 4. Finally, the conclusions are drawn in Section 5.

2. LSTM Neural Network for Fault Diagnosis

As a special type of recurrent neural network (RNN), the LSTM neural network was proposed by Hochreiter and Schmifhuber [24] to solve the vanishing or exploding gradient problem of RNNs [25], while retaining the ability of RNNs to process sequential data. In this section, we describe LSTM in more detail.

2.1. Structure of LSTM

The main component of an LSTM neural network is the LSTM cell, which can decide whether to update the state information of a memory cell. The structure of the LSTM cell is shown in Figure 1.

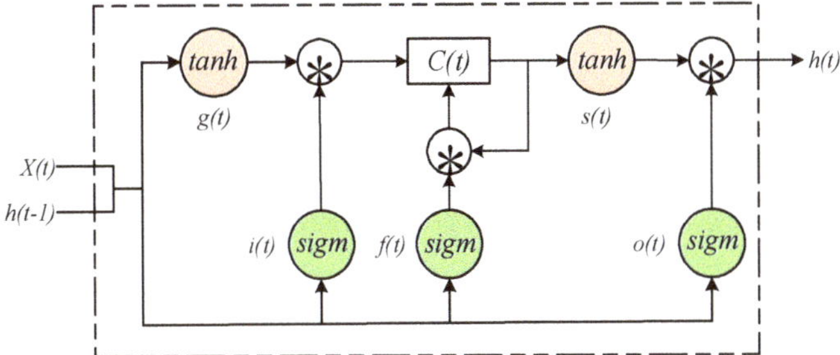

Figure 1. The Schematic diagram of an LSTM cell.

As shown in Figure 1, $h(t)$ and $x(t)$ are the output hidden states and inputs of the current time step, $h(t-1)$ represents the hidden state of the previous time step; sigm is the sigmoid function and *tanh* is the hyperbolic tangent function. $C(t)$ is a memory cell which is used for the preservation of information, and the flow of information into or out of $C(t)$ is regulated by three different gates:

1. The input gate $i(t)$, which decides whether the information can get in the memory element;
2. The forget gate $f(t)$, which decides whether the internal information needs to be forgotten;
3. The output gate $o(t)$, which decides what information can pass through the gate and get into the rest of the neural network.

The internal state node $s(t)$ and input node $g(t)$ are also integral parts of the LSTM cell. Here are the calculation procedures of the LSTM cell:

$$g(t) = \Phi(W_{gx}x(t) + W_{gh}h(t-1) + b_g), \tag{1}$$

$$i(t) = \sigma(W_{ix}x(t) + W_{ih}h(t-1) + b_i), \tag{2}$$

$$f(t) = \sigma(W_{fx}x(t) + W_{fh}h(t-1) + b_f), \tag{3}$$

$$o(t) = \sigma(W_{ox}x(t) + W_{oh}h(t-1) + b_o), \tag{4}$$

$$s(t) = g(t) * i(t) + C(t-1) * f(t), \tag{5}$$

$$h(t) = \Phi(s(t)) * o(t). \tag{6}$$

In the above equations, W_{jx}, W_{jh} and b_j, $j = g, j, f, o$ denote the input weight matrixes, hidden weight matrixes and bias vectors separately; $*$, σ and Φ are element-wise multiplications of two vectors, the *sigmoid* function and *tanh* function, respectively.

The LSTM neural network can learn when to open or close the gate to control the flow of information in LSTM cells automatically, so it can choose useful information to train the model.

2.2. Architecture of LSTM for Fault Diagnosis

The LSTM neural network is used for fault classification in fault diagnosis. The architecture for the LSTM network includes five layers: an input layer, an LSTM hidden layer, a fully connected layer, a softmax layer and a result output layer at the end. The architecture of the LSTM network is shown in Figure 2.

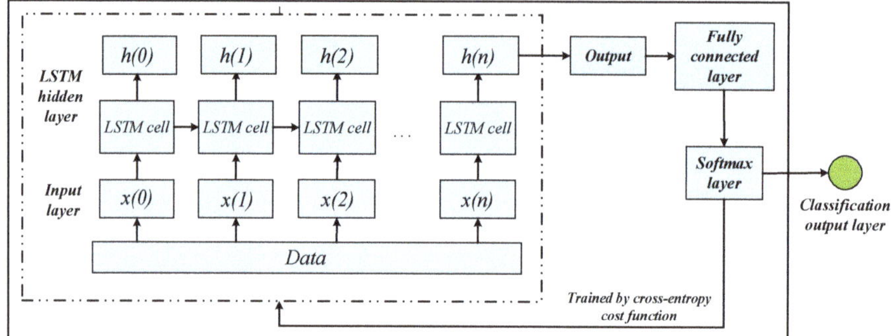

Figure 2. The architecture for LSTM network.

During the training process, the fault features are fed into the input layer first, then the data flow through LSTM cell and the result of LSTM cell is output to the LSTM hidden layer. The last output of the LSTM hidden layer is taken as the output of the LSTM network, and it is used to connect a fully connected layer to map outputs into the result space. The softmax layer follows the fully connected layer to calculate the probabilities for all the fault pattern. Finally, the fault diagnosis results are output to the classification output layer. After completing the training, the weights and bias will be adjusted to the optimal value, and then the test set is input into LSTM for fault diagnosis.

3. Cos-LSTM

The softmax cross entropy is often used as the loss function of the LSTM neural network; however, the softmax loss is insufficient to enable classification [26,27]. To solve this problem, the cosine loss function is adopted to optimize the LSTM neural network. This section provides details about the Cos-LSTM.

3.1. Cosine Loss

Based on the softmax loss, the cosine loss retains its advantage of enlarging the difference between classes [15], but reduces its sensitivity to different signal strengths and pays more attention to the difference of vectors in direction. The schematic of cosine loss is shown in Figure 3.

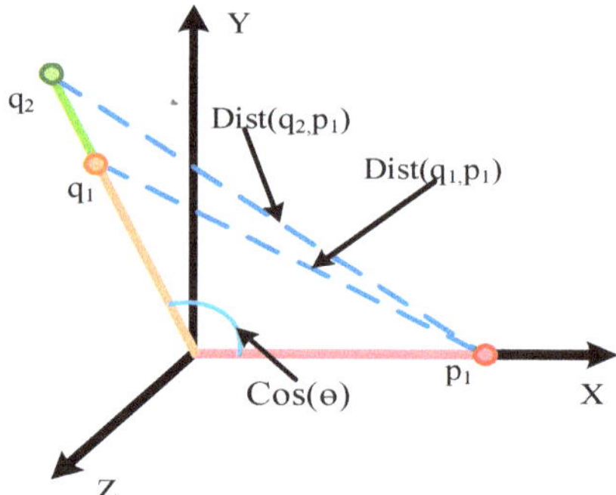

Figure 3. Schematic of Cosine Loss.

Suppose there are two signals q_1 and q_2 with the same fault, and the corresponding fault label is p_1. When softmax is taken as the loss function, the softmax loss can be formulated as follows,

$$\text{Loss}_{\text{soft}} = \frac{1}{B}\sum_{i=1}^{B} -\log\left(\frac{e^{\|W_i\|\|x\|\cos\theta_i}}{\sum_{j=1}^{N} e^{\|W_j\|\|x\|\cos\theta_j}}\right) \tag{7}$$

where B is the number of training samples and N is the number of classes, x and W represent the hidden layer output and the weight matrix respectively, and θ is the angle between W and x. Formula (2) suggests that softmax loss is related to signal strength, while cosine loss evaluates the size of the differences between classes according to cosine similarity between the two feature vectors. The cosine similarity is defined as follows:

$$similarity(A,B) = \frac{A \cdot B}{\| A \| * \| B \|} = \frac{\sum_{i=1}^{n} A_i * B_i}{\sqrt{\sum_{i=1}^{n} A_i^2} * \sqrt{\sum_{i=1}^{n} B_i^2}} \tag{8}$$

Taking 1—cosine similarity as the loss function, the cosine loss can be formulated as follows,

$$\text{Loss}_{\text{cos}} = \frac{1}{B}\sum_{i=1}^{B} 1 - \frac{y_i}{\sqrt{\sum_{j=1}^{N} y_j^2}} = \frac{1}{B}\sum_{i=1}^{B} \sqrt{1 - \frac{\| W_i \|^2 \| x \|^2 \cos\theta_i{}^2}{\sum \| W_j \|^2 \| x \|^2 \cos\theta_j{}^2}} \tag{9}$$

By Formula (5), the $\| x \|^2$ in this formula can be eliminated, so the cosine loss is independent of the signal strength. Therefore, taking cosine loss function as the loss function in gearbox fault diagnosis, the loss can be converted from Euclid space to angular space, thus eliminating the effect of signal strength and reduce the burden of network fitting.

3.2. The Process of Cos-LSTM for Fault Diagnosis

In this paper, there are two kinds of fault features extracted for evaluating the proposed method: the energy sequence feature and the wavelet energy entropy.

The energy sequence feature: The energy sequence features are extracted by wavelet packet decomposition (WPD). WPD is a signal decomposition tool that decomposes a signal to some nodes and every node represents a set of coefficients at a specified frequency band [28,29]. The wavelet packet is defined as follows:

$$\phi(t) = \sqrt{2}\sum_{k} h(k)\phi(2t-1) \tag{10}$$

$$\Psi(t) = \sqrt{2}\sum_{k} g(k)\phi(2t-1) \tag{11}$$

where $h(k)$ and $g(k)$ are a low-pass filter and a high-pass filter respectively. $\varnothing(t)$ and $\Psi(t)$ represent the scaling function and the wavelet function respectively. Additionally, $g(k)$ can be expressed by $h(k)$ using the formula $g(k) = (-1)^k h(1-k)$.

The signal is decomposed by Equations (12) and (13)

$$d_{j+1,2n}(t) = \sum_{l \in Z} h_{l-2k} d_{j,n}(t) \tag{12}$$

$$d_{j+1,2n+1}(t) = \sum_{l \in Z} g_{l-2k} d_{j,n}(t) \tag{13}$$

where j denotes the decomposition layer, $n \in \{0, 1, 2, \ldots, 2^j - 1\}$ is the number of nodes in layer j, l indicates the number of wavelet coefficients and $d_{j,n}$ represents the coefficient sequence at the jth layer, nth node.

Due to the large amount of data, we divided the vibration data into four segments and a three-layer WPD was performed on each segment of vibration data using Daubechies 3 (db 3) to obtain eight nodes [30–32]. The energy of each node $E_{j,n}$ could then be calculated through Formula (14)

$$E_{j,n} = \sum_k \left| d_{j,n}(k) \right|^2. \tag{14}$$

The total energy of the signal E is the sum of the energy of each node in layer three. It can be computed by (15):

$$E = \sum_{n=0}^{2^3-1} E_{j,n}. \tag{15}$$

and $P_{j,n}$ is defined by (16):

$$P_{j,n} = \frac{E_{j,n}}{E}. \tag{16}$$

Each of the signals can be decomposed to get eight nodes, and the energy sequences feature can be expressed as Equation (17) according to Equations (14)–(16).

$$x(i) = (P_{2,i}^{v1}, P_{2,i}^{v2}) \tag{17}$$

where $x(i)$ is the energy sequences feature and $i = 0, 1, \ldots, 7$, $P_{2,i}^{v1}$ and $P_{2,i}^{v2}$ indicate the $P_{2,i}$ for $s_{v1}(t)$ and $s_{v2}(t)$, which denote the vibration signals of the gearbox in the horizontal and vertical directions respectively.

Wavelet energy entropy: The signal is reconstructed according to the eight node coefficients obtained from the three-layer WPD above, and the reconstructed signal is divided into N segments on the basis of the time characteristics of the signal. The energy of each segment is calculated by Formula (14). The calculated energy is normalized by Formulas (15) and (16) to obtain the wavelet energy entropy. The wavelet energy entropy of the j-th layer n node of the WPD is defined as $H_{j,n}$, and can be formulated as follows:

$$H_{j,n} = -\sum_{i=1}^{N} P_{j,n}(i) \log P_{j,n}(i) \tag{18}$$

where $P_{j,n}(i)$ is the normalized value of the energy of each segment of the signal; $i = 0, 1, \ldots, N$. The value of N is 50 in this article.

According to the calculated wavelet energy entropy of each node, the wavelet energy entropy feature is formed by Equation (19):

$$T = [H_{3,1}, H_{3,2}, H_{3,3}, H_{3,4}, H_{3,5}, H_{3,6}, H_{3,7}, H_{3,8}] \tag{19}$$

The fault features obtained above are fed into the Cos-LSTM network to diagnose the gearbox fault. The flow chart of fault diagnosis based on the Cos-LSTM is shown in Figure 4.

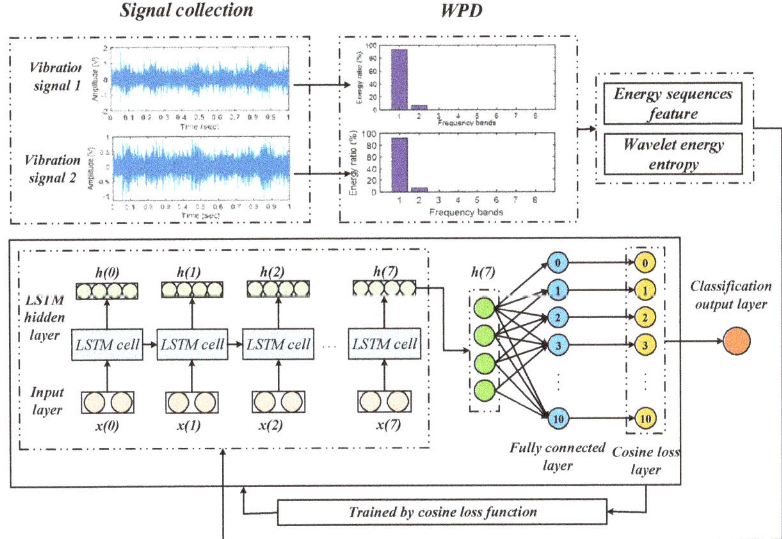

Figure 4. The flow chart of the Cos-LSTM method for gearbox fault diagnosis.

We used one LSTM hidden layer with eight LSTM cells to extract deeper features. The fault features are first normalized and then fed into the input layer. In this paper, we used N samples (N = 2200 samples) to train the model. Therefore, the size of the input layer is $N \times 8$ (time steps) $\times$ 2 (2-dimensional features), and the input size of each LSTM cell is $N \times 2$. The last output h (7) of the LSTM hidden layer connects a fully connected layer with 11 neurons, using cosine loss to calculate the probabilities for the 11-fault pattern.

The parameters of the LSTM neural network are presented as follows: time steps for LSTM = 8; the LSTM hidden layer neurons = 4; the fully connected layer neurons = 11; learning rate = 0.01; number of iterations of training = 10,000. The workflow of the Cos-LSTM is shown in Figure 5.

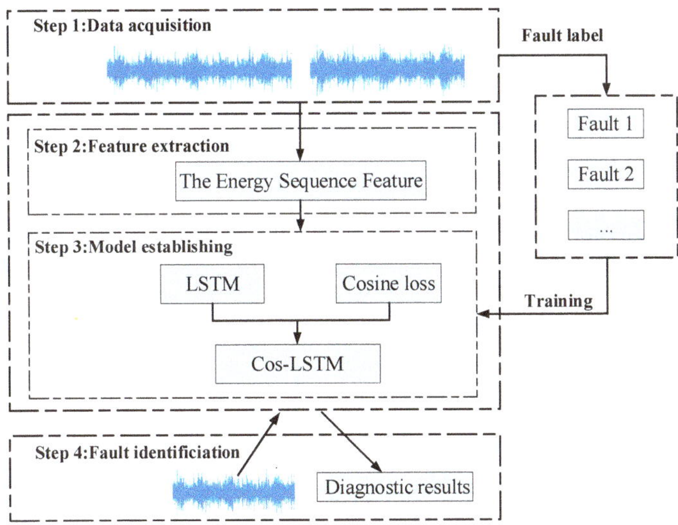

Figure 5. The workflow of the Cos-LSTM.

4. Experimental Validation

4.1. Experiment Description

The experimental test rig is illustrated in Figure 6a,b. The motor was controlled by an inverter and connected to the input shaft of the gearbox to transmit power by a coupling. An electromagnetic torque load was coupled with the output shaft of the gearbox through a V-belt. The electromagnetic torque load was controlled by a torque controller (TDK-Lambda, GEN 100-15-IS510; TDK-Lambda, Wuxi, China), which can adjust the torque of the load manually. Two accelerometers were mounted on the gearbox to collect signals, and the signals collected were transmitted to a laptop using the data acquisition card. Detailed information on the data acquisition system is provided in Table 1.

Table 1. Data acquisition settings.

Item	Parameter
Sensor	PCB ICP 353C03 accelerometer
Data acquisition box	NI cDAQ-9234
Software	LabVIEW
Sampling rate	50 kHz

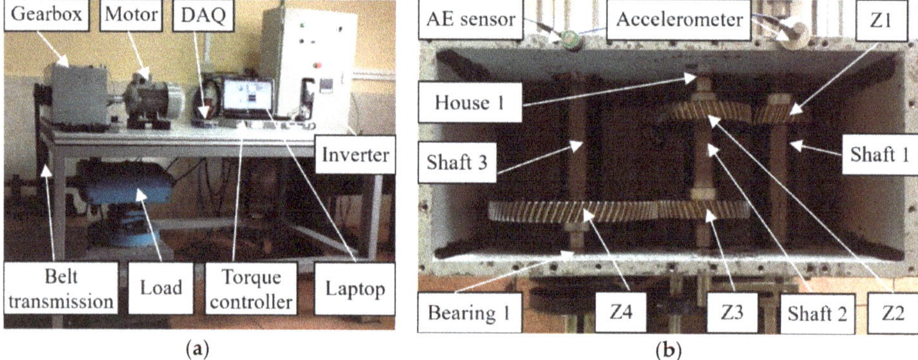

Figure 6. (**a**) Experimental test rig and (**b**) the structure of the gearbox.

The structure of the gearbox is displayed in Figure 6b. It consists of four gears, six bearings and three shafts. Shaft 1 was the input shaft, and gear Z1, with a module of 2.25 mm, a pressure angle of 20, a helical angle of 20, and 30 teeth, was installed on it. Shaft 1 transmitted the power to shaft 2 by a pair of gears (Z1 and Z2) in mesh. The output shaft (shaft 3) was driven by another helical gear Z4, with 80 teeth, which was meshed with the gear Z3. The helical gears Z2 and Z3 installed on shaft 2 both have 45 teeth and other parameters of them are the same to Z1. We installed one of the faulty components: bearing 1, bearing house 1, and gears Z1, Z2, Z3, Z4 every time on the gearbox to experiment. Table 2 shows all the condition patterns of the gearbox.

Table 2. Condition patterns of the gearbox.

Pattern Number	Faulty Component	Faulty Name	Input Speed (rpm)	Load (V)	View of the Failure
1	N/A	N/A	480, 720, 900	0, 10, 30	N/A
2	Gear Z_1	Worn tooth	480, 720, 900	0, 10, 30	
3	Gear Z_2	Chafing tooth	480, 720, 900	0, 10, 30	
4	Gear Z_3	Pitting tooth	480, 720, 900	0, 10, 30	
5	Gear Z_3	Worn tooth	480, 720, 900	0, 10 30	
6	Gear Z_4	Root crack tooth	480, 720, 900	0, 10, 30	
7	Gear Z_4	Chafing tooth	480, 720, 900	0, 10, 30	
8	Bearing 1	Inner race fault	480, 720, 900	0, 10, 30	
9	Bearing 1	Outer race fault	480, 720, 900	0, 10, 30	
10	Bearing 1	Ball fault	480, 720, 900	0, 10, 30	
11	House 1	Eccentric	480, 720, 900	0, 10, 30	

4.2. Experimental Results

Firstly, we verified the Cos-LSTM with the energy sequence features. We chose a test sample for explanation of the fault diagnosis process of our proposed method. The pattern number of this sample is 3 (chafing tooth), and the input speed and load of this sample are set to 480 rpm and zero respectively. The raw vibration signals and energy distribution map is shown in Figure 7. Figure 7a,c presents the raw signals $s_{v1}(t)$, $s_{v2}(t)$ of this sample collected on the gearbox and Figure 7b,d presents their energy distribution maps of the third layer WPD P^{v1} and P^{v2}. Putting the energy sequences feature of this sample into the Cos-LSTM, we got the probability of each fault pattern for the sample. The probability of the no. 3 fault pattern is 99.97% and the other 10 faults have a probability of 0.03%. The result shows that our proposed method considers that there is a fault numbered 3 (chafing tooth) in the gearbox. The result is correct for this test sample, so the method we proposed is effective.

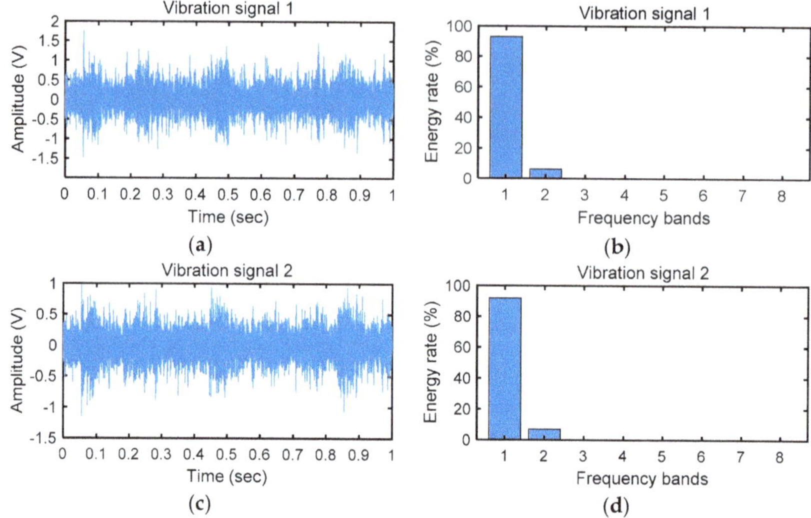

Figure 7. The raw vibration signals and energy distribution map.

From Table 2, it can be seen that three different input speeds and loads are set for all 11 fault patterns. Therefore, we have a total of 99 different tests, and each test is repeated five times. In each test, the signals are collected with 10 durations, and every duration covers 1 s. Therefore, we can get 9900 vibration signals. In order to train the model, we randomly choose 2200 samples as the training dataset. With the trained model, another 550 randomly chosen samples are used to test the effectiveness of the model. The effectiveness is measured by the accuracy rate. In this experiment, the accuracy rate is the number of correctly diagnosed samples divided by all the test samples, and the precision is the ratio of the number of samples correctly diagnosed with a fault pattern to the total number of samples diagnosed with such a fault pattern. The accuracy rate of the model is 98.55% in 550 samples. The accuracy rates and precision of our proposed model for the 11 fault patterns are shown in Figures 8 and 9 respectively.

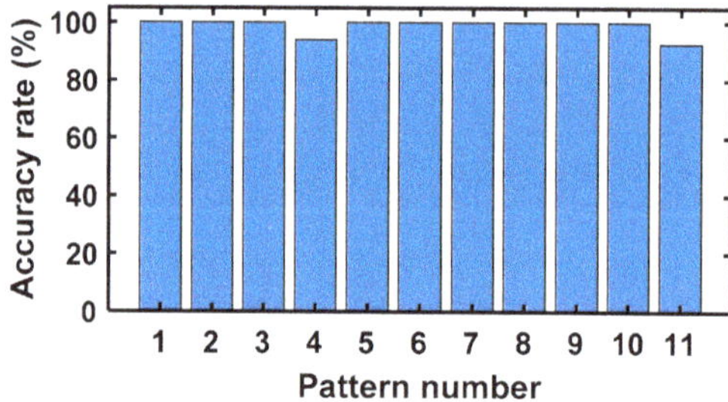

Figure 8. The accuracy rates for the 11 fault patterns.

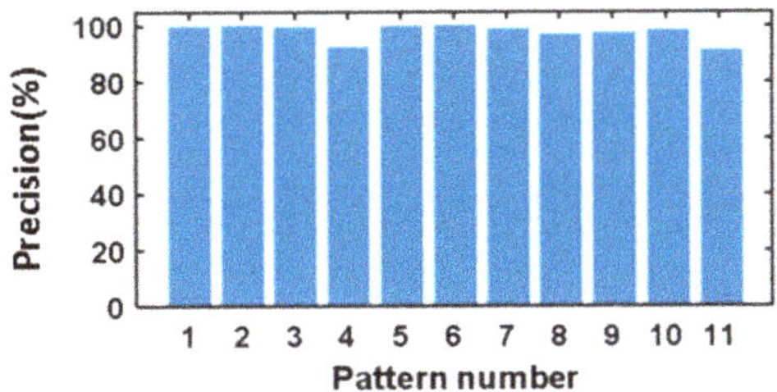

Figure 9. The precision for the 11 fault patterns.

4.3. Comparison Analysis

In this paper, the energy sequence features were used to verify the superiority of the Cos-LSTM by comparing with the traditional LSTM based on softmax loss and classic fault diagnosis methods, such as SVM, K-nearest neighbor (KNN) and backpropagation (BP) neural networks. In order to better evaluate the accuracy of the Cos-LSTM, we also used wavelet energy entropy feature for the fault diagnosis test. Table 3 shows the comparison results. Meanwhile, the different energy sequence features were extracted by changing the parameters of WPD such as wavelet basis function and data segment size, for evaluating the accuracy of the Cos-LSTM, and the results are displayed in Table 4.

Table 3. Comparisons with other classic fault diagnosis methods.

Feature	Fault Diagnosis Methods	Accuracy Rate
The energy sequence	Cos-LSTM	98.55%
	LSTM	96.72%
	SVM	65.48%
	KNN	83.93%
	BP neural network	69.64%
Wavelet Energy entropy	Cos-LSTM	98.08%

Table 4. Comparisons with different parameter of WPD.

Item	Parameter	Accuracy Rate
Wavelet basis function	Daubechies 3	98.55%
	Daubechies 2	96.36%
	Haar	93.82%
	Symlet	97.09%
Segment size	2	96.63%
	3	97.12%
	4	98.55%

According to Tables 3 and 4, the Cos-LSTM has the highest accuracy rate (98.55%) compared to other methods in the experimental results on the energy sequence features. After comparison and analysis, it can be found that: (1) comparison with traditional LSTM shows that the classification ability of cosine loss is better than that of softmax loss; (2) the accuracy rate of the LSTM neural network is better than KNN, SVM and BP neural networks, which indicates that the LSTM neural network has better feature-learning ability compared to classic fault diagnosis methods; (3) the evaluation results of Cos-LSTM using wavelet energy entropy are close to those using energy sequence features; (4) the accuracy rate of the Cos-LSTM is influenced by the energy sequence features extracted with different parameters of WPD, and the result shows that the energy sequence features extracted based on the wavelet basis function of Daubechies 3 (db3) and segment size 4 have better diagnostic accuracy rates; and (5) combined with the experimental results of energy sequence features and wavelet energy entropy, Cos-LSTM is able to diagnose the faults of the gearbox effectively.

5. Conclusions

This paper presented a fault diagnosis method for WT gearboxes based on the optimized LSTM network with cosine loss. The energy sequence features and the wavelet energy entropy were used to evaluate the Cos-LSTM network. The effectiveness of the Cos-LSTM was verified by a fault diagnosis experiment on a gearbox. The classification results show that the performance of the Cos-LSTM is better than that of the traditional LSTM and classic fault diagnosis techniques. Thus, the proposed method has superior performance in fault diagnosis. In the future, new studies will be conducted on feature learning directly from raw vibration signals using LSTM neural networks.

Author Contributions: Data curation, C.L.; Methodology, A.Y. and Z.Z.; Project administration, A.Y.; Resources, R.-V.S.; Supervision, A.Y.; Writing—original draft, Y.Y.; Writing—review & editing, Y.Y. and Z.Z. All authors have read and agreed to the published version of the manuscript.

Funding: This work was supported by the Key Science and Technology Research Project of Chongqing under grant cstc2018jszx-cyztzxX0032.

Conflicts of Interest: The authors declare no conflict of interest.

References

1. Zhao, X.; Yan, Z.; Zhang, X.-P. A Wind-Wave Farm System with Self-Energy Storage and Smoothed Power Output. *IEEE Access* **2016**, *4*, 8634–8642. [CrossRef]
2. Hu, A.; Yan, X.; Xiang, L. A new wind turbine fault diagnosis method based on ensemble intrinsic time-scale decomposition and WPT-fractal dimension. *Renew. Energy* **2015**, *83*, 767–778. [CrossRef]
3. Walford, C.A. *Wind Turbine Reliability: Understanding and Minimizing Wind Turbine Operation and Maintenance Costs*; Sandia National Laboratories: Albuquerque, NM, USA, 2006.
4. Jiang, G.; He, H.; Yan, J.; Xie, P. Multiscale Convolutional Neural Networks for Fault Diagnosis of Wind Turbine Gearbox. *IEEE Trans. Ind. Electron.* **2019**, *66*, 3196–3207. [CrossRef]
5. Feng, Z.; Qin, S.; Liang, M. Time-frequency analysis based on Vold-Kalman filter and higher order energy separation for fault diagnosis of wind turbine planetary gearbox under nonstationary conditions. *Renew. Energy* **2016**, *85*, 45–56. [CrossRef]
6. Zhang, Y.; Lu, W.; Chu, F. Planet gear fault localization for wind turbine gearbox using acoustic emission signals. *Renew. Energy* **2017**, *109*, 449–460. [CrossRef]
7. Bejger, A.; Chybowski, L.; Gawdzinska, K. Utilising elastic waves of acoustic emission to assess the condition of spray nozzles in a marine diesel engine. *J. Mar. Eng. Technol.* **2018**, *17*, 153–159. [CrossRef]
8. Bejger, A.; Drzewieniecki, J.B. The Use of Acoustic Emission to Diagnosis of Fuel Injection Pumps of Marine Diesel Engines. *Energies* **2019**, *12*. [CrossRef]
9. Jia, F.; Lei, Y.; Lin, J.; Zhou, X.; Lu, N. Deep neural networks: A promising tool for fault characteristic mining and intelligent diagnosis of rotating machinery with massive data. *Mech. Syst. Signal Process.* **2016**, *72–73*, 303–315. [CrossRef]
10. Lei, Y.; Jia, F.; Lin, J.; Xing, S.; Ding, S.X. An Intelligent Fault Diagnosis Method Using Unsupervised Feature Learning Towards Mechanical Big Data. *IEEE Trans. Ind. Electron.* **2016**, *63*, 3137–3147. [CrossRef]
11. Lei, Y.; He, Z.; Zi, Y.; Hu, Q. Fault diagnosis of rotating machinery based on multiple ANFIS combination with GAS. *Mech. Syst. Signal Process.* **2007**, *21*, 2280–2294. [CrossRef]
12. Wang, D.; Tse, P.W.; Guo, W.; Miao, Q. Support vector data description for fusion of multiple health indicators for enhancing gearbox fault diagnosis and prognosis. *Meas. Sci. Technol.* **2011**, *22*. [CrossRef]
13. Gao, Q.W.; Liu, W.Y.; Tang, B.P.; Li, G.J. A novel wind turbine fault diagnosis method based on intergral extension load mean decomposition multiscale entropy and least squares support vector machine. *Renew. Energy* **2018**, *116*, 169–175. [CrossRef]
14. Santos, P.; Villa, L.F.; Renones, A.; Bustillo, A.; Maudes, J. An SVM-Based Solution for Fault Detection in Wind Turbines. *Sensors* **2015**, *15*, 5627–5648. [CrossRef]

15. Abbasion, S.; Rafsanjani, A.; Farshidianfar, A.; Irani, N. Rolling element bearings multi-fault classification based on the wavelet denoising and support vector machine. *Mech. Syst. Signal Process.* **2007**, *21*, 2933–2945. [CrossRef]
16. Liu, W.Y.; Gao, Q.W.; Ye, G.; Ma, R.; Lu, X.N. A novel wind turbine bearing fault diagnosis method based on Integral Extension LMD. *Measurement* **2015**, *74*, 70–77. [CrossRef]
17. Lei, Y. *Intelligent Fault Diagnosis and Remaining Useful Life Prediction of Rotating Machinery*; Xi'an Jiaotong University Press: Xi'an, China, 2017.
18. Li, C.; Sanchez, R.-V.; Zurita, G.; Cerrada, M.; Cabrera, D. Fault Diagnosis for Rotating Machinery Using Vibration Measurement Deep Statistical Feature Learning. *Sensors* **2016**, *16*. [CrossRef]
19. Chen, Z.; Li, C.; Sanchez, R.-V. Gearbox Fault Identification and Classification with Convolutional Neural Networks. *Shock Vib.* **2015**. [CrossRef]
20. An, Z.; Li, S.; Wang, J.; Jiang, X. A novel bearing intelligent fault diagnosis framework under time-varying working conditions using recurrent neural network. *ISA Trans.* **2019**. [CrossRef]
21. Cao, L.; Zhang, J.; Wang, J.; Qian, Z. Intelligent fault diagnosis of wind turbine gearbox based on Long short-term memory networks. In Proceedings of the 2019 IEEE 28th International Symposium on Industrial Electronics, Vancouver, BC, Canada, 12–14 June 2019; pp. 890–895.
22. Medina, R.; Cerrada, M.; Cabrera, D.; Sanchez, R.-V.; Li, C.; de Oliveira, J.V. Deep Learning-Based Gear Pitting Severity Assessment using Acoustic Emission, Vibration and Currents signals. In Proceedings of the 2019 Prognostics and System Health Management Conference, Paris, France, 2–5 May 2019; pp. 210–216.
23. Wang, H.; Wang, Y.; Zhou, Z.; Ji, X.; Gong, D.; Zhou, J.; Li, Z.; Liu, W. CosFace: Large Margin Cosine Loss for Deep Face Recognition. In Proceedings of the 2018 IEEE/Cvf Conference on Computer Vision and Pattern Recognition, Salt Lake City, UT, USA, 18–23 June 2018; pp. 5265–5274.
24. Hochreiter, S.; Schmidhuber, J. Long short-term memory. *Neural Comput.* **1997**, *9*, 1735–1780. [CrossRef]
25. Graves, A. Supervised sequence labelling. In *Supervised Sequence Labelling with Recurrent Neural Networks*; Springer: Berlin/Heidelberg, Germany, 2012; pp. 5–13.
26. Wen, Y.; Zhang, K.; Li, Z.; Qiao, Y. A discriminative feature learning approach for deep face recognition. Proceedings of European Conference on Computer Vision, Amsterdam, The Netherlands, 8–16 October 2016; pp. 499–515.
27. Liu, W.; Wen, Y.; Yu, Z.; Li, M.; Raj, B.; Song, L. SphereFace: Deep Hypersphere Embedding for Face Recognition. In Proceedings of the 30th IEEE Conference on Computer Vision and Pattern Recognition, Honolulu, HI, USA, 21–26 July 2017; pp. 6738–6746.
28. Shao, Y.; Ge, L.; Fang, J. Fault diagnosis system based on smart bearing. In Proceedings of the 2008 International Conference on Control, Automation and Systems, Seoul, Korea, 14–17 October 2008; pp. 1084–1089.
29. Kedadouche, M.; Liu, Z. Fault feature extraction and classification based on WPT and SVD: Application to element bearings with artificially created faults under variable conditions. *Proc. Inst. Mech. Eng. Part C J. Mech. Eng. Sci.* **2017**, *231*, 4186–4196. [CrossRef]
30. Wang, D.C.; Ding, Y.F.; Zhu, C.X. A fault diagnosis method for gearbox based on neutrosophic K-Nearest Neighbor. *Shock Vib.* **2019**, *38*, 148–153. [CrossRef]
31. Wang, W.Q.; Yang, S. A method for choosing the wavelet decomposition level in structural fault analysis. *Struct. Environ. Eng.* **2009**. [CrossRef]
32. Wu, C.Z.; Jiang, P.C.; Feng, F.Z.; Chen, T.; Chen, X.L. Gearbox Faults diagnosis method for gearboxes based on 1-D convolutional neural network. *Shock Vib.* **2018**, *37*, 51–56. [CrossRef]

Article

Signal Denoising Method Using AIC–SVD and Its Application to Micro-Vibration in Reaction Wheels

Xianbo Yin, Yang Xu *, Xiaowei Sheng and Yan Shen

College of Mechanical Engineering, Donghua University, Shanghai 201620, China; yinxb_2008@163.com (X.Y.); shengxw@dhu.edu.cn (X.S.); shenyan1871@126.com (Y.S.)

* Correspondence: xuyang@dhu.edu.cn

Received: 21 October 2019; Accepted: 15 November 2019; Published: 18 November 2019

Abstract: To suppress noise in signals, a denoising method called AIC–SVD is proposed on the basis of the singular value decomposition (SVD) and the Akaike information criterion (AIC). First, the Hankel matrix is chosen as the trajectory matrix of the signals, and its optimal number of rows and columns is selected according to the maximum energy of the singular values. On the basis of the improved AIC, the valid order of the optimal matrix is determined for the vibration signals mixed with Gaussian white noise and colored noise. Subsequently, the denoised signals are reconstructed by inverse operation of SVD and the averaging method. To verify the effectiveness of AIC–SVD, it is compared with wavelet threshold denoising (WTD) and empirical mode decomposition with Savitzky–Golay filter (EMD–SG). Furthermore, a comprehensive indicator of denoising (CID) is introduced to describe the denoising performance. The results show that the denoising effect of AIC–SVD is significantly better than those of WTD and EMD–SG. On applying AIC–SVD to the micro-vibration signals of reaction wheels, the weak harmonic parameters can be successfully extracted during pre-processing. The proposed method is self-adaptable and robust while avoiding the occurrence of over-denoising.

Keywords: signal denoising; singular value decomposition; Akaike information criterion; reaction wheel; micro-vibration

---

1. Introduction

As a most common mechanical device, rotating machinery plays a vital role in modern industry. Unlike general equipment, rotating machinery is typically operated in harsh, high-speed, and heavy-load environments. These conditions can easily harm the key components of a mechanical system, such as gears, bearings, and rotors. With further expansion, the damage can cause equipment failure and even casualties. To ensure the safe operation of rotating machinery, fault detection techniques including vibration analysis, acoustic emission, temperature analysis, and wear debris analysis have been developed [1]. Among them, vibration analysis is widely used, owing to its signal testability and high correlation with structural dynamics. Simultaneously, in the fault diagnosis of rotating machinery, the corresponding signal processing technologies have been a part of the most useful approaches [2].

Considering the environmental and structural factors, the source signals are commonly mixed with random noise, which is problematic for the early fault detection of machinery [3]. For the purpose of extracting effective information, numerous reasonable methods are applied to reduce the noise from measured vibration signals. Affected by a series of non-linear factors, such as internal friction, loads, stiffness, and assembly gap, the vibration signals of rotating machinery have strong non-linear and non-stationary characteristics [4]. As powerful tools for non-stationary signal processing, time–frequency analysis methods are commonly used to analyze the characteristics of vibration signals. In general, time–frequency analysis methods include short-time Fourier transform (STFT), discrete wavelet transform (DWT), empirical mode decomposition (EMD) [5], local mean decomposition

(LMD) [6], and variational mode decomposition (VMD) [7]. Concurrently, in practical applications, the denoising methods based on time–frequency analysis have also made significant contributions. Currently, the methods based on wavelet analysis are the most well-known processing methods of signal denoising [8]. As a typical approach, based on the multi-resolution and self-similar characteristics of wavelet analysis, wavelet threshold denoising (WTD) reduces the noise in non-stationary signals [9]. In engineering applications, there are still a few limitations in WTD, such as the selection of the wavelet basis functions [10] and phase lag after denoising [11]. Similar to WTD, the quality of EMD threshold denoising strongly depends on the selection of threshold parameters [12]. To achieve ideal denoising, comparatively more advanced denoising methods are developed by the improvement of time–frequency analysis, such as EMD with Savitzky–Golay filter (EMD–SG) [13]. Apart from time–frequency analysis, significant research efforts have been made for realizing noise reduction, such as singular value decomposition (SVD) [14], matching tracking [15], and sparse representation [16].

SVD is a non-parametric technique first proposed by Beltrami in 1873 [17]. In engineering applications, signal processing based on SVD has been an effective approach to analyze non-linear and non-stationary signals. It has been utilized in various applications, including speech recognition [18], data compression [19], image processing [20], fault diagnosis [21], and signal denoising [22]. As a powerful signal processing technique, SVD exhibits excellent performance in mechanical fault diagnosis. Unlike the traditional decomposition algorithm, SVD ensures the stability of feature extraction based on the theory of matrix transformation [23]. For monitoring the condition of rotating machinery, Yang and Tse developed a denoising method of vibration signals by singular entropy; it studied the distribution characteristics of the noise and clean signals [24]. In addition, Golafshan and Sanliturk developed a novel SVD-based denoising method, which was successfully applied for ball bearing localized fault detection in both the time and frequency domains of the vibration signals [25].

However, there are two critical problems in SVD signal denoising: the selection of the construction matrix and determination of the effective singular values. Initially, a one-dimensional signal must be constructed in the trajectory matrix based on the matrix transformation principle of SVD. The common matrix forms include the Toeplitz matrix, cycle matrix, and Hankel matrix [26], of which the most widely used is the Hankel matrix. In reference [27], it was proven that an original signal could be decomposed into a linear superposition of a series of component signals by SVD using the Hankel matrix. Zhao and Ye pointed out that SVD based on the Hankel matrix was quite similar to the signal processing effect of wavelet transform [11]. In 2015, Jiang et al. used the singular values of Hankel–SVD as the characteristic parameters to diagnose bearings [28]. For the order determination of singular values, energy-based methods can appropriately select the active order under the premise of good prior knowledge, such as entropy increments [24] and cumulative contributions of the singular values [29]. In 2010, Zhao et al. used a curvature spectrum of singular values to choose the order of the valid singular values, thus reliably determining the total number of bearing raceway peeling pits [30]. Furthermore, numerous studies have been devoted to the analysis of the difference spectrum relying on the abrupt change of singular values to reduce noise [31]. Li et al. found a unique relationship between valid singular values and major frequencies, which assisted in the inverse verification of the singular value order [32]. In 2016, Zhang et al. completed order determination based on the difference of singular value variance, and thus extended SVD to the denoising of non-periodic signals [33]. When dealing with complex vibration signals in SVD-based denoising, the accuracy and robustness of the order determination are still the most significant properties.

To reduce noise effectively, a signal denoising method based on SVD and the Akaike information criterion (AIC) is proposed. This method can solve the problems of the selection of matrix structure and order determination of singular values. Based on the energy characteristics of the singular values, the optimal structure of the Hankel matrix is determined to act as the trajectory matrix of the signals. In the process of SVD, the effective singular values are accurately selected by adopting the improved AIC. After eliminating noise components, the remaining singular components are used to reconstruct an approximate matrix. Finally, the averaging method is utilized to obtain the denoising time series signal.

The remainder of this paper is organized as follows. Section 2 briefly reviews the principles of the SVD and AIC. Section 3 describes AIC–SVD to make it applicable for vibration signals containing colored noise. The effectiveness of the proposed method is verified by simulation analysis, as presented in Section 4, and the application of a reaction wheel, as described in Section 5. Finally, in Section 6, the conclusions are drawn.

2. Theoretical Background

2.1. Singular Value Decomposition of Signals

SVD is an orthogonal transformation. For a real matrix, $\mathbf{A} \in \mathbf{R}^{m\times n}$, there exist two orthogonal matrices, $\mathbf{U} \in \mathbf{R}^{m\times m}$ and $\mathbf{V} \in \mathbf{R}^{n\times n}$, that satisfy the equation given below [14]

$$\mathbf{A} = \mathbf{U\Sigma V}^T = \sum_{i=1}^{q} u_i\sigma_i {v_i}^T, \tag{1}$$

where the diagonal matrix, $\mathbf{\Sigma}$, is $[\mathbf{diag}(\sigma_1, \sigma_2, \cdots, \sigma_q), 0]$ or its transposition. The elements, $\sigma_i(\sigma_1 > \sigma_2 > \cdots > \sigma_q)$, are the singular values of the matrix $\mathbf{A}$, and $q = \min(m,n)$. $\mathbf{U}$ and $\mathbf{V}$ are the unitary matrices of $\mathbf{A}$, and their column vectors u_i and v_i are the eigenvectors of the covariance matrices, $\mathbf{AA}^T$ and $\mathbf{A}^T\mathbf{A}$, respectively.

The singular values correspond to the feature components of the decomposition matrix. Apart from their high stability, they also have the characteristics of proportional and rotational invariance. Therefore, SVD can ensure the robustness of the signal features represented by different singular values, in compliance with the properties required by the feature vectors in pattern recognition. In the SVD-based process of signals, the Hankel matrix is typically accepted as the trajectory matrix because of its characteristic of zero phase shift [11]. A signal containing a noise is indicated as a vector form, $\mathbf{s} = [s(1), s(2), \cdots, s(N)]$, and its corresponding $m \times n$ dimensional Hankel matrix form is expressed as

$$\mathbf{A} = (a_{ij})_{m\times n} = \begin{bmatrix} s(1) & s(2) & \cdots & s(n) \\ s(2) & s(3) & \cdots & s(n+1) \\ \vdots & \vdots & & \vdots \\ s(m) & s(m+1) & \cdots & s(N) \end{bmatrix}, \tag{2}$$

where $m = N - n + 1$ and $1 < n < N$.

The sampling signal can be expressed by Equation (2) as

$$\mathbf{s} = [\mathbf{A}(1, :), \mathbf{A}(2:m, n)]. \tag{3}$$

Defining $\mathbf{A}_i = u_i\sigma_i {v_i}^T$, the signal component, $\mathbf{P}_\mathrm{i}$, can be expressed as [28]

$$\mathbf{P}_\mathrm{i} = [\mathbf{A}_i(1,:), \mathbf{A}_i(2:m,n)]. \tag{4}$$

Based on Equations (1), (3), and (4), the original signal can be written as

$$\mathbf{s} = \sum_{i=1}^{q} \mathbf{P}_\mathrm{i}. \tag{5}$$

Based on Equation (5), by SVD using the Hankel matrix, the polluted signal can be decomposed into a simple linear superposition of a series of component signals [27]. For an additive noise signal,

$\mathbf{s} = \mathbf{x} + \mathbf{w_{noise}}$, an advantage of this decomposition is that the clean signal can be solved by the order of the effective singular values.

$$\mathbf{x} = \sum_{i=1}^{k} \mathbf{P_i}, \tag{6}$$

where $\mathbf{x}$ is the clean signal, and k is the order of the effective singular values.

2.2. Order Determination of Akaike Information Criterion

The AIC is an estimated measure of the fitting goodness of statistical models [34], and is currently used in the estimation of the source number. The decision functions of the AIC are as follows [35]:

$$AIC(d) = -2N(n-d)\log_{10}(L_d) + 2d(2n-d) \tag{7}$$

and

$$L_d = \frac{\prod\limits_{i=d+1}^{n} \lambda_i^{\frac{1}{n-d}}}{\frac{1}{n-d}\sum\limits_{i=d+1}^{n} \lambda_i}, \tag{8}$$

where $\lambda_i = \sigma_i{}^2$ denotes the eigenvalues of the unitary matrices, L_d is the maximum likelihood estimation of the eigenvalues, and $d = 1, 2, \cdots, n-1$ denotes the number of sources.

The AIC function consists of two parts. The former term is the maximum likelihood estimation of the model parameters, which reflects the parameter fitness of the principal components. The second term is the bias correction term inserted to convert the AIC into an unbiased estimator. The former term decreases with the increase in the number of sources, whereas the second term is contrary to the former. When the sum of the two terms is minimum, the best estimate of the effective order is obtained by balancing both the terms as

$$k = \underset{d}{\operatorname{argmin}}(AIC(d)). \tag{9}$$

3. Signal Denoising of Akaike Information Criterion–Singular Value Decomposition

3.1. Selection of Hankel Matrix Rows and Columns

To select the number of rows and columns of the Hankel matrix, the energy characteristics of the singular values are considered. The energy of the singular values indirectly reflects the information richness of the trajectory matrix [36], which is defined as

$$E(n) = \sum_{i=1}^{q} \sigma_i{}^2. \tag{10}$$

The relationship between the energy of the singular values and elements of the Hankle matrix can be derived from Equation (11).

$$\boldsymbol{AA}^T = [u_1 u_2 \cdots u_q] \cdot \begin{bmatrix} \lambda_1 & & & \\ & \lambda_2 & & \\ & & \ddots & \\ & & & \lambda_q \end{bmatrix} \cdot [u_1 u_2 \cdots u_q]^T, \tag{11}$$

$$E(n) = \lambda_1 + \lambda_2 + \cdots + \lambda_q = \sum_{j=1}^{n} \sum_{i=1}^{m} a_{ij}{}^2. \tag{12}$$

The difference in the number of rows and columns will modify the singular value energy. To easily distinguish the singular components and avoid feature coupling, the optimal number of matrix columns is selected based on the maximum energy of the singular values, i.e.,

$$\hat{n} == \underset{n}{\operatorname{argmax}} \left(E(n)\right) = \underset{n}{\operatorname{argmax}} \left(\sum_{j=1}^{n} \sum_{i=1}^{N-n+1} {a'_{ij}}^{2} \right). \tag{13}$$

According to Equation (13), the energy of the singular values is equal to the sum of the squares of all the matrix elements. When the structure of the Hankel matrix is a square or an approximate square, the corresponding energy of the singular values is maximum. Specifically, if N is even, the energy of the singular values is maximum at $n = N/2$ and $m = N/2 + 1$. If N is odd, the energy of the singular values is maximum at $n = m = (N+1)/2$. As the basis for selecting the optimal structure of Hankel matrix, the maximum criterion of singular value energy makes it convenient to identify the effective singular components.

3.2. Verification and Improvement of Order Determination

To verify the validity of the order determination based on the AIC, the different types of signals are designed. The expressions of the periodic, attenuation and sweep signal are given as

$$\begin{cases} x_1 = \sin(40\pi t) + 1.8\sin(100\pi t) + 0.5\sin(200\pi t) \\ x_2 = \exp^{-2t}[\sin(40\pi t) + 0.5\sin(200\pi t)] \\ x_3 = chirp(t, 10, 1, 100) \end{cases}. \tag{14}$$

Mixed with Gaussian white noise of different signal-to-noise ratios (SNRs), the initial signals turn into a series of polluted signals $s_{1(SNRs)}$, $s_{2(SNRs)}$ and $s_{3(SNRs)}$, respectively. At a sampling rate of 1 kHz and sampling time of 1 s, the polluted signals are constructed as 501 × 500 Hankel matrices to calculate by SVD. For simulation signals of different SNRs, the AIC is used to determine orders in comparison with cumulative contribution rate (CCR) and singular value curvature spectrum (CSM), as shown in Figure 1.

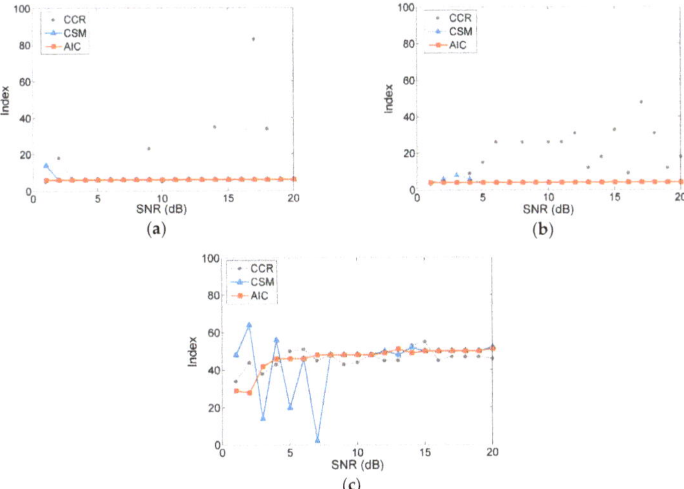

Figure 1. Comparison of the different methods in order determination: (**a**) signals $s_{1(SNRs)}$; (**b**) signals $s_{2(SNRs)}$; (**c**) signals $s_{3(SNRs)}$. CCR: cumulative contribution rate; CSM: singular value curvature spectrum; AIC: Akaike information criterion; SNR: signal-to-noise ratio.

Based on the main frequency analysis method, the effective orders of $s_{1(SNRs)}$ and $s_{2(SNRs)}$ can be rapidly determined as 6 and 4. In Figure 1a,b, the results calculated by the AIC are consistent with those by main frequency analysis method, remaining constant irrespective of the change in the SNR. Concurrently, violent jumps occur in the curves of both the CCR and CSM. As can be observed in Figure 1c, the effective orders by the AIC are more stable than the compared methods for sweep signals $s_{3(SNRs)}$. Therefore, the AIC improves the accuracy and robustness of the order determination, yielding results better than those obtained with other methods at different SNRs. The AIC can achieve viable noise separation, which is beneficial for reasonable noise reduction and feature extraction.

Apart from a white noise of uniform power, the actual vibration signals are also mixed with an uneven colored noise. To smooth the interference components in the background of the colored noise, the eigenvalues are modified by the diagonal loading technique [37] as follows:

$$\mu_i = {\sigma_i}^2 + \sqrt{\sum_{i=1}^{n} {\sigma_i}^2}. \tag{15}$$

Substituting the modified eigenvalues into the maximum likelihood estimation of the signals, the improved AIC function becomes as expressed in Equation (16). Therefore, the adaptive determination of the singular components can be achieved by minimizing the AIC objective function for the signals containing the colored noise.

$$AIC(d) = -2N(n-d)\log_{10}\left(\frac{\prod_{i=d+1}^{n} {\mu_i}^{\frac{1}{n-d}}}{\frac{1}{n-d}\sum_{i=d+1}^{n} \mu_i} \right) + 2d(2n-d). \tag{16}$$

3.3. Denoising of Akaike Information Criterion–Singular Value Decomposition

Combining the energy characteristics and AIC-based order determination of the singular values, a signal denoising method called AIC–SVD is proposed, as shown in Figure 2. The detailed steps of the method can be described as follows:

- Step 1. An $m \times n$ dimension Hankel matrix is chosen as the trajectory matrix of the sampling signal, $\mathbf{s} = [s(1), s(2), \cdots, s(N)]$, and then the optimal number of rows and columns of the matrix is selected according to the maximum energy criterion of the singular values;
- Step 2. SVD is performed on the optimal construction matrix to obtain a sequence of non-zero singular values, $\boldsymbol{\sigma} = (\sigma_1, \sigma_2, \cdots, \sigma_q)$. For signals containing the colored noise, the eigenvalues are corrected according to Equation (16). Next, the index of the minimum AIC value is determined by using the AIC, which is the order of effective singular values;
- Step 3. The inverse operation of SVD is applied to the singular components of the forward k order to obtain the approximate matrix, $\hat{\mathbf{A}}$;
- Step 4. According to the averaging method expressed in Equation (17), the denoised signal is obtained by the reconstruction of the time series signals from the approximate matrix.

$$\hat{x}(i) = \frac{1}{h-l+1}\sum_{j=l}^{h} \hat{A}(i-j+1, j), \tag{17}$$

where $i = 1, 2, \ldots, N$, $l = \max(1, i-n+1)$, and $h = \min(n, i)$.

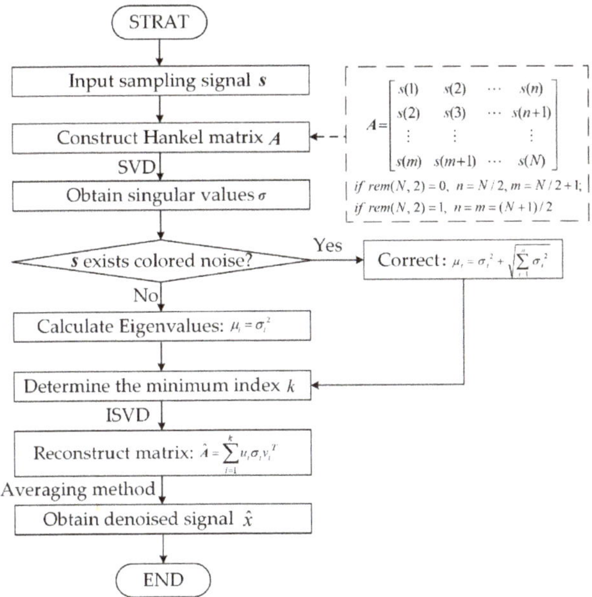

Figure 2. Flow chart of the signal denoising method using AIC–SVD. SVD: singular value.decomposition.

4. Simulation of Akaike Information Criterion–Singular Value Decomposition

4.1. Numerical Simulation

To verify the effectiveness of AIC–SVD in signal denoising, simulation experiments are performed with signal s_1, s_2 and s_3 mixed with a Gaussian white noise of 5 dB. At a sampling rate of 1 kHz and sampling time of 1 s, the corresponding waveform diagrams of the clean and polluted signals are shown in Figure 3. After selecting 501 × 500 Hankel matrices to construct the trajectory matrix of the signals, the singular values and the AIC values are calculated, as shown in Figure 4. Concurrently, the relevant parameters are extracted and listed in Table 1.

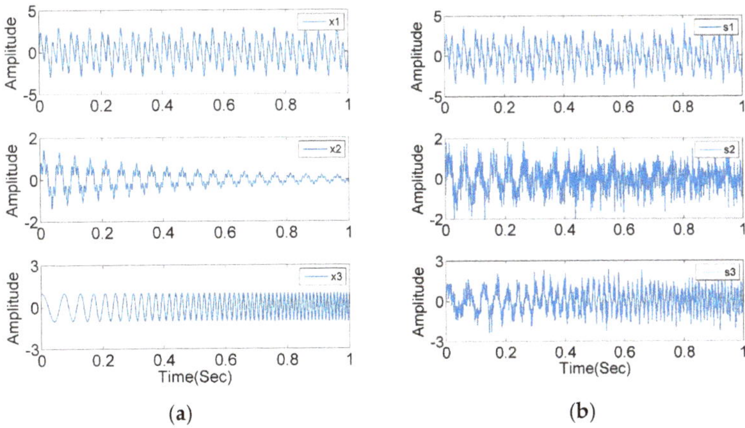

Figure 3. Waveform diagrams of simulation signals: (**a**) clean signals; (**b**) polluted signals with Gaussian white noise of 5 dB.

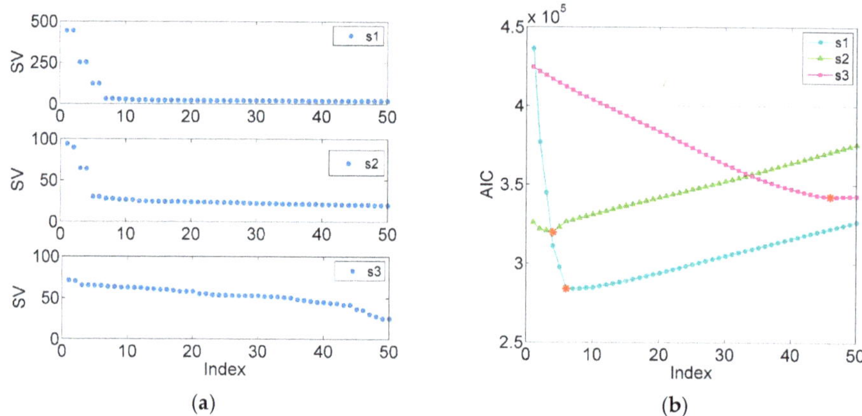

Figure 4. Calculation of simulation signals based on AIC–SVD: (**a**) singular values; (**b**) AIC values.

Table 1. Simulation parameters of signals based on AIC–SVD.

Signal	k	SV	AIC	Energy Ratio [1]	Valid Singular Spectrum	Error
s_1	6	127.8	2.485×10^5	84.49%	89.19%	5.92%
s_2	4	53.4	2.727×10^5	59.82%	54.82%	8.36%
s_3	46	32.9	3.426×10^5	63.84%	66.75%	4.56%

[1] Energy ratio is the energy ratio of clean signals to polluted signals.

As listed in Table 1, the minimum AIC value indices of the above-mentioned three signals are 6, 4, and 46, respectively. Concurrently, the corresponding effective singular spectral values are 89.19%, 54.82%, and 66.75%. The values of the valid singular spectrum are extremely close to the energy ratio of the initial pure signals, and the maximum error is 8.36%. This illustrates that the AIC exhibits a high performance for the order determination of singular values. To determine the reliability of the method, AIC–SVD is compared to WTD and EMD–SG by reconstructing the signals. The processed signals are shown in Figures 5–7.

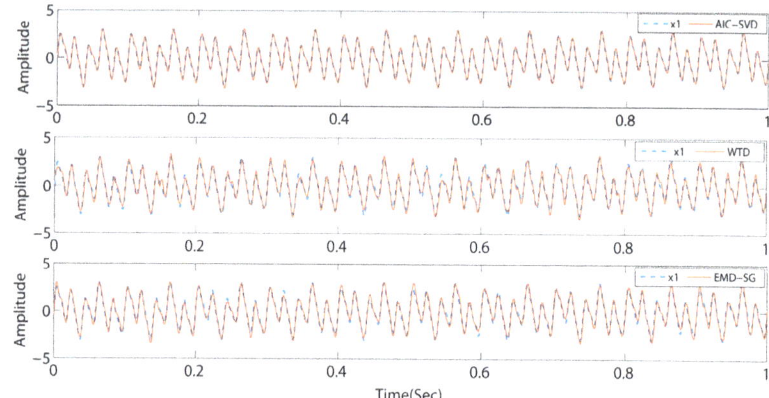

Figure 5. Comparison of denoising effects by different methods for signal s_1. WTD: wavelet threshold denoising; EMD-SG: empirical mode decomposition with Savitzky–Golay filter.

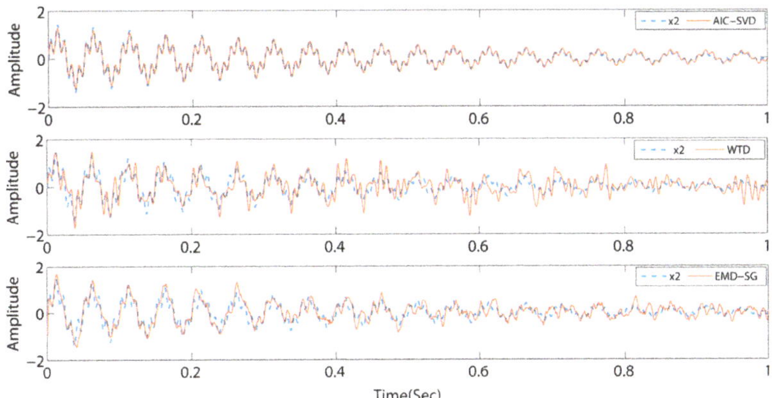

Figure 6. Comparison of denoising effects by different methods for signal s_2.

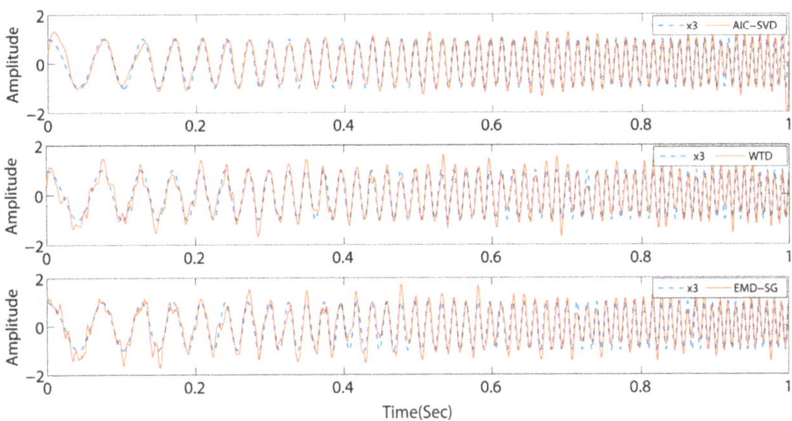

Figure 7. Comparison of denoising effects by different methods for signal s_3.

The comparison reveals that the signals processed by AIC–SVD are well restored by the morphology of the pure signal without a phase shift. For a periodic signal, the denoising effects of WTD and EMD–SG are similar overall to that of AIC–SVD. However, the attenuated signal and swept frequency signal have a notable issue. Specifically, the reconstructed signals exhibit a major waveform distortion, which is not conducive to the subsequent extraction and analysis of the features. The denoising method of AIC–SVD can prevent signal distortion while effectively removing noise. With zero phase shift characteristics, the method of AIC–SVD is suitable in the denoising of different types of signals.

4.2. Denoising Performance Evaluation

To describe the performance of denoising more intuitively and accurately, the simulation signals are further quantitatively analyzed by combining the SNR, root mean square error (RMSE), and waveform correlation coefficient (NCC). These evaluation indicators are defined as follows [38]:

$$SNR = 10\log_{10}\frac{\sum\limits_{i=1}^{N} x(i)^2}{\sum\limits_{i=1}^{N} [x(i)-\hat{x}(i)]^2}, \tag{18}$$

$$RMSE = \sqrt{\frac{1}{N}\sum_{i=1}^{N}[x(i) - \hat{x}(i)]^2}, \quad (19)$$

$$NCC = \frac{\sum_{i=1}^{N} x(i)\cdot\hat{x}(i)}{\sqrt{\left(\sum_{i=1}^{N} x(i)^2\right)\cdot\left(\sum_{i=1}^{N} \hat{x}(i)^2\right)}}. \quad (20)$$

The SNR and RMSE reflect the global characteristics of the denoising performance, whereas the NCC describes the local characteristics of the signals. To avoid the limitations of a single evaluation index, a comprehensive evaluation index (CID) of denoising is introduced by integrating the SNR, RMSE and NCC. It can be defined as

$$CID = \frac{SNR\cdot NCC}{RMSE}. \quad (21)$$

According to Equation (21), a large value of CID corresponds to a good performance in signal denoising. For the simulation signals, the denoising performance parameters of different methods are calculated and listed in Table 2. Subsequently, Gaussian white noise with different SNRs (2 dB, 5 dB, and 10 dB) is added to the pure signals. The CID values of the denoising at the different SNRs are shown in Figure 8.

Table 2. Denoising performance parameters at SNR of 5 dB.

Evaluation Parameters	WTD			EMD–SG			AIC–SVD		
	s_1	s_2	s_3	s_1	s_2	s_3	s_1	s_2	s_3
SNR	31.548	7.196	19.007	34.548	9.908	18.334	51.407	38.295	23.496
RMSE	0.309	0.273	0.274	0.266	0.239	0.283	0.115	0.058	0.219
NCC	0.979	0.800	0.933	0.985	0.845	0.933	0.997	0.990	0.954
CID	100	21	65	128	35	60	447	657	103
Computing time (s)	0.9	1.2	1.4	3.6	2.5	2.4	3.7	2.7	2.7

RMSE: root mean square error; NCC: waveform correlation coefficient; CID: comprehensive evaluation index.

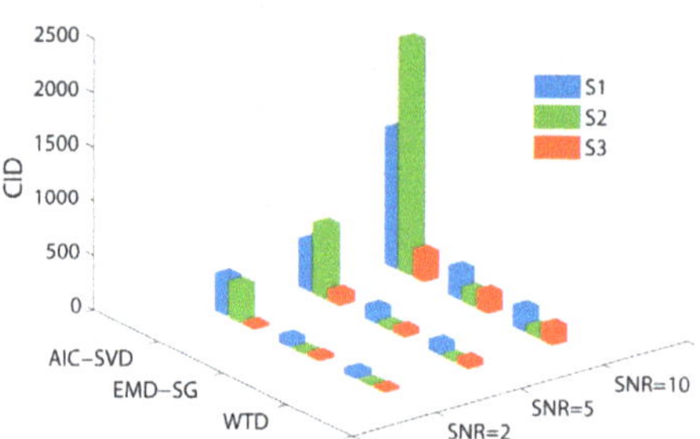

Figure 8. Comparison of the CID for the different denoising methods.

The data in Table 2 prove that the SNRs of the signals are improved after denoising by both the methods, of which AIC–SVD leads to the largest increase. The minimum NCC value of AIC–SVD is 0.954, which can preserve the local waveform characteristics of the initial signal well, avoiding signal distortion. In Figure 8, the CID values of the different denoising methods increase with the improvement in the SNR of the initial signal, and the overall denoising performance of AIC–SVD is significantly better than those of the compared methods. Specifically, for the attenuated signal, the corresponding CID value of AIC–SVD at a 5 dB SNR is 657, which is much larger than those of the other methods. The powerful denoising performance of AIC–SVD for the attenuated signal shows that it is an effective pre-processing tool for vibration signals with pulse characteristics.

5. Study on Micro-Vibration Signal Denoising of Reaction Wheels

5.1. Micro-Vibration Test

As important attitude control components of a satellite, reaction wheels have the general characteristics of rotating machinery. The specific structure of a reaction wheel is depicted in Figure 9. It primarily consists of a rotor supported by ball bearings encased in a housing and driven by a brushless direct current (DC) motor. Influenced by some factors such as the internal rotor imbalance, bearing imperfections, and structural modes, a reaction wheel generates disturbance forces and moments during running. The negative impact of the disturbances is unacceptable for the normal operation of the payloads in satellites [39]. To ensure the successful implementation of the operations in space, it is necessary to analyze the micro-vibration characteristics of reaction wheels.

An on-ground micro-vibration test is frequently performed as an approach to study the micro-vibration characteristics of reaction wheels. It is conducted using the Kistler micro-vibration test device, as depicted in Figure 10. During the operation of a reaction wheel, the disturbance response is transmitted to the force measurement platform through the transfer tool. Then, the micro-vibration signals collected by piezoelectric sensors are transmitted to a data acquisition (DAQ) system via a charge amplifier, which are displayed and processed on a computer.

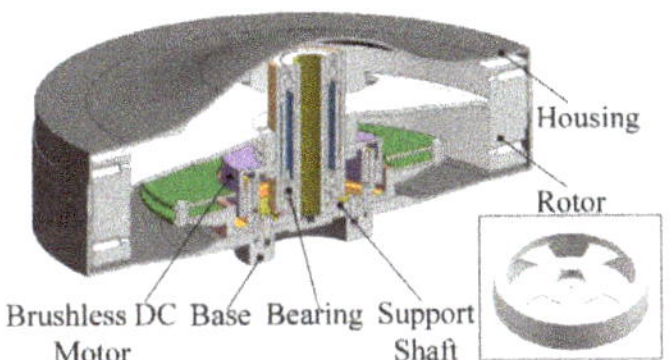

Figure 9. Structure of a reaction wheel. DC: direct current.

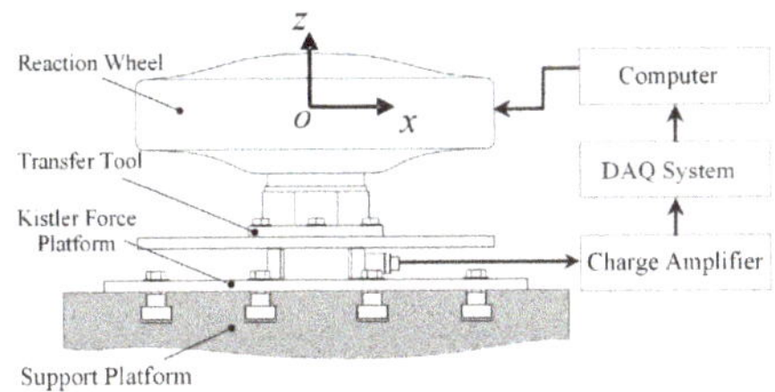

Figure 10. Micro-vibration test of a reaction wheel. DAQ: data acquisition.

The micro-vibration signals of the reaction wheel are collected at different rotational speeds (0–2000 rpm). Performing the fast Fourier transform on the time domain signals, three-dimensional waterfall diagrams of the radial and axial disturbances are obtained, as shown in Figure 11. The vibration of the reaction wheel mainly concentrates on the radial disturbance forces F_x, axial disturbance forces F_z, and radial disturbance torque M_x. Relatively, the magnitude of the axial disturbing moment M_z is small, which can be ignored. Therefore, the analysis of the micro-vibration signals is carried out in F_x, F_z and M_x.

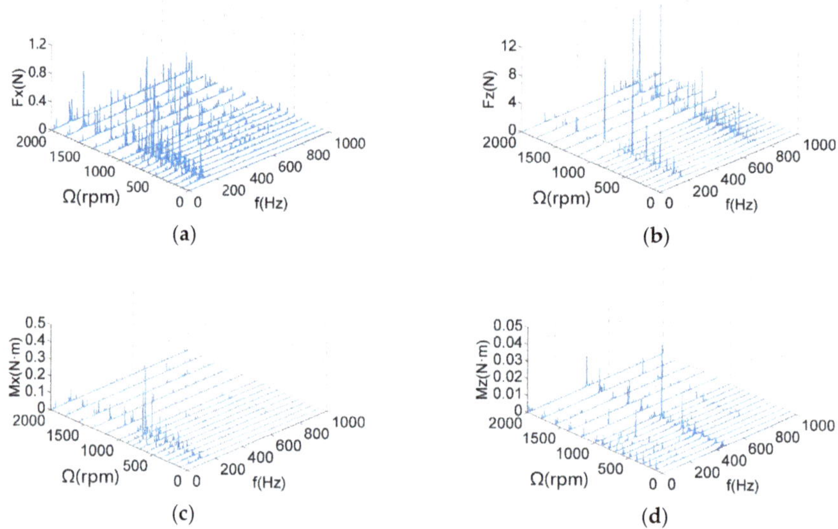

Figure 11. Waterfall diagrams of a reaction wheel: (**a**) F_x; (**b**) F_z; (**c**) M_x; (**d**) M_z.

5.2. Analysis of Micro-Vibration Denoising

Excluding the environmental factors, the noise of micro-vibration signals is also derived from the internal torque fluctuations and frictional interference. To separate the noise component from the micro-vibration signals, a general processing method called peak threshold denoising is used in reaction wheels at present. Based on the amplitude statistical characteristics of the noise, the threshold value to remove the noise from the original signal is determined. It is described as [40]

$$DT = \mu + N_\delta \cdot \delta, \tag{22}$$

where μ and δ are the mean and standard deviation of the spike amplitude, respectively, and N_δ is a user-defined tolerance level, which also depends on the SNR of the sampling signals. Generally, the value of N_δ can be 2 or 3.

In the study of a reaction wheel under the ultimate working conditions, the micro-vibration signals at 1800 and 2000 rpm are selected for the denoising analysis. The frequency of interest is set within 500 Hz, which is the main frequency band that causes satellite jitter. The threshold values are calculated according to different tolerance levels, as shown in Figure 12. Similarly, WTD and EMD–SG are used to suppress noise in micro-vibration signals, as shown in Figure 13.

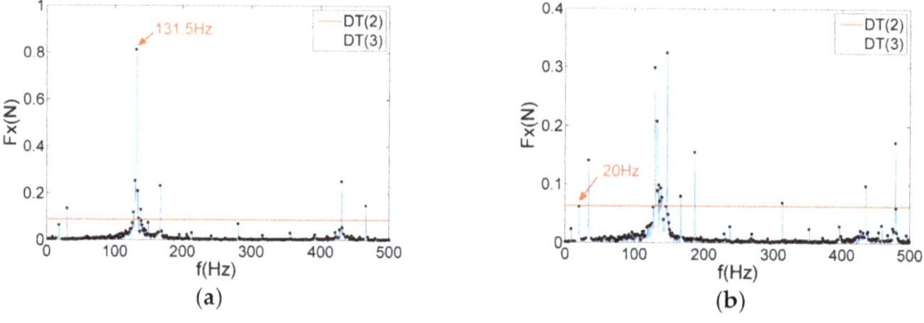

Figure 12. Peak threshold denoising of the micro-vibration signals: (**a**) 1800 rpm; (**b**) 2000 rpm.

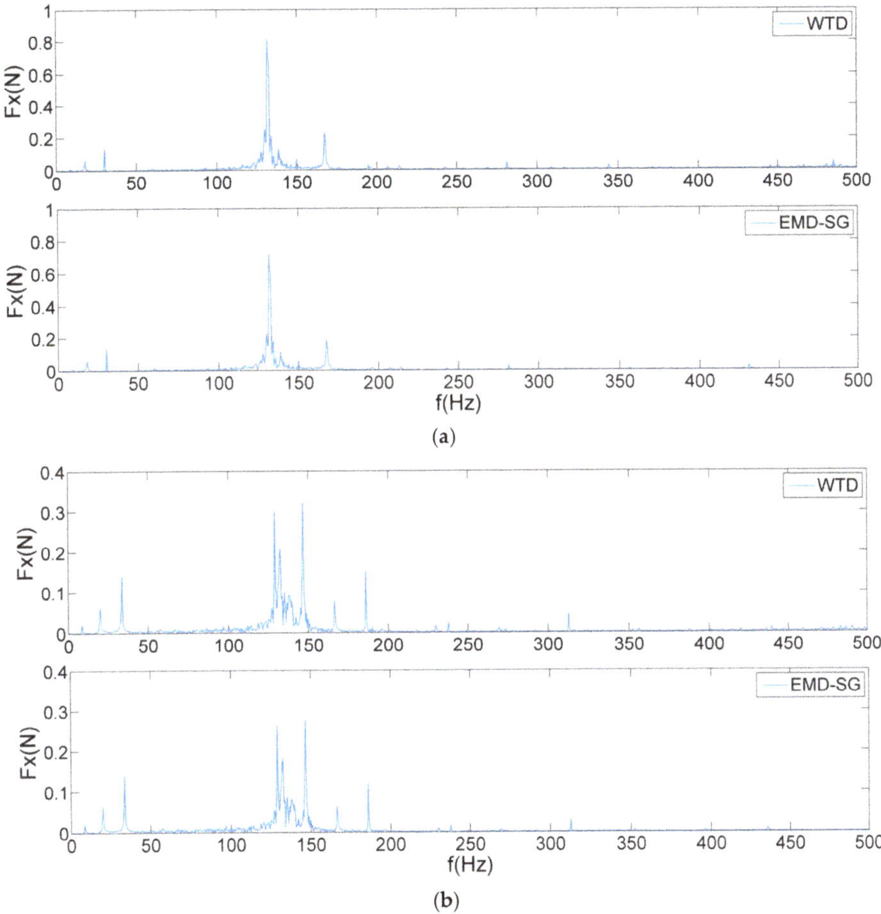

Figure 13. Spectra of micro-vibration signals by WTD and EMD–SG: (**a**) 1800 rpm; (**b**) 2000 rpm.

As exhibited in Figure 12, the magnitude of the user tolerance level directly affects the final effect of the signal denoising. In Figure 12a, a reaction wheel generates a large disturbance force at 131.5 Hz owing to the coupling of the harmonic responses and structural modes, which increases the threshold value to filter out some critical frequency features. Some obvious feature frequencies are equally easy to be removed in Figure 12b, such as 20 Hz. As shown in Figure 13, WTD and EMD–SG mainly act on high-frequency of test signals, which appear under-denoising in the low-frequency range and lose super-harmonics. The filtered details frequently indicate that the system experiences a significant motion mechanism, which is not conducive to the subsequent characteristic analysis and fault diagnosis.

Owing to inappropriate parameter setting and resonance coupling, these denoising methods can easily cause phenomena of over-denoising and under-denoising. Therefore, AIC–SVD is introduced into the pre-processing of the micro-vibration signals of the above-mentioned reaction wheel. By constructing the Hankel matrix of micro-vibration signals, the singular values are solved by SVD. Owing to the presence of colored noise in the micro-vibration signals, the improved AIC is used to determine the order of the effective singular value by correcting the eigenvalues. According to the calculation results as shown in Figure 14, the indices of minimum AIC value is selected to reconstruct the approximate matrixes. Once the time series signals are restored by the averaging method, denoised frequency spectra are obtained, as shown in Figure 15.

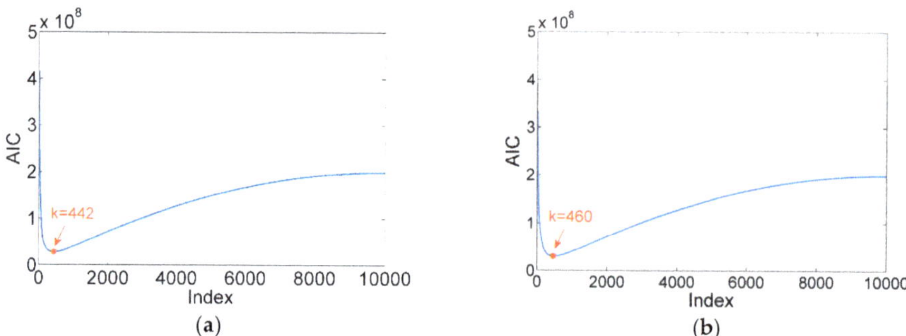

Figure 14. AIC diagrams at different rotational speeds: (**a**) 1800 rpm; (**b**) 2000 rpm.

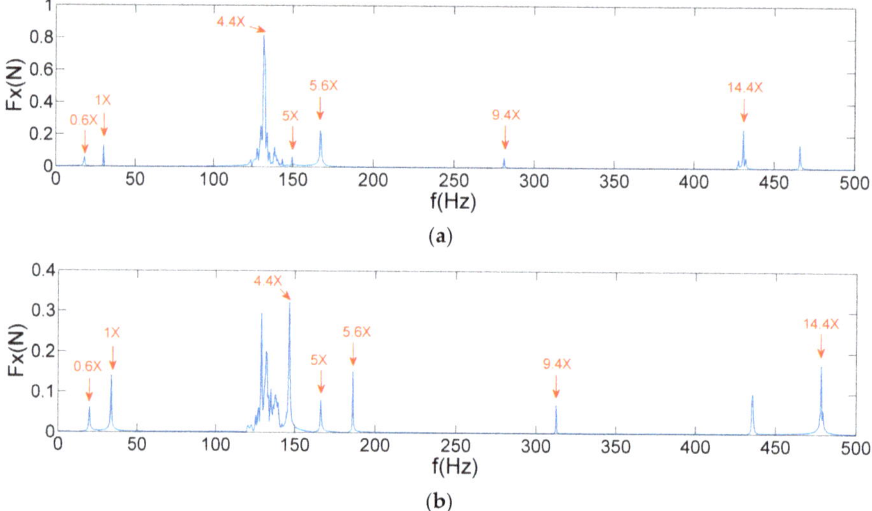

Figure 15. AIC–SVD denoised frequency spectrum of F_x: (**a**) 1800 rpm; (**b**) 2000 rpm.

By comparing Figures 12, 13 and 15, it is observed that AIC–SVD can effectively eliminate the noise from the micro-vibration signals. The denoised signals are convenient in the extraction of harmonic features. As shown in Figures 16 and 17, the micro-vibration signals of F_z and M_x are processed by AIC–SVD. And the related parameters of the reaction wheel are listed in Table 3.

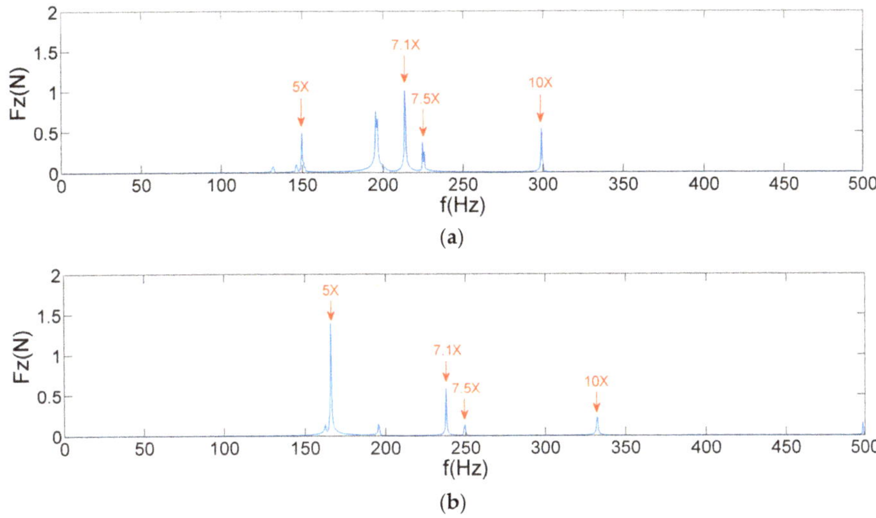

Figure 16. AIC–SVD denoised frequency spectra of F_z: (**a**) 1800 rpm; (**b**) 2000 rpm.

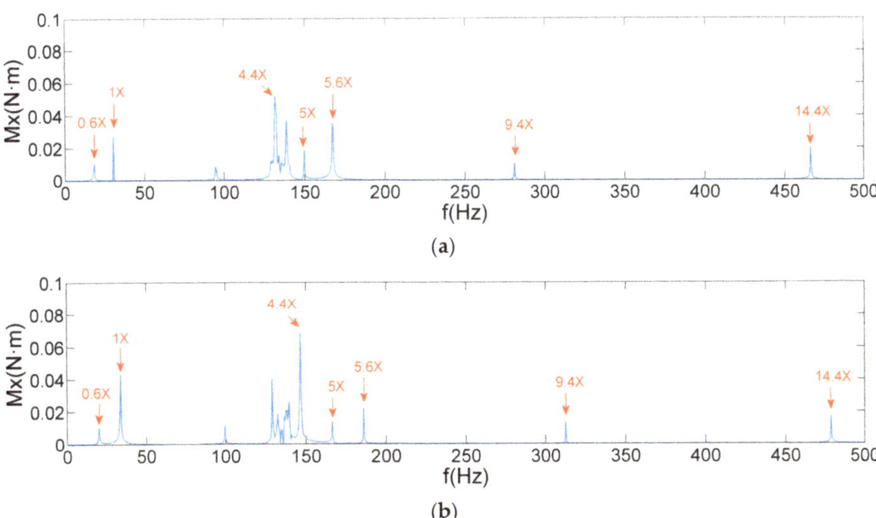

Figure 17. AIC–SVD denoised frequency spectra of M_x: (**a**) 1800 rpm; (**b**) 2000 rpm.

Table 3. Characteristic parameters of micro-vibration signals by AIC–SVD.

Disturbing Component	Speed (rpm)	k	Valid Singular Spectrum	Computing Time (s)	Harmonic Coefficient
F_x	1800	442	94.9%	436	0.6, 1, 4.4, 5, 5.6, 9.4, 14.4
	2000	460	92.4%	448	
F_z	1800	422	99.3%	444	5, 7.1, 7.5, 10
	2000	498	98.9%	450	
M_x	1800	64	86.7%	440	0.6, 1, 4.4, 5, 5.6, 9.4, 14.4
	2000	68	88.1%	442	

The data listed in Table 3 provide all the harmonic coefficients and related frequencies. The average running time of AIC–SVD is 443 s, which is mainly caused by SVD of matrices at the high sampling frequency. Combined with the analysis of the disturbance mechanism, it reveals that the denoised signals include a fundamental harmonic caused by the rotor imbalance, a sub-harmonic of 0.6 times frequency caused by the bearing cage defects, and super-harmonics. Super-harmonics contain 4.4, 5, 5.6, 9.4, and 14.4 times frequency in both F_x and M_x, 5, 7.1, 7.5, and 10 times frequency in F_z, which are caused by the coupling of bearing imperfections.

6. Conclusions

This paper presents a powerful denoising method based on SVD and the improved AIC. Simulation analysis and an engineering application are undertaken to demonstrate the effectiveness of the proposed AIC–SVD, and the following conclusions can be drawn:

(1) In the signal processing of SVD based on Hankel matrix, the energy of the singular values is maximum when the matrix structure is a square or an approximate square. Currently, the feature components provide the largest degree of distinction, which is convenient for the order determination of the effective singular values.
(2) The method of order determination based on the AIC possesses high accuracy and robustness. Furthermore, AIC–SVD is significantly better than WTD and EMD–SG in the denoising performance for the signals containing Gaussian white noise.
(3) In the micro-vibration signal pre-processing of reaction wheels, AIC–SVD achieves a reasonable denoising effect for the signals containing Gaussian white noise and colored noise. This solves the problem of over-denoising and under-denoising caused by inappropriate parameter selection and modal resonance factor. The proposed method has strong adaptability to vibration signal processing under different working conditions, which is beneficial in the extraction of harmonic features.

By extracting the harmonic parameters, a reasonable disturbance model of reaction wheels will be established to describe the characteristics of micro-vibration. This will be of far-reaching significance in the study of orbital-operation monitoring and vibration-reduction space satellites. In addition, there is still room for improvement in the running efficiency of AIC–SVD, which would be optimized in the following study.

Author Contributions: Development of theoretical approach, X.Y. and Y.X.; numerical analysis, X.Y., X.S. and Y.S.; writing—original draft preparation, X.Y.; writing—review and editing, Y.X.

Funding: This research was funded by the National Natural Science Foundation of China (U1831123), and the Fundamental Research Funds for the Central Universities (2232017A3-04).

Acknowledgments: The authors want to acknowledge supports given by Shanghai Engineering Center for Microsatellites in experiments.

Conflicts of Interest: The authors declare no conflict of interest.

References

1. Zhao, M.; Jia, X.; Lin, J.; Lei, Y.; Lee, J. Instantaneous speed jitter detection via encoder signal and its application for the diagnosis of planetary gearbox. *Mech. Syst. Signal. Process.* **2018**, *98*, 16–31. [CrossRef]
2. Wei, Y.; Li, Y.; Xu, M.; Huang, W. A review of early fault diagnosis approaches and their applications in rotating machinery. *Entropy* **2019**, *21*, 409. [CrossRef]
3. Zhang, S.; Wang, Y.; He, S.; Jiang, Z. Bearing fault diagnosis based on variational mode decomposition and total variation denoising. *Meas. Sci. Technol.* **2016**, *27*, 075101. [CrossRef]
4. Yuan, J.; Wang, Y.; Peng, Y.; Wei, C. Weak fault detection and health degradation monitoring using customized standard multiwavelets. *Mech. Syst. Signal. Process.* **2017**, *94*, 384–399. [CrossRef]

5. Huang, N.E.; Shen, Z.; Long, S.R.; Wu, M.C.; Shih, H.H.; Zheng, Q.; Yen, N.-C.; Tung, C.C.; Liu, H.H. The empirical mode decomposition and the Hilbert spectrum for nonlinear and non-stationary time series analysis. *Proc. R. Soc. Lond. Ser. A Math. Phys. Eng. Sci.* **1998**, *454*, 903–995. [CrossRef]
6. Smith, J.S. The local mean decomposition and its application to EEG perception data. *J. R. Soc. Interface* **2005**, *2*, 443–454. [CrossRef]
7. Dragomiretskiy, K.; Zosso, D. Variational mode decomposition. *IEEE Trans. Signal Process.* **2013**, *62*, 531–544. [CrossRef]
8. Beenamol, M.; Prabavathy, S.; Mohanalin, J. Wavelet based seismic signal de-noising using Shannon and Tsallis entropy. *Comput. Math. Appl.* **2012**, *64*, 3580–3593. [CrossRef]
9. Ahn, J.-H.; Kwak, D.-H.; Koh, B.-H. Fault detection of a roller-bearing system through the EMD of a wavelet denoised signal. *Sensors* **2014**, *14*, 15022–15038. [CrossRef]
10. Lu, N.; Xiao, Z.; Malik, O. Feature extraction using adaptive multiwavelets and synthetic detection index for rotor fault diagnosis of rotating machinery. *Mech. Syst. Signal. Process.* **2015**, *52*, 393–415. [CrossRef]
11. He, Q.; Wang, X.; Zhou, Q. Vibration sensor data denoising using a time-frequency manifold for machinery fault diagnosis. *Sensors* **2014**, *14*, 382–402. [CrossRef] [PubMed]
12. Zhao, X.; Ye, B. Similarity of signal processing effect between Hankel matrix-based SVD and wavelet transform and its mechanism analysis. *Mech. Syst. Signal. Process.* **2009**, *23*, 1062–1075. [CrossRef]
13. Boudraa, A.-O.; Cexus, J.-C. Denoising via empirical mode decomposition. *Proc. IEEE ISCCSP* **2006**, *4*, 2006.
14. Kang, M.; Kim, J.-M. Singular value decomposition based feature extraction approaches for classifying faults of induction motors. *Mech. Syst. Signal. Process.* **2013**, *41*, 348–356. [CrossRef]
15. Wang, S.; Chen, X.; Cai, G.; Chen, B.; Li, X.; He, Z. Matching demodulation transform and synchrosqueezing in time-frequency analysis. *IEEE Trans. Signal. Process.* **2013**, *62*, 69–84. [CrossRef]
16. Hao, Y.; Song, L.; Ke, Y.; Wang, H.; Chen, P. Diagnosis of compound fault using sparsity promoted-based sparse component analysis. *Sensors* **2017**, *17*, 1307. [CrossRef]
17. Beltrami, E. Sulle funzioni bilineari. In *Giornale di Matematiche ad Uso degli Studenti Delle Università Italiane*; Battaglini, G., Fergola, E., Eds.; Libreria Scientifica e Industriale: Naples, Italy, 1873; Volume 11, pp. 98–106.
18. Lilly, B.; Paliwal, K. Robust Speech Recognition Using Singular Value Decomposition Based Speech Enhancement. In Proceedings of the IEEE TENCON'97 Brisbane-Australia, Region 10 Annual Conference, Speech and Image Technologies for Computing and Telecommunications (Cat. No. 97CH36162). Brisbane, Queensland, Australia, 4 December 1997; pp. 257–260.
19. Samraj, A.; Sayeed, S.; Raja, J.E.; Hossen, J.; Rahman, A. Dynamic clustering estimation of tool flank wear in turning process using SVD models of the emitted sound signals. *World Acad. Sci. Eng. Technol.* **2011**, *80*, 1322–1326.
20. Sadek, R.A. SVD based image processing applications: State of the art, contributions and research challenges. *Int. J. Adv. Comput. Sci. Appl.* **2012**, *3*, 26–34.
21. Liu, T.; Chen, J.; Dong, G. Singular spectrum analysis and continuous hidden Markov model for rolling element bearing fault diagnosis. *J. Vib. Control.* **2015**, *21*, 1506–1521. [CrossRef]
22. Shi, J.; Liang, M.; Guan, Y. Bearing fault diagnosis under variable rotational speed via the joint application of windowed fractal dimension transform and generalized demodulation: A method free from prefiltering and resampling. *Mech. Syst. Signal. Process.* **2016**, *68*, 15–33. [CrossRef]
23. Han, T.; Jiang, D.; Zhang, X.; Sun, Y. Intelligent diagnosis method for rotating machinery using dictionary learning and singular value decomposition. *Sensors* **2017**, *17*, 689. [CrossRef] [PubMed]
24. Yang, W.-X.; Tse, P.W. Development of an advanced noise reduction method for vibration analysis based on singular value decomposition. *NDT E Int.* **2003**, *36*, 419–432. [CrossRef]
25. Golafshan, R.; Sanliturk, K.Y. SVD and Hankel matrix based de-noising approach for ball bearing fault detection and its assessment using artificial faults. *Mech. Syst. Signal. Process.* **2016**, *70*, 36–50. [CrossRef]
26. Zhao, M.; Jia, X. A novel strategy for signal denoising using reweighted SVD and its applications to weak fault feature enhancement of rotating machinery. *Mech. Syst. Signal. Process.* **2017**, *94*, 129–147. [CrossRef]
27. Zhao, X.; Ye, B. The Similarity of Signal Processing Effect between SVD and Wavelet Transform and Its Mechanism Analysis. *Acta Electron. Sin.* **2008**, *36*, 1582–1589.
28. Jiang, H.; Chen, J.; Dong, G.; Liu, T.; Chen, G. Study on Hankel matrix-based SVD and its application in rolling element bearing fault diagnosis. *Mech. Syst. Signal Process.* **2015**, *52*, 338–359. [CrossRef]

29. Zhao, X.Z.; Ye, B.Y.; Chen, T.-J.A. Influence of Matrix Creation Way on Signal Processing Effect of Singular Value Decomposition. *J. South. China Univ. Technol. (Nat. Sci. Ed.)* **2008**, *36*, 86–93. (In Chinese)
30. Zhao, X.Z.; Ye, B.Y.; Chen, T.-J.A. Selection of Effective Singular Values Based on Curvature Spectrum of Singular Values. *J. South. China Univ. Technol. (Nat. Sci. Ed.)* **2010**, *38*, 11–18, 23. (In Chinese)
31. Zhao, X.; Ye, B. Selection of effective singular values using difference spectrum and its application to fault diagnosis of headstock. *Mech. Syst. Signal. Process.* **2011**, *25*, 1617–1631. [CrossRef]
32. Li, Z.; Li, W.; Zhao, X. Feature frequency extraction based on singular value decomposition and its application on rotor faults diagnosis. *J. Vib. Control.* **2019**, *25*, 1246–1262. [CrossRef]
33. Zhang, X.; Tang, J.; Zhang, M.; Ji, Q. Noise subspaces subtraction in SVD based on the difference of variance values. *J. Vibroeng.* **2016**, *18*, 4852–4861.
34. Akaike, H. A new look at the statistical model identification. *IEEE Trans. Autom. Control.* **1974**, *19*, 716–723. [CrossRef]
35. Cheng, W.; Lee, S.; Zhang, Z.; He, Z. Independent component analysis based source number estimation and its comparison for mechanical systems. *J. Sound Vib.* **2012**, *331*, 5153–5167. [CrossRef]
36. Geng, Y.; Zhao, X. Optimization of Morlet wavelet scale based on energy spectrum of singular values. *J. Vib. Shock* **2015**, *34*, 133–139.
37. Ma, N.; Goh, J.T. Efficient Method to Determine Diagonal Loading Value. In Proceedings of the 2003 IEEE ICASSP'03 International Conference on Acoustics, Speech, and Signal Processing, Hongkong, China, 6–10 April 2003; pp. V-341–V-344.
38. Jin, T.; Li, Q.; Mohamed, M.A. A Novel Adaptive EEMD Method for Switchgear Partial Discharge Signal Denoising. *IEEE Access* **2019**, *7*, 58139–58147. [CrossRef]
39. Kim, D.-K. Micro-vibration model and parameter estimation method of a reaction wheel assembly. *J. Sound Vib.* **2014**, *333*, 4214–4231. [CrossRef]
40. De Lellis, S.; Stabile, A.; Aglietti, G.; Richardson, G. A semiempirical methodology to characterise a family of microvibration sources. *J. Sound Vib.* **2019**, *448*, 1–18. [CrossRef]

Article

Cognitive Frequency-Hopping Waveform Design for Dual-Function MIMO Radar-Communications System

Yu Yao [1,*], Xuan Li [1] and Lenan Wu [2]

[1] School of Information Engineering, East China Jiaotong University, Nanchang 330031, China; lixuan0799@outlook.com

[2] School of Information Science and Engineering, Southeast University, Nanjing 210096, China; wuln@seu.edu.cn

* Correspondence: shell8696@hotmail.com; Tel.: +86-1317-789-6959

Received: 30 November 2019; Accepted: 8 January 2020; Published: 11 January 2020

Abstract: A frequency-hopping (FH)-based dual-function multiple-input multiple-output (MIMO) radar communications system enables implementation of a primary radar operation and a secondary communication function simultaneously. The set of transmit waveforms employed to perform the MIMO radar task is generated using FH codes. For each transmit antenna, the communication operation can be realized by embedding one phase symbol during each FH interval. However, as the radar channel is time-variant, it is necessary for a successive waveform optimization scheme to continually obtain target feature information. This research work aims at enhancing the target detection and feature estimation performance by maximizing the mutual information (MI) between the target response and the target returns, and then minimizing the MI between successive target-scattering signals. The two-step cognitive waveform design strategy is based upon continuous learning from the radar scene. The dynamic information about the target feature is utilized to design FH codes. Simulation results show an improvement in target response extraction, target detection probability and delay-Doppler resolution as the number of iterations increases, while still maintaining high data rate with low bit error rates between the proposed system nodes.

Keywords: multiple-input multiple-output (MIMO); frequency-hopping code; dual-function radar-communications; information embedding; mutual information (mi); waveform optimization

1. Introduction

The multiple-input multiple-output (MIMO) radar [1–3] is a significant technique owing to the enhancement in performance it provides over the traditional radar systems with a single transmit antenna. The transmit waveforms are detected by a matched filter bank in the MIMO radar receiver. Making use of the knowledge of the propagation channels, a superior spatial resolution can be achieved. Furthermore, a MIMO radar has several advantages such as prominent interference rejection capability, enhanced parameter estimation performance, and better flexibility for transmit waveform optimization [4–6]. Several works in the literature discussed the subject of MIMO waveform optimization [7–11]. In [7], the authors considered the joint design of both transmit waveforms and receive filters for a collocated MIMO radar with the existence of signal-dependent interference and white noise. The design problem is formulated into a maximization of the signal-to-interference-plus-noise ratio (SINR), including various constraints on the transmit waveforms. Stoica and Li designed the covariance matrix of the transmit waveforms to control the spatial power [8]. However, the cross correlation between the transmit waveforms at a specific target range is minimized. Several research works [12–15] designed the transmit waveform directly instead of just the covariance matrix. The work of [13] considered prior knowledge of the target impulse response (TIR) and utilized the information to select the transmit waveforms which optimize the mutual information (MI) between the target echoes and the TIR. The work of [15] investigated the problem of the spectrally compatible waveform design for MIMO radar in the presence of multiple targets and signal-dependent interference. A new method was proposed to deal with a more general problem, i.e., designing a spectrally compatible waveform for multiple targets, by minimizing the waveform energy of the overlayed space-frequency bands. The waveform optimization which employs prior knowledge of TIR is also implemented in the single-input multiple-output (SIMO) radar system [16].

The work of [17] derived MIMO radar ambiguity functions. San Antonio and Fuhrmann discussed some properties of the MIMO radar ambiguity function, which provide several ideas for MIMO waveform optimization [17]. The frequency-hopping (FH) sequences presented in [18] were used in MIMO radar configuration [19,20]. The FH orthogonal transmit waveforms discussed in [18] are initially considered for multiuser radar system. Furthermore, [18] employed the FH codes to minimize the peaks in the cross correlation functions of the transmit waveforms as much as possible. However, in the multiuser radar scenario, each operator activates its individual system. This is different from the MIMO radar system where the receiver antennas can cooperate to resolve the target responses. The FH sequences, which were designed by Chen and Yang, optimized the MIMO ambiguity function.

Based on their characteristics of being easily produced and modulus constant, FH codes are considered a good choice for the MIMO radar waveforms. A new scheme for optimizing the MIMO radar waveforms was provided in [19]. The scheme makes the energy of the MIMO ambiguity function spread in the range and angular dimensions evenly, as well as decreases the sidelobes in the MIMO radar ambiguity function. The work of [19] also designed optimal FH waveforms, which had separate FH codes and amplitudes, for a collocated MIMO radar system. The joint design problem can be solved by using game theory, provided by [20]. The authors considered the two objective functions, corresponding to FH codes and amplitudes separately, as two interacting players. By this concept, the joint optimization scheme obtained better integrated code and amplitude matrices that can improve performance much better than separate designs.

The works of [21–26] indicated a possibility of employing the radar-communications integration concept to solve the lack of radio frequency (RF) spectrum. Efficient utilization of shared bandwidth between wireless communications and radar can be achieved by using dynamic frequency allocation. For example, using dynamic frequency allocation is a way to make shared bandwidth between wireless communications and radar possess more efficient utilization. The work of [23] proposed a novel dual-function radar-communications (DFRC) strategy to embed quadrature amplitude modulation (QAM) based communication information in the radar waveforms by exploiting sidelobe control and waveform diversity. In [27,28], the authors proposed a novel distributed DFRC MIMO system capable of simultaneously performing radar and communication tasks. The distributed DFRC MIMO system performs both objectives by optimizing the power allocation of the different transmitters in the DFRC system. The proposed strategy can serve multiple communication receivers located in the vicinity of the distributed DFRC MIMO system. Numerous recent studies [29–31] considered that the developing concept of DFRC is secondary to the main radar task. Communication source embedding into the illumination of MIMO radar system is realized using waveform diversity, sidelobe control, or the time-modulated array technique, which was studied in [30]. Hassanien and Himed presented a signaling strategy for communication source embedding into the illumination of FH-based MIMO radar system [32]. The main principle behind the signaling strategy is to embed phase modulation (PM)-based symbols by using phase rotating the FH pulses. The phase shift is implemented to each transmit FH waveform of the MIMO radar system. The PM-based symbol embedding does not influence the function of the MIMO radar system, which uses the FH waveforms.

The investigation on cognitive radar waveform optimization has received a lot of interest [33–35]. To further enhance the performance of the TIR estimation in a time-varying target scene, the transmitted signal parameters should be constantly adjusted. Then, updated knowledge about the time-varying target scene is employed to allocate fundamental resources like transmitted signal parameters in a cognitive mode [34]. A new strategy for optimizing the waveforms of a cognitive radar was presented in [34]. The aim is to enhance the performance of target estimation by minimizing the mean-square error (MSE) of the estimates of target scattering coefficients (TSC) based on Kalman filtering and then minimizing MI between the radar target echoes at successive time instants. However, there is also an increase in the computational load due to the Kalman filtering step in the waveform optimization. Such a cognitive radar system cannot be used in real applications to address the environmental sensing issues. To improve the performance of the target parameters estimation and classification, the pioneering study by Bell in [36] developed an information-theoretic method for the radar waveforms optimization. The authors in [37–39] extended the information-theoretic method by maximizing the MI between the target response and the target-scattering signals as a waveform design criterion in the MIMO radar system. The work of [40] proposed an innovative method to designing the transmit signal of cognitive MIMO radar system, which combines the signal optimization and selection processes. The works of [41–44] present a signaling scheme for information embedding into the illumination of the radar using FH pulses. An FH-based joint radar-communication system enables implementing a primary radar operation and a secondary communication function simultaneously. Then, the authors consider the problem of radar codes optimization under a peak-to-average-power ratio (PAR) and an energy constraint. However, a time-variant radar scenario is not considered.

Based on the points discussed above, it is interesting to discuss the performance of an adaptive dual-function MIMO radar-communications system that combines the adaptive FH waveform optimization scheme stated in [38] and PM-based information embedding strategy stated in [31]. To further adapt to the dynamic radar environment, we consider the problem of adaptive waveforms design and propose a two-step waveform optimization scheme, which provides better target detection performance and high data-rate communication capability between the proposed dual-function systems. The proposed scheme is summarized as follows:

Step 1. Waveform Design: this part includes the design of the FH waveform for the individual MIMO transmit antennas. The primary goal is to maximize the MI between the target-scattering signal and the estimated target response. The method ensures that the target-scattering signal at each time instant is dependent on the target response.

Step 2. Waveform Selection: after the best waveform ensemble is gained, part two is to choice the suitable PM-based FH sequences for emission. The principle of this module is to minimize the MI between consecutive target scattering signals. This part ensures that we constantly obtain target returns that become independent of each other in time, with the purpose of achieving more information about the target characteristics at each time instant of reception.

The premise of the FH waveform optimization scheme is channel estimation. The target feature estimation can be performed by the MIMO radar receiver through observations implemented in the previous time instant. A feedback loop enables the delivery of the estimates to the dual-function MMO radar-communications transmitter. At a result, the optimization strategy allows the MIMO radar transmitter to constantly adjust FH codes to suit the time-varying channel scene.

We choose the FH waveform for the following reasons:

(1) The channel environment is complicated to wireless communications due to densely populated scatterers. However, FH waveforms are immune to multipath channel fading under the circumstances.
(2) The FH waveforms are robust to antagonistic environments by offering low interception probability. Furthermore, FH waveforms are immune to clutter interference.
(3) The constant-modulus waveforms have the property of high transmission power efficiency. FH waveforms enjoy the constant-modulus feature and are easy to generate.

The primary innovations of our work are summarized as follows:

(1) We define the PM-based FH waveforms in dual-function MIMO radar-communications configuration and derive the associated MIMO ambiguity function;
(2) We develop a two-step waveform optimization scheme in the adaptive PM-based dual-function MIMO radar-communications framework;
(3) We evaluate the performance of the proposed scheme in terms of target response estimation, delay-Doppler resolution and communication symbol error rate (SER).
(4) We compare the proposed scheme with other radar systems through analysis of the target detection and receiver operating characteristics (ROC) in an interference noise environment.

The organization of this paper is as follows. In Section 2, we describe the dual-function MIMO radar-communications system and the PM-based FH signal model. In Section 3, we derive the MIMO radar ambiguity function of the PM-based FH waveform. In Section 4, we present an adaptive approach to optimizing the proposed information embedding waveform. The transmit waveforms are designed at step 1 of the algorithm and selected based on the criterion presented in step 2. The simulation results demonstrating the proposed scheme are presented in Section 5. Finally, our conclusions and directions for possible future work are drawn in Section 6.

Throughout this paper, the following notations will be used. We use boldface lowercase letters and boldface uppercase letters to denote vectors and matrices, respectively; $(.)^*$ to denote the complex conjugate operation; $(.)^T$ to denote the transpose operation; $(.)^H$ to denotes the Hermitian operation; $\otimes$ to denote the Kronecker product; $\mathbf{I}_{MN}$ to denote the identity matrix of size $MN \times MN$; $Angle(\cdot)$ to denote the angle of a complex number.

2. System Configuration and Signal Model

2.1. Phase Modulation (PM)-Based Frequency Hopping (FH) Waveforms

A PM-based method for embedding information into the radar emission was recently proposed in [31]. To deliver a finite number of binary bits per radar pulse, the PM-based method maps the binary data into a phase symbol that belongs to a phase dictionary of an appropriate size. During each radar pulse, the PM-based method embeds one phase symbol into the radar emission toward the intended communication direction. At the communication receiver, a phase detector is used to detect the embedded symbol and, subsequently, deciphers the corresponding binary sequence. Unlike the amplitude modulation (AM) and amplitude shift keying (ASK) methods [30], the PM-based method offers the ability to embed information toward communication receivers, regardless of whether they are located within the sidelobe or the main lobe.

Since target detection is the main task of the dual-function radar-communications system, the transmit waveform should be considered primarily based on the requirements of the radar function. One fundamental requirement of the radar function is the high efficiency of transmitted power. So constant-modulus waveforms are selected. FH waveform enjoys the constant-modulus feature and is easy to generate. Furthermore, FH pulse waveform is immune to multipath channel fading and clutter interference under the circumstances.

The configuration for a joint radar-communications system and the PM-based FH signal model was developed in [44]. In this section, we follow the methods of [44] and further develop a PM-based dual-function MIMO radar-communications system. We express the MIMO FH waveforms as

$$\phi_m(t) = \sum_{q=1}^{Q} e^{j2\pi c_{m,q}\Delta f t} u(t - \Delta t), m = 1, \ldots, M \tag{1}$$

where $c_{m,q}, m = 1, \ldots, M; q = 1, \ldots, Q$ describes the FH code, M and Q denote the number of transmit antennas and the length of FH code, respectively. Δf and Δt respectively denote the frequency step and the hopping interval duration, and

$$u(t) = \begin{cases} 1, & 0 < t < \Delta t \\ 0, & otherwise \end{cases} \tag{2}$$

The duration of a pulse is expressed as $T_0 = Q\Delta t$. We assume that the code is $c_{m,q} \in \{1, \ldots, J\}$, where J denotes a predefined value. As a result, the bandwidth of the radar pulse can be approximately denoted by $J\Delta f$. To realize transmit waveform orthogonality, FH code should be designed to meet the following requirement $c_{m,q} = c_{m',q}, \forall q, m = m'$. Several papers proposed FH waveform optimization for MIMO systems (see [19]; and references therein). Then, we propose that a dual-function MIMO radar-communications system with phase symbols yields the extended virtual data model at the radar receiver. We also present the information-embedding scheme and the associated transmission rate. Let $\{\Omega_{m,q} \in [0, 2\pi]\}, m = 1, \ldots, M, q = 1, \ldots, Q$ be a set of MQ PM-based symbols. Hence, the set of FH waveforms is expressed as:

$$x_m(t) = \sum_{q=1}^{Q} e^{j\Omega_{m,q}} e^{j2\pi c_{m,q}\Delta f t} u(t - \Delta t), m = 1, \ldots, M \tag{3}$$

In (3), Δt should be designed to meet the requirement:

$$\int_0^{\Delta t} e^{j2\pi c_{m,q}\Delta f t} e^{-j2\pi c_{m',q'}\Delta f t} dt = 0, m = m', q = q' \tag{4}$$

By using (1) and the orthogonality between the FH waveforms stated in (4), it is easy to prove that PM-based FH waveforms $x_m(t), m = 1, \ldots, M$ are also orthogonal. The set of transmitted waveforms is presented in Figure 1.

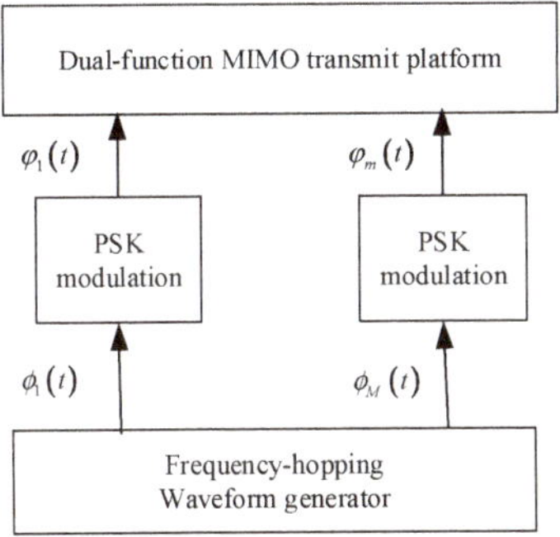

Figure 1. Illustrative diagram of a dual-function multiple-input multiple-output (MIMO) transmit platform.

The $N\times1$ vector of the target scattering signals at the output of the radar receive antenna array can be expressed as:

$$\mathbf{y}(t, i) = \mathbf{H}^T(i)\mathbf{x}(t, i) + \mathbf{n}(t, i) \tag{5}$$

where i describes the index. $\mathbf{H}(i) = \left[h^i_{m,n}\right]_{M\times N}$ indicates the target response matrix during the i-th radar scan. $h_{m,n}$ stands for the channel coefficient between the m-th transmit element and the n-th receive element. $\mathbf{x}(t, i) = [x_1(t, i), \ldots, x_M(t, i)]^T$ represents the $M\times1$ vector of PM-based waveforms during the i-th radar scan. $x_m(t, i)$ is the waveform transmitted from the m-th transmit element, and $\mathbf{n}(t, i)$ is an $N\times1$ vector of zero-mean white Gaussian noise.

The target-scattering signal components associated with the individual PM-based FH waveforms are achieved by using a matched filter bank. As a result, the target scattering signals observed at the output of the MIMO radar receiver (5) are matched-filtered to the proposed PM-based sequences (3), yielding the $MN \times 1$ vector of the virtual signal. The extended vector can be expressed as:

$$\begin{aligned} \mathbf{r}(i) &= vec(\int_{T_0} \mathbf{y}(t,i)\mathbf{x}^H(t)dt) \\ &= vec(\mathbf{H}^T) + \hat{\mathbf{n}}(i) \end{aligned} \tag{6}$$

In (6), $vec(\cdot)$ describes the vectorization operator that stacks the columns of a matrix into one long column vector, and $\hat{\mathbf{n}}(i)$ denotes the $MN \times 1$ zero-mean white Gaussian noise with covariance $\delta_z^2 \mathbf{I}_{MN}$.

2.2. Information-Embedding Scheme

It is assumed that a phase symbol denotes B bits of binary sequence. During the i-th radar pulse, the binary sequence that needs to be embedded is used to select phase symbols $\Omega_{m,q}(i), m = 1, \ldots, M, q = 1, \ldots, Q$ from a pre-defined constellation of $K = 2^B$ symbols. It is considered that the constellation is uniformly distributed between $[0, 2\pi]$, which can be expressed as:

$$\mathbb{C}_{PSK} = \left\{0, \frac{2\pi}{K}, \ldots, \frac{(K-1)2\pi}{K}\right\} \tag{7}$$

The set of PM-based FH waveforms at time i can be rewritten as:

$$x_m(t,i) = \sum_{q=1}^{Q} e^{j\Omega_{m,q}(i)} e^{j2\pi c_{m,q}\Delta f t} u(t-\Delta t), m = 1, \ldots, M \tag{8}$$

In (8), $\Omega_{m,q}(i) \in \mathbb{C}_{PSK}$. To simplify discussion and implementation, it is assumed that a communication receiver equipped with a single antenna is located at a known direction θ_c. Therefore, the received signal at the output of the communication receiver can be expressed as:

$$y(t,i) = \alpha_c \mathbf{a}^T(\theta_c)\mathbf{x}(t,i) + n(t,i) \tag{9}$$

where α_c denotes the channel attenuation coefficient. It is considered that α_c keeps constant during the whole processing interval. $\mathbf{a}(\theta_c)$ indicates the steering vector of the transmit antenna array, and $n(t,i)$ denotes the zero-mean white Gaussian noise with covariance δ_w^2.

It is assumed that the communication receiver has full knowledge of the code $c_{m,q}$ and the step Δf. Hence, the communication received signals (5) are matched-filtered to the orthogonal FH sub-pulses yields:

$$\begin{aligned} r_{m,q}(i) &= \int_0^{\Delta t} y(t,i) e^{-j2\pi c_{m,q}\Delta f t} u(t-\Delta t)dt \\ &= \alpha_c e^{-j\pi d_m \sin\theta_c} e^{\Omega_{m,q}(\tau)} + n_{m,q}(i) \\ & m = 1, \ldots, M, q = 1, \ldots, Q \end{aligned} \tag{10}$$

In (10), d_m denotes the distance between the first and the m-th antennas of the transmit array. The communication receiver has the ability to undo $e^{-j\pi d_m \sin\theta_c}$ before it estimates the symbol $\Omega_{m,q}(i)$. $r_{m,q}(i)$ is viewed as a phase-rotated and noisy version of the m-th entry of $\mathbf{a}(\theta_c)$. Hence, $\Omega_{m,q}(i)$ that need to be embedded is estimated as:

$$\hat{\Omega}_{m,q}(i) = Angle(r_{m,q}(i)) - Angle(\alpha_c) + 2\pi d_m \sin\theta_c \tag{11}$$

Then, the communication receiver compares $\hat{\Omega}_{m,q}(i)$ to the predefined $\mathbb{C}_{PSK}$. $\Omega_{m,q}(i)$ is restored from $r_{m,q}(i)$ at the output of the (m,q)-th matched filter. It allows the communication receiver to determine $\Omega_{m,q}(i)$ and convert $\Omega_{m,q}(i)$ into the original sequence. The advantages of the communication information embedding scheme have been presented in [44].

3. Multiple-Input Multiple-Output (MIMO) Radar Ambiguity Function

In this section, we consider a radar target at $\chi(\tau, v, f)$ where τ describes the delay of the target range, v denotes the Doppler frequency, $f = 2\pi \frac{d_R}{\lambda} \sin\theta$ indicates the spatial frequency. Here θ and λ are the angle of the target and the wavelength, respectively. d_R and d_T are the distance between the transmit antennas and between the receiver antennas, respectively. To simplify the discussion, we assume $d_T = d_R$ in this paper. $x_m(t), m = 1, \ldots, M$ denotes the waveform radiated by the m-th transmit antenna. In [17], the MIMO radar ambiguity function can be defined as:

$$\chi(\tau, v, f, f') = \sum_{m=1}^{M} \sum_{m'=1}^{M} \chi_{m,m'}(\tau, v) e^{j2\pi(fm - f'm')} \tag{12}$$

where

$$\chi_{m,m'}(\tau, v) = \int_{-\infty}^{\infty} x_m(t) x^*_{m'}(t+\tau) e^{j2\pi vt} dt \tag{13}$$

In (13), $\chi_{m,m'}(\tau, v)$ describes the cross ambiguity function, which implicates two radar waveforms $x_m(t)$ and $x_{m'}(t)$. $\chi_{m,m'}(\tau, v)$ is analogous to the SIMO ambiguity function presented in [19]. We now discuss the MIMO radar ambiguity function for the case when $x_m(t)$ is composed of the shifted forms of a rectangular pulse $u_m(t)$.

$$x_m(t) = \sum_{l=1}^{L} u_m(t - T_l) \tag{14}$$

where l denotes the number of the rectangular pulse and T describes pulse repetition interval (PRI). The cross ambiguity function of the waveform $x_m(t)$ is defined as:

$$\chi_{m,m'}(\tau, v) = \sum_{l'=1}^{L} \sum_{l=1}^{L} \chi^u_{m,m'}(\tau + T_l - T_{l'}, v) e^{j2\pi vTl} \tag{15}$$

where $\chi^u_{m,m'}(\tau, v)$ describes the cross ambiguity function of the rectangular pulses $u_m(t)$ and $u_{m'}(t)$. We assume that the pulse duration T_0 and the Doppler frequency v are small enough so that $T_0 v \approx 0$. At a result, the envelope of the Doppler frequency keeps unchanged within an entire pulse period. The cross ambiguity function $\chi^u_{m,m'}(\tau, v)$ reduces to the cross correlation function $r^u_{m,m'}(\tau)$, which is not a function of v any more, we obtain:

$$\chi^u_{m,m'}(\tau, v) \approx r^u_{m,m'}(\tau) \tag{16}$$

We further assume that little reflections take place at these second trip ranges. Therefore, the cross ambiguity function of $x_m(t)$ can be rewritten as:

$$\chi_{m,m'}(\tau, v) \simeq r^u_{m,m'}(\tau) \sum_{l=1}^{L} e^{j2\pi v T_l} \tag{17}$$

The MIMO ambiguity function of $x_m(t)$ can be expressed as:

$$\chi(\tau, v, f, f') = \sum_{m=1}^{M} \sum_{m'=1}^{M} r^u_{m,m'}(\tau) e^{j2\pi(fm - f'm')} \sum_{l=1}^{L} e^{j2\pi v T_l} \tag{18}$$

It is worth noting that the MIMO ambiguity function $\chi(\tau, v, f, f')$ depends on the cross correlation functions $r^u_{m,m'}(\tau)$. Furthermore, the shifted forms of a rectangular pulse $\{u_m(t)\}$ just have an impact on the range and spatial resolution. The pulses have no effect on the Doppler resolution. Consequently, to gain a sharp MIMO ambiguity function, these waveforms should be designed such that the function $\Omega(\tau, f, f')$ is denoted as:

$$\Omega(\tau, f, f') = \sum_{m=1}^{M} \sum_{m'=1}^{M} r^{u}_{m,m'}(\tau) e^{j2\pi(fm - f'm')} \tag{19}$$

Next, we discuss the MIMO radar ambiguity function of the PM-based information embedding FH waveforms. The proposed waveforms can be expressed as (3). We intend to obtain the expression for the function $\Omega(\tau, f, f')$ in terms of $\{c_{m,q}\}$ and $\{\Omega_{m,q}\}$. To derive $\Omega(\tau, f, f')$, we therefore begin with the cross correlation function of the PM-based FH waveform. Making use of (3) and (16), the cross correlation function can be expressed as:

$$r^{x}_{m,m'}(\tau) = \sum_{q=1}^{Q} \sum_{q'=1}^{Q} \chi^{rect}(\tau - (q' - q)\Delta t, (c_{m,q} - c_{m',q'})\Delta f) e^{j\Omega_{m,q}} e^{j2\pi\Delta f(c_{m,q} - c_{m',q'})q\Delta t} e^{j2\pi\Delta f c_{m',q'}\tau} \tag{20}$$

and the function $\Omega(\tau, f, f')$ can be expressed as:

$$\Omega(\tau, f, f') = \sum_{m,m'=1}^{M} \sum_{q,q'=1}^{Q} \chi^{rect}(\tau - (q' - q)\Delta t, (c_{m,q} - c_{m',q'})\Delta f) e^{j\Omega_{m,q}} e^{j2\pi\Delta f(c_{m,q} - c_{m',q'})q\Delta t} e^{j2\pi\Delta f c_{m',q'}\tau} e^{j2\pi(fm - f'm')} \tag{21}$$

where $\chi^{rect}(\tau, v)$ describes the ambiguity function of $u(t)$, which can be denoted by:

$$\chi^{rect}(\tau, v) = \begin{cases} \frac{\Delta t - |\tau|}{\Delta t} \sin c(v(\Delta t - |\tau|)) e^{j\pi v(\tau + \Delta t)}, & \text{if} |\tau| < \Delta t \\ 0, & \text{otherwise} \end{cases} \tag{22}$$

For $M = 1$, the function is the special case of the SIMO radar. For the general case of the MIMO radar $M > 1$, not only the auto-correlation functions but also the cross-correlation functions between the waveforms should be taken into account such that the function (22) is sharp around $\{(\tau, f, f') \big| \tau = 0, f = f'\}$.

4. Waveform Optimization

In this section, we aim to enhance the target detection and feature estimation performance by maximizing the MI between the target response and the target returns in the first step, and then minimizing the MI between successive target scattering signals in the second step. These two stages correspond to the design of the ensemble of excitations and the selection of a suitable signal out of the ensemble, respectively. The two-step cognitive waveform design strategy is based upon continuous learning from the radar scene. The dynamic information about the target feature is utilized to design PM-based FH codes. In this way the transmitter adjusts its probing signals to suit the dynamically changing environment.

Step 1: this step involves the design of PM-based FH waveforms for the dual-function transmit array. The main idea of waveform design is to maximize the MI between the target scattering signal and the target response, subject to the transmit power constraint.

Step 2: once an ensemble of optimal transmit waveforms has been designed, then we select the most reasonable waveform for emission from the ensemble. The key concept of waveform selection is to minimize the MI between the target scattering signals at present and the next target-scattering signals. The step ensures that we continually obtain target scattering signals that are independent of each other in time, in order to achieve more feature information of the target at each time instant of reception. Figure 2 describes the architecture of an adaptive dual-function MIMO radar communication system.

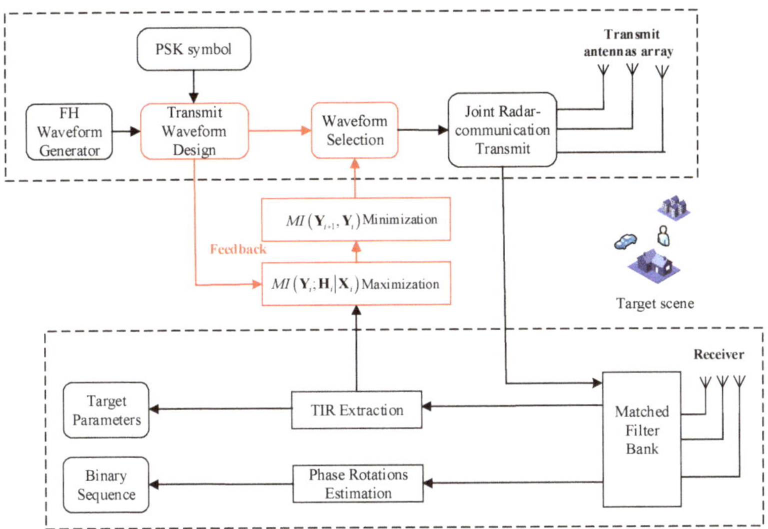

Figure 2. The architecture of an adaptive dual-function MIMO radar communication system.

Step 1: during the i-th radar scan, the set of PM-based FH waveforms $x_m(t,i), m = 1, \ldots, M$ can be expressed as a matrix $\mathbf{X}_i \in \mathbb{C}^{M \times K}$ after discrete sampling, where $\mathbb{C}$ describes the number domain and K indicates the sample number. $\mathbf{N} \in \mathbb{C}^{N \times K}$ is zero-mean white noise matrix. Therefore, the $N \times K$ matric of the target scattering signals is written as:

$$\mathbf{Y}_i = \mathbf{H}_i^T \mathbf{X}_i + \mathbf{N} \tag{23}$$

Here $\mathbf{H}_i \sim \mathbb{C}\left(0, \mathbf{R}_{\mathbf{H}_i}\right)$ and $\mathbf{N} \sim \mathbb{C}(0, \mathbf{R}_{\mathbf{N}})$, and $\mathbf{R}_{\mathbf{H}_i} = E\left\{\mathbf{H}_i \mathbf{H}_i^H\right\}$ and $\mathbf{R}_{\mathbf{N}} = E\left\{\mathbf{N}^H \mathbf{N}\right\}$ indicate the covariance matrices of the channel response $\mathbf{H}_i$ and the zero-mean white noise $\mathbf{N}$, respectively. We intend to maximize the MI between the target-scattering signal and the target response given the transmit waveforms. This involves that the target-scattering signals would be more dependent upon the actual target feature information. According to the definition of MI, we have:

$$I(\mathbf{Y}_i; \mathbf{H}_i | \mathbf{X}_i) = H(\mathbf{Y}_i | \mathbf{X}_i) - H(\mathbf{N}) \tag{24}$$

In (24), $I(\mathbf{Y}_i; \mathbf{H}_i | \mathbf{X}_i)$ represents the MI between two random variates $\mathbf{Y}_i$ and $\mathbf{H}_i$ given the transmit matrix $\mathbf{X}_i$, and $H(\mathbf{Y}_i | \mathbf{X}_i)$ indicates the conditional entropy that $\mathbf{X}_i$ conveys about $\mathbf{Y}_i$. The main objective of this step is to maximize $I(\mathbf{Y}_i; \mathbf{H}_i | \mathbf{X}_i)$ between $\mathbf{Y}_i$ and $\mathbf{H}_i$ given $\mathbf{X}_i$. According to the definition of entropy, we have:

$$H(\mathbf{Y}_i | \mathbf{X}_i) = \int -p(\mathbf{Y}_i | \mathbf{X}_i) In[p(\mathbf{Y}_i | \mathbf{X}_i)] d\mathbf{Y}_i \tag{25}$$

Here $p(\mathbf{Y}_i | \mathbf{X}_i)$ indicates the conditional probability density function (PDF) of the received matrix $\mathbf{Y}_i$ given transmit matrix $\mathbf{X}_i$. The conditional PDF $p(\mathbf{Y}_i | \mathbf{X}_i)$ can be expressed as follows:

$$\begin{aligned} p(\mathbf{Y}_i | \mathbf{X}_i) &= \prod_{n=1}^{N} p\left(\mathbf{y}_{n,i} \middle| \mathbf{X}_i\right) \\ &= \prod_{n=1}^{N} \frac{1}{\pi^K \det\left(\mathbf{X}_i^H \mathbf{R}_{\mathbf{H}_i} \mathbf{X}_i + \mathbf{R}_{\mathbf{N}}\right)} e^{-\mathbf{y}_{n,i}^H \left(\mathbf{X}_i^H \mathbf{R}_{\mathbf{H}_i} \mathbf{X}_i + \mathbf{R}_{\mathbf{N}}\right)^{-1} \mathbf{y}_{n,i}} \\ &= \frac{1}{\pi^{NK} \left[\det\left(\mathbf{X}_i^H \mathbf{R}_{\mathbf{H}_i} \mathbf{X}_i + \mathbf{R}_{\mathbf{N}}\right)\right]^N} e^{-tr\left[\left(\mathbf{X}_i^H \mathbf{R}_{\mathbf{H}_i} \mathbf{X}_i + \mathbf{R}_{\mathbf{N}}\right)^{-1} \mathbf{Y}_i^H \mathbf{Y}_i\right]} \end{aligned} \tag{26}$$

Solving (25) and (26) gives rise to the following result for the conditional entropy [40]:

$$H(\mathbf{Y}_i|\mathbf{X}_i) = NKIn(\pi) + NK + NIn\left[\det\left(\mathbf{X}_i^H \mathbf{R}_{\mathbf{H}_i} \mathbf{X}_i + \mathbf{R}_\mathbf{N}\right)\right] \tag{27}$$

Similarly, the result for the entropy of the noise can be derived as follows:

$$H(\mathbf{N}) = NKIn(\pi) + NK + NIn[\det(\mathbf{R}_\mathbf{N})] \tag{28}$$

Making use of Equations (24), (27) and (28), the MI between the target-scattering signal and the channel response given transmit waveforms can be rewritten as follows:

$$I(\mathbf{Y}_i; \mathbf{H}_i|\mathbf{X}_i) = NIn\left[\det\left(\mathbf{X}_i^H \mathbf{R}_{\mathbf{H}_i} \mathbf{X}_i + \mathbf{R}_\mathbf{N}\right)\right] - NIn[\det(\mathbf{R}_\mathbf{N})] \tag{29}$$

Therefore, we can formulate the MI maximization criterion as follows:

$$\begin{aligned} &\max_{\mathbf{X}_i}\left\{NIn\left[\det\left(\mathbf{X}_i^H \mathbf{R}_{\mathbf{H}_i} \mathbf{X}_i + \mathbf{R}_\mathbf{N}\right)\right] - NIn[\det(\mathbf{R}_\mathbf{N})]\right\}_{\frac{n!}{r!(n-r)!}} \\ &s.t.\ \mathrm{tr}\left[\mathbf{X}_i^H \mathbf{X}_i\right] \le P_0 \\ &\quad c_{m,q} \ne c_{m',q},\ \text{for } m = m', \forall q \quad \Delta t \Delta f = 1 \end{aligned} \tag{30}$$

where P_0 indicates the transmit power. The work of [40] has given a rigorous solution of the above optimization problem (30). Then, we can obtain the ensemble $\mathbb{C}_{\hat{\mathbf{X}}_i}$ out of the whole set of PM-based FH sequences, and the corresponding power allocation vector over diverse dual-function transmit antennas. We start the design procedure with the transmit waveforms from the PM-based FH matrix, and allocate the power on a pulse level as well as across the transmit antennas based on the MI maximization criterion.

Step 2: we then proceed to the waveform selection procedure, in which the successive target scattering signals are different from each other. The step ensures that we achieve more information of the target feature at each time instant of reception. We denote the MI between the successive target scattering signals at time i and at time $i+1$ as:

$$I(\mathbf{Y}_{i+1}, \mathbf{Y}_i) = H\left(\mathbf{Y}_{i+1}\middle|\mathbf{X}_{i+1}\right) + H(\mathbf{Y}_i|\mathbf{X}_i) - H\left(\mathbf{Y}_{i+1}, \mathbf{Y}_i\middle|\mathbf{X}_{i+1}, \mathbf{X}_i\right) \tag{31}$$

In (31), the term $H(\mathbf{Y}_i|\mathbf{X}_i)$ (or $H\left(\mathbf{Y}_{i+1}\middle|\mathbf{X}_{i+1}\right)$) denotes the entropy of $\mathbf{Y}_i$ (or $\mathbf{Y}_{i+1}$) at time i (or $i+1$) given the knowledge of $\mathbf{X}_i$ (or $\mathbf{X}_{i+1}$). The term $H\left(\mathbf{Y}_{i+1}, \mathbf{Y}_i\middle|\mathbf{X}_{i+1}, \mathbf{X}_i\right)$ is defined similarly. According to the literature [6,34], the above Equation (31) can be rewritten as follows:

$$\begin{aligned} I(\mathbf{Y}_i, \mathbf{Y}_{i+1}) &= -NIn\left(\det\left(\mathbf{I}_{(M\times M)} - \mathbf{D}_{i,i+1}^2\right)\right) \\ &= -N\sum_{m=1}^{M} In\left(1 - \left(d_{i,i+1}^m\right)^2\right) \end{aligned} \tag{32}$$

where $d_{i,i+1}^m\left(d_{i,i+1}^1 \ge d_{i,i+1}^2 \ge \ldots \ge d_{i,i+1}^M\right)$ is the diagonal element of the diagonal matrix $\mathbf{D}_{i,i+1}$. $\mathbf{D}_{i,i+1}$ is achieved by singular value decomposition (SVD) of the covariance matrix, which can be denoted as:

$$\mathbf{R}_{\mathbf{Y}_i,\mathbf{Y}_{i+1}} = E\left\{\mathbf{Y}_i^H \mathbf{Y}_{i+1}\right\} = \mathbf{X}_i^H \mathbf{R}_{\mathbf{H}_i} \mathbf{X}_{i+1} \tag{33}$$

In (33), $\mathbf{R}_{\mathbf{Y}_i,\mathbf{Y}_{i+1}}$ is the cross-covariance of the expressions for $\mathbf{Y}_i$ and $\mathbf{Y}_{i+1}$. Hence, the MI minimization criterion can be expressed as:

$$\begin{aligned}
&\min_{\mathbf{X}_{i+1}\in\mathbb{C}_{\hat{\mathbf{X}}_i}} \left\{-N\sum_{m=1}^{M} In\left\{1-\left(d_{i,i+1}^{m}\right)^{2}\right\}\right\} \\
&s.t.\ \mathrm{tr}\left[\mathbf{X}_{i+1}^{H}\mathbf{X}_{i+1}\right] \le P_0 \\
&c_{m,q} \ne c_{m',q},\ \text{for } m = m', \forall q\ \ \Delta t\Delta f = 1
\end{aligned} \tag{34}$$

We can estimate the values for $\mathbf{Y}_{i+1}$ over all possible values of $\hat{\mathbf{X}}_{i+1} \in \mathbb{C}_{\hat{\mathbf{X}}_i}$ using (23). Thus, we can also form an estimate of all the values of the corresponding $\mathbf{D}_{i,i+1}$ $(d_{i,i+1}^{m}(d_{i,i+1}^{1} \ge d_{i,i+1}^{2} \ge \ldots \ge d_{i,i+1}^{M}))$ and choose the value for $\hat{\mathbf{X}}_{i+1}$ that minimizes (34). The above waveform optimization problem (34) is convex. We can obtain the optimal solution directly by using a MATLAB optimization toolbox, such as CVX.

Step 1 designs the optimal ensemble $\mathbb{C}_{\hat{\mathbf{X}}_i}$ based on MI maximization criterion over the spatial domain, and step 2 selects the optimized PM-based FH waveform for each dual-functional transmit antenna from the ensemble $\mathbb{C}_{\hat{\mathbf{X}}_i}$ based on MI minimization criterion over the temporal domain. The proposed adaptive waveform design and selection procedures can be summarized as Algorithm 1.

Algorithm 1. The adaptive waveform design and selection algorithm

Initialize the transmit matrix $\mathbf{X}_0$ and the covariance matrix of the noise $\mathbf{R}_{\mathbf{N}}$.
1. At the initial time $t = 0$, measure the target scattering signal $\mathbf{Y}_0$ and calculate the covariance of the target-scattering signal $\mathbf{R}_{\mathbf{Y}_0}$. The covariance matrix of the channel response $\mathbf{R}_{\mathbf{H}_0}$ can be obtained through successive measurements with uniform power allocation over the transmit antennas by solving (36).
2. The optimized waveform ensemble $\mathbb{C}_{\hat{\mathbf{X}}_0}$ can be obtained through the waveform design process by solving (30).
3. At time $t = 1$, measure $\mathbf{Y}_1$ and calculate $\mathbf{R}_{\mathbf{Y}_1}$ and the cross-covariance matrix $\mathbf{R}_{\mathbf{Y}_0,\mathbf{Y}_1}$. The actual channel response $\mathbf{H}_1$ is obtained by de-convolving the target-scattering signal with the transmit signal by using (23).
4. Calculate the corresponding singular values $d_{0,1}^{m}, m = 1, \ldots, M$ of $\mathbf{R}_{\mathbf{Y}_0,\mathbf{Y}_1}$ and $\hat{\mathbf{X}}_1 \in \mathbb{C}_{\hat{\mathbf{X}}_0}$ can be acquired through the waveform selection process by solving (38).
5. At time $t = 1$, emission $\hat{\mathbf{X}}_1$ and observe the corresponding received signal $\hat{\mathbf{Y}}_1$ to achieve $\mathbf{R}_{\mathbf{H}_1}$ by using (36).
6. If $i = I_{\max}$, the iterative procedure ends; or else, repeat steps 2–5 iteratively.

It is worth noting that adaptation is included in the waveform design and selection procedures through the feedback process and numerous interactions with the radar channel. In summary, the adaptive waveform optimization strategy for target detection is implemented according to the block diagram in Figure 3.

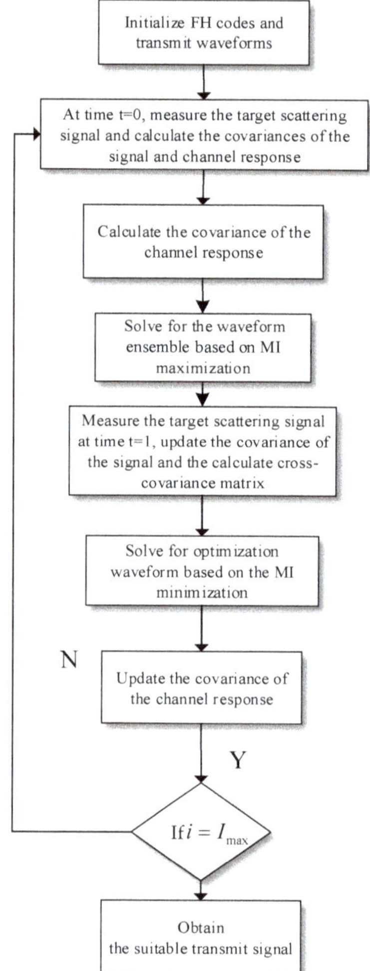

Figure 3. The adaptive waveform design and selection scheme for target detection.

5. Simulation

In this section, numerical results based on Monte Carlo simulations have been provided to validate the effectiveness of the proposed method. Without loss of generality, each entry of the channel matrices follows the standard complex Gaussian distribution. The simulation parameters are based on radar application with a high Pulse Repetition Frequency (PRF), such as in X-band radar. A data rate in the range of dozens of Mbps can be achieved. We provide a comparison between the proposed scheme and the method of [32]. To implement the method in [32], we consider a dual-function MIMO system operating in the X-band with carrier frequency $f_c = 8.2$ GHz and bandwidth $B = 500$ MHz. The sampling frequency is $f_s = 10^9$ sample/sec, which is taken as the Nyquist rate. The PRI is $T_0 = 10$ µs. We assume an arbitrary linear transmit array consisting of $M = 16$ elements. We further assume that the minimum transmit/receive antenna spacing is sufficiently larger than half wavelength (distributed MIMO configuration). Hence, the correlation introduced by finite antenna element spacing is low enough that the fades associated with two different antenna elements can be considered independent. To implement the radar function, we further assume that the FH step is $\Delta f = 10$ MHz, the length of the FH code is $Q = 20$ and the FH interval duration is $\Delta t = 0.1$ μs. We generate a set of 16 FH pulse waveforms. The parameter $J = 50$ is used. Therefore, the 320 FH code is generated randomly from the set $\{1, 2, \ldots, J\}$, where $J = 50$.

We employ orthogonal sequences of the FH pulse over the transmit antenna elements. The backscatter signals are matched filtered at the receivers and the transmitted signals are later modified by the waveform optimization module as shown in Figure 2. The optimized transmission sequence at one particular transmit antenna after the two-step optimization process. The target response extracted from the received target echoes after matched filtering at the end of 20 iterations of the algorithm, where an excellent performance of the target response estimation can be observed. At each iteration, the scattering coefficients for the target and non-target scatterers in $\mathbf{H}_i$ vary as described by the Swerling III model [45,46]. This causes the amplitude returns of the backscatter signals to vary at each instance. However, the amplitudes of the echoes from the target are always assumed to be stronger than those from the clutter sources.

5.1. Target Detection Performance

Figure 4a illustrates the detection probability offered by the proposed scheme versus the signal-to-noise ratio (SNR) for different iterations. The iteration process is run 20 times. All optimal waveforms are generated by the proposed two-step algorithm. The value of requested SNR increases as the probability increase for a particular number of iterations. The SNR decreases as the number of iterations increase for a certain detection probability. The detection performance offered by the proposed scheme improves as the number of iterations increases. Simulation results show that, on average, 20 iterations of the waveform optimization algorithm are required in order to achieve convergence of the target response estimation for a wide range of radar scenes.

Figure 4b illustrates the ROC for the following four approaches while the value of received SNR equals to 8 dB. (1) 4×4 MIMO system based on conventional maximum a posterior (MAP) approach; (2) 4×4 MIMO system using the Kalman filtering [42]; (3) 4×4 MIMO system based on MI optimization (step 2) scheme; (4) 4×4 dual-function MIMO radar-communications system using the proposed scheme.

The plot is run at the end of 24 iterations. For the probability of false alarm $P_{fa} = 0.005$, the of target detection probability generated by the proposed scheme is 0.95 as compared with 0.75 provided by the Kalman filtering method, 0.7 by MI optimization (step 2) criterion and 0.55 by conventional MAP approach. As the two-step scheme can use the temporal correlation of target characteristic during the radar scan interval, the dual-function MIMO system constantly adapts transmit mode to suit the dynamic radar scene. Furthermore, the successive target returns are considered as independent of each other. The property guarantees that information about the radar scene is learned at each instant of reception. In this case, the detection performance offered by the proposed scheme is best.

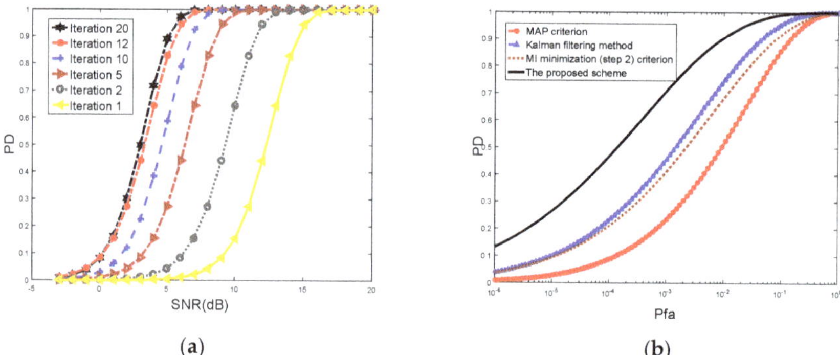

Figure 4. (**a**) The detection probability for different iterations provided by the proposed scheme; (**b**) The receiver operating characteristics (ROC) for four approaches.

5.2. Target Response Estimation Performance

Figure 5a displays the MSE achieved by the proposed scheme with regard to the estimation of target response. This plot demonstrates an improved MSE performance for the two-step optimization as compared with the conventional MAP approach, the Kalman filtering method and the MI optimization (step 2) modules, particularly for the first few iterations.

It is evident from Figure 5 that the MSE performance offered by the proposed scheme is superior to the conventional MAP approach and the Kalman filtering method. The MSE performance offered by the proposed scheme is superior to MI optimization (step 2) criterion. Similarly, Figure 5b shows the MSE performance offered by the proposed scheme with respect to the estimate of target response. The MSE of target response estimation provided by the two-step scheme and other several approaches are compared to reveal the benefit of the proposed scheme.

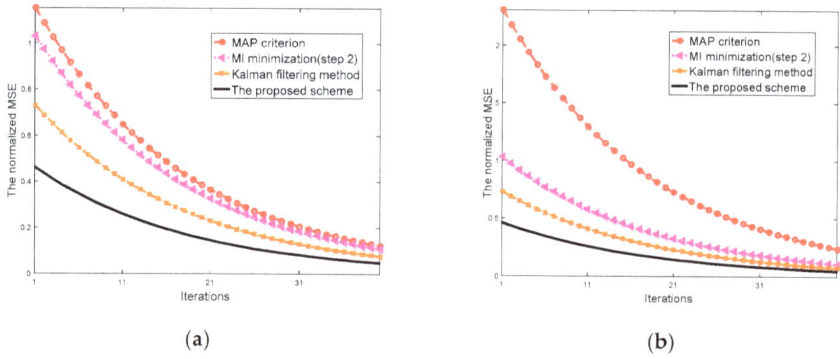

Figure 5. (**a**) The mean-square error (MSE) of target response estimation subject to power constraint; (**b**) the MSE of target response estimation subject to power and probability constraint.

5.3. Delay-Doppler Resolution Performance

Figure 6a displays the 2×2 MIMO radar ambiguity function contours of the original waveform. The delay-Doppler resolution deteriorates owing to the presence of environment noise as shown in Figure 6a. This phenomenon will became worse if the interference scatterers are placed in the vicinity along the line linking the radar target and the antenna. The radar target is considered to be located at the origin of the plan. Figure 6b displays the MIMO radar ambiguity function contours of the optimization waveform by the proposed scheme at the end of 20 iterations. As can be seen from Figure 6a, the target discrimination capability becomes significantly enhanced by the increasing iteration number. Specifically, the noise is suppressed by about 2.5 dB.

The delay-Doppler resolution, which is the output of matched filter at the radar receiver, is connected to the MIMO radar ambiguity function of the transmit waveforms. The near-ideal thumbtack response would appear if the statistically independent waveforms are employed for emission. We constantly use the optimization waveforms provided by the proposed scheme to match the estimated target response. It is worth noting that the estimated target response is continuously updated at each iteration. At a result, the matching process ensures that the noise interference can be suppressed over the radar channels and further enhance the SNR of the target-scattering signals. Figure 6b illustrates the enhancement in SNR with multiple iteration number. The phenomenon shows the enhanced ability of the proposed dual-function MIMO system to discriminate the target from radar environment and resolve the target's range and velocity.

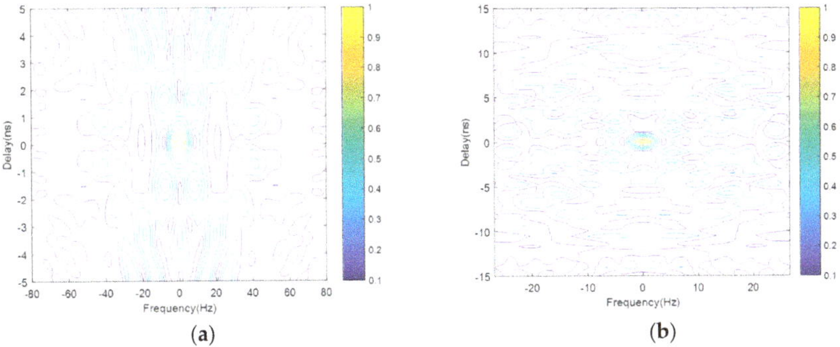

Figure 6. Ambiguity Function (AF) contours showing delay-Doppler resolution at (**a**) iteration 1 and (**b**) iteration 20, which demonstrates smaller focal area in (**b**) as compared with (**a**).

5.4. Symbol Error Rate Performance

We study the SER performance of the communication source embedding scheme using Binary Phase Shift Keying (BPSK), Quadrature Phase Shift Keying (QPSK), 16-Phase Shift Keying (16-PSK) and 256-Phase Shift Keying (256-PSK) constellations and compare the original waveform with the encoded waveform. The original waveform corresponds to data rate of $R = 32; 64; 128$ and 256 Mbps, respectively. To obtain encoded waveform, a convolutional encoder of rate 2/3 is employed in the original waveform. We use a Viterbi decoder to decode the received encoded waveform at the communication receiver. The communication channel coefficient is taken as $|\alpha_c| = 1$, and the phase of α_c is uniformly distributed within the interval $[0, 2\pi]$. To test the SER performance, we generate a number of 10×10^{17} random BPSK, QPSK, 16-PSK and 256-PSK symbols. The SERs versus SNR for all constellation sizes is illustrated in Figure 7.

The results demonstrate that the smaller the constellation size is, the better the SER performance will be. As the constellation size increases, it is more difficult to detect the symbols. It can be explained that the defective cross-correlation between the non-orthogonal transmit sequences influences the detection performance. Obviously, the encoded waveform achieves greater SER performance as compared to the original waveform. Therefore, intersymbol interference is a significant source of detection error resulting in performance degradation. It is expected that the SER performance gets worse if longer FH waveforms are used. It is worth noting that, for all techniques tested, the SER performance offered by the encoded waveform outperforms the performance provided by the original waveform. Therefore, the SER superiority comes at the price of lower data rate.

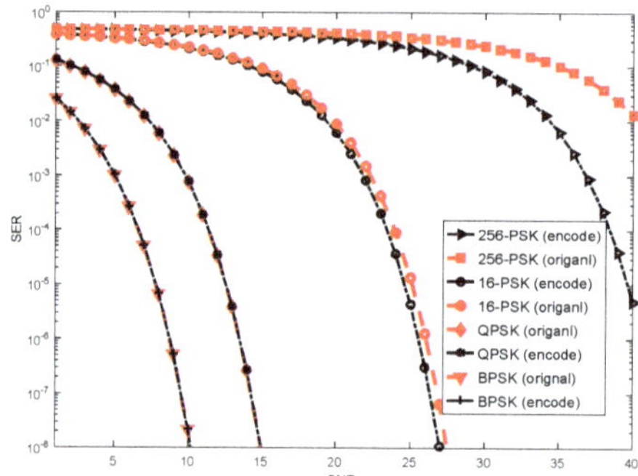

Figure 7. The symbol error rates (SERs) versus signal-to-noise ratio (SNR) for Binary Phase Shift Keying (BPSK), Quadrature Phase Shift Keying (QPSK), 16-Phase Shift Keying (16-PSK) and 256-Phase Shift Keying (256-PSK).

Figure 8 illustrates the throughput result provided by the proposed optimization waveform versus distance for BPSK, QPSK, 16-PSK, and 256-PSK constellation. 256-PSK waveform provides a data rate of approximately 8 Mbps at a distance of 10 m, which is better than that generated by BPSK, QPSK, 16-PSK constellations. The highest data rate is acquired by the 256-PSK waveform within a distance of 60 m, as the distance between the system nodes increases the data rate decreases.

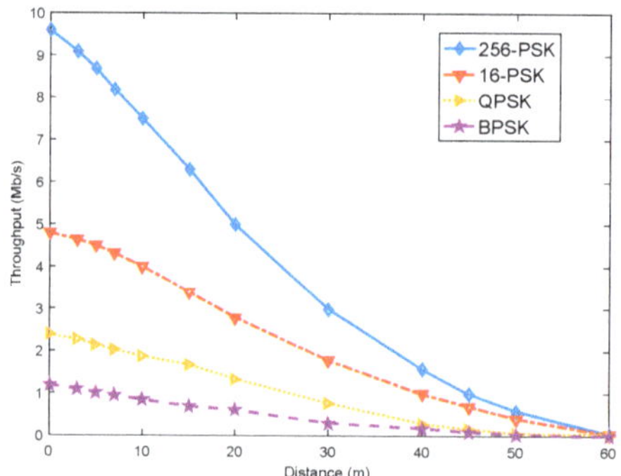

Figure 8. Comparative throughput of BPSK, QPSK, 16-PSK, 256-PSK.

5.5. Detection Variation Performance

The detection constraint optimization has recently been studied in works such as [38,39], where the authors address the problem of radar code design for target recognition in the presence of colored Gaussian disturbance. The objective function in [38] aims to maximize the weighted average Euclidean distance between the ideal echoes from different target hypotheses. Furthermore, additional practical constraints are considered in [38]. For example, the modulus of the waveform is restricted to be a constant and the detection constraints require that the achievable SNR for each target hypothesis is larger than a desired threshold.

Figure 9 indicates the detection variation offered by the proposed scheme subject to the detection constraint. We assume a radar scene, which has three range-separated targets. The target-scattering signals derived from the radar scene is normalized and the proposed dual-function MIMO system intends to discriminate the scatterers by using a particular detection threshold.

With subsequent iterations of the proposed algorithm, the detection performance of the multiple targets is enhanced. As can be seen from Figure 9, by suppressing noise, the dual-function MIMO system could discriminate three scatterers effectively at the end of 20 iterations. The result proves the performance of target detection presented in Figure 6b as well.

The detection performance is enhanced by providing the waveform design (step 1) procedure in the proposed scheme. The waveform design procedure ensures the maximum of the Euclidean distance between the target echoes from different scatterers based on the MI maximization criterion in (30). The proposed design procedure is similar to the optimization method in [43]. By increasing the number of iterations, we can obtain more accurate estimates of the target responses, which are used to improve detection of the multiple targets.

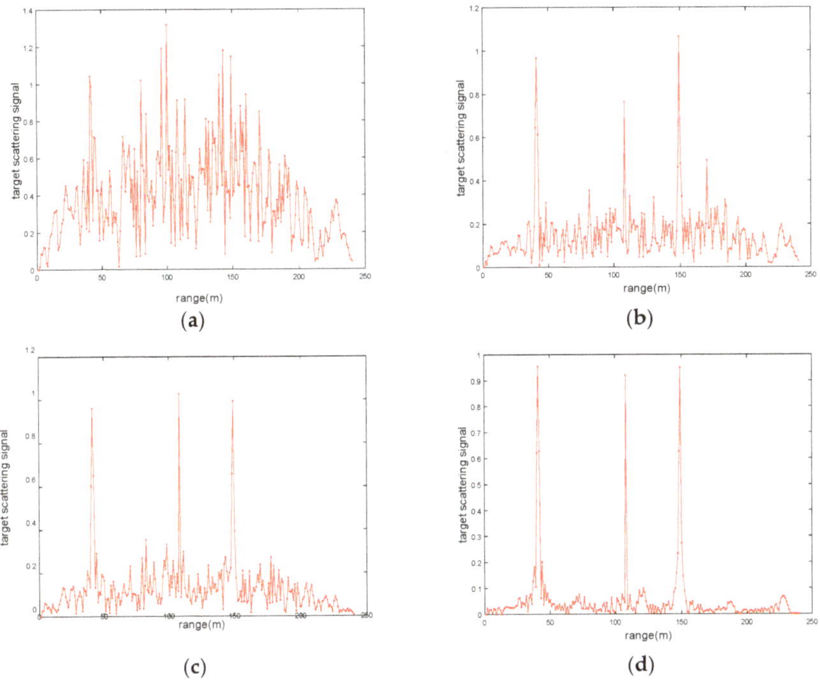

Figure 9. Target-scattering signal profiles at (**a**) iteration 1, (**b**) iteration 5, (**c**) iteration 10, and (**d**) iteration 20.

5.6. Comparison with Other Methods

We have assess the performance of the new approach in terms of the Bit Error Rate (BER) and compare it with the method in [29]. Note that the latter approach uses a single sequence in tandem with 4 sidelobe levels towards the communication direction to deliver 2 bits of information. On the other hand, we use a PM-based FH waveform to deliver two bits per pulse. To test the BER, a sequence of symbols unencoded (two bits each) is used. Furthermore, a convolutional encoder of rate 2/3 is used in the original waveform leading to encoded waveform. Both the unencoded and the encoded waveforms are embedded independently using the approach of [29] and the new approach. The received encoded signal is decoded using a Viterbi decoder. The BERs versus the signal-to-noise ratio (SNR) for the two approaches is presented in Figure 10 for both the unencoded as well as the encoded data sequences. Obviously, the proposed approach achieves better BER performance compared to the method of [29]. Note that the latter approach embeds 25% of the information via each of the four beams. Thus, intersymbol interference is a non-negligible source of detection error leading to capability degradation. The behavior can be expected to be worse if longer pulses are transmitted. Note that, for two approaches tested, the BER with respect to the encoded waveform outperforms the BER with respect to the unencoded waveform. However, this BER superiority comes at the price of slower data transmission rate.

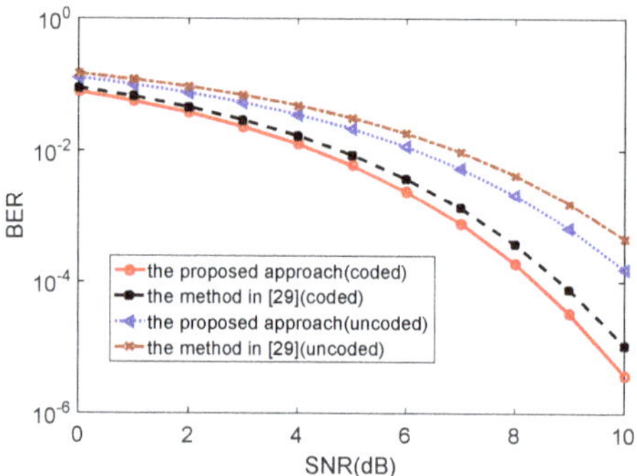

Figure 10. The SERs versus SNR for BPSK.

In Figure 11, we compare the performance for adaptive waveform provided by the proposed method to the performance for a static waveform [32] over multiple snapshots. As the proposed method selects specific waveforms, which generate target returns having low correlation over time, the system adapts its transmit waveform better to the fluctuating target Radar Cross-Section (RCS). On the other hand, the static waveform [32] is unable to match the time-varying target response. Therefore, the performance of the static waveform [32] is worse than the proposed adaptive waveform.

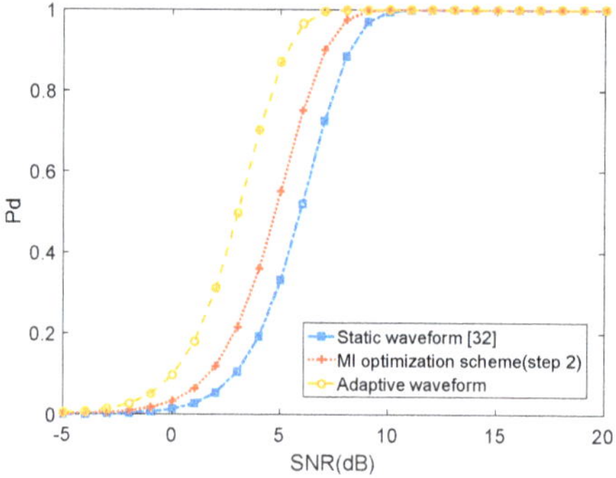

Figure 11. Probability of detection for adaptive waveform and static waveform.

6. Conclusions

The problem of dual-functionality system design for joint radar and communications operation was considered. We further presented a two-step waveform optimization algorithm for the proposed system, which combines the waveform optimization and selection processes. The proposed algorithm is based upon constant learning of the radar scene at the receiver and adaptation of the probing codes to suit the dynamic target feature. The adaptive process ensures maximum information extraction from the target of interest. The effectiveness of the proposed technique and its superiority over existing techniques in terms of the BER performance and the target detection performance were demonstrated through extensive simulations. The proposed system would form a joint platform for future intelligent transportation applications for which both environmental perception and establishment of data links are crucial. Future research will look into the tradeoff between the performance improvement offered by the proposed approach and the computational complexity involved.

Author Contributions: Conceptualization, Y.Y. and X.L.; Methodology, Y.Y.; Software, X.L.; Validation, Y.Y., X.L. and L.W.; Formal Analysis, Y.Y.; Investigation, Y.Y.; Resources, X.L.; Data Curation, X.L.; Writing-Original Draft Preparation, Y.Y.; Writing-Review & Editing, X.L.; Visualization, Y.Y.; Supervision, L.W.; Project Administration, L.W.; Funding Acquisition, Y.Y. All authors have read and agreed to the published version of the manuscript.

Funding: This work was supported by the National Natural Science Foundation of China (61761019, 61861017, 61861018, 61862024) and the Natural Science Foundation of Jiangxi Province (Jiangxi Province natural Science Fund) (20181BAB211014, 20181BAB211013) and Foundation of Jiangxi Educational Committee of China (GJJ180352).

Conflicts of Interest: The authors declare no conflict of interest.

References

1. Li, J.; Stoica, P. (Eds.) *MIMO Radar Signal Processing*; Wiley: Hoboken, NJ, USA, 2009.
2. Li, B.; Petropulu, A.P.; Trappe, W. Optimum co-design for spectum sharing between matrix completion based MIMO radars and a MIMO communication system. *IEEE Trans. Signal Process.* **2016**, *64*, 4562–4575. [CrossRef]
3. Qian, J.; Lops, M.; Zheng, L.; Wang, X.; He, Z. Joint System Design for Coexistence of MIMO Radar and MIMO Communication. *IEEE Trans. Signal Process.* **2018**, *66*, 3504–3519. [CrossRef]
4. Liu, S.; Zhang, Y.D.; Shan, T.; Tao, R. Structure-Aware Bayesian Compressive Sensing for Frequency-Hopping Spectrum Estimation with Missing Observations. *IEEE Trans. Signal Process.* **2018**, *66*, 2153–2166. [CrossRef]
5. Fan, W.; Liang, J.; Li, J. Constant Modulus MIMO Radar Waveform Design with Minimum Peak Sidelobe Transmit Beampattern. *IEEE Trans. Signal Process.* **2018**, *66*, 4207–4222. [CrossRef]
6. Yao, Y.; Zhao, J.; Wu, L. Adaptive Waveform Design for MIMO Radar-Communication Transceiver. *Sensors* **2018**, *18*, 1957. [CrossRef]
7. Wu, L.; Babu, P.; Palomar, D.P. Transmit Waveform/Receive Filter Design for MIMO Radar with Multiple Waveform Constraints. *IEEE Trans. Signal Process.* **2017**, *66*, 1526–1540. [CrossRef]
8. Stoica, P.; Li, J.; Xie, Y. On Probing Signal Design for MIMO Radar. *IEEE Trans. Signal Process.* **2007**, *55*, 4151–4161. [CrossRef]
9. Tang, B.; Li, J. Spectrally Constrained MIMO Radar Waveform Design Based on Mutual Information. *IEEE Trans. Signal Process.* **2018**, *67*, 821–834. [CrossRef]
10. Yang, Y.; Blum, R. MIMO radar waveform design based on mutual information and minimum mean-square error estimation. *IEEE Trans. Aerosp. Electron. Syst.* **2007**, *43*, 330–343. [CrossRef]
11. Wang, L.; Zhu, W.; Zhang, Y.; Liao, Q.; Tang, J. Multi-Target Detection and Adaptive Waveform Design for Cognitive MIMO Radar. *IEEE Sens. J.* **2018**, *18*, 9962–9970. [CrossRef]
12. Fuhrmann, D.R.; San Antonio, G. Transmit beamforming for MIMO radar systems using signal crosscorrelation. *IEEE Trans. Aerosp. Electron. Syst.* **2009**, *44*, 171–186. [CrossRef]
13. Song, X.; Zhou, S.; Willett, P. Reducing the Waveform Cross Correlation of MIMO Radar with Space–Time Coding. *IEEE Trans. Signal Process.* **2010**, *58*, 4213–4224. [CrossRef]
14. Jajamovich, G.H.; Lops, M.; Wang, X. Space-Time Coding for MIMO Radar Detection and Ranging. *IEEE Trans. Signal Process.* **2010**, *58*, 6195–6206. [CrossRef]
15. Cheng, Z.; Liao, B.; He, Z.; Li, Y.; Li, J. Spectrally Compatible Waveform Design for MIMO Radar in the Presence of Multiple Targets. *IEEE Trans. Signal Process.* **2018**, *66*, 3543–3555. [CrossRef]
16. Yang, Y.; Blum, R.S.; He, Z.S.; Fuhrmann, D.R. MIMO radar waveform design via alternating projection. *IEEE Trans. Signal Process.* **2010**, *58*, 1440–1445. [CrossRef]
17. Antonio, G.S.; Fuhrmann, D.R.; Robey, F.C. MIMO Radar Ambiguity Functions. *IEEE J. Sel. Top. Signal Process.* **2007**, *1*, 167–177. [CrossRef]
18. Chen, C.-Y.; Vaidyanathan, P. MIMO Radar Ambiguity Properties and Optimization Using Frequency-Hopping Waveforms. *IEEE Trans. Signal Process.* **2008**, *56*, 5926–5936. [CrossRef]
19. Gogineni, S.; Nehorai, A. Frequency-Hopping Code Design for MIMO Radar Estimation Using Sparse Modeling. *IEEE Trans. Signal Process.* **2012**, *60*, 3022–3035. [CrossRef]

20. Han, K.; Nehorai, A. Jointly optimal design for mimo radar frequency-hopping waveforms using game theory. *IEEE Trans. Aerosp. Electron. Syst.* **2016**, *52*, 809–820. [CrossRef]
21. Griffiths, H.; Cohen, L.; Watts, S.; Mokole, E.; Baker, C.; Wicks, M.; Blunt, S. Radar spectrum engineering and management: Technical and regulatory issues. *Proc. IEEE* **2015**, *103*, 85–102. [CrossRef]
22. Chiriyath, A.R.; Paul, B.; Jacyna, G.M.; Bliss, D.W. Inner Bounds on Performance of Radar and Communications Co-Existence. *IEEE Trans. Signal Process.* **2015**, *64*, 464–474. [CrossRef]
23. Ahmed, A.; Zhang, Y.D.; Gu, Y. Dual-function radar-communications using QAM-based sidelobe modulation. *Digit. Signal Process.* **2018**, *82*, 166–174. [CrossRef]
24. Bliss, D.W. Cooperative radar and communications signaling: The estimation and information theory odd couple. In *Proceedings of the 2014 IEEE Radar Conference, Cincinnati, OH, USA, 19–23 May 2014*; Institute of Electrical and Electronics Engineers (IEEE): New York, NY, USA, 2014; pp. 0050–0055.
25. Geng, Z.; Deng, H.; Himed, B. Adaptive radar beamforming for interference mitigation in radar-wireless spectrum sharing. *IEEE Signal Process. Lett.* **2015**, *22*, 484–488. [CrossRef]
26. Junhui, Z.; Tao, Y.; Yi, G.; Jiao, W.; Lei, F. Power control algorithm of cognitive radio based on non-cooperative game theory. *China Commun.* **2013**, *10*, 143–154. [CrossRef]
27. Khawar, A.; Abdelhadi, A.; Clancy, C.; Clancy, T. Target Detection Performance of Spectrum Sharing MIMO Radars. *IEEE Sens. J.* **2015**, *15*, 1. [CrossRef]
28. Ahmed, A.; Zhang, Y.D.; Himed, B. Distributed Dual-Function Radar-Communication MIMO System with Optimized Resource Allocation. In *Proceedings of the 2019 IEEE Radar Conference (RadarConf), Boston, MA, USA, 22–26 April 2019*; Institute of Electrical and Electronics Engineers (IEEE): New York, NY, USA, 2019; pp. 1–5.
29. Hassanien, A.; Amin, M.G.; Zhang, Y.D.; Ahmad, F. Dual-function radar-communications: Information embedding using sidelobe control and waveform diversity. *IEEE Trans. Signal Process.* **2016**, *64*, 2168–2181. [CrossRef]
30. Hassanien, A.; Amin, M.G.; Zhang, Y.D.; Ahmad, F. Signaling strategies for dual-function radar communications: An overview. *IEEE Aerosp. Electron. Syst. Mag.* **2016**, *31*, 36–45. [CrossRef]
31. Hassanien, A.; Amin, M.G.; Zhang, Y.D.; Ahmad, F. Phase-modulation based dual-function radar-communications. *IET Radar, Sonar Navig.* **2016**, *10*, 1411–1421. [CrossRef]
32. Hassanien, A.; Himed, B.; Rigling, B.D. A dual-function MIMO radar-communications system using frequency-hopping waveforms. In *Proceedings of the 2017 IEEE Radar Conference (RadarConf), Seattle, WA, USA, 8–12 May 2017*; Institute of Electrical and Electronics Engineers (IEEE): New York, NY, USA, 2017; pp. 1721–1725.
33. Ahmed, A.; Gu, Y.; Silage, D.; Zhang, Y.D. Power-Efficient Multi-User Dual-Function Radar-Communications. In *Proceedings of the 2018 IEEE 19th International Workshop on Signal Processing Advances in Wireless Communications (SPAWC), Kalamata, Greece, 25–28 June 2018*; Institute of Electrical and Electronics Engineers (IEEE): New York, NY, USA, 2018; pp. 1–5.
34. Yao, Y.; Zhao, J.; Wu, L. Cognitive Radar Waveform Optimization Based on Mutual Information and Kalman filtering. *Entropy* **2018**, *20*, 653. [CrossRef]
35. Zhao, J.; Guan, X.; Li, X.P. Power allocation based on genetic simulated annealing algorithm in cognitive radio networks. *Chin. J. Electron.* **2013**, *22*, 177–180.
36. Bell, M. Information theory and radar waveform design. *IEEE Trans. Inf. Theory* **1993**, *39*, 1578–1597. [CrossRef]
37. Romero, R.A.; Bae, J.; Goodman, N.A. Theory and Application of SNR and Mutual Information Matched Illumination Waveforms. *IEEE Trans. Aerosp. Electron. Syst.* **2011**, *47*, 912–927. [CrossRef]
38. Tang, B.; Tang, J.; Peng, Y. MIMO Radar Waveform Design in Colored Noise Based on Information Theory. *IEEE Trans. Signal Process.* **2010**, *58*, 4684–4697. [CrossRef]
39. Aubry, A.; De Maio, A.; Huang, Y.; Piezzo, M.; Farina, A. A new radar waveform design algorithm with improved feasibility for spectral coexistence. *IEEE Trans. Aerosp. Electron. Syst.* **2015**, *51*, 1029–1038. [CrossRef]
40. Chen, Y.; Nijsure, Y.; Yuen, C.; Chew, Y.H.; Ding, Z.; Boussakta, S. Adaptive Distributed MIMO Radar Waveform Optimization Based on Mutual Information. *IEEE Trans. Aerosp. Electron. Syst.* **2013**, *49*, 1374–1385. [CrossRef]

41. Naghibi, T.; Behnia, F. MIMO Radar Waveform Design in the Presence of Clutter. *IEEE Trans. Aerosp. Electron. Syst.* **2011**, *47*, 770–781. [CrossRef]
42. Yao, Y.; Zhao, J.; Wu, L. Waveform Optimization for Target Estimation by Cognitive Radar with Multiple Antennas. *Sensors* **2018**, *18*, 1743. [CrossRef]
43. Wang, X.; Hassanien, A.; Amin, M.G. Dual-Function MIMO Radar Communications System Design Via Sparse Array Optimization. *IEEE Trans. Aerosp. Electron. Syst.* **2019**, *55*, 1213–1226. [CrossRef]
44. Yao, Y.; Zhao, J.; Wu, L. Frequency-Hopping Code Design for Target Detection via Optimization Theory. *J. Optim. Theory Appl.* **2019**, *183*, 731–756. [CrossRef]
45. Nijsure, Y.; Chen, Y.; Boussakta, S.; Yuen, C.; Chew, Y.H.; Ding, Z. Novel System Architecture and Waveform Design for Cognitive Radar Radio Networks. *IEEE Trans. Veh. Technol.* **2012**, *61*, 3630–3642. [CrossRef]
46. Goodman, N.A.; Venkata, P.R.; Neifeld, M.A. Adaptive Waveform Design and Sequential Hypothesis Testing for Target Recognition with Active Sensors. *IEEE J. Sel. Top. Signal Process.* **2007**, *1*, 105–113. [CrossRef]

Article

Proposal of a Geometric Calibration Method Using Sparse Recovery to Remove Linear Array Push-Broom Sensor Bias

Jun Chen *, Zhichao Sha, Jungang Yang and Wei An

College of Electronic Science, National University of Defense Technology, Changsha 410073, China; shazhichao163@163.com (Z.S.); yangjungang@nudt.edu.cn (J.Y.); anwei@nudt.edu.cn (W.A.)

* Correspondence: chenjun11@nudt.edu.cn; Tel.: +86-731-8700-3326

Received: 14 August 2019; Accepted: 9 September 2019; Published: 16 September 2019

Abstract: The rational function model (RFM) is widely used in the most advanced Earth observation satellites, replacing the rigorous imaging model. The RFM method achieves the desired calibration performance when image distortion is caused by long-period errors. However, the calibration performance of the RFM method deteriorates when short-period errors—such as attitude jitter error—are present, and the insufficient and uneven ground control points (GCPs) can also lower the calibration precision of the RFM method. Hence, this paper proposes a geometric calibration method using sparse recovery to remove the linear array push-broom sensor bias. The most important issue regarding this method is that the errors related to the imaging process are approximated to the equivalent bias angles. By using the sparse recovery method, the number and distribution of GCPs needed are greatly reduced. Meanwhile, the proposed method effectively removes short-period errors by recognizing periodic wavy patterns in the first step of the process. The image data from Earth Observing 1 (EO-1) and the Advanced Land Observing Satellite (ALOS) are used as experimental data for the verification of the calibration performance of the proposed method. The experimental results indicate that the proposed method is effective for the sensor calibration of both satellites.

Keywords: geometric calibration; long- and short-period errors; equivalent bias angles; sparse recovery; linear array push-broom sensor

1. Introduction

With the accelerating development of technical aeronautics and space exploration, many countries have launched advanced Earth observation satellites in recent years. The linear array push-broom sensor is the payload which has been carried by the most satellites. This significant quantity of Earth observation has had a great impact on overall production and human life. However, the imaging process of sensors is easily influenced by a variety of errors, including thermal deformation error, optical distortion error, satellite position, attitude measurement errors, and so on [1,2]. Commonly, the raw images contain many image distortions and cannot be directly applied to the processing of subsequent applications. Typically, errors which result from image distortion can be divided into two sources: long-period errors, including assembling error, optical distortion error, and thermal distortion error, which are constant, non-varying, or slowly varying [3]; and short-period errors, including satellite position, attitude measurement errors, and attitude jitter error. If these errors are insufficiently characterized or uncorrected, significant distortion of the raw image can result. Therefore, it is necessary to complete geometric calibration of the sensor to eliminate image distortion before raw image application.

The imaging process involves projecting a point on the surface of the Earth toward the sensor's focal plane [4]. The imaging models are used to describe this imaging process consist of two types [5].

The first type is the rigorous imaging model, which involves a set of coordinate transformations. Many rigorous image models have been proposed by different scholars for a variety of sensors in previous research [1,2,6,7]. However, most imagery vendors often do not provide details of advanced Earth observation satellites, such as the precise parameters and work mode of sensor or the satellite orbit, to the user. Therefore, this type of model has many limitations in practical use. The other type is the generalized imaging model. There are a lot of generalized imaging models, with the main ones being the rational function model (RFM), the direct linear transformation model, and the polynomial model. The RFM is widely applied in sensor calibration due to its simple form and high precision compared to the other models [5]. However, the RFM has strict requirements regarding the distribution and number of ground control points (GCPs) in each frame, and it also needs to calculate the coefficients frame-by-frame. The RFM method effectively calibrates image distortion caused by the long-period errors—such as thermal distortion error, optical distortion error, and assembling error—but the calibration performance of the RFM method deteriorates in the presence of short-period errors, such as attitude jitter [8]. The majority of advanced Earth observation satellites undergo attitude jitter, which results from attitude control operations, the dynamic structure, and so on [9]. Referring to the previous studies [8,9], we found three classic ways to handle satellite attitude jitter. The first way is to apply an advanced hardware device, such as an angular displacement sensor, to achieve high-resolution measurement data. However, this is infeasible for many satellites due to restrictions regarding the economy and technology. The second way is to estimate attitude jitter by using multispectral parallax images. This method is unsuited for the estimation of attitude jitter of satellites which are not equipped with multispectral sensors. The third way is to use high precision GCPs in scenes; the effectiveness of this method is decided by the distribution and number of GCPs [8]. However, it is difficult to extract many GCPs with high precision in each scene, especially if water or cloud coverage exists in the scene.

In order to solve the above-mentioned problems, this paper puts forward a geometric calibration method by using sparse recovery to remove linear array push-broom sensor bias. The errors relating to the imaging process are approximated to the equivalent bias angles [10], and the equivalent bias angles of the image sequence are recovered by compressive sensing in this method. Hence, the traditional problem regarding error-solving in the sensor calibration process is transformed into a new problem regarding signal recovery in this paper.

2. Sensor Geometric Calibration Model

2.1. Sensor Geometric Calibration Modeling

Over the past decade, the linear array push-broom sensor has been used as a main payload assembled by the most advanced Earth observation satellites. The imaging process of the linear array push-broom sensor projects a point on the surface of Earth, such as a GCP, to the sensor's focal plane [4]. The basic flow diagram of this process is shown in Figure 1. The rigorous imaging model for the linear array push-broom sensor is presented as follows [6,7,11].

$$\begin{bmatrix} X-X_F \\ Y-Y_F \\ Z-Z_F \end{bmatrix} = m\times \mathbf{R}_{ECI}^{ECF}\times \mathbf{R}_{orb}^{ECI}(\theta_\Omega,\theta_i,\theta_\omega)\times\mathbf{R}_{orb}^{body}(\varphi,\varepsilon,\psi)\times\mathbf{R}_{body}^{sen}(\phi_X,\phi_Y,\phi_Z)\times\mathbf{M}_{mir}(\theta_0,\theta_c)\times\begin{bmatrix} 0 \\ y \\ -f \end{bmatrix} \quad (1)$$

where (X,Y,Z) represents the projective position in the Earth-centered fixed (ECF) coordinate system, y represents the image column position in the focal plane coordinate system, (X_F,Y_F,Z_F) represents the satellite position in the ECF, m represents a scale factor, f represents the focal length, $\mathbf{R}_{ECI}^{ECF}$ denotes the rotation from the Earth-centered inertial (ECI) to the ECF coordinate system, $\theta_\Omega,\theta_i,\theta_\omega$ represent the orbital elements of the satellite (right ascension of the ascending node, inclination, and true perigee angle), $\mathbf{R}_{orb}^{ECI}$ denotes the orbital elements that give the rotation from the satellite orbit to the ECI coordinate system, φ,ε,ψ represent the Euler angles of the satellite attitude, $\mathbf{R}_{orb}^{body}$ denotes the satellite attitude angles that give the rotation from the satellite body to the satellite orbit coordinate system,

ϕ_X, ϕ_Y, ϕ_Z represent the assembling angles of the sensor, $\mathbf{R}_{body}^{sen}$ denotes the assembling angles that give the rotation from the sensor to the satellite body coordinate system, θ_0, θ_c represent the sensor pointing angles (the mirror assembling angle and the mirror rotation angle), and $\mathbf{M}_{mir}$ denotes the sensor pointing angles that give the rotation from the pointing to the sensor coordinate systems.

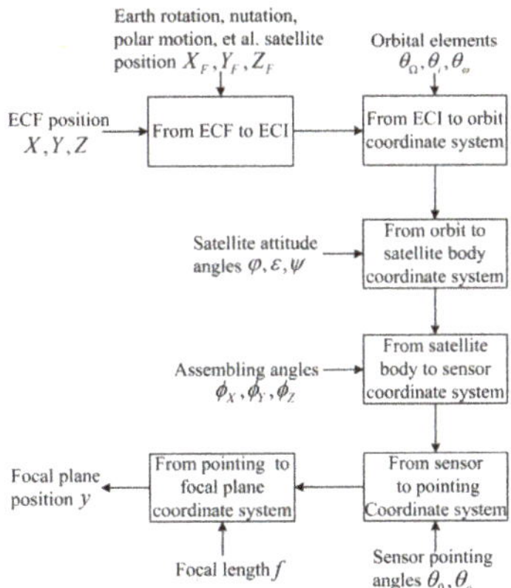

Figure 1. Basic flow diagram of the imaging process for the linear array push-broom sensor.

Compared with the true values, the imaging parameters usually solve some undistinguished errors in the imaging process. The long-period errors, such as thermal distortion error, optical distortion error, assembling error, and the short-period errors—such as satellite orbit error and attitude error—of the imaging process are approximated to the equivalent bias angles. The sensor geometric calibration model is presented as [1,2,10]

$$\begin{bmatrix} X - X_F \\ Y - Y_F \\ Z - Z_F \end{bmatrix} = m \times \mathbf{R}_{Eq}(\alpha, \beta, \theta) \times \mathbf{R}_{ECI}^{ECF} \times \mathbf{R}_{orb}^{ECI} \times \mathbf{R}_{orb}^{body} \times \mathbf{R}_{body}^{sen} \times \mathbf{M}_{mir} \times \begin{bmatrix} 0 \\ y \\ -f \end{bmatrix} \quad (2)$$

where α, β, θ represent the equivalent bias angles and $\mathbf{R}_{Eq}(\alpha, \beta, \theta)$ denotes the equivalent rotation matrix

$$\mathbf{R}_{Eq}(\alpha, \beta, \theta) = \begin{pmatrix} 1 & 0 & 0 \\ 0 & \cos\alpha & \sin\alpha \\ 0 & -\sin\alpha & \cos\alpha \end{pmatrix} \begin{pmatrix} \cos\beta & 0 & -\sin\beta \\ 0 & 1 & 0 \\ \sin\beta & 0 & \cos\beta \end{pmatrix} \begin{pmatrix} \cos\theta & -\sin\theta & 0 \\ \sin\theta & \cos\theta & 0 \\ 0 & 0 & 1 \end{pmatrix} \quad (3)$$

2.2. Error Modeling

All on-orbit satellites suffer from an environment that varies cyclically between cooling and heating due to the circular orbit around Earth. The thermal distortion error resulting from the satellite in the Earth's orbit causing a solar eclipse is the biggest error in the imaging process [12]. The period of this cycle is the same as the orbit period of the satellite. Thermal distortion errors present in three directions, with the error in the direction toward the sun being the largest. The optical distortion error and assembling error are constant bias errors or long-period variation errors [4]. Attitude jitter is also a primary source of decreasing sensor performance. The satellite position and attitude measurement errors, which are a type of Gaussian error, can be effectively removed by using compressive sensing, therefore, they are not included in the bias angle model. According to the characteristics of these errors, the equivalent bias angles can be described by the constant errors and a set of period components varying in sinusoidal waveform, demonstrated as [4,12]

$$\begin{cases} \alpha(t) = \varsigma_\alpha + \sum_i^n A_{\alpha i} \sin(2\pi f_{\alpha i} t + \zeta_{\alpha i}) \\ \beta(t) = \varsigma_\beta + \sum_i^n A_{\beta i} \sin(2\pi f_{\beta i} t + \zeta_{\beta i}) \\ \theta(t) = \varsigma_\theta + \sum_i^n A_{\theta i} \sin(2\pi f_{\theta i} t + \zeta_{\theta i}) \end{cases} \quad (4)$$

where $\varsigma_\alpha, \varsigma_\beta, \varsigma_\theta$ represent the constant errors, $A_{\alpha i}, A_{\beta i}, A_{\theta i}$ represent the amplitudes of the period components, $f_{\alpha i}, f_{\beta i}, f_{\theta i}$ represent the frequencies of the period components, and $\zeta_{\alpha i}, \zeta_{\beta i}, \zeta_{\theta i}$ represent the phases of the period components.

3. Geometric Calibration by Using Sparse Recovery

3.1. Compressive Sensing

As a new theory in signal processing, compressive sensing was proposed in 2004 and rapidly developed in the following years. The core idea of this theory mainly involves two aspects. The first is the sparsity of the signal, which means that the majority of the elements are have a value of zero or are otherwise very small [13]. The other is sampling irrelevance, in other words, the measurement matrix required to meet the restricted isometry property (RIP) [14]. The traditional Nyquist–Shannon sampling theorem indicated that the sampling frequency must exceed the Nyquist sampling frequency to restore the signal without distortion. However, the sparse signal is exactly recovered by compressive sensing from its incomplete measurements far below the Nyquist sampling frequency [13]. In practical problems, the majority of signals are not initially sparse. Fortunately, these signals can be converted into sparse signals using sparse transformation. Sparse transformation can be done by discrete Fourier transformation (DFT), discrete cosine transformation, and wavelet transformation. Hence, these signals are recovered by compressive sensing. Because of its incomparable advantage in handling large-scale compressible data, compressive sensing is widely used in the fields of radio communication, array signal processing, medical imaging, and so on.

3.2. Procedure of Proposed Method

The geometric calibration method consists of five primary steps: (a) periodic wavy pattern recognition; (b) sparse GCP image shift calculation; (c) dense GCP image shift calculation; (d) equivalent bias angles recovery; and (e) image calibration. The procedure of the proposed method is shown in Figure 2.

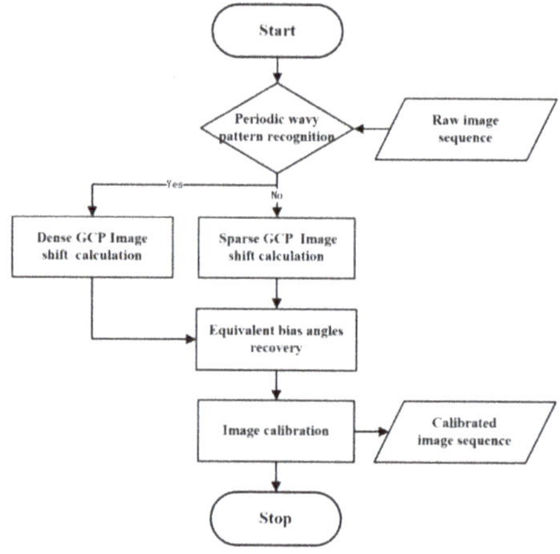

Figure 2. Flowchart of the proposed method.

3.2.1. Periodic Wavy Pattern Recognition

The first step is to recognize the periodic wavy pattern of the raw image sequence frame-by-frame. Once the raw image exists with periodic wavy patterns, which are caused by short-period error, such as attitude jitter error, the subsequent process of Step 3 is necessary for the frame. There are three processes that need to be completed for periodic wavy pattern recognition. First, a line or edge detection algorithm is used to extract the road, dike, and coastline from the raw images. Then, the smoothness of the extracted lines or edges is extracted. Finally, the existence of periodic wavy patterns in the images is determined. The details of this step can be seen in [15].

3.2.2. Sparse GCP Image Shift Calculation

There are three main processes in this step. First, a handful of images with good imaging conditions are randomly selected from the raw image sequence, excluding images with periodic wavy patterns, as the measurement scenes. Then, a small number of GCPs with high precision from each measurement scene are extracted and matched. The sampling interval of GCPs is set to larger, and these GCPs are defined as sparse GCPs in this paper. The latitude and longitude ranges of each measurement scene are calculated to determine whether there are GCPs in each measurement scene; the templates of GCPs used in this GCPs extraction process are obtained from the Shuttle Radar Topography Mission data. The details of this process are seen in [16]. Finally, we calculate the ideal position of the GCPs in the focal plane coordinates using Equation (2), and the equivalent bias angles are set to zero in this process. The real position of the GCPs are measured from the raw images and the image shift of sparse GCPs is calculated as

$$\begin{cases} \Delta x_i = x_i - x_{ci} \\ \Delta y_i = y_i - y_{ci} \end{cases} \quad (5)$$

where $(\Delta x_i, \Delta y_i)$ is the image shift between the real position (x_i, y_i) and the ideal position (x_{ci}, y_{ci}) of ith GCP.

3.2.3. Dense GCP Image Shift Calculation

The raw images with periodic wavy patterns are selected as the measurement scenes, and a number of GCPs with intensive distribution from each measurement scene were extracted and matched with periodic wavy patterns. The sampling interval of the GCPs is set to smaller; these GCPs are defined as dense GCPs. The exact number of GCPs is determined by the sparsity and length of the recovered signal, the recovery algorithm, and so on. Generally speaking, the length of the measurements is far below the length of the recovered signal [14]. The rest of this step follows the same process as Step 2.

3.2.4. Equivalent Bias Angle Recovery

The most important thing in this crucial step is the signal fusion. It is known that the compressive sensing method can only recover 1-D column vectors, therefore, 3-D equivalent bias angles signals must be merged into a 1-D time varying signal according to the time order. The problem regarding solving 3-D signal errors is transformed into the problem relating to 1-D signal recovery. After obtaining the image shift of GCPs, the remaining tasks of this step are to select a desired sparse basis and an appropriate recovery algorithm and to construct a suitable measurement matrix to exactly recover the 1-D time varying signal. The 3-D signals are separated from the 1-D recovered signal. The details of this important work are presented in the following sections.

3.2.5. Image Calibration

There are two processes in this step. One involves calculating the calibrated position, and the other is to do with resampling the raw image. Once the equivalent bias angles are recovered, the calibrated image positions are easily calculated using Equation (2) with the imaging parameters. The image resampling can be completed using the bilinear interpolation method. The gray values of the calibrated image are obtained according to the calibrated image positions and the raw image.

3.3. Criticism of the Proposed Method

If the equivalent bias angles are considered to be errors in an ideal situation, the estimation of equivalent bias angles is a typical adjustment of indirect observations. The function model of the adjustment of these indirect observations is expressed as

$$\mathbf{L} = \mathbf{B}\mathbf{X} + \mathbf{d} \tag{6}$$

where **L** is the observation, **B** is the measurement matrix, **X** is the estimated parameter, and **d** is an offset constant.

Generally, the estimated parameter is expressed as

$$\mathbf{X} = \mathbf{X}_0 + \hat{\mathbf{x}} \tag{7}$$

where $\mathbf{X}_0$ is the ideal value of the estimated parameter and $\hat{\mathbf{x}}$ is the correction of the estimated parameter.

Let

$$\mathbf{L}_1 = \mathbf{L} - \mathbf{L}_0 = \mathbf{L} - (\mathbf{B}\mathbf{X}_0 + \mathbf{d}) \tag{8}$$

where $\mathbf{L}_0$ is the ideal observation value.

According to Equations (6) and (7), Equation (8) can be simplified as

$$\mathbf{L_1} = \mathbf{B}\hat{\mathbf{x}} \tag{9}$$

where $\hat{\mathbf{x}}$ is the equivalent bias angle, $\mathbf{L}_1$ is the image shift of the GCP, and **B** is the measurement matrix. When there is nonlinear relation between **L** and $\mathbf{X}(\mathbf{L} = F(\mathbf{X}))$, partial derivative of function $F(\mathbf{X})$ must be sought to obtain the measurement matrix **B**.

3.3.1. Measurement Matrix and Measurement Equation

(a) Measurement variables

According to Equation (8), $\mathbf{L}_1$ is determined by **L** and $\mathbf{L}_0$ is the real position of the GCP and $\mathbf{L}_0$ is the ideal position of the GCP when M GCPs are extracted and matched from the raw image sequence. The **L** and $\mathbf{L}_0$ are expressed as

$$\begin{aligned} \mathbf{L} &= \begin{bmatrix} x_1 & y_1 & x_2 & y_2 & \cdots & x_M & y_M \end{bmatrix}^{\mathrm{T}} \\ \mathbf{d} &= \begin{bmatrix} x_{c1} & y_{c1} & x_{c2} & y_{c2} & \cdots & x_{cM} & y_{cM} \end{bmatrix}^{\mathrm{T}} \end{aligned} \tag{10}$$

According to Equation (5), L_1 is expressed as

$$\begin{aligned} \mathbf{L}_1 &= \begin{bmatrix} \Delta x_1 & \Delta y_1 & \Delta x_2 & \Delta y_2 & \cdots & \Delta x_M & \Delta y_M \end{bmatrix}^{\mathrm{T}} \\ &= \begin{bmatrix} x_1 - x_{c1} & y_1 - y_{c1} & x_2 - x_{c2} & y_2 - y_{c2} & \cdots & x_M - x_{cM} & y_M - y_{cM} \end{bmatrix}^{\mathrm{T}} \end{aligned} \tag{11}$$

where L_1 is a $2M \times 1$ vector.

(b) Measurement equation

According to Equation (9), the measurement equation of one GCP is expressed as

$$\begin{bmatrix} \Delta x_m & \Delta y_m \end{bmatrix}^{\mathrm{T}} = \mathbf{B}_{mn} \begin{bmatrix} \alpha_n & \beta_n & \theta_n \end{bmatrix}^{\mathrm{T}} \tag{12}$$

where $\begin{bmatrix} \alpha_n & \beta_n & \theta_n \end{bmatrix}^T$ are the equivalent bias angles of the mth GCP and $\mathbf{B}_{mn}$ is the 2×3 measurement matrix.

The measurement equation of M GCP is expressed as

$$\mathbf{L}_1 = \mathbf{B}\hat{\mathbf{x}} \Rightarrow \begin{bmatrix} \Delta x_1 \\ \Delta y_1 \\ \Delta x_2 \\ \Delta y_2 \\ \vdots \\ \Delta x_M \\ \Delta y_M \end{bmatrix} = \begin{bmatrix} \mathbf{B}_{11} & 0 & \cdots & 0 \\ 0 & \mathbf{B}_{22} & \cdots & 0 \\ \vdots & \vdots & \ddots & \vdots \\ 0 & 0 & \cdots & \mathbf{B}_{MN} \end{bmatrix} \begin{bmatrix} \alpha_1 \\ \beta_1 \\ \theta_1 \\ \alpha_2 \\ \beta_2 \\ \theta_2 \\ \vdots \\ \alpha_N \\ \beta_N \\ \theta_N \end{bmatrix} = \begin{bmatrix} b_{11}^1 & b_{12}^1 & b_{13}^1 & 0 & 0 & 0 & \cdots & 0 & 0 & 0 \\ b_{21}^1 & b_{22}^1 & b_{23}^1 & 0 & 0 & 0 & \cdots & 0 & 0 & 0 \\ 0 & 0 & 0 & b_{11}^2 & b_{12}^2 & b_{13}^2 & \cdots & 0 & 0 & 0 \\ 0 & 0 & 0 & b_{21}^2 & b_{21}^2 & b_{21}^2 & \cdots & 0 & 0 & 0 \\ \vdots & \vdots & \vdots & \vdots & \vdots & \vdots & \ddots & \vdots & \vdots & \vdots \\ 0 & 0 & 0 & 0 & 0 & 0 & \cdots & b_{11}^M & b_{12}^M & b_{13}^M \\ 0 & 0 & 0 & 0 & 0 & 0 & \cdots & b_{21}^M & b_{22}^M & b_{23}^M \end{bmatrix} \begin{bmatrix} \alpha_1 \\ \beta_1 \\ \theta_1 \\ \alpha_2 \\ \beta_2 \\ \theta_2 \\ \vdots \\ \alpha_N \\ \beta_N \\ \theta_N \end{bmatrix} \quad (13)$$

where $\mathbf{L}_1$ is a $2M \times 1$ measurement vector, $\mathbf{B}$ is a $2M \times 3N$ measurement matrix, and $\hat{\mathbf{x}}$ is a $3N \times 1$ merged equivalent bias angle vector.

(c) Linearization of measurement equation

The other important issue regards obtaining the element of the measurement matrix $\mathbf{B}$. To reduce the complexity of the algorithm, the linearization of measurement equations for each GCP must be completed before recovering the signal. According to Equation (2), the relationship between the estimated parameters and the measurements is nonlinear, therefore a partial derivative of Equation (2) should be sought to obtain the measurement matrix $\mathbf{B}$. However, seeking a partial derivative of Equation (2) is difficult, and the numerical analysis method is used to estimate the elements of the measurement matrix $\mathbf{B}$. Next, the elements b_{11}^m and b_{21}^m of $\mathbf{B}_{mn}$ are taken as examples to introduce the details of the solution.

1. The upper limit of the equivalent bias angles A, the increment $\Delta\gamma$ in each step, and the cycle index $I = \frac{A}{\Delta\gamma}$ are determined;
2. The first parameter α_n of nth equivalent bias angles is set to $i \times \Delta\gamma i \in [0, I]$ and the other parameters are set to zero;
3. The image positions of the mth GCP by using the imaging parameters and the equivalent bias angles $\begin{pmatrix} i \times \Delta\gamma & 0 & 0 \end{pmatrix} i \in [0, I]$ are calculated'
4. After obtaining the $I + 1$ image positions $\begin{pmatrix} x_0 & y_0 \end{pmatrix}$,$\begin{pmatrix} x_1 & y_1 \end{pmatrix}$, and $\begin{pmatrix} x_I & y_I \end{pmatrix}$, the variations of image positions Δx_i and Δy_i are calculated using

$$\begin{cases} \Delta x_i = x_i - x_0 \\ \Delta y_i = y_i - y_0 \end{cases} i \in [1, I] \quad (14)$$

5. b_{11}^m and b_{21}^m are calculated as follows. The other elements of $\mathbf{B}_{mn}$ can be calculated in the same way

$$\begin{cases} b_{11}^m = \frac{1}{I} \sum\limits_{i=1}^{I} \frac{\Delta x_i}{i \times \Delta\gamma} \\ b_{21}^m = \frac{1}{I} \sum\limits_{i=1}^{I} \frac{\Delta y_i}{i \times \Delta\gamma} \end{cases}. \quad (15)$$

3.3.2. Sparse Basis and Sparse Representation

The equivalent bias angles signals based on Equation (4) are not sparse in the time domain. According to previous studies [8,17], the DFT converts equivalent bias angles signals into the sparse representation. The simulated signals based on Equation (4) are displayed in Figure 3a. The three curves in Figure 3a represent the simulated three-dimensional equivalent bias angles signals with different constant errors and period components. The red curve indicates the X direction, the blue curve indicates the Y direction, and the black curve indicates the Z direction. The results of the 1-D DFT are displayed in Figure 3b.

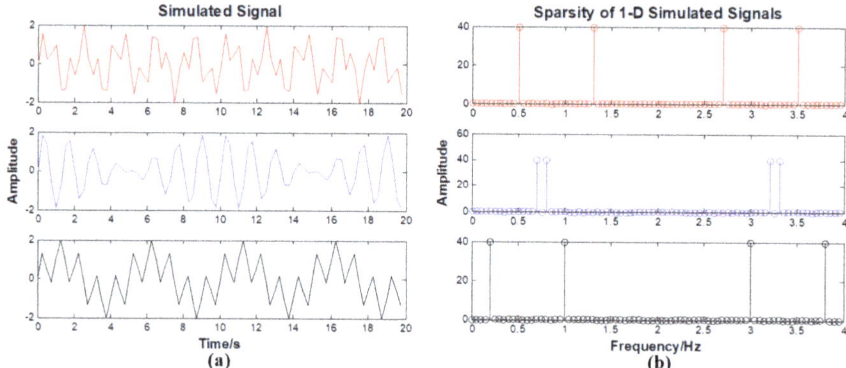

Figure 3. (**a**) Simulated 3-D equivalent bias angles signals. (**b**) Results of 1-D DFT for the 1-D equivalent bias angle signal.

To easily recover the equivalent bias angles signals, the 3-D equivalent bias angle signals need to be merged into a 1-D signal according to time order. The 1-D merged signal and the results of the 1-D DFT are displayed in Figure 4.

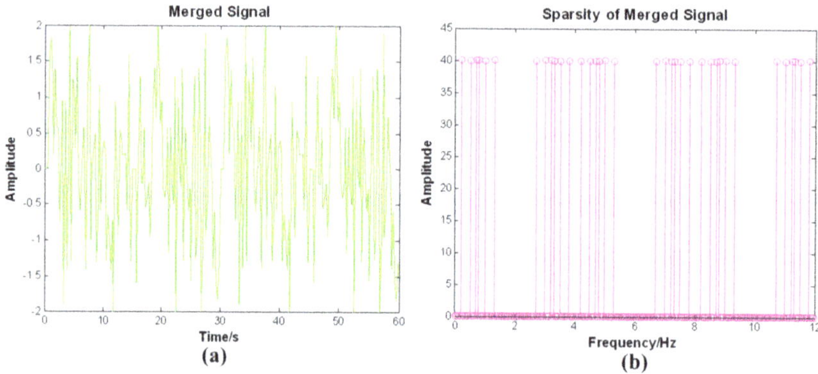

Figure 4. (**a**) 1-D merged signal. (**b**) Results of the 1-D DFT.

From Figure 3b, it is seen that the 3-D equivalent bias angle signals are sparse in the frequency domain and the 1-D merged signal is sparse in the frequency domain, as seen in Figure 4b. Therefore, we select the 1-D DFT basis as the sparse basis $\mathbf{\Psi}$ in the proposed method. The performances of the other sparse bases are all worse than the 1-D DFT basis. The definition of sparse basis $\mathbf{\Psi}$ is

$$\mathbf{\Psi} = \frac{1}{N}\begin{bmatrix} e^{j2\pi(0\times0)/N} & e^{j2\pi(1\times0)/N} & \cdots & e^{j2\pi((N-1)\times0)/N} \\ e^{j2\pi(0\times1)/N} & e^{j2\pi(1\times1)/N} & \cdots & e^{j2\pi((N-1)\times1)/N} \\ \vdots & \vdots & \ddots & \vdots \\ e^{j2\pi(0\times(N-1))/N} & e^{j2\pi(1\times(N-1))/N} & \cdots & e^{j2\pi((N-1)\times(N-1))/N} \end{bmatrix} \tag{16}$$

3.3.3. Signal Recovery

It is obvious from Equation (12) that the 2-D measurement equation used to solve the 3-D equivalent bias angles of one GCP is a pathological problem. The traditional methods assume that the equivalent bias angles are unchanged in the frame period and use classical optimal estimation methods—such as least squares estimation, Bayes estimation, or maximum likelihood estimation—to combine a number of GCPs to estimate the equivalent bias angles of each frame. However, this assumption is unreasonable when equivalent bias angles contain short-period errors. Due to $M << N$, the solution of Equation (13) is highly undetermined; therefore, the traditional methods do not work in this situation. Fortunately, it is possible to exactly recover the values of $\hat{\mathbf{x}}$ using compressive sensing if $\hat{\mathbf{x}}$ is represented by the sparse basis.

The measurement equation of M GCP in Equation (13) is rewritten as

$$\mathbf{L}_1 = \mathbf{B}\hat{\mathbf{x}} = \mathbf{B}\mathbf{\Psi}\mathbf{f} = \widetilde{\mathbf{B}}\mathbf{f} \tag{17}$$

where $\mathbf{\Psi}$ is a $3N \times 3N$ sparse basis, $\mathbf{f}$ is a $3N \times 1$ sparse representation, $\hat{\mathbf{x}} = \mathbf{\Psi}\mathbf{f}$, and $\widetilde{\mathbf{B}} = \mathbf{B}\mathbf{\Psi}$ is a $2M \times 3N$ sensing matrix. In Equation (17), $\mathbf{L_1}$ is considered to be the product of $\mathbf{f}$ times $\widetilde{\mathbf{B}}$ [13]. The 1-D DFT method achieves RIP [17]. The GCP is selected randomly when the measurement matrix $\mathbf{B}$ achieves RIP, therefore, the sensing matrix $\widetilde{\mathbf{B}}$ also achieves RIP. We use the matching pursuit algorithm to recover the exact values of $\mathbf{f}$ due to its lower computational complexity [18,19]. The merged equivalent bias angle signal $\hat{\mathbf{x}}$ is obtained as

$$\hat{\mathbf{x}} = \mathbf{\Psi}\mathbf{f} \tag{18}$$

4. Experiments and Analysis

Two experiments were used to compare the proposed method with the RFM method in this section. The first method employs the image data from the Hyperion of Earth Observing 1 (EO-1) and shows the performances of two methods designed to estimate the long-period components of equivalent bias angles. The second method, which uses the image data from the Panchromatic Remote-sensing Instrument for Stereo Mapping (PRISM) of the Advanced Land Observing Satellite (ALOS), shows that the proposed method has a superior performance regarding the estimation of the short-period components. The details of the experiments are described as follows.

4.1. *Hyperion Data Experiment*

4.1.1. Data Description

Hyperion, as the main payload of EO-1, is a typical linear array push-broom sensor. A total of 100 Hyperion frame images from 7 June 2002 were selected as the experimental data. In the proposed method, we selected 40 frames with some evenly distributed GCPs from the experimental data as the measurement scenes. The measurement scene used in the proposed method is represented in Figure 5a, with the red crosses denoting the GCPs. The rest of the frames were selected as verification scenes, which evenly distribute 50 random check points (RCPs) in each scene. One verification scene used in the proposed method is represented in Figure 5b, with the blue triangles denoting the RCPs. For the RFM method, we extracted some GCPs in each frame; the verification scenes were the same as in the proposed method. The measurement and verification scene of the RFM method are represented in Figure 5c. The GCPs of the measurement scenes were used as the inputs for both methods. The RCPs in the verification scenes were selected as the valuators to exactly evaluate the calibration performances of both methods. To adequately compare the methods, we designed 10 testing cases with different numbers of GCPs (5–50) and counted the calibration results. GCPs appeared to distribute evenly in this experiment. The original size (3129 × 256) of the Hyperion images was changed in order to be presented in Figure 5.

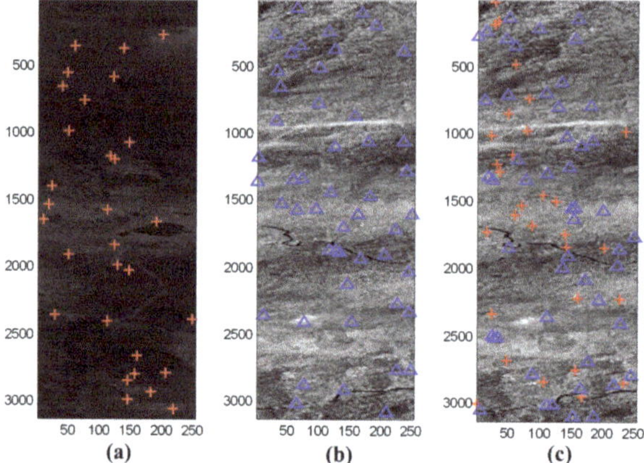

Figure 5. (**a**) Measurement scene and (**b**) verification scene of the proposed method. (**c**) Measurement and verification scene of the rational functional model (RFM) method (30 GCPs and 50 random check points (RCPs)).

4.1.2. Experimental Results

The calibration results of both methods for one frame showing 30 GCPs are displayed in Figure 6. The calibration results of the two methods in this case could be distinguished by the naked eye, even when details were closely examined (Figure 6).

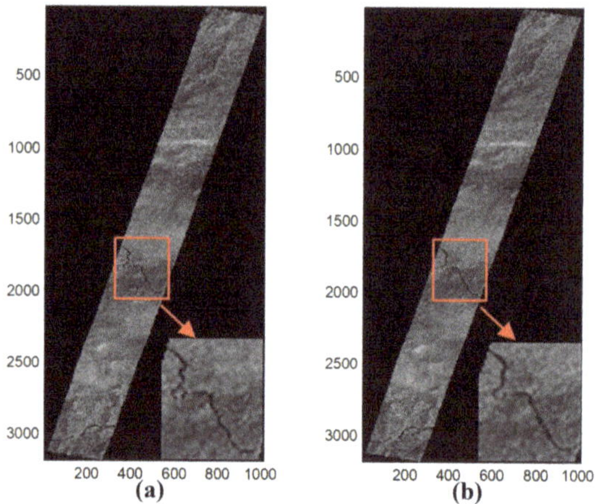

Figure 6. The calibration results of (**a**) the proposed method and (**b**) the RFM method.

To make accurate comparisons between the two methods, the calibrated residuals of the RCPs were employed in the statistical analysis. The mean value and root mean square error (RMSE) of the calibrated residuals were used as the measurable indicators. The variation trends of the mean value and the RMSE for the total GCPs across the two methods are shown in Figure 7. At least 19 GCPs were required to solve the coefficients of RFM; the RFM method did not work with 5, 10, or 15 GCPs. As the number of GCPs increased, the mean value of the two methods gradually decreased, as shown in Figure 7a. Similarly, the RMSE of the two methods gradually decreased in Figure 7b as the number of GCPs increased, i.e., the calibration performance of the two methods improved when the number of GCPs increased. The RFM method was superior in calibration performance when presented with enough GCPs. However, the proposed method did well when insufficient GCPs were available. Therefore, the proposed method achieved a better calibration performance in situations with insufficient GCPs compared with the RFM method.

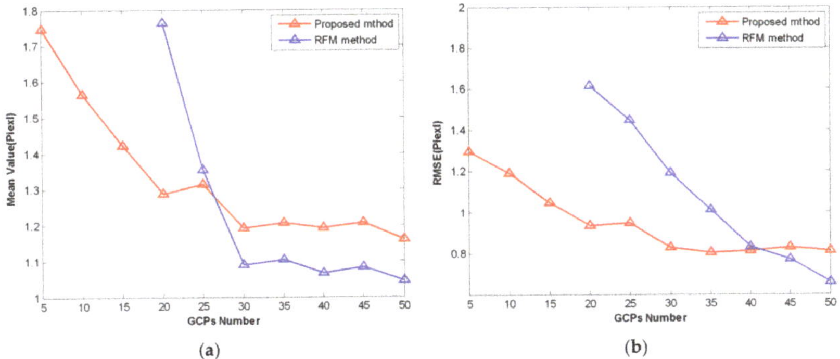

Figure 7. The variation trends of (**a**) mean value and (**b**) root mean square error (RMSE) of the number of GCPs across the two methods.

4.2. Experiment on ALOS Data

4.2.1. Data Description

PRISM, the primary sensor carried by ALOS, was designed to obtain a high-resolution digital surface model to derive orthoimages on a global scale. PRISM has three panchromatic linear array push-broom sensors (forward, nadir, and backward) with a 2.5 m spatial resolution and a width of 35 km swaths. The raw image data of PRISM was observed on 20 August 2006 and covered 50 scenes; these were used as the experimental data. To verify the performances of the two methods regarding the estimation of short-period components, one scene from the experimental data was used as the typical scene, as shown in Figure 8. We extracted 100 evenly distributed GCPs as inputs for the two methods and 50 evenly distributed RCPs to evaluate the calibration performances of the two methods in this typical scene. Also, we randomly chose 20 scenes with 30 evenly distributed GCPs as the measurement data from the rest of the experimental data.

Figure 8. Image data of a typical scene.

4.2.2. Experimental Results

We used the line or edge detection algorithm to extract the road in a typical scene; the detection results are shown in Figure 9.

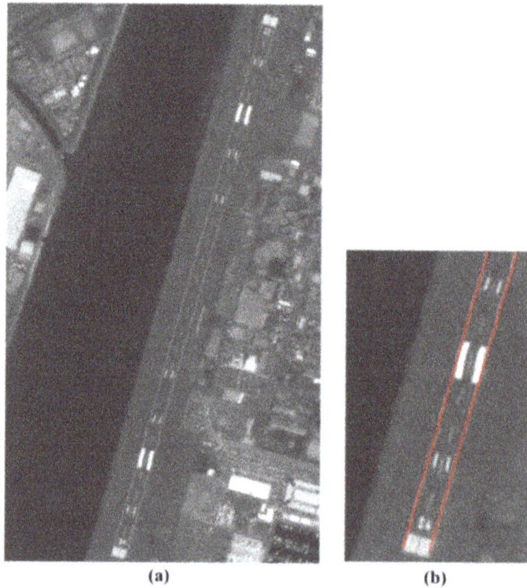

Figure 9. The line detection results of a typical scene. (**a**) Hiroshima West Airfield and (**b**) enlarged details.

Figure 9a shows the Hiroshima West Airfield. As seen in Figure 9b, the airfield runway had obvious periodic wavy patterns which were caused by short-period errors. The optical system of PRISM was affected by the motion of another sensor's mirror when this image was in the process of being captured, therefore, Step 3 of the proposed method was performed to deal with this image data. The partial details of the calibration results of a typical scene using the different methods are shown in Figure 10. The periodic wavy patterns in the calibration results of the proposed method are hardly recognizable in Figure 10a, but the periodic wavy patterns are easily recognized in the calibration results of the RFM method in Figure 10b.

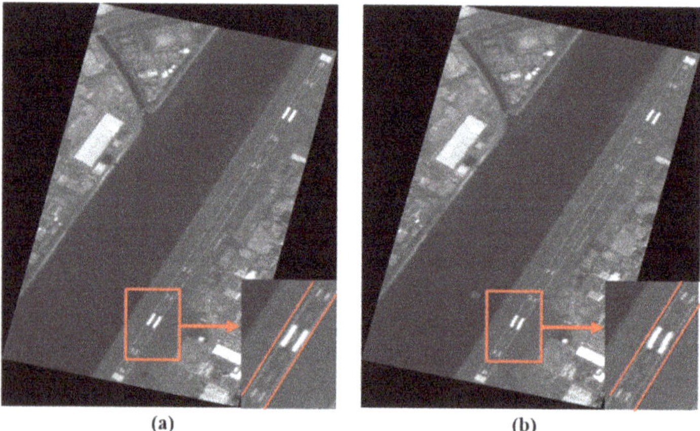

Figure 10. Partial details of the calibration results for a typical scene. (**a**) Proposed method and (**b**) RFM method.

The calibration results of the two methods for 50 RCPs in typical scene are shown in Table 1, which shows that the mean value and RMSE of the calibrated results for the proposed method were less than those of the RFM method. The results of this experiment indicated that the proposed method did well in estimating the short-period components of equivalent bias angles when compared with the RFM method.

Table 1. Calibration results of the two methods using 50 RCPs in a typical scene.

	Mean Value (urad)	RMSE (Pixels)
Proposed Method	0.43	0.68
RFM Method	1.67	2.03

5. Conclusions

Geometric calibration must be carried out before the application of raw images. This paper proposed a geometric calibration method using sparse recovery to remove linear array push-broom sensor bias. The errors in the imaging process were approximated to the equivalent bias angles in this method. By using the sparse recovery method, the proposed method exactly estimated long-period errors with a small number of GCPs available. Also, the proposed method effectively removed short-period errors by recognizing periodic wavy patterns in advance. The preliminary experimental results indicated the practicality and superior calibration performance of the proposed method when used for image data captured by the EO-1 and ALOS satellites. Compared with the traditional methods, the proposed method did well in situations with insufficient GCPs and short-period error calibration.

Future research will focus on the effects of GCP distribution on the proposed method. It is also important to apply the proposed method to other types of sensors.

Author Contributions: Z.S., J.Y., and W.A. conceived and designed the experiments; J.C. performed the experiments; J.C. and Z.S. analyzed the data; J.C. wrote the paper.

Acknowledgments: This work was supported in part by the National Natural Science Foundation of China under grant no. 61605242 and the Hunan Provincial Natural Science Foundation under grant no. 2016JJ3025. These foundations offered the cost of publishing this paper in an open access journal.

Conflicts of Interest: The authors declare no conflict of interest.

References

1. Iwasaki, A.; Fujisada, H. ASTER geometric performance. *IEEE Trans. Geosci. Remote. Sens.* **2005**, *43*, 2700–2706. [CrossRef]
2. Takaku, J.; Tadono, T. PRISM on-orbit geometric calibration and DSM performance. *IEEE Trans. Geosci. Remote. Sens.* **2009**, *47*, 4060–4073. [CrossRef]
3. Clemons, T.M.; Chang, K.C. Sensor calibration using in-situ celestial observations to estimate bias in space-based missile tracking. *IEEE Trans. AES* **2012**, *48*, 1403–1427. [CrossRef]
4. Xue, Y.H.; An, W. A novel target LOS calibration method for IR scanning sensor based on control points. In Proceedings of the SPIE, Beijing, China, 5 December 2012; pp. 314–318.
5. Sun, J.B.; Ni, L.; Zhou, J.Q. *Principle and Applications of Remote Sensing*; Wuhan University Press: Wuhan, China, 2009; pp. 123–137.
6. Poli, D.; Toutin, T. Review of developments in geometric modeling for high resolution satellite pushbroom sensors. *Photogramm. Rec.* **2012**, *27*, 58–73. [CrossRef]
7. Chen, Y.F.; Xie, Z.; Qiu, Z.; Zhang, Q.; Hu, Z. Calibration and validation of ZY-3 optical sensors. *IEEE Trans. Geosci. Remote. Sens.* **2015**, *53*, 4616–4626. [CrossRef]
8. Wang, P.; An, W.; Deng, X.P.; Ma, C. Geometric correction method to correct influence of attitude jitter on remote sensing imagery using compressive sampling. *J. Appl. Remote Sens.* **2015**, *9*, 095077. [CrossRef]

9. Tong, X.H.; Xu, Y.S.; Ye, Z.; Liu, S.J.; Tang, X.M.; Li, L.Y.; Xie, H.; Xie, J.F. Attitude oscillation detection of the ZY-3 satellite by using multispectral parallax images. *IEEE Trans. Geosci. Remote. Sens.* **2015**, *53*, 3522–3534. [CrossRef]
10. Chen, J.; An, W.; Yang, J.G.; Wang, P. Geometric correction method for linear array pushroom infrared imagery using compressive sampling. *J. Appl. Remote Sens.* **2016**, *10*, 042010. [CrossRef]
11. Jiang, Y.H.; Zhang, G.; Tang, X.M.; Li, D.R.; Huang, W.C.; Pan, H.B. Geometric calibration and accuracy assessment of ZiYuan-3 multispectral images. *IEEE Trans. Geosci. Remote. Sens.* **2014**, *52*, 4161–4172. [CrossRef]
12. Kistosturian, H.G. On-orbit calibration of satellite antenna-pointing errors. *IEEE Trans. AES.* **1990**, *26*, 88–1121. [CrossRef]
13. Candès, E.; Wakin, M.B. An introduction to compressive sampling. *IEEE Signal. Process. Mag.* **2008**, *25*, 21–30. [CrossRef]
14. Donoho, D.; Huo, X. Uncertainty principle and ideal atomic decomposition. *IEEE Trans. Inf. Theory.* **2001**, *47*, 2845–2862. [CrossRef]
15. Jensen, R.R.; Jackson, M.W.; Lulla, V. Single line correction method to remove aircraft roll errors in hyperspectral imagery. *J. Appl. Remote Sens.* **2008**, *2*, 023529. [CrossRef]
16. Eppler, W.G.; Paglieroni, D.W. GOES landmark positioning system. In Proceedings of the SPIE, Denver, CO, USA, 18 October 1996; pp. 789–804.
17. Rudelson, M.; Vershynin, R. Sparse reconstruction by convex relaxation: Fourier and Gaussian measurements. In Proceedings of the Conference on Information Sciences & Systems, Princeton, NJ, USA, 22–24 March 2006; pp. 207–212.
18. Tropp, J.A.; Gilbert, A.C. Signal recovery from random measurements via orthogonal matching pursuit. *IEEE Trans. Inf. Theory* **2007**, *53*, 4655–4666. [CrossRef]
19. Ling, Q.; Guo, Y.L.; Ling, Z.P.; An, W. A constrained sparse representation model for hyperspectral anomaly detection. *IEEE Trans. Geosci. Remote. Sens.* **2019**, *57*, 2358–2371. [CrossRef]

Article

Modulation Classification Using Compressed Sensing and Decision Tree–Support Vector Machine in Cognitive Radio System

Xiaoyong Sun, Shaojing Su, Zhen Zuo *, Xiaojun Guo and Xiaopeng Tan

College of Intelligent Science and Technology, National University of Defense Technology, Changsha 410073, China; sunxiaoyong14@nudt.edu.cn (X.S.); susj-5@163.com (S.S.); jeanakin@nudt.edu.cn (X.G.); tanxiaopeng14@nudt.edu.cn (X.T.)

* Correspondence: z.zuo@nudt.edu.cn; Tel.: +86-731-84573353

Received: 15 January 2020; Accepted: 3 March 2020; Published: 6 March 2020

Abstract: In this paper, a blind modulation classification method based on compressed sensing using a high-order cumulant and cyclic spectrum combined with the decision tree–support vector machine classifier is proposed to solve the problem of low identification accuracy under single-feature parameters and reduce the performance requirements of the sampling system. Through calculating the fourth-order, eighth-order cumulant and cyclic spectrum feature parameters by breaking through the traditional Nyquist sampling law in the compressed sensing framework, six different cognitive radio signals are effectively classified. Moreover, the influences of symbol length and compression ratio on the classification accuracy are simulated and the classification performance is improved, which achieves the purpose of identifying more signals when fewer feature parameters are used. The results indicate that accurate and effective modulation classification can be achieved, which provides the theoretical basis and technical accumulation for the field of optical-fiber signal detection.

Keywords: modulation classification; high-order cumulant; cyclic spectrum; compressed sensing; decision tree–support vector machine

---

1. Introduction

As one of the booming communication technologies in the information era, modulation classification (MC) technology [1] has a very important application value in the field of wireless communication. For example, it can play an important role in communication investigation, electronic countermeasures, signal authentication, interference identification, spectrum management, etc. At present, the wireless communication network has maintained a steady and rapid development trend, the network construction is increasingly integrated, and network applications are everywhere. At the same time, the inherent contradiction between the centralized static network and the dynamic change of the environment also causes serious problems like the low utilization of spectrum resources in the wireless communication network. Therefore, cognitive radio (CR) technology [2–5] is proposed and considered as a promising technology to solve these problems. MC plays an important role in CR based on spectrum sensing and feature analysis.

Cognitive radio has been widely accepted as a new technology in the field of wireless communication in the new era. In the cognitive radio network (CRN), in order to avoid interference with the transmission of the primary users, it is essential to accurately sense the presence for any contemporaneous transmission of the primary users in the observed spectrum [6]. The primary user signal error detection will cause the secondary user to waste the spectrum opportunity. Noise, shadow and multipath fading lead to a serious degradation of signal characteristics in conventional wireless communication scenarios. This makes signal detection very difficult in a low signal-to-noise ratio

(SNR) environment [7,8]. In addition, because the primary user (authorized user) and the cognitive user (unauthorized user) cannot communicate with each other, accurate MC can not only avoid mutual interference between them, but also provide the multi-dimensional spectrum information of the surrounding wireless environment, which helps to improve the inefficient use of spectrum resources in the CRN. With the different modulation parameters and methods used in the wide-band communication signal, MC has gradually been studied in depth and has become one of the main methods of signal recognition and classification.

MC has been playing an important role in the field of wireless communication for a long time, especially in dynamic spectrum management and interference recognition. A variety of methods and classifiers have been proposed in the literature, but most of them only identify a few modulation formats, such as low-order modulation format, or require some knowledge of parameters of the signal. MCs of a CR system are roughly divided into 4 categories: (a) Multiple quadrature amplitude modulation (MQAM) and multiple phase shift keying (MPSK) signals are classified based on signal envelope variance and wavelet transform, but the recognition rate is low at low SNR [9,10]; (b) artificial neural networks (ANN) based on machine-learning algorithms for automatic signal type recognition, which requires the most appropriate ANN and will lead to an increase in calculation time and risk of over-fitting [11,12]; (c) identification from higher-order cumulant (HOC) using fourth-order cumulant, which cannot identify some signals with the same fourth-order cumulant [13,14]; (d) feature parameters are extracted from the time domain, frequency domain and power spectrum of signals to classify and identify a modulation signal, but some feature parameter extraction processes are complex and easily interfered with by noise [15–17]. The proposed method mainly focuses on the recognition of single-feature parameters, and most classifiers adopted an increase the complexity of the system.

In this paper, we propose a new modulation classification method that combines high-order cumulants and cyclic spectrum feature extraction methods with a decision tree–support vector machine (DT–SVM) classifier. In the feature extraction phase, the compressed sensing (CS) method is used to obtain the compressed sample size of the feature parameters, and the influence of key factors on the classification accuracy in the modulation classification process is analyzed. CS is a signal processing technique called "sampling compression combo". The CS method can map signals from high-dimensional space to low-dimensional space through a small number of observations (non-adaptive linear projection) of sparse signals, and maintain the original structure of the signal [18]. The sparse signal reconstruction is actually reconstructing the original signal from the signal observations with high probability by solving the non-linear optimization problem, which breaks through the limitations of the traditional Shannon–Nyquist sampling theorem and solves the performance requirements of a sampling system when processing cognitive radio signal. It also relieves the pressure of storage, transmission and processing for large amounts of the traditional sampled data. The combination of HOC and the cyclic spectrum can distinguish the same cumulant of different signals, and achieve the MC through the DT–SVM classifier. Combining the advantages of HOC and cyclic spectrum features, the algorithm directly obtains the compressed values of feature parameters through the CS theory, and analyzes the influence of symbol length and compression ratio on the recognition accuracy. The simulation results show that the algorithm has a better classification performance in low SNR and the validity of the method is verified.

The rest of this paper is structured as follows. Section 2 introduces the feature extraction method and its characteristics in detail. Section 3 introduces the compression sampling values of feature parameters obtained by combining the compression sensing theory. Section 4 describes the structure of the decision tree–support vector machine classifier. In Section 5, some simulation results are presented. Finally, Section 6 sums up the conclusions.

2. Feature Extraction

2.1. Feature Extraction Based on Higher-Order Cumulant (HOC)

For wireless channel model, we studied the property of HOC and the insensitivity of its second-order terms to Gaussian noise, the kth-order cumulant $\mathbf{C}_{k,n}(m_1, m_2, \cdots, m_{k-1})$ of a complex-valued stationary random process $x(t)$, can be defined as:

$$\mathbf{C}_{k,n}(m_1, m_2, \cdots, m_{k-1}) = cum(x(t), x(t+m_1), \cdots, x(t+m_{k-1})) \quad (1)$$

where $x(t+m_k)$ denotes a function of different time delays and regardless of t, $cum(\bullet)$ means taking the cumulant. Therefore, its fourth-order cumulant is:

$$\begin{aligned} \mathbf{C}_{4,n}(m_1, m_2, \cdots, m_3) = \; & E[x(n)x(n+m_1)x(n+m_2)x(n+m_3)] - \mathbf{C}_{2,n}(m_1)\mathbf{C}_{2,n}(m_2-m_3) \\ & -\mathbf{C}_{2,n}(m_2)\mathbf{C}_{2,n}(m_3-m_1) - \mathbf{C}_{2,n}(m_3)\mathbf{C}_{2,n}(m_1-m_2) \end{aligned} \quad (2)$$

Based on the above theory, the fourth-order, sixth-order and eight-order cumulants of the zero-mean $x(t)$, are shown as:

$$\begin{aligned} \mathbf{C}_{4,0} &= cum(x,x,x,x) = \mathbf{M}_{4,0} - 3\mathbf{M}_{2,0}{}^2 \\ \mathbf{C}_{4,1} &= cum(x,x,x,x^*) = \mathbf{M}_{4,1} - 3\mathbf{M}_{2,1}\mathbf{M}_{2,0} \\ \mathbf{C}_{4,2} &= cum(x,x,x^*,x^*) = \mathbf{M}_{4,2} - \left|\mathbf{M}_{2,0}\right|^2 - 2\mathbf{M}_{2,1}{}^2 \\ \mathbf{C}_{6,0} &= cum(x,x,x,x,x,x) = \mathbf{M}_{6,0} - 15\mathbf{M}_{4,0}\mathbf{M}_{2,0} + 30\mathbf{M}_{2,0}{}^3 \\ \mathbf{C}_{6,3} &= cum(x,x,x,x^*,x^*,x^*) = \mathbf{M}_{6,3} - 9\mathbf{C}_{4,2}\mathbf{C}_{2,1} - 6\mathbf{C}_{2,1}{}^3 \\ \mathbf{C}_{8,0} &= cum(x,x,x,x,x,x,x,x) = \mathbf{M}_{8,0} - 28\mathbf{M}_{6,0}\mathbf{C}_{2,0} - 35\mathbf{M}_{4,0}{}^2 + 420\mathbf{M}_{4,0}\mathbf{M}_{2,0}{}^2 - 630\mathbf{M}_{2,0}{}^4 \end{aligned} \quad (3)$$

where $\mathbf{M}_{pq} = E[x(t)^{p-q}x^*(t)^q]$ denotes the pth-order mixing moment [19].

In the practical application of MC, we need to estimate the HOC value of the signal from the received symbol sequence in the shortest possible time. Sample estimations of the correlations are given by:

$$\begin{aligned} \mathbf{C}_{4,0} &= \tfrac{1}{N}\sum_{n=1}^{N}(x(t))^4 - 3\mathbf{C}_{2,0}^2 \\ & \quad \cdots\cdots \\ \mathbf{C}_{8,0} &= \tfrac{1}{N}\sum_{n=1}^{N}(x(t))^8 - 28\mathbf{C}_{2,0}\tfrac{1}{N}\sum_{n=1}^{N}(x(t))^6 - 35\mathbf{M}_{4,0}^2 + 420\mathbf{M}_{4,0}\mathbf{M}_{2,0}^2 - 630\mathbf{M}_{2,0}^2 \end{aligned} \quad (4)$$

Substituting the estimated values into Equation (4), we can obtain all of the features for the considered six wireless signal types. Table 1 shows some of these features for a number of these signals. These values are computed under the constraint of unit variance in noise free conditions. It can be seen that by computing of these values, we can classify the wireless signal types.

Table 1. Theoretical values of higher-order cumulant (HOC) for six wireless signal modulations.

	$\lvert\mathbf{C}_{4,0}\rvert$	$\lvert\mathbf{C}_{4,1}\rvert$	$\lvert\mathbf{C}_{4,2}\rvert$	$\lvert\mathbf{C}_{6,0}\rvert$	$\lvert\mathbf{C}_{6,3}\rvert$	$\lvert\mathbf{C}_{8,0}\rvert$
OOK	2	2	2	16	13	272
DPSK	2	2	2	16	13	272
QPSK	1	0	1	0	4	34
OQPSK	1	0	1	0	4	34
16QAM	0.68	0	0.68	0	2.08	13.9808
64QAM	0.619	0	0.619	0	1.7972	11.5022

Table 1 shows that OOK (on-off keying), DPSK (differential phase shift keying), QPSK (quadrature phase shift keying), OQPSK (offset quadrature phase shift keying) have the same theoretical values of HOC. In addition, 16QAM (16 quadrature amplitude modulation) and 64QAM (64 quadrature amplitude modulation) have similar HOC values. Therefore, we can define a feature parameter $\mathbf{T1} = |\mathbf{C}_{8,0}|/|\mathbf{C}_{4,0}|$ that is calculated in Table 2 and divides signals into three categories including (OOK, DPSK), (QPSK, OQPSK) and (16QAM, 64QAM). It is worth noting that the absolute value and ratio form are used to eliminate the effect of phase jitter and amplitude [20].

Table 2. Theoretical values of T1 for six wireless signal modulations.

	OOK,DPSK	**QPSK,OQPSK**	**16QAM**	**64QAM**
T1	136	34	20.56	18.5819

Owing to the difference between the phase jump rules of QPSK and OQPSK, the sampling sequence of both can be performed with a differential operation, i.e.,

$$\Delta x(t) = x(t+1) - x(t) = (a_{t+1} - a_t)\exp[j(2\pi f_c + \Delta\theta_c)] \quad (5)$$

where $x(t)$ denotes the signals of QPSK and OQPSK, a_k is the transmitted symbol sequences, f_c denotes the carrier frequency and θ_c denotes the phase jitter. For the sake of discussion, we assume that f_c and θ_c have been completed timing synchronization. The values of HOC under difference operation are calculated in Table 3. Then we define another feature parameter $\mathbf{T2} = |\mathbf{C}_{d8,0}|/|\mathbf{C}_{d4,0}|^2$ is calculated in Table 4 to classify QPSK and OQPSK, where $\mathbf{C}_{d8,0}$ and $\mathbf{C}_{d4,0}$ represent the cumulants after differential operation.

Table 3. Theoretical values of HOC after difference between QPSK and OQPSK.

	$\lvert C_{d4,0}\rvert$	$\lvert C_{d4,1}\rvert$	$\lvert C_{d4,2}\rvert$	$\lvert C_{d6,3}\rvert$	$\lvert C_{d8,0}\rvert$
QPSK	2	0	2	8	68
OQPSK	2	0	0.89	2	131.4

Table 4. Theoretical values of T2 for QPSK and OQPSK.

	QPSK	**OQPSK**
T2	17	32.85

2.2. Feature Extraction Based on Cyclic Spectrum

Since the T1 of (OOK, DPSK) and (16QAM, 64QAM) are the same or similar, a cyclic spectral density function for noise suppression is proposed for identification. Assuming $x(t)$ is the cyclostationary signal, and its mean value and autocorrelation function are periodic with T_0 shown as:

$$m_x(t+T_0) = m_x(t) \quad (6)$$

$$\mathbf{R}_x(t+T_0+\frac{\tau}{2}, t+T_0-\frac{\tau}{2}) = \mathbf{R}_x(t+\frac{\tau}{2}, t-\frac{\tau}{2}) \quad (7)$$

where τ is the delay variable. Because the autocorrelation function has periodicity, its Fourier series can be written as:

$$\mathbf{R}_{x\alpha}(t+\frac{\tau}{2}, t-\frac{\tau}{2}) = \sum_{\alpha} \mathbf{R}_{x\alpha}(\tau)e^{j2\pi x} \quad (8)$$

where α stands for the frequency corresponding to the instantaneous autocorrelation and is often called the cyclic frequency. In addition, $\mathbf{R}_{x\alpha}$ is the coefficient of the Fourier series which is given by:

$$\mathbf{R}_{x\alpha}(\tau) = \frac{1}{T_0}\int_{-\frac{T_0}{2}}^{\frac{T_0}{2}} \mathbf{R}_{x\alpha}(t+\frac{\tau}{2}, t-\frac{\tau}{2})e^{-j2\pi\alpha t}dt \quad (9)$$

The Fourier transform of the cyclic autocorrelation function can be written as:

$$\mathbf{S}_{x\alpha}(f) \triangleq \int_{-\infty}^{\infty} \mathbf{R}_{x\alpha}(\tau) e^{-j2\pi f t} d\tau \tag{10}$$

where $\mathbf{S}_{x\alpha}(f)$ is called power spectral density and f is the spectral frequency.

The $\mathbf{R}_{x\alpha}(\tau)$ can be seen as the cross-correlation of two complex frequency shift components $u(t)$ and $v(t)$ of $x(t)$, i.e.,

$$\mathbf{R}_{x\alpha}(\tau) = \mathbf{R}_{uv\alpha}(\tau) = \frac{1}{T_0} \int_{-\frac{T_0}{2}}^{\frac{T_0}{2}} u(t + \frac{\tau}{2}) v^*(t - \frac{\tau}{2}) dt. \tag{11}$$

where $u(t) = x(t)e^{j2\pi\alpha t}$, $v(t) = x(t)e^{-j2\pi\alpha t}$.

From Equation (10) we can obtain $\mathbf{S}_{x\alpha}(f) = \mathbf{S}_{uv\alpha}(f)$. Through the cross-spectrum analysis, we can obtain:

$$\mathbf{S}_{x\alpha}(f) \triangleq \lim_{T_0 \to \infty} \lim_{\Delta t \to \infty} \mathbf{S}_{uvT_0}(f)_{\Delta t} = \lim_{T_0 \to \infty} \lim_{\Delta t \to \infty} \frac{1}{\Delta t} \int_{-\frac{\Delta t}{2}}^{\frac{\Delta t}{2}} \mathbf{S}_{X_{T_0}\alpha}(t, f) dt \tag{12}$$

$$\mathbf{S}_{X_{T_0}\alpha}(t, f) = \frac{1}{T_0} \mathbf{X}_{T_o}(t, f + \frac{\alpha}{2}) \mathbf{X}_T^*(t, f + \frac{\alpha}{2}) \tag{13}$$

$$\mathbf{X}_{T_0}(t, f + \frac{\alpha}{2}) = \int_{t-\frac{T_0}{2}}^{t+\frac{T_0}{2}} x(u) e^{-j2\pi f u} du \tag{14}$$

where Equation (12) is used to estimate the cyclic spectral density, Equation (13) is the cyclic periodic diagram, Equation (14) is the short-time Fourier transform (STFT) formula, Δt is the length of received data, T_0 is the window length for the STFT, and $(\bullet)^*$ is the complex conjugate.

According to the insensitivity of the cyclic spectrum to noise and the above theory, the characteristic parameter **T3** $= \max(\mathbf{S}_{x\alpha})$ is defined to distinguish the signal set of (OOK, DPSK) and (16QAM, 64QAM).

3. Compressed Values of Feature Parameters Based on Compressed Sensing

3.1. Compressed Value of HOC

CS techniques perform successfully whenever applied to so-called compressible and/or K-sparse signals, i.e., signals that can be represented by $K \ll N$ significant coefficients over an N-dimensional basis [21,22]. The K-sparse signal $s(t) \in \mathbb{R}^N$ of dimension N is accomplished by computing a measurement vector $y(t) \in \mathbb{R}^M$ that consists of $M \ll N$ linear projections of the vector $s(t)$. The compression sampling rate $f_{cs} = (M/N)f_s$, where f_s is traditional sampling rate and $\delta = M/N \in (0,1)$ is called the compression ratio. The linear compression sampling process can be described as:

$$y = \Phi s + n \tag{15}$$

where Φ represents a $M \times N$ matrix, usually over the field of real numbers. It is noted that the measurement matrix Φ is a random matrix satisfying the restricted isometry property (RIP) [23], and its form is various, such as Gaussian matrix [24] and local Hadamard matrix [25].

According to the theory of feature extraction and compression sensing, the linear square compression sampling process can be defined as:

$$[\![x]\!]^2 = \Phi [\![s]\!]^2 \tag{16}$$

where $[\![\bullet]\!]$ represents the product operation of the corresponding elements between the vectors. From Equation (15), we can through CS to simplify the reconstruction.

First, we define a relationship between the autocorrelation matrix $\mathbf{R}_{[\![x]\!]^2}$ and $\mathbf{R}_{[\![s]\!]^2}$:

$$\mathbf{R}_{[\![x]\!]^2} = [\![x]\!]^2([\![x]\!]^2)^T = (\Phi[\![s]\!]^2)(\Phi[\![s]\!]^2)^T = \Phi\mathbf{R}_{[\![s]\!]^2}\Phi^T \tag{17}$$

Second, we obtain a relationship between $[\![x]\!]^2$ and $\mathbf{R}_{[\![x]\!]^2}$:

$$[\![x]\!]^2 = \mathbf{P}_{[\![x]\!]^2} vec(\mathbf{R}_{[\![x]\!]^2}) \tag{18}$$

where $vec(AXB) = (B^T \otimes A)vec(X)$ ($\otimes$ denotes Kronecker product) and $\mathbf{P}_{[\![x]\!]^2} \in \{0,1\}^{n \times n^2}$ that maps the linearly products to the vectorized counterparts $[\![x]\!]^2$ and $\mathbf{R}_{[\![x]\!]^2}$ [26].

According to Equation (4), the compressed value of fourth-order cumulant can be defined as:

$$\mathbf{C}_{(4,0)\alpha} = F[\![s]\!]^4 - 3[\![F[\![s]\!]^2]\!]^2 \tag{19}$$

where $F = \frac{1}{N}[\exp(-j2\pi\alpha n/N)]_{(\alpha,n)} \in \mathbb{R}^{M_\alpha \times N}$ is the discrete Fourier transform (DFT) matrix.

Therefore, the linear representation process of the vector-form $\mathbf{C}_{(4,0)\delta}$ is shown as:

$$\begin{aligned}\mathbf{C}_{(4,0)\delta} &= FP_s vec(R_{[\![s]\!]^2}) - 3P_{Fs} vec(\mathbf{R}_{F[\![s]\!]^2}) = FP_s vec(\mathbf{R}_{[\![s]\!]^2}) - 3P_{Fs} vec(F\mathbf{R}_{[\![s]\!]^2}F^T) \\ &= [FP_s - 3P_{Fs}(F \otimes F)]vec(\mathbf{R}_{[\![s]\!]^2}) = [FP_s - 3P_{Fs}(F \otimes F)][P_{[\![x]\!]^2}(\Phi \otimes \Phi)]^{\dagger}[\![x]\!]^2\end{aligned} \tag{20}$$

where $[\bullet]^{\dagger}$ stands for the pseudo-inverse operation.

Finally, by deriving the linear compressed sampling process of the fourth-order cumulant, we can obtain the compressed value of the eighth-order as follows:

$$\begin{aligned}\mathbf{C}_{(8,0)\delta} = \; & \{FP_s[(P_s - 28P_s^{1/2}P_{Fs}^{1/2}(F \otimes F)^{1/2} - 35FP_s + 420P_{Fs}(F \otimes F)] - \\ & 630P_{Fs}^2(F \otimes F)^2\}\bullet[P_{[\![x]\!]^2}(\Phi \otimes \Phi)]^{\dagger}[\![x]\!]^2\end{aligned} \tag{21}$$

Therefore, the first and second characteristic parameters after CS is $\mathrm{T1} = |\mathbf{C}_{(8,0)\delta}|/|\mathbf{C}_{(4,0)\delta}|$ and $\mathrm{T2} = |C_{d(8,0)\delta}|/|C_{d(4,0)\delta}|^2$.

3.2. Compressed Value of Cyclic Spectrum

It can be seen from Equations (10) and (16) that there is no direct linear relationship between the compressed sampled value x in the time domain and the cyclic spectrum $S_{x\alpha}$, so the existing reconstruction algorithm cannot be used to implement the cyclic spectrum estimation. It is necessary to use some explicit linear relations between the second-order statistic to derive its transformation, and indirectly establish the linear relationship between them, so as to use the existing reconstruction algorithm to complete the estimation of the cyclic spectrum [27].

In order to obtain the compressed values of the cyclic spectrum, we first need to obtain the relationship between the cyclic spectrum matrix $\mathbf{S}_{s\delta}$ and the cyclic autocorrelation matrix $\mathbf{R}_{s\delta}(u,v)$. Let $\mathbf{R}_{s\delta}(u,v)$ denote the form of the time-varying covariance matrix R. When x is a real value, R is a symmetric semi-positive definite matrix. For the convenience of calculation, we convert it into an auxiliary covariance matrix:

$$\mathbf{R} = \begin{bmatrix} \mathbf{R}_{s\delta}(0,0) & \mathbf{R}_{s\delta}(0,1) & \mathbf{R}_{s\delta}(0,2) & \cdots & \mathbf{R}_{s\delta}(0,N-1) \\ \mathbf{R}_{s\delta}(1,0) & \mathbf{R}_{s\delta}(1,1) & \mathbf{R}_{s\delta}(1,2) & \cdots & 0 \\ \mathbf{R}_{s\delta}(2,0) & \mathbf{R}_{s\delta}(2,1) & \mathbf{R}_{s\delta}(2,2) & \cdots & 0 \\ \cdots & \cdots & \cdots & \cdots & \cdots \\ \mathbf{R}_{s\delta}(N-1,0) & 0 & 0 & \cdots & 0 \end{bmatrix} \tag{22}$$

where N is the sampling point.

The matrix R contains all the elements in the $\mathbf{R}_{s\delta}$ vector except for the zero elements. The relationship between $\mathbf{R}_{s\delta}$ and R can be expressed as:

$$vec\{\mathbf{R}\} = \mathbf{H}_N\mathbf{R}_{s\delta} \tag{23}$$

where $\mathbf{H}_N \in \{0,1\}^{N^2\times(N(N+1)/2)}$, $vec\{\bullet\}$ means the vectorization operation, and the $\mathbf{R}_{s\delta}$ vector is mapped to $vec\{\mathbf{R}\}$.

So the relationship between the cyclic spectrum matrix and the cyclic autocorrelation matrix is shown as:

$$\begin{aligned} \mathbf{R}_{s\delta} &= \sum_{v=0}^{N-1} \mathbf{G}_v\mathbf{R}\mathbf{D}_v \\ \mathbf{S}_{s\delta} &= \mathbf{R}_{s\delta}\mathbf{F} \end{aligned} \tag{24}$$

where $\mathbf{G}_n = [\frac{1}{N}\exp(-j\frac{2\pi}{N}a(n+\frac{v}{2}))]_{(a,n)} \in \mathbb{R}^{N\times N}$, $\mathbf{F} = [\exp(-j2\pi vb/N)]_{(v,b)}$ is the N-point DFT matrix and $\mathbf{D}_v$ is an $N\times N$ matrix with only its (v,v)th diagonal element being 1 and all other elements being 0. In addition, n and v are time delay, $a, b \in [0, N-1]$ denotes digital cyclic frequency and $\alpha = fa/N$ stands for the cyclic frequency.

The time-varying covariance matrix $\mathbf{R}_x = E\{x_m x_m^T\}$ of the compressed value x is also a symmetric semi-definite matrix, which can be rearranged into a vector $\mathbf{R}_{x\delta}$ of length $M(M+1)/2$ to represent as:

$$\begin{aligned} \mathbf{R}_{x\delta} = \ & [\mathbf{R}_{x\delta}(0,0), \mathbf{R}_{x\delta}(1,0), \cdots, \mathbf{R}_{x\delta}(M-1,0) \\ & \mathbf{R}_{x\delta}(0,1), \mathbf{R}_{x\delta}(1,1), \cdots, \mathbf{R}_{x\delta}(M-2,1) \\ & \mathbf{R}_{x\delta}(0,M-1)]^T \end{aligned} \tag{25}$$

Through the linear formula conversion, we can define two projection matrices $\mathbf{P}_m \in \{0,1\}^{N^2\times(N(N+1)/2)}$ and $\mathbf{Q}_m \in \{0,1/2,1\}^{(M(M+1)/2)\times M^2}$ map the entries of x, s to those in $vec(\mathbf{R}_x)$ and $vec(\mathbf{R}_s)$, it can be shown that:

$$\begin{aligned} vec\{\mathbf{R}_x\} &= \mathbf{P}_m x \\ s &= \mathbf{Q}_m vec\{\mathbf{R}_s\} \end{aligned} \tag{26}$$

where $\mathbf{P}_m$ and $\mathbf{Q}_m$ are special mapping matrices.

Because of $x(t) = \Phi s(t)$, we can obtain:

$$x = \mathbf{Q}_m vec(\mathbf{R}_x) = \mathbf{Q}_m(\Phi\otimes\Phi)vec(\mathbf{R}_s) = \mathbf{Q}_m(\Phi\otimes\Phi)\mathbf{P}_m s = \Theta s \tag{27}$$

where $\Theta = \mathbf{Q}_m(\Phi\otimes\Phi)\mathbf{P}_m \in \mathbb{R}^{\frac{M(M+1)}{2}\times\frac{N(N+1)}{2}}$.

Following the equation (24), we can obtain $vec(\mathbf{R}_{s\delta})$ is:

$$vec(\mathbf{R}_{s\delta}) = \sum_{v=0}^{N-1} (\mathbf{G}_v^T \otimes \mathbf{D}_v) vec(\mathbf{R}_s) = \Omega s \tag{28}$$

where $\Omega = \sum_{v=0}^{N-1} (g_v^T \otimes \mathbf{D}_v)\mathbf{P}_m \in \mathbb{R}^{N^2\times(N(N+1)/2)}$.

Through the Equations (24), (27) and (28), we can derive the measurement vector x as a linear function of the vector-form cyclic spectrum $\mathbf{S}_{s\delta}$ as:

$$\mathbf{S}_{s\delta} = \Xi\Omega\Theta^{\dagger} x \tag{29}$$

where $\Xi = (\mathbf{F}^{-T}\otimes\mathbf{I}_N)^{-1}$, $\mathbf{I}_N$ is the N dimension unit matrix. Therefore, the third characteristic parameter after CS is T3 = $\max(S_{s\delta})$.

4. The Structural Process of Decision Tree–Support Vector Machine Classifier

4.1. The Principle of Support Vector Machine

Support vector machine (SVM) is based on the principle of structural risk minimization [28–30]. Its final solution can be transformed into a quadratic convex programming problem with linear constraints. There is no local minimum problem. By introducing the kernel function, the linear SVM can be simply extended to the non-linear SVM, and there is almost no additional computation for high-dimensional samples.

The main idea can be seen from Figure 1 that for a linear separable case, the idea of maximizing the classification boundary is used to seek the optimal hyperplane H, while H1 and H2 are hyperplanes passing through the closest sample to the H and parallel to h, respectively, and the distance between them is called the classification interval. In the case of linear indivisibility, the linear indivisible samples in the low-dimensional input space are transformed into the high-dimensional feature space by the non-linear mapping algorithm, so that they can be linearly separable, in order that the high-dimensional feature space can be solved by the linear analysis method.

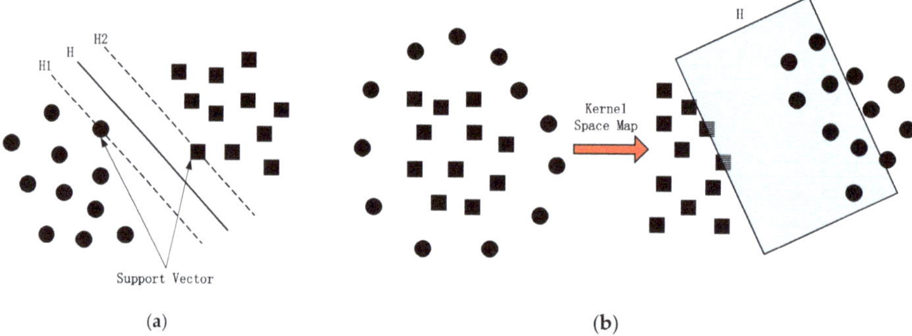

Figure 1. Linear (**a**) and non-linear (**b**) classification of support vector machine (SVM).

Suppose that the training set is $\{(x_i, x_i), i = 1, 2, \ldots, L\}$ and the expected output is $y_i \in \{+1, -1\}$, where +1 and −1 represent two kinds of class representation respectively. If $x_i \in R^n$ belongs to the first category, the corresponding output is $y_i = +1$; if it belongs to the second category, the corresponding output is $y_i = -1$. The linear separability of the problem shows that there is a hyper-plane $(w * x) + b = 0$, which makes the positive and negative inputs of the training points located on both sides of the hyper-plane, respectively. When the training sets are not completely linearly separable, we can introduce the relaxation variable $\xi_i \geq 0 \quad i = 1, 2, \ldots, L$, then the objective function is transformed into:

$$\begin{aligned} &\min \quad \phi(w) = \tfrac{1}{2}||w||^2 + C\sum_{i=1}^{L} \xi_i \\ &s.t. \quad y_i[(w \cdot x_i) + b] \geq 1 - \xi_i \\ &\xi_i \geq 0 \quad i = 1, 2, \ldots, L \end{aligned} \tag{30}$$

where $C \geq 0$ is the penalty parameter. A larger C indicates a larger penalty for misclassification and is the only parameter that can be adjusted in the algorithm. By choosing the proper kernel function $K(x, x^T)$ and using the Lagrange multiplier method to solve Equation (30), the corresponding dual problem can be obtained as follows:

$$\begin{aligned} &\min_{\alpha} \frac{1}{2} \sum_{i=1}^{L} \sum_{j=1}^{L} y_i y_j \alpha_i \alpha_j K(x_i, x_j) - \sum_{j=1}^{L} \alpha_j \\ &\text{s.t.} \quad \sum_{i=1}^{L} y_i \alpha_i = 0 \\ &0 \le \alpha_i \le C \qquad i = 1, 2, \ldots, L \end{aligned} \tag{31}$$

Equation (31) obtains the optimal solution $\alpha^* = (\alpha_1^*, \alpha_2^*, \ldots, \alpha_L^*)$ and selects a positive component $0 \le \alpha_j^* \le C$ of α^*, and calculates $b^* = y_j - \sum_{i=1}^{L} y_i \alpha_i^* K(x_i, x_j)$ accordingly. Finally, the policy function $f(x) = \text{sgn}(\sum_{i=1}^{L} y_i \alpha_i^* K(x_i, x_j) + b^*)$ is obtained.

4.2. The Structure of Decision Tree–Support Vector Machine Classifier

Through the above calculation and analysis of the compressed values of the feature parameters based on CS, the feature parameters are input into the decision tree–support vector machine (DT–SVM) structure with efficient calculation to realize signal classification. Adding support vector machine (SVM) to every node of the decision tree can comprehensively utilize the efficient computing power of the decision tree structure and the high classification performance of the SVM to achieve the high-precision classification of MC. In particular, for the K-class classification problem, the method of DT–SVM only needs to construct K-1 SVM sub classifier. The classification process of decision tree is shown in Figure 2.

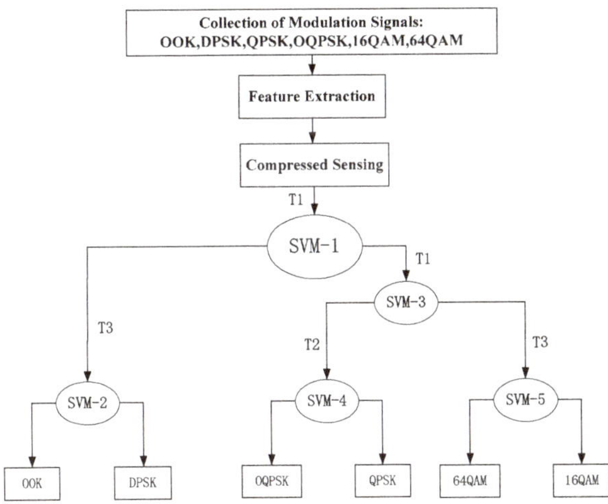

Figure 2. The identification process of decision-tree (DT) classifier.

Using the three features in Figure 1 to complete the MC, the specific steps of the process are as follows:

(1) Three feature vectors are obtained from six kinds of wireless modulation signal data through feature extraction module;
(2) The feature vector is input into the compression-sensing module to obtain the respective compression sampling values, as shown in Figures 3 and 4;
(3) Six kinds of wireless signals are roughly classified by T1. The (OOK, DPSK) signals can be separated by SVM-1, and the remaining signals are classified into one class;
(4) For (OOK, DPSK) signals, SVM-2 and T3 are used to realize classification;

(5) By SVM-3 and T1, the residual signals can be divided into two categories: (QPSK, OQPSK) and (16QAM, 64QAM);

(6) The T2 after differential operation and SVM-4 are used to classify QPSK and OQPSK;

(7) Finally, the classification of 16QAM and 64QAM signals is realized by the T3 and SVM-5.

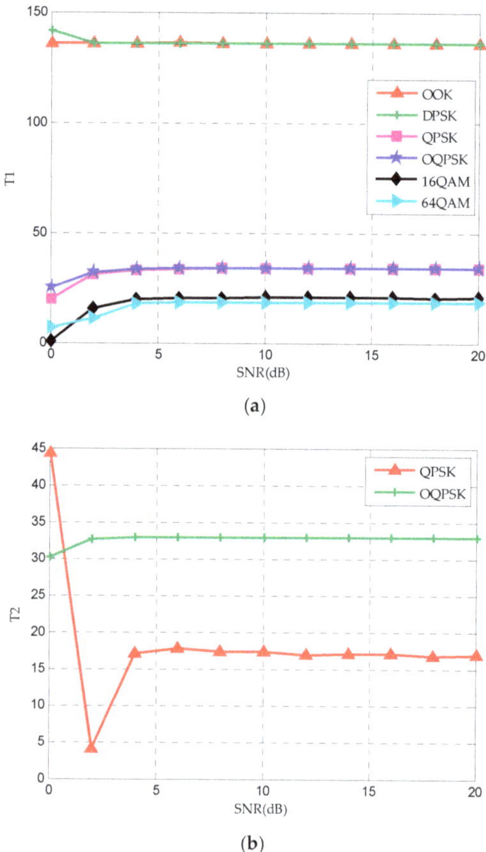

Figure 3. The values of feature parameters (**a**) T1 and (**b**) T2 under different signal-to-noise ratios (SNRs).

It can be seen from Figure 3a that the compressed value of the feature parameter T1 tends to be stable with the increase of SNR and conforms to the theoretical value. In addition, it is obvious from the figure that T1 can classify the six signals into three categories includes (OOK, DPSK), (QPSK, OQPSK) and (16QAM, 64QAM). Similarly, it can be seen from Figure 3b that the feature parameter T2 after difference can distinguish QPSK from OQPSK. The circular spectrum and the cross-sectional diagrams of cycle frequency of OOK, DPSK, 16QAM and 64QAM are shown in Figure 4. It can be seen from Figure 4 that the maximum value of the cyclic spectrum of different signals is different, so the remaining signals can be distinguished by T3.

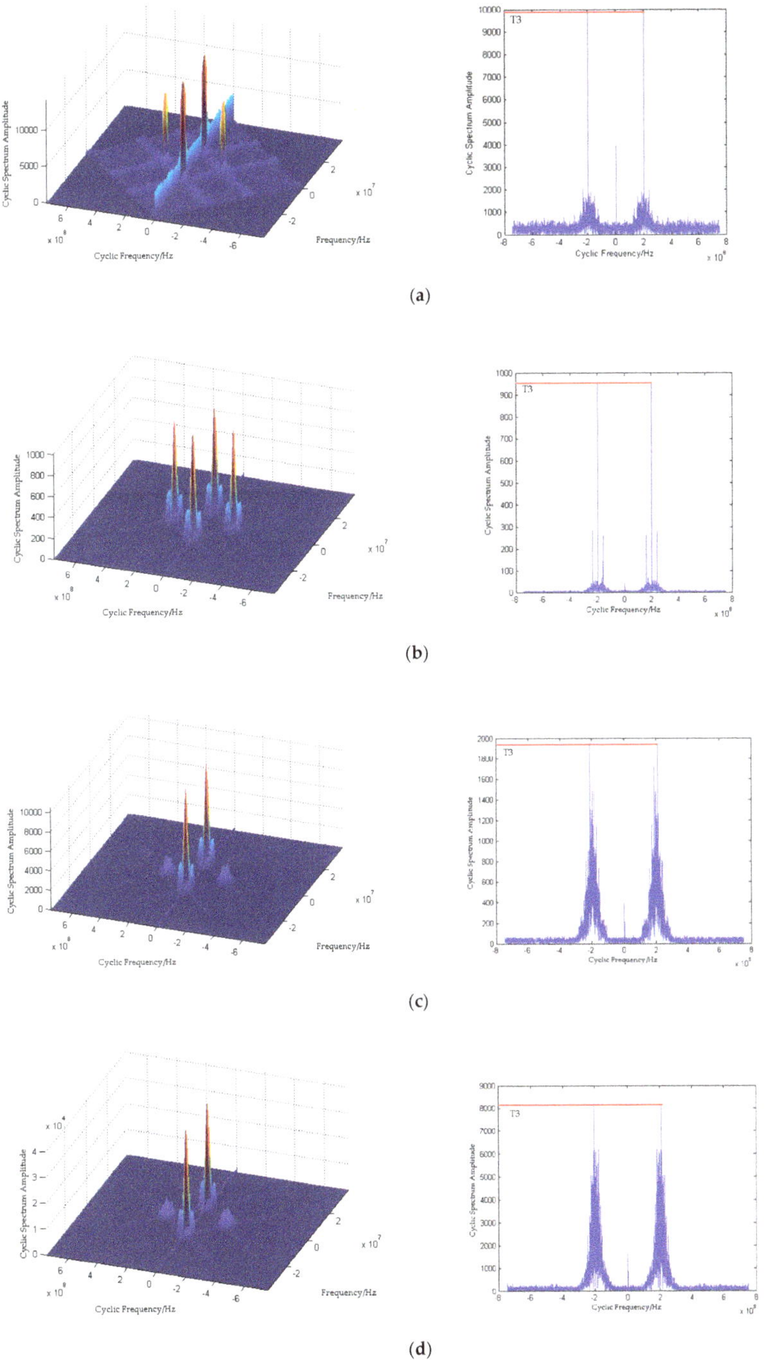

Figure 4. Cyclic spectrum and cross-sectional diagrams of (**a**) OOK (on-off keying), (**b**) DPSK (differential phase shift keying), (**c**) 16QAM (16 quadrature amplitude modulation) and (**d**) 64QAM (64 quadrature amplitude modulation).

5. Simulation Results and Discussion

For signal modulation classification task, the modulation set is {OOK, DPSK, QPSK, OQPSK, 16QAM, 64QAM}. In this simulation process, all modulation signals adopt the same modulation parameters, that is, the carrier frequency is 100 kHz, the symbol rate is 40 kbps, and the sampling frequency is 800kHz. For each kind of modulation signal, the simulation generates 500 characteristic samples under different SNR (SNR from −5 dB to 15 dB, interval 5 dB). The K-fold cross validation (K-CV) method is used to evaluate the generalization ability of the model, which can not only improve the data utilization, but also solve the over fitting problem to a certain extent, so as to select the model. In this paper, K is chosen as 10. The basic principle of K-CV is to divide the training data set into K equal subsets. Each time, K-1 data are used as training data, and other data are used as test data. In this way, we repeat K times, estimate the expected generalization error according to the mean square error (MSE) average value after K times iteration, and finally select a group of optimal parameters.

To evaluate the influence of different symbol lengths on the classification performance, we select the symbol length N = 512, 1024, 2048, 4096 for the OOK, QPSK and 16QAM signals (as examples), respectively, to analyze the classification accuracy in Figure 5. As can be seen from Figure 5, with the increase of symbol length, the trend of classification accuracy is gradually increasing and finally tends to 100%. However, the increase of symbol length affects the classification accuracy mainly in the case of low SNR. Considering the influence of the computation cost in the classification process, 2048 is chosen as the symbol length in this paper.

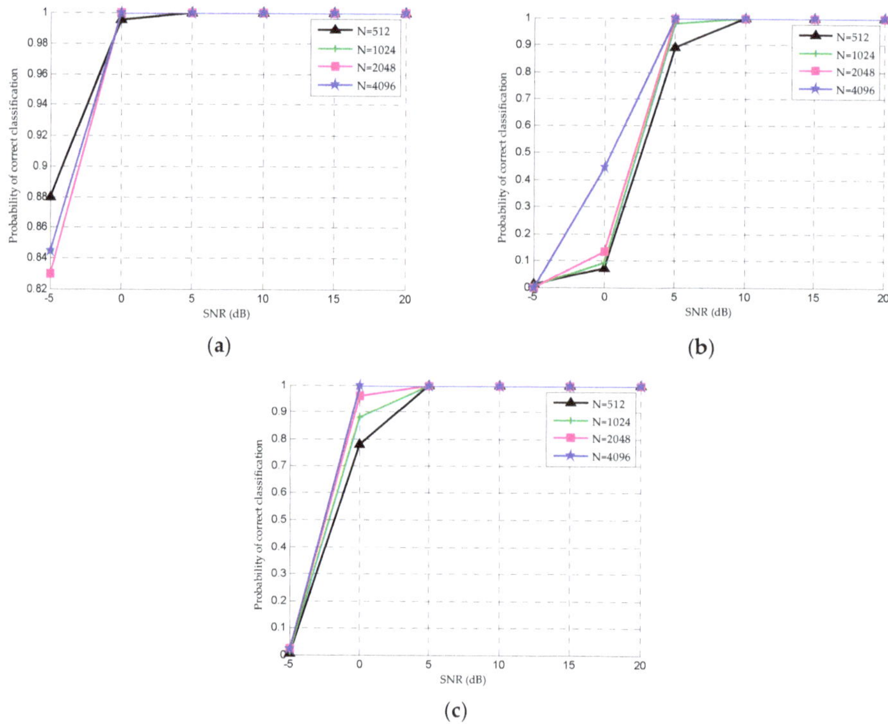

Figure 5. Correct classification rate with different symbol length for (**a**) OOK, (**b**) QPSK and (**c**) 16QAM under different SNRs.

In order to evaluate the impact of different compression ratios on the classification performance, we analyzed the classification accuracy in Figure 6 with the compression ratios of 25%, 37.5%, 50% and

75% for OOK, QPSK and 16QAM signals (as examples) when the symbol length is 2048. As can be seen from Figure 6, with the increase of compression ratio, the classification accuracy increases slightly and finally tends to 100%. The increase of symbol rate has little effect on recognition rate. Therefore, in order to reduce the sampling rate and system complexity as much as possible, 25% is selected as the compression rate value.

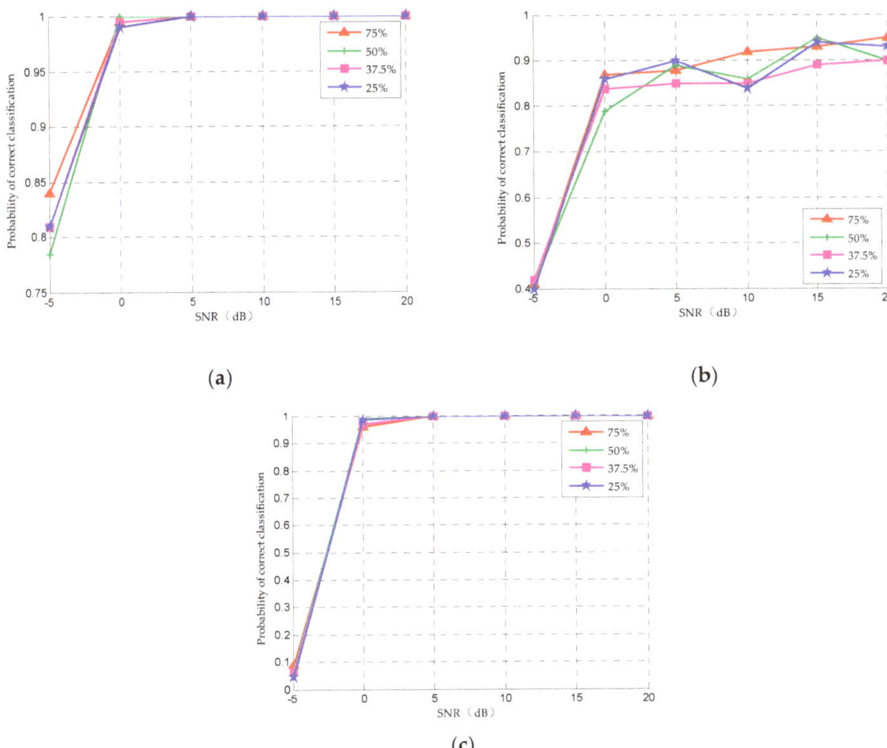

Figure 6. Correct classification rate with different compression rate for (**a**) OOK, (**b**) QPSK and (**c**) 16QAM under different SNRs.

In order to optimize the parameters of kernel function and improve the classification accuracy, the grid search method is used. Table 5 shows the optimization results of penalty parameter and kernel parameter of each node of DT–SVM and the classification accuracy of sub-SVM under the RBF kernel function. It can be seen from the table that under different SNR conditions, with the improvement of SNR, the classification accuracy of sub nodes has improved. When the SNR is 0 dB or above, the average classification accuracy has reached 100%.

Table 5. The optimization results and identification accuracy of the sub-SVM of each node in DT–SVM.

SNR	SVM-1	SVM-2	SVM-3	SVM-4	SVM-5	AVERAGE
	Acc/% (c,γ)	Acc/% (c,γ)	Acc/% (c,γ)	Acc/% (c,γ)	Acc/% (c,γ)	Acc/%
−5 dB	88.33 $(2^{11.2},2^{13.8})$	100 $(2^{0},2^{0})$	81.25 $(2^{0.5},2^{3})$	95 $(2^{-2.5},2^{15})$	100 $(2^{0},2^{0})$	92.92
0 dB	100 $(2^{-8},2^{2})$	100 $(2^{0},2^{0})$	100 $(2^{-5},2^{8.5})$	100 $(2^{-5},2^{7.5})$	100 $(2^{0},2^{0})$	100
5 dB	100 $(2^{-8},2^{-2})$	100 $(2^{0},2^{0})$	100 $(2^{0},2^{0})$	100 $(2^{0},2^{0})$	100 $(2^{0},2^{0})$	100

In order to prove the superiority of this method in recognition accuracy, Table 6 shows the classification accuracy of six different cognitive radio signals using multidimensional HOC, cyclic spectrum and DT–SVM classifier when the kernel function is RBF. In addition, the sizes of training and testing subsets are selected as 80% and 20% of the whole set of eigenvectors. The results show that with the increase of SNR, the classification accuracy of six kinds of signal is improved. When the SNR is 0 dB, the classification accuracy is 100%. It is proved that this method still has high recognition accuracy under low SNR. In addition, a new modulation signal can be introduced to expand the flexibility of the method and shows better compatibility, which will certainly increase the complexity of the algorithm and the classification time of the whole classification system.

Table 6. The classification accuracy of the cognitive radio signals using multidimensional HOC, cyclic spectrum and DT–SVM classifier.

SNR	Classification Accuracy of Cognitive Radio Signals (%)						
	OOK	DPSK	QPSK	OQPSK	16QAM	64QAM	AVERAGE
−5 dB	72.5	72.5	74.69	74.69	83.25	83.25	76.81
0 dB	100	100	100	100	100	100	100
5 dB	100	100	100	100	100	100	100
10 dB	100	100	100	100	100	100	100
15 dB	100	100	100	100	100	100	100

6. Conclusions

We have proposed a method through simulation to identify the cognitive radio signals based on compressed sensing combined with HOC and cyclic spectrum, which has been proved to perform well in noisy situations. It successfully achieves reconstructing the feature parameters of HOC and cyclic spectrum directly from the sub-Nyquist rate rather than reconstructing the original signal. The simulation results indicate that this method can effectively achieve modulation classification for six kinds of cognitive radio signals with three feature parameters. In this paper, the proposed method is relatively simple and the feature parameters used are few, and they are also less affected by noise. By analyzing the effect of symbol length and compression rate on the classification rate, the classification performance is improved and the classification accuracy can reach 100% when the SNR is 0 dB. This technique utilizes a sampling rate of CS much lower than the Nyquist sampling rate and noise-insensitive feature extraction algorithm, realizes blind classification without any prior information from the transmitter, and has low computational complexity.

Author Contributions: X.S. and Z.Z. conceived and designed the experiments; X.S. and X.T. performed the experiments; X.S. and X.G. analyzed the data; X.S., Z.Z. and X.G. contributed the simulation software and experimental facilities; X.S. and Z.Z. wrote the paper; S.S. participated in the funding acquisition and investigation. All authors have read and agreed to the published version of the manuscript.

Funding: This work was supported by the Frontier Science and Technology Innovation Project in the National Key Research and Development Program under Grant No. 2016QY11W2003, Natural Science Foundation of

Hunan Province under Grant No. 2018JJ3607, Natural Science Foundation of China under Grant No. 51575517 and National Technology Foundation Project under Grant No. 181GF22006.

Conflicts of Interest: The authors declare no conflict of interest.

References

1. Zhu, X.; Fujii, T. A modulation classification method in cognitive radios system using stacked denoising sparse autoencoder. In Proceedings of the 2017 IEEE Radio and Wireless Symposium (RWS), Phoenix, AZ, USA, 15–18 January 2017; pp. 218–220.
2. Chen, S.; Shen, B.; Wang, X.; Yoo, S. A strong machine learning classifier and decision stumps based hybrid adaBoost classification algorithm for cognitive radios. *Sensors* **2019**, *19*, 5077. [CrossRef] [PubMed]
3. Sutton, P.; Nolan, K.E.; Doyle, L. Cyclostationary signatures in practical cognitive radio applications. *IEEE J. Sel. Area. Comm.* **2008**, *26*, 13–24. [CrossRef]
4. Hu, H. Cyclostationary Approach to Signal Detection and Classification in Cognitive Radio Systems. In *Cognitive Radio Systems*; Wang, W., Ed.; Beijing University of Posts and Telecommunications: Beijing, China, 2009.
5. Ganesan, G.; Li, Y.G. Cooperative spectrum sensing in cognitive radio Part I: Two users networks. *IEEE Trans. Wirel. Commun.* **2007**, *6*, 2204–2213. [CrossRef]
6. Ma, J.; Li, G.Y.; Juang, B.H. Signal Processing in Cognitive Radio. *Proc. IEEE* **2009**, *97*, 805–823.
7. Zhao, Q.; Sadler, B.M. A Survey of Dynamic Spectrum Access. *IEEE Signal Proc. Mag.* **2007**, *24*, 79–89. [CrossRef]
8. Tandra, R.; Sahai, A. SNR walls for signal detection. *IEEE J-STSP* **2008**, *2*, 4–17. [CrossRef]
9. Li, C.; Xiao, J.; Xu, Q. A novel modulation classification for PSK and QAM signals in wireless communication. In Proceedings of the IET International Conference on Communication Technology and Application (ICCTA), Beijing, China, 14–16 October 2011.
10. Liu, J.; Luo, Q. A novel modulation classification algorithm based on daubechies5 wavelet and fractional fourier transform in cognitive radio. In Proceedings of the IEEE 14th International Conference on Communication Technology, Chengdu, China, 9–11 November 2012; pp. 115–120.
11. Liu, A.; Zhu, Q. Automatic modulation classification based on the combination of clustering and neural network. *J. China Univ. Posts Telecommun.* **2011**, *18*, 13–19. [CrossRef]
12. Xu, Y.; Li, D.; Wang, Z.; Liu, G.; Lv, H. A deep learning method based on convolutional neural network for automatic modulation classification of wireless signals. In Proceedings of the International Conference on Machine Learning and Intelligent Communications (MLICOM), Weihai, China, 5–6 August 2017; Volume 226, pp. 373–381.
13. Liu, L.; Xu, J. A novel modulation classification method based on high order cumulants. In Proceedings of the International Conference on Wireless Communications, Networking and Mobile Computing, Wuhan, China, 22–24 September 2006.
14. Chen, X.; Wang, H.; Cai, Q. Performance analysis and optimization of novel high-order statistic features in modulation classification. In Proceedings of the 4th International Conference on Wireless Communications, Networking and Mobile Computing, Dalin, China, 12–14 October 2008.
15. Yuan, H.; Sun, X.; Li, H. The modulation recognition based on decision-making mechanism and neural network integrated classifier. *High Technol. Lett.* **2013**, *19*, 132–136.
16. Liu, N.; Liu, B.; Guo, S.; Luo, R. Investigation on signal modulation recognition in the low SNR. In Proceedings of the International Conference on Measuring Technology and Mechatronics Automation, Changsha, China, 13–14 March 2010.
17. Yoo, Y.; Baek, J. A novel image feature for the remaining useful lifetime prediction of bearings based on continuous wavelet transform and convolutional neural network. *Appl. Sci.* **2018**, *8*, 1102. [CrossRef]
18. Wang, S.; Sun, Z.; Liu, S.; Chen, X.; Wang, W. Modulation classification of linear digital signals based on compressive sensing using high-order moments. In Proceedings of the European Modeling Symposium, Pisa, Italy, 21–23 October 2014.
19. Zhang, X. *Modern Signal Processing*, 3rd ed.; Tsinghua University: Beijing, China, 2015; pp. 219–221.
20. Hui, B.; Tang, X.; Gao, N.; Zhang, W.; Zhang, X. High order modulation format identification based on compressed sensing in optical fiber communication system. *Chin. Opt. Lett.* **2016**, *14*, 14–18.

21. Candes, E.J.; Romberg, J.; Tao, T. Robust uncertainty principles: Exact signal reconstruction from highly incomplete frequency information. *IEEE Trans. Inform. Theory* **2006**, *52*, 489–509. [CrossRef]
22. Baraniuk, R.G. Compressive Sensing. *IEEE Signal Proc. Mag.* **2007**, *24*, 118–121. [CrossRef]
23. Candes, E.J.; Tao, T. Decoding by linear programming. *IEEE Trans. Inform. Theory* **2005**, *51*, 4203–4215. [CrossRef]
24. Candes, E.J.; Romberg, J.K.; Tao, T. Stable signal recovery from incomplete and inaccurate measurement. *Commun. Pur. Appl. Math.* **2006**, *59*, 1207–1223. [CrossRef]
25. Tsaig, Y.; Donoho, D.L. Extensions of compressed sensing. *Signal Process.* **2006**, *86*, 549–571. [CrossRef]
26. Tian, Z.; Tafesse, Y.; Sadler, B.M. Cyclic feature detection with sub-Nyquist sampling for wideband spectrum sensing. *IEEE J-STSP* **2012**, *6*, 58–69. [CrossRef]
27. Kirolos, S.; Laska, J.; Wakin, M.; Duarte, M.; Baron, D.; Ragheb, T.; Massoud, Y.; Baraniuk, R. Analog-to-information conversion via random demodulation. In Proceedings of the IEEE Dallas/CAS Workshop on Design, Application, Integration and Software, Richardson, TX, USA, 29–30 October 2006.
28. Awe, O.P.; Deligiannis, A.; Lambotharan, S. Spatio-temporal spectrum sensing in cognitive radio networks using beamformer-aided SVM algorithms. *IEEE Access* **2018**, *6*, 25377–25388. [CrossRef]
29. Yokota, S.; Endo, M.; Ohe, K. Establishing a classification system for high fall-risk among inpatients using support vector machines. *CIN Comput. Inform. Nurs.* **2017**, *35*, 408–416. [CrossRef] [PubMed]
30. Zhang, W. Automatic modulation classification based on statistical features and support vector machine. In Proceedings of the URSI General Assembly and Scientific Symposium (URSI GASS), Beijing, China, 16–23 August 2014.

MDPI
St. Alban-Anlage 66
4052 Basel
Switzerland
Tel. +41 61 683 77 34
Fax +41 61 302 89 18
www.mdpi.com

Sensors Editorial Office
E-mail: sensors@mdpi.com
www.mdpi.com/journal/sensors

www.ingramcontent.com/pod-product-compliance
Lightning Source LLC
LaVergne TN
LVHW071609170726
843515LV00008B/2353

* 9 7 8 3 0 3 6 5 0 0 1 2 6 *